Sustainability Solutions

An Interdisciplinary and Transdisciplinary Nexus

Sustainability Solutions addresses a wide range of global challenges with applied solutions in the interdisciplinary and transdisciplinary sustainability nexus. Titles in the series present a strong sustainable development focus, and topics which connect environment, economic and social applications are encouraged. Topics could include environmental impacts, vulnerability, risk assessment and reduction, human-environmental systems (e.g. food, water, health) and innovative solutions to climate change. Volumes in the series present novel solutions to societal challenges using innovative application, for example demonstrating a solutions-based approach using machine learning, artificial intelligence, decision support systems, advance geospatial technology. Case studies and shorter report volumes which present knowledge transfer for policymaking and governance are also encouraged.

Keywords:

Interdisciplinary environmental solutions, transdisciplinary knowledge transfer, geospatial technologies, natural resource management, decision support system, artificial intelligence, deep learning, smart technology, environmental science, economic, social and spatial, human-environment interactions, sustainable development.

Gopal Krishna Panda · Rama Mohapatra ·
Sailesh Samanta · Kishor Dandapat · Suman Paul ·
Uday Chatterjee
Editors

Socio-Ecological Crisis of Development

Community Resilience and Sustainable Development

Springer

Editors
Gopal Krishna Panda
Department of Geography
School of Tribal Resource Management
Kalinga Institute of Social Science
(Deemed to be University)
Bhubaneswar, Odisha, India

Sailesh Samanta
School of Surveying and Land Studies
Papua New Guinea Universtiy
of Technology
Lae, Papua New Guinea

Suman Paul
Department of Geography
Sidho-Kanho-Birsha University
Purulia, West Bengal, India

Rama Mohapatra
Department of Geography and Geographic
Information Science
Minnesota State University Mankato
Mankato, MN, USA

Kishor Dandapat
Department of Geography,
Jogamaya Devi College (Affiliated to
University of Calcutta)
Kolkata, West Bengal, India

Uday Chatterjee
Department of Geography
Bhatter College, Dantan (Autonomous)
Medinipur, West Bengal, India

ISSN 3059-278X ISSN 3059-2798 (electronic)
Sustainability Solutions
ISBN 978-3-032-22170-4 ISBN 978-3-032-22171-1 (eBook)
https://doi.org/10.1007/978-3-032-22171-1

This Springer imprint is published by the registered company Springer Nature Switzerland AG
The registered company address is: Gewerbestrasse 11, 6330 Cham, Switzerland

Dedicated to social scientists, environmentalists, planners, and development practitioners, whose critical inquiry, ethical commitment, and action-oriented research continue to illuminate pathways toward resilient communities and sustainable socio-ecological futures.

Preface

Development has always been seen as a pathway to human progress and prosperity. While the world has witnessed unprecedented economic growth, technological innovation, and urban expansion, these very processes have also deepened ecological degradation, social inequalities, and cultural disruptions. The same growth that was meant to uplift us has often come at the cost of the natural systems that sustain life straining the very ecological foundations. The result is a growing socio-ecological crisis where environmental degradation, social inequality, and cultural dislocation intertwine to the challenges of sustainable development. This book, *Socio-Ecological Crisis of Development: Community Resilience and Sustainable Development*, presents a comprehensive exploration of empirical case studies from across the countries of India, Indonesia and Vietnam with the application of advanced geostatistical and geospatial modeling's as well as machine learning and artificial intelligence focusing on the complex social and ecological crisis of development and innovative solutions toward achieving community resilience and sustainable solutions. This volume aims to contribute to ongoing academic and policy discourses on sustainable development, resilience thinking, and environmental governance. The chapters in this book explore how development can be reimagined—how we can move from exploitation and degradation to restoration, from dependency to self-reliance, and from crisis to creativity. It is believed that true progress cannot exist without harmony between people and the planet.

The book comprises of three distinct parts. Part I is focusing on the eternal conflict of development versus environment and society. In this part, Chap. 1 provides an in-depth understanding on indigenous responses to socio-ecological crises of developmental initiatives integrating community resilience and sustainable solutions. Chapter 2 explores the implications of Neo-Urbanization and Ecological Modernization on social and environmental development. Part II of the book seeks to understand how development reshapes ecosystem, livelihood, and social fabric, and how communities adapt, innovate, and build resilience amid these transformations.

Part II explores critical intersections between ecology, economy, and equity, drawing insights from real-world experiences, policy frameworks, and indigenous knowledge systems. This part contains eight chapters relating to Socio-Ecological

Issue of Urbanization, Earthquake vulnerability and Community coping, soil contamination of metalloids and its impact on human health. Other themes of this part are Spatial Analysis of Social Vulnerability to Flood Hazards, Environmental Shifts and community resilience in Sundarbans, Societal Implication and Community Response to COVID-19 and Indian Governance system toward climate change offering valuable insight toward community resilience and sustainable solutions.

Part III is shifting the focus on socio-ecological crisis of development and sustainable solutions. Drawing from research, case studies, and grounded experiences, this part seeks to bridge the theory and practice of sustainability to show that meaningful solutions emerge when scientific insight meets community wisdom. This part includes eight chapters, Chap. 11 focuses on integration of agriculture and tourism for sustainable livelihood from Indonesia. Chapter 12 highlighted on empirical studies on Socioeconomic Livelihoods and Environmental Sustainability. Chapter 13 explores how climate village program in Indonesia has been implemented for climate crisis adaptation and mitigation. Chapter 14 focuses on Climate Village Program for Adaptation and Mitigation, and Chap. 15 explores community based environmental management strategy as an inclusive solution for sustainable local communities from Indonesia. Chapter 17 explores Urban Green Spaces for Neighborhood Sustainability and the final chapter Chap. 18 highlighted on climate resilience development policy, strategies, and practices for sustainable future of Indonesia.

This book aims to contribute to ongoing academic and policy discourses on sustainable development, resilience thinking, and environmental governance. It brings in to focus those scholars, practitioners, and policymakers should reconsider development paradigms through the lens of ecological ethics. It is our sincere hope that this book inspires reflection, dialogue, and action toward a more equitable, resilient, and sustainable world.

Bhubaneswar, India	Gopal Krishna Panda
Mankato, USA	Rama Mohapatra
Morobe, Papua New Guinea	Sailesh Samanta
Kolkata, India	Kishor Dandapat
Purulia, India	Suman Paul
Kharagpur, India	Uday Chatterjee

Acknowledgments

We are grateful to the communities, researchers, and colleagues whose experiences and encouragement from their reviewed literatures shaped this work. Their stories reaffirm the solutions that the socio-ecological crisis lays not only in technology or policy but in the collective wisdom and resilience of people who sustain life within fragile ecosystems. We remain deeply indebted to the innumerable authors, researchers, and institutions whose insights and resilience have inspired this work. Their enduring relationship with the environment reminds us that sustainability is not only a policy goal but a shared moral and cultural responsibility that value human well-being and planetary health in equal measure. We are very much thankful to all our authors and contributors for their commitment to timely submission and insightful work to enrich this book with ideas and concepts relevant to the contemporary situation of the new millennium. We express our heartfelt thanks to our publishing Editor **Margaret Deignan** and series Editor **Dr. Uday Chatterjee** for their guidance, monitoring, and supervision for this publication. We are also thankful to all the authors whose work has been cited in this publication which have been the valuable sources of information in formulating the key concepts, ideas, and approaches in this publication.

Bhubaneswar, India	Gopal Krishna Panda
Mankato, USA	Rama Mohapatra
Morobe, Papua New Guinea	Sailesh Samanta
Kolkata, India	Kishor Dandapat
Purulia, India	Suman Paul
Medinipur, India	Uday Chatterjee

Contents

Editors and Contributors

About the Editors

Prof. Gopal Krishna Panda is a physical geographer, presently working as Emeritus Professor in the Post Graduate Department of Geography, School of Tribal Resource Management at KISS (Kalinga Institute of Social Science) Deemed University, Bhubaneswar and former Professor and Head of the Department of Geography of Utkal University, Bhubaneswar. He has been awarded the Emeritus Fellowship of UGC in the year 2016 and honored with the Senior Fellowship of ICSSR, New Delhi in 2019. During his tenure of service in the Department of Geography at Utkal University, he has served in different capacities as Professor and Head, Chairman of the Post Graduate Council, Member of the Syndicate, Senate and Chief Executive of the Population Research Center of Utkal University. He has nearly 35 years of teaching and research in the areas of physical geography, coastal geomorphology, climate change, natural hazards and disaster management and geospatial technology focusing on capacity building, programme evaluation and value addition. He was a National Scholarship holder and Gold Medallist of his time in Utkal University in the year 1977. He was a Visiting Fellow of the Liverpool School of Tropical Medicine availing British Council Fellowship in 1995. He was the Principal Investigator for a major research project sponsored by the UNDP and OSDMA to prepare the "Odisha State Natural Hazards and Disaster Vulnerability Analysis and Risk Assessment Report" for the state of Odisha. Under his guidance 16 students have been awarded Ph.D. degree in geography and environmental science. He was awarded Smt. Parbati Mishra and Prof. M. M. Mohanty Memorial Environmentalist of the Year award in the years 2017 and 2021 by the Odisha Environment Society. He has been the editor of two important books published from the Springer Nature i.e., Indigenous People and Nature, and Indigenous knowledge in disaster risk reduction. Professor Gopal Krishna Panda is an acclaimed physical geographer and Emeritus Professor at Kalinga Institute of Social Sciences (KISS) Deemed University, Bhubaneswar. Previously, he served as Professor and Head of Geography at Utkal University. His distinguished career spans 35 years, focusing on physical geography, coastal geomorphology, climate change, disaster management, and geospatial technology. Professor Panda holds prestigious fellowships from UGC (2016) and ICSSR (2019). He was a National Scholarship holder and Gold Medallist (1977) and Visiting Fellow at Liverpool School of Tropical Medicine (1995). He led the UNDP and OSDMA-sponsored project on Odisha's natural hazards and disaster vulnerability analysis. Under his guidance, 16 students earned Ph.D. in geography and environmental science. He received the Environmentalist of the Year award (2017, 2021) from Odisha Environment Society and edited two Springer Nature books: *Indigenous People and Nature* and *Indigenous Knowledge in Disaster Risk Reduction.*

Prof. Rama Mohapatra is an expert in applications of GISc (Geographic Information Science), remote sensing technologies, econometrics, and machine learning to model geographical data. Overall, he has more than 25 years of teaching and research experience in the field of GISc. This is his 14th year at Minnesota State University, Mankato. His primary role at Minnesota State University involves teaching numerous GISc classes, serving as thesis adviser, guiding students on their internships, and supervising graduate and undergraduate student research projects. Apart from his teaching and research, Prof. Mohapatra is also involved in many committees and professional organizations. Professor Mohapatra obtained his Ph.D. degree in Geography from the University of

Wisconsin-Milwaukee in 2012. His dissertation title was "Monitoring and modeling urban growth using remote sensing and GIS technologies." Professor Mohapatra completed his bachelors and masters (M.A. and M.Phil.) degrees in geography from the Utkal University, Odisha, India. He has several journal publications, presented in numerous conferences, and is involved in many research projects. Professor Mohapatra's teaching focuses on Geoanalytics, Geospatial modeling, 3D GIS, Temporal GIS, Web GIS, Utility GIS, Business GIS, etc. His courses provide students with the skills necessary to investigate Spatio-temporal patterns that result from human environment interactions. In his classes, students get involved in independent research work such as tracking hurricanes, mapping crime hot spots, analyzing disease outbreaks, locating new businesses, targeting customers for marketing, setting a new bee pollinator garden, managing water distribution network of a city, creating story maps, building Web GIS apps, etc. Professor Mohapatra's teaching philosophy emphasizes on project-based learning and he has successfully guided several students to conduct high level in-class research based on the content taught in classes. Some of his students have presented these research findings in national and local conferences as paper or poster presentations.

Dr. Sailesh Samanta is working as a Professor and Section Head of the GIS section under the School of Surveying and Land Studies at The Papua New Guinea University of Technology, Papua New Guinea. He is also coordinating the undergraduate distance program in Geomatics at this university. His maiden Ph.D. was on Climatological modeling of Temperature and Rainfall through GIS Techniques. He has more than 50 peer-reviewed publications in national/international journals and more than 1150 citations up-to-date. Dr. Samanta achieved a pinnacle position in research and development in Geoinformatics, Climate modeling, Hydro-meteorology, Earth Sciences, Disaster Management, natural resources management, sustainable development, etc. He contributed to different international collaboration projects as a principal consultant or co-team member. His current involvement is with the Commonwealth Scientific and Industrial Research

Organisation (CSIRO), Australian Centre of Agricultural Research (ACAR), and Pacific Islands Universities Research Network (PIURN). His research profile can be visited at https://scholar.google.com/citations?user=vd3ddccAAAAJ&hl=en.

Dr. Kishor Dandapat is working as an Assistant Professor, Department of Geography at Jogamaya Devi College, Kolkata, under University of Calcutta, West Bengal, India. He had a master's degree in Geography from Pt. Ravi Shankar Shukla University, Raipur and a Ph.D. from Utkal University, Bhubaneswar, under the guidance of Emeritus Professor Dr. Gopal Krishna Panda, Bhubaneswar and he qualified UGC-NET examination in the subject of Geography. His area of interest is in Geography, mostly focuses on human ecology, environmental issues, natural hazards and disaster management, and fluvial geomorphology. Dr. Dandapat published a book from international publishers *Flood Hazards Management: A Community Approach*, Notion Press, Chennai, India. Currently completed a Major Research Project entitled by "*Vulnerability of tribal communities towards food and nutrition insecurity: A study in Jhargram District of West Bengal*" sponsored by Indian Council of Social Science Research, ICSSR, New Delhi, India. He is a reviewer of several reputed national and international journals including those published from Springer Nature and Elsevier. He has reviewed various national and international journals, including *Sustainable Water Resources Management*, *Environmental Management*, *Disaster Risk Studies Scientific Reports*, and *Arabian Journal of Geoscience*. Dr. Dandapat has actively shared his research findings through presentations at international, national, and state-level seminars and conferences. His research has also been published in numerous articles in reputable peer-reviewed international and national journals, as well as in edited books.

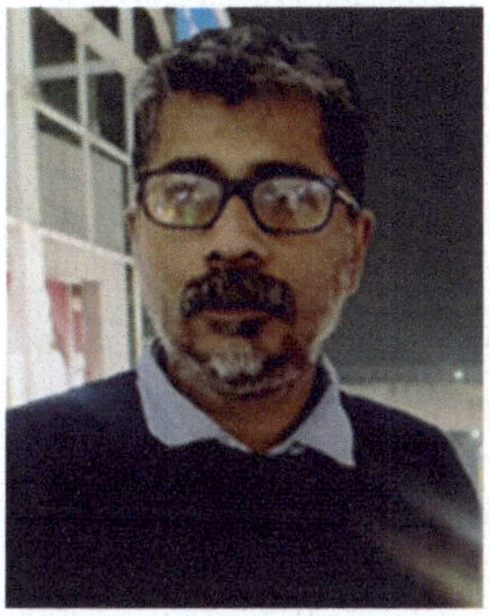

Dr. Suman Paul was born in Kolkata, West Bengal, India on July 31st, 1978. He received his B.Sc. and M.Sc. degrees from University of Calcutta, Calcutta India in 2000 and 2002 respectively. He was associated as an Assistant Professor at Post Graduate Department of Geography with more than nine years of experience at Krishnagar Government College and Acharya B. N. Seal Government College. Presently he is the HOD and **Professor of Geography Department, Sidho-Kanho-Birsha University, Purulia, West Bengal, India**. He achieved National Scholarship and stood 1st position (1st Class FIRST) at Under Graduate level from Calcutta University. His areas of specialization are urban geography, environmental hazards and disaster analysis, GIS and remote sensing. He completed his Ph.D. on Urban Geography from the Department of Geography, University of Calcutta, West Bengal, India entitled "Emerging Urban Centres—Status and Possibilities: A Case Study of the District of North 24 Parganas, West Bengal." He has presented several research papers in international, national and state levels seminars and conferences. He has published a number of articles in reputed peer-reviewed international, national journal and in edited books.

Dr. Uday Chatterjee is an Assistant Professor at the Department of Geography, Bhatter College, Dantan, affiliated with Vidyasagar University, West Bengal, India. He earned his Doctoral Degree in Applied Geography from Ravenshaw University, Cuttack, Odisha. His research interests encompass Urban Planning, Social and Human Geography, Applied Geomorphology, Indigenous Knowledge and Hazards and Disasters, Risk Assessment and Resilience, Community Response and Environmental Issues. Dr. Chatterjee has been actively involved in editing scholarly works, including the Special Issue on Urbanism, Smart Cities, and Modeling for GeoJournal, Springer, Special issue in Global Social Welfare, Springer and Frontiers Environmental Science journal, Geological journal Wiley, Editorial Board Member of Scientific Report (Nature Portfolio) journal and serving as a Book Series Editor for Developments in Environmental Science, Elsevier and Sustainability Solutions-An Interdisciplinary and

Transdisciplinary Nexus, Springer Nature. His extensive research and editorial contributions highlight his commitment to advancing geographical knowledge and its applications. His research work has been funded by the West Bengal Pollution Control Board (WBPCB) Government of West Bengal, India Indian Council of Social Science Research (ICSSR), Government of India. He has published research papers in reputed journals, edited books with reputed publisher like Springer, Elsevier, CRC Press, and author books, book chapters, and conference proceedings. Currently, Dr. Chatterjee doing an international project, in collaboration with Indonesia, Malaysia, and Japan funded by APN (Asia Pacific-Global Change Research, Grant https://doi.org/10.30852/p.22128), and visited Saudi Arabia for International collaboration and conference session chair and Keio University, Tokyo, Japan as a Visiting Professor.

Contributors

Md. Juhayr Akbar Department of Geography and Environment, University of Dhaka, Dhaka, Bangladesh

Rizal Akbar Aldyan National Research and Innovation Agency (BRIN), Jakarta, Indonesia

Purnama Alamsyah National Research and Innovation Agency, South Jakarta, Indonesia

Reza Amarta Prayoga Research Center for Social Welfare, Village and Connectivity-National Research and Innovation Agency, Jakarta, Indonesia; Faculty of Social and Political Sciences, University of Indonesia, Depok, Indonesia

Rosita Novi Andari Research Center for Social Welfare, Villages and Connectivity, National Research and Innovation Agency (BRIN), Jakarta, Indonesia

Iwan Setiajie Anugrah Research Center for Social Welfare, Villages and Connectivity, National Research and Innovation Agency (BRIN), Jakarta, Indonesia

Shayantani Banerjee Amity Institute of English Studies and Research (AIESR), Amity University Jharkhand, Ranchi, India

Rio Heykhal Belvage Reserch Center for Social Welfare, Village, and Connectivity, National Research and Innovation Agency, Jakarta, Indonesia

Gouri Sankar Bhunia Independent Researcher, Kharagpur, Paschim Medinipur, West Bengal, India

Sheuli Biswas Department of Geography, Sadhu Ram Chand Murmu University, Jhargram, West Bengal, India

Lina Bose Department of History, Amity University Jharkhand, Ranchi, India

Carolina Research Center for Social Welfare, Villages and Connectivity, National Research and Innovation Agency (BRIN), Jakarta, Indonesia

Danarti National Research and Innovation Agency (BRIN), Jakarta, Indonesia

Kishor Dandapat Assistant Professor, Department of Geography, Jogamaya Devi College (Affiliated to University of Calcutta), Kolkata, West Bengal, India

Kader Demiröz Altintaş Department of Emergency Aid and Disaster Management, Artvin Coruh University, Artvin Merkez, Türkiye

Sevda Demiröz Yıldırım Department of Emergency Aid and Disaster Management, Burdur Mehmet Akif Ersoy University Bucak School of Health, Bucak, Burdur, Türkiye

Sachin J. Deore Department of Geography, Post Graduate Studies and Research, SNDT Women's University, Pune, MS, India

Amar M. Dhere Department of Science, SNDT College of Home Science, Pune, MS, India

Siti Fatimah Research Center for Social Welfare, Village and Connectivity-National Research and Innovation Agency, Jakarta, Indonesia

I. Gusti Ngurah Jayanti Research Center for Environmental Archaeology, Maritime Archeology, and Cultural Sustainability, National Research and Innovation Agency, Jakarta, Indonesia

Md. Iqbal Hasan Department of Geography and Environment, University of Dhaka, Dhaka, Bangladesh

Herlina Reserch Center for Social Welfare, Village, and Connectivity, National Research and Innovation Agency, Jakarta, Indonesia

Rifat Bin Hossain Department of Geography and Environment, University of Dhaka, Dhaka, Bangladesh

Kurnia Suci Indraningsih Research Centre for Social Welfare, Village, and Connectivity, National Research and Innovation Agency, Jakarta, Indonesia

Andi Ishak National Research and Innovation Agency, South Jakarta, Indonesia

Srinivas Jangirala Jindal Global Business School, O. P. Jindal Global University, Sonipat, Haryana, India

Ria Jayanthi National Research and Innovation Agency, South Jakarta, Indonesia

Herma Juniati National Research and Innovation Agency (BRIN), Jakarta, Indonesia

Siti Khoiriyah Doctoral Program of Environmental Science, Universitas Sebelas Maret, Surakarta, Indonesia

Mehtap Kılıç Yildiz Department of Emergency Aid and Disaster Management, Ardahan University, Ardahan Merkez, Türkiye

Biplab Mahali Department of Geography, University of Calcutta, Kolkata, West Bengal, India

Prasanna Mahata Department of Geography, Sadhu Ram Chand Murmu University, Jhargram, West Bengal, India;
The University of Burdwan, Bardhaman, West Bengal, India

Abdul Manan Faculty of Economic, University of Mataram, Mataram, Indonesia

Tri Margono National Research and Innovation Agency (BRIN), Jakarta, Indonesia

Gabriela Michael Indian Maritime University, Chennai, India

Harshita Mishra Rashtriya Raksha University, Shivamogga, Karnataka, India

Swati Mittal Department of English, School of Media Studies and Humanities (SMeH), Manav Rachna International Institute of Research and Studies (MRIIRS), Faridabad, Haryana, India

Sri Najiyati National Research and Innovation Agency (BRIN), Jakarta, Indonesia

Nafisa Nandini Department of Geography and Environment, University of Dhaka, Dhaka, Bangladesh

Rukmini Nugroho Dewi National Research and Innovation Agency (BRIN), Jakarta, Indonesia

Fatwa Nurul Hakim Research Center for Social Welfare, Village and Connectivity-National Research and Innovation Agency, Jakarta, Indonesia

Gopal Krishna Panda Department of Geography, School of Tribal Resource Management, Kalinga Institute of Social Science, Deemed to be University, Odisha, Bhubaneswar, India

Meenakshi Prasad P.G. Department of Geography, Magadh University, Bodh Gaya, Bihar, India

Djoko Puguh Wibowo Research Center for Social Welfare, Village and Connectivity-National Research and Innovation Agency, Jakarta, Indonesia

Lis Purbandini Research Center for Social Welfare, Village and Connectivity-National Research and Innovation Agency, Jakarta, Indonesia

Tri Bastuti Purwantini Research Centre for Social Welfare, Village, and Connectivity, National Research and Innovation Agency (BRIN), Jakarta, Indonesia

Dian Purwasantana National Research and Innovation Agency (BRIN), Jakarta, Indonesia

Intan Hapsari Surya Putri National Research and Innovation Agency (BRIN), Jakarta, Indonesia

Rajashekhar Rao Bangady Killur School of Agriculture, Faculty of Natural Resources, Papua New Guinea University of Technology, Lae, Papua New Guinea

Muslim Sabarisman Research Center for Social Welfare, Villages and Connectivity, National Research and Innovation Agency (BRIN), Jakarta, Indonesia

Rachmini Saparita Research Center for Social Welfare, Villages and Connectivity, National Research and Innovation Agency (BRIN), Jakarta, Indonesia

Hari Harjanto Setiawan Research Center for Social Welfare, Villages and Connectivity, National Research and Innovation Agency (BRIN), Jakarta, Indonesia

Ulfa Sevia Azni Research Center for Social Welfare, Village and Connectivity-National Research and Innovation Agency, Jakarta, Indonesia

Tasneem Sharmin Department of Geography and Environment, University of Dhaka, Dhaka, Bangladesh

Sandeep Singh Jindal Global Business School, O. P. Jindal Global University, Sonipat, Haryana, India

Saraswati Soegiharto National Research and Innovation Agency (BRIN), Jakarta, Indonesia

Nuzul Solekhah Research Center for Social Welfare, Village and Connectivity-National Research and Innovation Agency, Jakarta, Indonesia; Department of Anthropology, Gadjah Mada University, Yogyakarta, Indonesia

Badrun Susantyo Research Center for Social Welfare, Villages and Connectivity, National Research and Innovation Agency (BRIN), Jakarta, Indonesia

Chandrani Tiwari Siddharth College, Bilaspur, Chhattisgarh, India

Shivani Vashist Department of English, School of Media Studies and Humanities (SMeH), Manav Rachna International Institute of Research and Studies (MRIIRS), Faridabad, Haryana, India

Deepika Chandra Verma Indian Institute of Management Calcutta, Kolkata, West Bengal, India

Eko Wahyono Reserch Center for Social Welfare, Village, and Connectivity, National Research and Innovation Agency, Jakarta, Indonesia

I. Wayan Rupa Research Center for Religious Heritage and Civilization, National Research and Innovation Agency, Jakarta, Indonesia

Arditya Wicaksono National Research and Innovation Agency, South Jakarta, Indonesia

Febtri Wijayanti National Research and Innovation Agency, South Jakarta, Indonesia

Janardan Krishna Yadav Jindal Global Business School, O. P. Jindal Global University, Sonipat, Haryana, India

Yumantoko Reserch Center for Social Welfare, Village, and Connectivity, National Research and Innovation Agency, Jakarta, Indonesia

Part I
Introduction

Chapter 1
Integrating Community Resilience and Sustainable Solutions for an Eco-Conscious Future: Indigenous Responses to Socio-Ecological Crises and Environmental Vulnerability

Swati Mittal and Shivani Vashist

Abstract The rapid and non-sustainable development process has increased socio-ecological disasters and resulted in the loss of biodiversity, climate vulnerability, and worsening inequalities. In this chapter, Indigenous societies are framed as offering resilience and sustainability based on knowledge systems that integrate ecology, culture, and governance. Based on a qualitative comparative analysis, six case studies are described: the Zapatista rebellion in Mexico, the Kayapo in the Xingu Indigenous Park, Brazil, the Maasai in Kenya's Mara ecosystem, Sámi reindeer people in northern Europe, Himalayan people in India, and Amazonian people in Brazil. These cases reveal strategies like agroecology, rotational grazing, sacred forest protection, and community-based governance that simultaneously ensure local livelihood and avoid environmental degeneration. In comparing these disparate examples with recent literature on conservation and climate adaptation, the chapter demonstrates the universal relevance of Indigenous methods in preventing deforestation, accumulating biological diversity, and enhancing adaptation capacity. The aim is to highlight the potential in integrating Indigenous ecological knowledge with modern policy instruments in the development of replicable models of resilience. This will take a step toward an eco-sensitive future in which environmental stewardship and social justice become one and the same. The novelty of the study lies in the cross-regional comparison, where Indigenous leadership emerges as an overlooked but central mechanism for sustaining sustainability.

Keywords Socio-ecological crisis · Indigenous knowledge · Community resilience · Climate adaptation · Biodiversity conservation · Sustainability

S. Mittal (✉) · S. Vashist
Department of English, School of Media Studies and Humanities (SMeH), Manav Rachna International Institute of Research and Studies (MRIIRS), Faridabad, Haryana, India
e-mail: Swatimittal.smeh@mriu.edu.in

S. Vashist
e-mail: shivani.smeh@mriu.edu.in

G. K. Panda et al. (eds.), *Socio-Ecological Crisis of Development*, Sustainability Solutions, https://doi.org/10.1007/978-3-032-22171-1_1

1.1 Introduction

1.1.1 Global Environmental Crisis and Socio-Ecological Vulnerability

Indigenous communities do not see themselves as owners of the Earth's resources. They believe in nature worship and also have faith that the resources available on earth should be used without exploitation. But those whose ancestral wisdom promotes reverence for the Earth have become the ones who are often marginalized in environmental discussions. They are frequently labeled as "in need of development," their voices are silenced, and their ideas about environmental conservation are ignored. Instead, external authorities impose their own notions and often dismiss Indigenous knowledge as "backwards." According to Mishra et al. (2021), "Indigenous people constitute an important section of society in many countries. Despite being a numerically smaller section, they are culturally diverse and distributed mostly in valuable, natural resources-rich regions worldwide. In the era of globalisation, industrialisation, and trade liberalisation, indigenous communities have become more vulnerable to displacement, land alienation, cultural erosion, and social exclusion."

Yet research shows that some of the most insightful and sustainable approaches to environmental conservation come directly from these communities. They teach us that since nature always nurtures mankind, mankind also, in return, has to protect nature. According to Abas et al. (2022), "Indigenous people have long histories with their land, which has provided sustenance in very direct and intimate ways." The twenty-first century is a witness to unparalleled environmental degeneration. The Intergovernmental Science-Policy Platform on Biodiversity and Ecosystem Services (IPBES) tells humankind that nearly one million species are condemned in the next twenty years due to habitat clearance, climate warming, pollution, and destructive agriculture. Modernity comes with its own burdens and disadvantages. For example, one can see how deforestation alone is proceeding at an alarming rate with an estimated annual deforestation of up to 10 million hectares of forests, as calculated by the Food and Agriculture Organization (FAO, 2020), primarily in the tropics.

These ecological and social crises are concentrated unevenly. According to Maurya et al. (2020), populations that reside in ecologically vulnerable regions—and often Indigenous and farming cultures—bear an unequal burden of the Impacts of climate variability, deforestation, and soil erosion. What once gave life—air, water, forests—is increasingly poisoned, depleted, and unstable. It is high time to step toward sustainable solutions before nature becomes a mirror, in a way it starts reflecting back the consequences of human neglect and exploitation. Vulnerability is heightened by structural inequalities such as political exclusion, insecure property regimes, and non-participation in institutional decision-making (Adger et al., 2005; Sikor & Lund, 2009). According to Turner et al. (2003), scientists have increasingly perceived sustainability as a commodity that is unattainable alone with technological progress or top-down government policy. According to UNEP (2021) Climate and ecological shock-resilience, on the other hand, requires the co-integration of various knowledge systems. Particularly those founded on long-standing human environment interactions, like Indigenous communities, can contribute more to environmental conservation (Berkes et al., 2000). According to the United Nations (2021), Indigenous people have rightful claims over the Earth's resources, yet these rights are frequently overlooked or exploited by those outside their communities. According to Vásquez Coronado et al. (2024), development should be directed toward the benefit of society to provide justice to everyone, but in reality, development leads to environmental degradation and displacement of the marginalized communities.

This is not just a case in the meantime, the Intergovernmental Panel on Climate Change (IPCC, 2022) has warned mankind that the planet is indeed pre-warmed by 1.1 °C over pre-industrial temperatures as a consequence of sharply increasing extreme weather conditions, disturbing ecological systems, and raising socio-economic inequalities. It is quite shocking that the nature which nurtures man, in reality, is itself becoming a victim of Socio-Ecological Crises and Environmental Vulnerability.

According to Zipperer et al. (2020), environmental degradation is one of the most urgent challenges we face today. Communities must set aside their differences and come together to protect the planet, just as the Earth provides shelter and sustenance to all species without discrimination. Amidst environmental degradation, moving toward sustainable solutions has become the need of the hour. As the planet struggles under the weight of climate change and reckless exploitation, one must look to those who have stayed true to their roots. Those who have carried forward ancestral wisdom hold some of the most practical and time-tested answers. And such sustainable solutions can come from their ways of managing forests, water, land, and food. Because there's not an approach that sees sustainability as an abstract idea but a lived practice passed down through generations (Agrawal & Ostrom, 2001).

1.1.2 Indigenous People as Environmental Caretakers

Exploring the meaning of sustainable development, Rist and Dahdough-Guebad (2006) argued that "Considering the needs of future generations is only meaningful when the conception of humans goes beyond a merely ego-centric and materialist understanding of life and living beings. This implies recognition that without a specific view on the world, which allows the establishment of a meaningful—and thus action guiding—relationship between humans, society and nature, sustainable development will remain literally without sense."

Indigenous people throughout the planet live on about 28% of the planet's territory but control regions with over 80% of the planet's biological richness (Garnett et al., 2018; Sobrevila, 2008). This exceptional correlation between Indigenous tenure and environmental integrity is now gaining wider consideration from conservationists, climate scientists, and policymakers. Indigenous peoples' knowledge systems—drawn upon relational cosmologies between ecological, spiritual, and human domains—are radically different from reductionist Western science paradigms (Kimmerer, 2013; Ogar et al., 2020). Contrary to asserting nature as a commodity, Indigenous paradigms consistently position the environment as a living one with reciprocal responsibilities.

Empirical data supports the ecological effectiveness of such paradigms. Walker et al. (2020) argue that Indigenous ecological knowledge (IEK) is valuable for good adaptive capacity through incorporating the integration of social norms of reciprocity, self-restraint, and stewardship in resource management practice (Berkes et al., 2000). These conclusions not merely identify Indigenous knowledge's intrinsic value. They also show that it has instrumental value for interglobal sustainability challenges. But even with such evidence, Indigenous people are poorly represented in the formal spaces of decision-making. Their local knowledge is downplayed as anecdotal or "local" rather than as systematic and empirically grounded (Whyte, 2018). According to Cutter et al. (2008), this epistemological exclusion, in turn, has upheld environmental destruction and cultural erasure. It requires research that gives Indigenous leadership voices centrality and accords it primacy, that supports their knowledge systems, and documents their value in universal locations of climate adaptation and conservation (Brondizio et al., 2009). With an international survey, Fa et al. (2021) concluded that environmental outcomes for biodiversity in lands managed by Indigenous peoples are similar to or better compared to formally protected areas.

1.1.3 Contemporary Research and Literature Review

Sustainable and environmental governance research over the last two decades has emphasized participatory and community-oriented responses. Scholars of resilience (Folke, 2006; Walker et al., 2004) argued that socio-ecological systems are non-linear, adaptive, and complex and can never therefore adequately be tackled through technocratic, top-down methods. Instead, as framed as the capacity of systems to recover disturbances while maintaining core functions, resilience is strengthened where there is direct community engagement in decision-taking. Community-based natural resource management (CBNRM) is one such prevailing paradigm. In southern Africa, for instance, CBNRM programs have consolidated wildlife conservation achievements while achieving financial returns for citizens (Murphree, 2009). In similar proportion, Ostrom's (2012) seminal study on the common-pool theory established that societies, with facilitating conditions, are capable of sustainable management of common assets without external compliance. These cases have since been followed up with comparative international work showing Indigenous and local systems of governance are more often superior relative to state-based conservation in ecological sustainability and equity (Chhatre & Agrawal, 2009; Reyes-García et al., 2019). Nonetheless, there are gaps. With the exception of work on one region or case-specific processes, the majority of research is based on one region, and comparative, cross-regional research that synthesizes lessons between ecological and cultural contexts is lacking. Policy argumentation even yet still promotes technocratic "solutions" such as carbon markets, geoengineering, or fortress-form conservation that do not permit Indigenous participation (Büscher & Fletcher, 2020). They risk reproducing ecological inequities through excluding the very peoples most interested in protecting systems. There is also a pressing need for integration of science and Indigenous ecological knowledge. Though co-production of knowledge is increasingly popular in climate adaptation science (Tengö et al., 2014), there are epistemological and institutional impediments. Western science emphasizes quantitative data and repeatability of experiments, while Indigenous systems privilege relational and site-specific knowledge over centuries. Looking up to and a combination of these epistemologies requires greater than token representation. It can be said that it requires balanced collaboration that considers power disparities and Indigenous sovereignty (Nadasdy, 1999; Whyte, 2018). Their study identifies and researches such gaps. It does it through offering a comparative analysis of six geographically and culturally diverse Indigenous communities. In an effort to put the work next to recent scientific publications on conservation and climate resilience, the study also follows both convergences with and novel additions with respect to such publications.

1.1.4 Indigenous Case Studies in Context

The study has six Indigenous communities under consideration. According to the Coordination of Indigenous Organisations of the Brazilian Amazon (2010), they possess unique ecological histories, yet as a whole shed light upon common patterns in Indigenous persistence. These are the Zapatistas in Mexico, Kayapo in Brazil, Maasai in Kenya, Sámi in northern Europe, societies in the Himalayas in India, and those in the Amazon in Brazil.

Zapatista Movement (Mexico): Even though they had been pushed to the edges for a long time, the Chiapas Zapatista movement set up independent municipalities based on Indigenous values of making decisions as a group. By adopting agroecology—diversified, small-scale agriculture that enhances soil fertility and minimizes chemical use—the Zapatistas are pursuing an alternative development strategy compared to large-scale farming. Studies show that Zapatista agroecosystems have more biodiversity and better soil use than monoculture farms, and that they help food sovereignty (Maison, 2023; Toledo et al., 2003).

Brazil's Indigenous Park Xingu: There are over 11 million hectares of rainforest inhabited by the Kayapo people in the Brazilian Amazon. They have been able to prevent invasions of logging and mining with the help of community monitoring, alliances with non-governmental organizations (NGOs), and sustainable harvesting. Studies show that deforestation rates inside the territory of the Kayapo are significantly lower compared with adjacent non-Indigenous territories (BenYishay et al., 2017). Their model shows Indigenous territory rights acting directly to contribute to climate mitigation worldwide with carbon sinks.

Kenya's Mara Ecosystem Maasai: The Maasai pastoralists historically applied rotational grazing for keeping semi-arid savannahs with livestock herds. It allows the rebuilding of grasslands, enhances diversity, and prevents soil erosion (Reid et al., 2014). Climate variability and privatization of grazing lands are threatening such livelihoods. The Maasai innovatively integrated customary knowledge with modern paradigms of conservation through the development of community conservancies, and eco-tourism revenue is generated without compromising ecological integrity (Bedelian & Ogutu, 2017).

Northern European Sámi Reindeer Herders: In Scandinavia and Russia, the Sámi have traditionally been reindeer herders, adapting mobility patterns with the seasonally shifting Arctic ecological systems. Climate change today disrupts migration corridors in altered snow patterns and plant transitions (Forbes et al., 2016). Sámi adaptation is based on blending traditional ecological knowledge with satellite tracking and scientific knowledge in a hybrid adaptation model. Their experiences are central in international Arctic sustainability discourses.

Communities of the Himalayas of India: The communities of the Indian Himalayas live in a geographically diverse and culturally vibrant region. Indigenous people such as the Lepcha and the Bhotiya irrigated with traditional irrigation systems known as the kuls and retained sacred groves as repositories of biodiversity (Tiwari, 2010). Ongoing glacial retreat and erratic monsoons render the practices as good exemplars of Indigenous knowledge in regulating water and managing ecosystems.

Tribes of the Amazon (Brazil): Many other Amazonian tribes, besides the Kayapo, use ethnobotanical knowledge to take care of trees. Changing how we farm and using plants that heal helps keep the environment in balance and makes people healthier. Recent studies indicate that Indigenous Amazonian regions serve as defenses against deforestation and sequester billions of tons of carbon (Walker et al., 2020).

These examples, while different, show that Indigenous ecological knowledge is not something that has been around for a long time; it is a living, changing system that can be used by everyone. Redford and Fearn (2007) say that indigenous communities work very hard to keep the environment safe. They show how Indigenous societies come up with new ideas while dealing with political and economic marginalization.

1.1.5 Research Problem and Significance

Even as research on Indigenous environmental management progresses scientifically, climate and nature conservation policies globally are shaped by top-down, centralized approaches that marginalize Indigenous perspectives. For example, "fortress conservation," the exclusive development of closed domains, continues to displace Indigenous peoples under the assumption that biodiversity is best preserved in their absence (Brockington et al., 2008). These methods violate Indigenous sovereignty and contradict empirical studies indicating that Indigenous-managed landscapes produce enhanced biodiversity results (Fa et al., 2021; Garnett et al., 2018). Climate change meanwhile turbocharges ecological vulnerability impacting Indigenous people foremost. Arctic warmings jeopardize Sámi grazing and sheep-herding traditions, and sea-level rise jeopardizes Pacific Islander territory. In Latin America and sub-Saharan Africa, resource appropriation and privatization deepen socio-economic marginalisation. These processes constitute a double disaster: ecological breakdown driven by unsustainable systems and Indigenous sovereignty erosion. Why research on this problem matters is the understanding that Indigenous ecological knowledge is not alternative practice alone yet sustainability principles. Rotational grazing among Maasai, conservation of sacred Himalayan Mountain forests, and Zapatista self-governance are whole-of-systems integration exemplars with ecological, cultural, and societal systems each supporting each other. Contrary to reductional epistemologies treating systems as isolable biological systems, Indigenous systems sciences propose systems as socio-ecological wholes—an appreciation now buttressed with theory on the topic of resilience and with sustainability science (Berkes, 2018; Folke, 2006). Yet comparative research compiling and comparing between continents is lacking. Most research is geography-specific, and policymakers do not have an integrated framework on which to generalize Indigenous practice while adjusting for cultural differentiation. It is out of this void that the present study emerged, and it compares six very disparate Indigenous societies in Asia, Europe, Latin America, and Africa. In comparing the cases, the research helps highlight similar strategies—including agroecology, sacred resource conservation, and rotational resource use—that cut across ecological and cultural conditions, and so create transferable wisdom internationally.

Despite increasing evidence of their value, Indigenous people remain marginalized in climate and conservation policy. Extractivist economies, dispossession of lands, and exclusive regimes of conservation displace Indigenous peoples while stripping away their ecological intentionality. "Green growth" policies instead often embed and reify injustice as policies privilege carbon markets and eco-tourism over Indigenous property regimes (Stevens, 2014). The research gap targeted in the present paper is the product of the disconnect: while Indigenous territories are globally acknowledged as sites of exceptional biodiversity richness, research and policy rarely juxtapose cross-regional Indigenous protocols for doing so into comparative frames.

Scholarship and policy are episodic—limited to single ethnographies or single-site analysis—entailing an understanding gap about how different Indigenous systems of governing cumulatively illuminate sustainability and resilient trajectories. In response, the study uses a qualitative, cross-regional comparative research method. Six case studies are open to analysis: Chiapas Zapatista movement, Mexico; the Xingu Indigenous Park Kayapo, Brazil; Maasai of Kenya's Mara ecosystem; northern European Sámi reindeer people; Indigenous people of the Himalayas, India; and Brazilian Amazonian tribes. In an examination of ecological practice alongside agroforestry, conservation of sacred forests, rotational grazing, and self-governance of locations, the research supports shared principles and site-specific innovations. In comparative analysis, the study shows that the knowledge derived from Indigenous people's ecological lives (TEK) is systematically contrasted and synthesized with other knowledge systems. It also offers transferable exemplars of resilient ways combining biodiversity conservation, climate adaptation, and community flourishing.

1.1.6 Novelty of the Study

The chapter's uniqueness lies in its cross-regional and comparative methodology. Other research typically documents Indigenous ecological knowledge for nation-states or regions; however, this study systematically compares six Indigenous societies across four continents. By doing this, it is possible to find both new ideas that are specific to a certain situation and general rules for how to be resilient. Second, the chapter changes the focus from seeing Indigenous knowledge as an add-on to making Indigenous leadership and governance the main focus of sustainability. Indigenous systems are frequently characterized as "traditional knowledge" to be appropriated for wisdom, disregarding the political autonomy and leadership that sustain these systems.

By focusing on Indigenous governance, from Zapatista self-government to Maasai conservancies, the study highlights the interdependence of ecological stewardship and social justice. Thirdly, the analysis clearly connects Indigenous traditions with more recent scientific writing on climate adaptation, protecting biodiversity, and socio-ecological resilience. Instead of showing Indigenous systems as old or separate, the chapter shows them as flexible and adaptable ways of thinking that can help us understand new policy ideas. This integrative perspective aids in reconciling epistemological disparities between Indigenous and Western knowledge systems. The study provides a replicable analytical framework for understanding Indigenous resilience by identifying practices, assessing ecological outcomes, and evaluating governance mechanisms. Such a framework may assist policymakers, researchers, and activists in converting localized experiences into broadly applicable sustainability strategies while preserving cultural specificity.

1.1.7 Objective of the Study

The chapter's overall objective is an analysis of Indigenous societies from different ecological and cultural backgrounds contributing to sustainability and resilience amidst socio-ecological disasters. The study initially examines Indigenous ecological knowledge and practices—such as agroecology, rotational grazing, sacred forests and conservation, and ethnobotany—across six case studies. It subsequently juxtaposes the practices with contemporary scientific research on climate adaptation, biodiversity conservation, and socio-ecological resilience, emphasizing both points of convergence and unique contributions.

A second objective is an evaluation of the replicability and policy relevance of the Indigenous strategies with an integration into wider debates on sustainability, environmental justice, and governance. Third, the chapter points out Indigenous leadership and governance as primary entry points for an integration of ecological stewardship and social justice in sustainability discourse at the global level. Overall, the objectives position the study between Indigenous studies, sustainability science, and environmental governance and show an eco-conscious future not as one based on technological innovation alone but one based on the recognition and elevation of Indigenous leadership.

1.2 Rationale of the Study

According to Steffen et al. (2015), the expanding socio-ecological calamity resulting from unsustainable development trends calls for immediate intervention harmonizing environmental sustainability and the pursuit of social justice. Indigenous communities have historically been key actors in environmental protection with strongly-rooted knowledge systems appreciating ecological integrity and sustainability of natural systems. Even so, against their immense body of contributions, Indigenous knowledge systems are broadly excluded from prevailing environmental policies and development paradigms. In response, the research is motivated by the necessity to integrate Indigenous resilience-building as a means to advance sustainability initiatives globally, recognizing their capacity to prevent environmental degradation and maintain long-term ecological integrity.

An analysis of case studies from various geographic regions underscores Indigenous peoples' creativity and adaptability in addressing environmental vulnerabilities, including deforestation, climate variability, and the decline of biological diversity. The research also aims to challenge dominant development paradigms that prioritize economic growth over ecological integrity and Indigenous peoples' land ownership and control. In response, there must be a transition to environmentally inclusive governance processes that recognize and incorporate Indigenous knowledge systems. Accordingly, the research aspires to contribute to sustainable development discourses with an advocacy for policies based on the empowerment of Indigenous peoples with a guarantee of their leadership in environmental decision-making and the accomplishment of an environmentally-conscious as well as brighter future.

1.3 Materials and Methods

This research employs a qualitative case study research technique for an examination of adaptive, sustainable development policies thought of with consideration of Indigenous people while reinserting traditional ecological knowledge back into modern-day conservation. The research examines six case studies depending on differing geographical locations: Chiapas Zapatista, Mexico; Xingu Indigenous Park and the Kayapo people, Brazil; the Maasai Mara system, Kenya; Sámi Northern European reindeer people; the Himalayan Indigenous people, India; and the Amazonian people, Brazil. These case studies are selected depending on sustainable environmental management and ecological antagonism accounts. The research selects six case studies depending on differing geography and exemplary Indigenous responses depending on environmental disasters. Each case exhibits differing yet correlating patterns of resilient responses, sustainable resource utilization, and resistance against ecological disaster with an informative examination of Indigenous-led environmental conservation depending on differing ecosystems. The research employs a study base of research literature, environmental briefing papers, and statistical records depending on forest clearance rates, climate antagonism, and erosion of biodiversity.

Sources of knowledge are peer-reviewed journal articles, books, and NGO briefing papers as primary knowledge sources, and secondary knowledge sources are governmental policies, Indigenous statements, and press coverage. The research employs thematic analysis according to key patterns depending on Indigenous-led conservation, climate adaptation, and sustainable resource utilization. The research employs qualitative results over statistical records so as to enable an informative investigation of Indigenous knowledge systems and knowledge transferability in modern-day environmental governance. With the comparison of results between differing Indigenous people, the research aims to have a comparative study of exemplary sustainability policies using traditional ecological knowledge.

The study is primarily based on a large body of literature and case study examination of what is already out there. No primary field research was undertaken as there was a time and resource constraint. It, however, examines case studies, reports, and datasets already in print. The data sources comprise secondary data and available literature. There are case studies in the chapter based on availed data or reports, specifically where field research was not possible. There is, however, considerable research undertaken in order to examine case data and argue with evidence.

1.3.1 Methodological Flowchart: Overview of the Research Process

Figure 1.1 shows the flow of the adopted methodological framework in the chapter. We used a qualitative, case-based methodology via secondary sources, thematic coding, and comparative analysis for examining Indigenous-led conservation frameworks globally and in the region.

The research employs a qualitative, case-study research approach designed to examine Indigenous-led conservation strategies at the international and regional levels. It begins with the selection of the research subject with diligence, followed by the selection of relevant environmental challenges from the international to the regional domain. Six case studies were chosen taking into consideration their

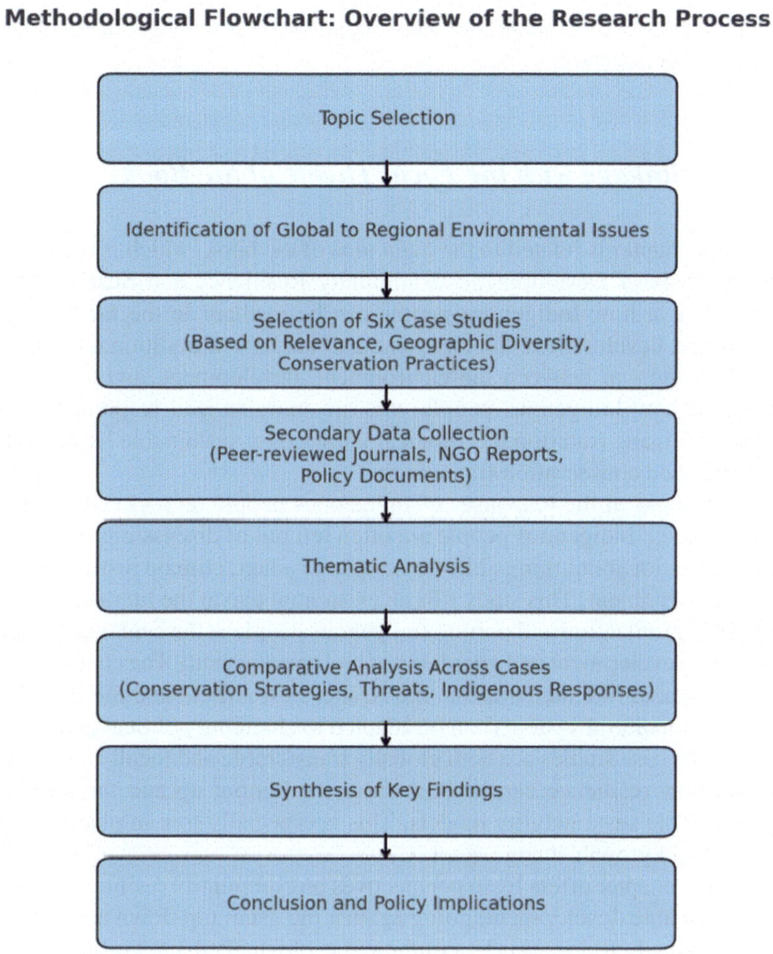

Fig. 1.1 Methodological flowchart: an overview of the research process

ecological significance, geographic coverage, and Indigenous conservation strategies. The secondary data were collected from peer-reviewed journal articles and non-governmental report documents. Also, data collection was done from policy papers and acted as a rich and variable evidence base.

Thematic analysis was then applied in identifying patterns, strategies, and responses that are recurring between and within selected cases. It was followed by a comparative analysis, variations and similarities in conservation strategies, ecological hazards, and Indigenous responses being examined. Results were incorporated with the potential of showcasing key best practices and lessons upon which study conclusions as well as policy responses were founded. This research framework allows for a systematic and stringent research process with an overriding focus on the interactions between Indigenous knowledge systems, ecological sustainability, and community resilience.

1.3.2 Interlinkage with the Focal Theme of the Book

This research chapter is related to the main idea of the book, which is called "Socio-Ecological Crisis of Development: Community Resilience and Sustainable Solutions." It looks at how Indigenous people can be resilient in the face of environmental damage, development displacement, and the loss of traditional heritage. The study bridges the gap between the environment, development, and social disaster by looking at how Indigenous people who are marginalized respond to environmental hazards using traditional ecological knowledge, sustainable livelihoods, and community-based conservation frameworks.

The study looks at the responses of Indigenous people on a global, continental, and regional level. Indigenous people are often left out of discussions about growth, but they know a lot about things like being able to adapt, climate justice, and taking care of the environment. This study fills an existential gap in the literature by turning the script on victimization and putting Indigenous people at the center as the creators of sustainable development and guardians of biological wealth. The chapter uses case studies from places like the Amazon, the Arctic, East Africa, and the Himalayas to show how sociocultural systems can be adapted to changing political and ecological conditions. The case studies are both globally transferable and locally adapted. Each case shows how resilience can be built from the bottom up and through culture, using transferable sustainability models. This is especially true in places where the environment and growth change quickly.

Overall, the chapter offers fresh perspectives on community resilience as a way to achieve sustainable development, going against the usual top-down and supposedly exploitative growth. It does this by combining ecological ethics, grassroots politics, and fairness into a single framework, which is exactly what is needed to fix the socio-ecological problem that the book is trying to fix.

1.4 Results and Discussion

1.4.1 Case Study 1: The Zapatista Movement and Indigenous Environmental Governance in Chiapas, Mexico

The Zapatista Revolution, led by the Ejército Zapatista de Liberación Nacional (EZLN), was born as a direct response against the systematic marginalization and environmental devastation faced by Indigenous people of Chiapas, Mexico. Chiapas is one of Mexico's richest but poorest revenue-gathering provinces, and it is long home to conflict over lands between Indigenous people and large-scale agriculture enterprises, sawmills, and the Chiapas province government. At first and foremost, the rebellion was meant to give indigenous voices voice over national policy agenda in Mexico (Toledo et al., 2003). For most of the preceding five centuries, the voices of the indigenous people have been violently suppressed or quietly overlooked. Indigenous people are regularly mistreated with some of the poorest work conditions in Mexico, and their lands and natural resources have been taken. The Zapatista movement, an alliance of mostly Indigenous Tzotzil, Tzeltal, Tojolabal, and Chol people, has been central in standing up against environmental exploitation while promoting sustainable lands development and self-government (Harvey, 1998). The Zapatista movement was central in supporting Indigenous people's lands and environmental sustainability (Van der Haar, 2004). Cleaver (1998) writes that, for a long time before the rebellion began in 1994, people in Chiapas mainly used legal means of protest, such as marches and rallies that often traveled as far as Mexico City and to demand things such as the recuperation of lands taken away from people. The Mexican government replied against such legal appeals with favors (to empower tools of power locally) and incessant stalling, especially where it concerned lands appeals. As appeals for lands reform kept flooding in, the government did not want the power of the region's elites weakened, so it urged people to relocate into forests that weren't being farmed. Because of that, people from different regions of Chiapas and other regions in Mexico hewed new farmlable lands and towns out of the forests.

1.4.2 Historical Context and Emergence of the Zapatista Movement

The Zapatista uprising formally began on January 1, 1994, with the agreement constituting the implementation of the North American Free Trade Agreement (NAFTA) (Stahler-Sholk, 2010). The agreement accelerated the adoption of neoliberal economic policies supporting privatization and deregulation, putting Indigenous communal ownership of lands at risk and paving the ways for expansion of extractive businesses (Collier & Quaratiello, 2005). Russell (2005) explains that the movement's declaration of war against Mexico's government stemmed from long histories of dispossession of Indigenous lands and the ecological effects of extractive capitalism. The Zapatistas established autonomous towns in Chiapas, debunking the development paradigms of the nation-state in favor of self-determining systems of self-governance combining historical Indigenous systems with modern sustainable systems. These are towns with primary focus on common ownership of lands, sustainable farming using agroecological farming, and the protection of rainforest regions with rich biological diversity. The Lacandon Jungle, one of the most biodiverse locations in Mexico, is one of the focal points of confrontation between Indigenous people, agriculture modeled after large-scale businesses, and eco-preservation policies. Expansion of grazing of cattle, commercial planting of forests, and plantations based on monoculture agriculture—typically with the support of the nation-state and large corporate bodies—has led to deforestation, erosion of lands, and alienation of Indigenous farming lands (Bobrow-Strain, 2007). As observed by Toledo et al. (2003), these agronomic techniques ensure not merely agrarian sovereignty over food but also carbon storage and biological diversity preservation. The Zapatista movement resists such environmental disasters with the occupations of lands, grassroots ecological preservation policies, and judicial challenges over Indigenous ownership of lands. According to Toledo et al. (2003), among the Zapatistas' major achievements is the communitarian management of recuperated lands where Indigenous farmers work lands using sustainable agronomic techniques, including agroforestry, crop rotation, and organic farming.

1.4.3 Zapatista Autonomous Governance and Environmental Sustainability

The Good Government Councils (Juntas de Buen Gobierno) of 2003 are an important part of Zapatista self-government because they make sure that environmental policy is both good for the environment and meets people's needs. The groups make rules that limit harmful activities like cutting down trees, mining, and using chemical pesticides. They also require reforestation and traditional water management (Rus et al., 2003). Cerqueira (2024) looks at how the Zapatista movement understands the relationship between people and nature and focuses on how their struggle is based on a Mayan-derived Indigenous cosmology. She says that the movement's fights for environmental and political justice are connected because their desire for independence is shown through sustainable agroecological practices. By putting care, respect, and equality at the top of their farming, the Zapatistas don't just fight against the dominant system of extractivist modernity; they also change the way people interact with the environment in ways that break down the long-dominant system of capitalist modernity.

Their opposition to using genetically modified seeds and chemical fertilizers in farming is in line with larger Indigenous reasons for being able to survive in harsh environments and adjust to climate change. The Zapatista system of self-government gives people on the ground control over their surroundings. This is an example for conservation efforts around the world. The Zapatista system of food sovereignty also puts a lot of emphasis on native crops, communal gardening, and saving seeds. This is very different from the monoculture-industrial agriculture that is common in the rest of Mexico.

1.4.4 Global Implications of the Zapatista Movement on the World and Lessons Learned

The Zapatista struggle for environmental justice is also important for the global movement of Indigenous resistance. Their model for self-determined growth goes against the environmental policies that are most popular and are based on extractivism. The Zapatistas offer an option to environmental government that is based on making money from the top down. Their plan is based on long-term economic growth and protecting nature through community-based efforts (McMichael, 2012). Also, the fact that they put a lot of stress on territory and sovereignty shows how protecting Indigenous lands could help fight climate change. Compared to state-managed conservation areas, forests that are handled by indigenous people have lower rates of deforestation and higher amounts of biodiversity (Stevens, 2014). Russell (2005) shows how the Zapatista Movement used the internet to help globalization work. Small protests in different countries got attention and support from people all over the world. With the help of digital communication, dominating discourses could be questioned and support could be built across borders. A lot of people can even connect their own computers straight to networks like EcoNet from the Association for Progressive Communications. The Zapatista Movement is a great example of how Indigenous people can stay strong and protect the environment. Through autonomous towns, they reclaimed their land, set up sustainable farming, and fought against practices led by corporations that were bad for the environment. Their plan, which is based on self-government, ecological stewardship, and strong communities, is interesting and could be useful for the global environmental movement. Recognizing and supporting Indigenous-led conservation is a key part of tackling the disasters of climate change, wildlife loss, and social and environmental inequality that are all linked.

Their methods, like agroforestry and not using chemical fertilizers, are not only long-lasting, but they can also handle the changes brought on by climate change, like uneven rains and soil erosion. These are specifically designed to help the Zapatista villages in Chiapas deal with the unique environmental problems they face, but they may be used by everyone. The Juntas de Buen Gobierno (Good Government Councils) protect the rights of indigenous people to their land by passing laws that support environmental justice and survival. Zapatistas have fought back against corporate and government-led efforts to export their land through grassroots democracy and legal lobbying. They have also pushed for policies that support community self-governance over economic pressures from outside the community. It also talked about similar case studies that happened in other places and showed how community-led protection worked in those places. Their model shows how to fight extractivist, top-down development models and come up with more inclusive, bottom-up policies. For example, it shows how to include Indigenous knowledge in the international agenda for climate justice. Researchers Rus et al. (2003) say that the Zapatista uprising is similar to other Indigenous uprisings around the world. They say that the Zapatistas are resisting extractivist practices and want an end to them. They also want self-government over lands that are over-claimed and occupied by Indigenous people, which is different from what is claimed and occupied today. Agroecological farming, managing shared lands, and advocating for change are all examples of similar practices that show how the Indigenous global movement for self-government and environmental justice is growing.

The Zapatista movement's success in supporting sustainability and fighting against environmental exploitation is a great example of how Indigenous peoples can start and lead solutions to climate change. Grassroots groups go against the common belief that environmental policy should be based on making money and set out a plan for more sustainable and fair action in other places. These and other examples from around the world, like the Amazon and African nature protection, show that Indigenous peoples are very important in fighting climate change and the loss of biodiversity.

1.4.5 Case Study 2: The Kayapo People in Brazil's Xingu Indigenous Park

The Kayapo are an indigenous group that lives in the Xingu Indigenous Park in the Brazilian Amazon. This park is protected. The Xingu is a very important place for the environment because it has big rainforests, rivers, and many different kinds of plants and animals. There are a lot of Indigenous people who live there. The Kayapo have a long history of religious and cultural ties to the land. They are most famous for being warriors and for their beautiful body painting. They get food, medicine, and building materials from the rainforest, which is how they make a living. Mining, deforestation, and large-scale farming are taking over more and more of their land. This is bad for their way of life and the environment in the Xingu area. Illegal logging, farming with cattle, and gold mining have always been threats to the Kayapo. All of these activities destroy forests and pollute the land where the Kayapo live (Schwartzman et al., 2013). More than illegal logging, cattle farming, and gold prospecting have always been bad for the Kayapo because they destroy forests and pollute their land (Schwartzman et al., 2013). In the past few decades, building things like roads and dams has made it even harder for them to clear their land (Fearnside, 2017).

According to Pretty et al. (2009), these development projects are ruinous on river systems, displace individuals, and are harmful to biological diversity. Even though the Brazilian government legally outlined Indigenous lands, there is weak enforcement, and corporate interests override Indigenous property rights. The Kayapo have responded with planned resistance against the threat. Their most notable activism was in the latter end of the 1980s, where they protested against the Altamira Dam venture that was set to flood large tracts of their lands. The resistance of the Kayapo even includes direct confrontation with illicit loggers and planned collaboration with environmental civil societies, with an efficient advocacy voice internationally. Their advocacy work with the courts, including resorting to international press, was central in winning lands and forcing the Brazilian government to halt environmentally degrading enterprises such as the Belo Monte Dam. With an open protest, Kayapo leaders, including Chief Raoni Metuktire, appealed using the international press so as to force the Brazilian government to abandon the project (Turner, 1993).

1.4.6 Sustainable Practices and Environmental Stewardship

Sustainable land management is used by the people of Kayapo for long-term rainforest health. They apply customary agroforestry practice, rotational cultivation, and controlled burning for long-term soil fertility and richness of biodiversity (Zimmerman et al., 2001). In contrast with large-scale farming that depletes the land, the agricultural activity of the people of Kayapo restores the ecosystem. Among their best conservation methods is the adoption of sustainable economic enterprises and community-based eco-tourism, including handicrafts and the harvesting of Brazil nuts (Schwartzman & Zimmerman, 2005).

These kinds of businesses not only keep their culture alive, but they also help people make money without using too much land. These methods make sure that people always have food and also make the rainforest less likely to be hurt by natural disasters that happen outside of it. They can now use satellite images and drones to check on illegal tree cutting more easily because they work with groups like the Rainforest Foundation and the Amazon Conservation Team (Nepstad et al., 2006). It is important for the rights of Indigenous people to land that the Brazilian government change the way it handles the environment. Many states chose mining and farming over protecting the environment (Ferrante & Fearnside, 2019). Because people need more food, soy, and beef, there has been more deforestation and land theft, which makes life harder for Indigenous people. Nevertheless, there are some legal mechanisms protecting Indigenous land in the international agreements such as the UN Declaration on the Rights of Indigenous Peoples.

People of Kayapo continue fighting against their rights and demonstrate, take direct action and even file lawsuits, although the situation is getting more complicated. Indigenous people and environmental justice are now being combated by people across the globe. The Kayapo people are also struggling to save their land as a model to all who would wish to combat climate change. This will enable the entire world to combat climate change immediately by ensuring trees that trap carbon are not cut down (Walker et al., 2020). The Kayapo residents of the Xingu Indigenous Park can serve as a good example to explain the necessity of Indigenous people to preserve the nature and combat the elements that destroy it. Kayapo has long been practicing agro forestry, rotating crop farming as well as controlled burning. This demonstrates the role of the significance of the Indigenous people in the conservation of wildlife. These traditional methods of doing things maintain the earth healthy and save the numerous plants and animals, which are found in the Amazon rainforest.

Their community-led protection is a great example of how to protect the environment all over the world. The Kayapo's fight against a world search for climate justice is to protect the land, which stops cutting down trees, which is a big reason why the climate is changing. Conservation efforts led by indigenous people, such as the Zapatistas in Mexico or the First Nations in Canada, show how important it is for people to be able to control their own land and use it in a way that doesn't hurt it in the long run. This is good for the health of the environment both locally and globally.

1.4.7 Case Study 3: The Maasai Mara Ecosystem in Kenya

Maasai are semi-nomadic Indigenous people whose traditional pastoralist livelihood, with sustainable grazing of their animals, is key in maintaining ecological balance of the Maasai Mara ecosystem. According to Wang et al. (2020), their rich knowledge about animals and lands is key in maintaining the region's biodiversity. Maasai are a people of the Nilotic with exceptional semi-nomadic pastoralist culture and dress, and are predominantly found in northern, central, and south Kenya, and north Tanzania in the African Great Lakes. Maasai traditionally are animists with belief in the existence of spiritual energy in nature. Maasai are found in north, middle, and south Kenya and north Tanzania regions and the immediate environs and around Maasai Mara National Reserve and Serengeti National Park. Maasai Mara in Kenya is one of the most famous wildlife destination and is home to the Maasai people as key players in the conservation activity in the region and in the sector of tourism. Maasai Mara is found in south-western Kenya and is one of the most diverse regions in Africa in diversification of the range of ecosystems and in diversification of wildlife. Distinguished are the Maasai Mara's savannah landscape and the Great Migration of wildebeests and zebras and other animals. Maasai Mara is about an approximate size of about 1,510 km^2 and is part of the greater Serengeti-Mara ecosystem covering large extents in Tanzania. There exist some safari tour operators promoting sustainable eco-tourism and Maasai culture. In the face of mounting threat of danger from tourism and private ownership of lands, Maasai have resorted to utilizing the systems of community-based conservation and eco-tourism systems that are not merely valuable for sustainable protection of wildlife but also ensure Maasai voices are core components of the lands and development of the community. Maasai Mara people created 24 conservancies covering an area of about 180,000 ha (450,000 acres) of lands, basically doubling the wildlife habitats away from the Maasai Mara National Reserve. Maasai Mara tribe sustainability strategy is attaining balance between customary ways and maintaining the unique Maasai region's special ecosystem and wildlife, frequently with sustainable Maasai tourism and community conservancies (Rimisho & Matei, 2024). Literature shows that sustainable tourism, community-based conservation, and human-wildlife conflict management are key drivers for long-term environmental and Maasai people's welfare (Onchwati et al., 2010).

1.4.8 Challenges Facing the Maasai Mara Ecosystem

In the past, there was common ownership of lands among the Maasai that accommodated rotational pasturing and sustainable resource utilization. In the 1960s and 1970s, the Kenyan government, however, initiated reforms in tenure in lands with privatization and individualisation of lands and subdivision of lands into single ownership plots. The move was targeted at sparking development, with the end product being the sale of lands to non-Maasai citizens and commercial investors and subsequent subdivision into small ordinary rangelands it previously inhabited (Galvin et al., 2008). Maasai pastoralists lost grazing lands, and the end product was overgrazing in small pasturable pieces of land, low productivity in animals and increased conflict in the utilization of land. Encroachment and expansion into wildlife habitats are widespread and are herein destroying habitats and fragmenting habitats (Haddad et al., 2015). The Maasai Mara houses one of the highest predator densities on the continent of Africa, including cheetahs, leopards, and lions. With changes in lands and the loss of wildlife habitats, there is increased conflict between pastoralists and predators (Ogutu et al., 2011). Lion and hyena predation on animals resulted in reprisal killings of predators by Maasai pastoralists, which threaten one of the continent's most valuable predator populations. Livestock loss also violates the economic stability and poverty reduction agenda of the Maasai.

Climate change made the environmental damage in the Maasai Mara even worse. Long periods of drought, strange weather patterns, and strange rainfall patterns can change the way animals graze and make the pasture less plentiful and less good (Homewood et al., 2012). Overgrazing, made worse by changes in the weather, caused land degradation, with soils eroding and vegetation coverage decreasing. This made the threat to both wildlife and livestock even worse. The changes resulted in less availability of pastures, with the Maasai adjusting their utilization of lands favorable to supporting the population of both wildlife and livestock. Tourism is one of the Maasai Mara's main economic enterprises, generating millions of dollars into the Kenyan economy annually. Unbridled tourism and intensified development of infrastructure, however, came with it habitat destruction and wildlife perturbation (Bedelian & Ogutu, 2017). Moreover, wildlife conservation policies, including setting up conservancies and protected zones, ostensibly while protecting wildlife at the expense of Maasai livelihoods, have shut out grazing lands, isolating the Indigenous people even further. Even as some of the conservancies involve revenue-sharing between tourists and host peoples, not all Maasai equally share in similar policies.

1.4.9 Indigenous Resilience and Adaptive Strategies

To meet mounting demands for conservation, Maasai started Community-Based Conservation (CBC) programs, such as the Mara Conservancies model, that allow for economic gains for sustainable land usage and wildlife protection while ensuring

Maasai people maintain ownership of ancestral lands. Methodology allows for an economic gain for conservation and helps the Maasai people maintain ownership over ancestral lands (Reid et al., 2014). In an effort to reduce over-dependency on grazing livestock, the majority of Maasai diversify sources of livelihood. Some are now in the tourism industry as safari tours, cultural performances, and working with the conservancies. Some have ventured into beekeeping, handicrafts, and small-scale agriculture, creating alternative livelihood sources and less grazing pressure on lands (Goldman & Riosmena, 2013). The Maasai people continue practising Traditional Ecological Knowledge (TEK) in land management. Seasonal migration, rotational grazing, and planned burning are among the techniques that have allowed the Maasai people historically to manage rangeland resources sustainably. Integrating TEK with modern conservation science may build integrated, holistic methods of land management beneficial for wildlife and pastoralists.

Maasai women are increasingly assuming leadership in conservation and sustainable development initiatives. Environmentally sustainable goods, including honey and traditional craftsmanship, are produced and marketed by women's cooperatives on foreign and home markets. Environmental education and afforestation activities for rehabilitating denuded lands and building up biodiversity are undertaken by women's organizations as well (Brown et al., 1987). The Maasai Mara ecosystem is one of the international case studies of achieving conservation goals with Indigenous people's rights and sustainable livelihoods. Like other international case studies, including the Zapatistas of Mexico, the Maasai model of community-based conservation exhibits the potential for marrying Indigenous knowledge and leadership with environmental policy and climate justice. Even as conservation policies mostly have been non-inclusive with regard to the Maasai, new community-based projects demonstrate that sustainable conservation is possible with Indigenous voices in decision-making. The Maasai's experiences are indicative of an increasing trend at the international level in support of inclusive practices of conservation. Like other similar trends, including that of the Kayapo of Brazil, community-based conservation models founded on Indigenous knowledge safeguard livelihoods and biodiversity as well and remain relevant for climate and conservation policies globally.

1.4.10 Case Study 4: Sámi Reindeer Herders in Northern Europe

Sámi people are an Indigenous people settling in the north of Norway, Sweden, Finland, and Russia's Kola Peninsula. They traditionally found livelihoods based on reindeer herding as a culture and socio-economic foundation of Sámi society. Their livelihood is Arctic-environment-based, currently threatened by climate and fast industrialization and uncompromising policies of the state. Regardless of such adverse conditions, Sámi reindeer herders have been brilliant in embracing old ways with an objective of preserving both the culture and endangered Arctic habitats.

Climate change is today the biggest threat encountered by Sámi reindeer herding. Increased temperatures have transformed the Arctic environment so that it is less accessible for grazing. Increased winters are cause for freeze–thaw cycles that develop layers of ice on grazing pastures and prevent entry by the reindeer to lichen, the food it feeds on during winters (Tyler et al., 2007). Compulsory state borders and state control over Sámi customary usage patterns of usage make it much harder and difficult to maintain customary usage patterns of lands, with some regimes of herding having to modify corridors and scale down number of animals in response to demands of the law. Climatic unpredictability also disrupts corridors and calving intervals and threatens survival of herds. Widening of industry imposes an additional barrier against Sámi feeding system adaptability. Mines, forestry activity, hydroelectricity development, and development of wind power on customary Sámi pastures ranges separates pastures and restricts trans-seasonal migrations (Johnsen et al., 2015).

1.4.11 Sámi Resilience and Adaptive Strategies

To counter such socio-ecological stresses, Sámi herders have adapted in numerous ways. Among such adaptations is the utilization of GPS tracking and drone monitoring to monitor reindeer migration in order better to manage herds in the face of changing environmental conditions. Various Sámi nations have also tried supplemental feeding in cold winters as an alternative for withered pasture availability. Sámi organizations have been actively engaged in political advocacy as a strategy of acquiring rights over lands and observing policies against Sámi lifeways. Sámi Parliaments established in Norway, Sweden, and Finland have brought Indigenous representation so as to increase greater participation in environmental decision-making (Lantto & Mörkenstam, 2007). Court victories, such as the Norwegian Supreme Court decision against the construction of wind farms over Sámi lands in 2021, established Indigenous sovereignty over the ground and against exploitive industrial practice. In a cultural context, the Sámi have been actively engaged in preserving and reviving the Sámi language and knowledge platforms. Traditional ecological knowledge (TEK) with knowledge on migration patterns, weather forecasting, and sustainable pasture usage is being documented and incorporated into customary and scholarly practice. Sámi young people are introduced to modern reindeer herding learning platforms combining customary practice with modern technological tools so as to ensure continuity of their heritage amidst the changing times. There are a variety of Sustainable Indigenous-Led Solutions lessons from the case. According to Johnsen and Eira (2015), the Sámi reindeer herders' case is one of Indigenous knowledge value in climate adaptation and sustainable resource utilization (Johnsen et al., 2015). Their case is a lesson in securing tenure on the ground, customary and modern technological integration and enhancing Indigenous political representation.

The example of the Sami resistance to ecological and economic pressures can be discussed as the important case study of the ability of Indigenous people to be on the forefront in creating sustainable and environmentally friendly solutions to environmental problems. Centuries-old experience on the balance between human activity and environmental protection can contribute to the sustainability agendas on a continent-level, as the Sami offer to advocate Indigenous land tenure, to advance climate justice, and appreciate traditional ecological knowledge. Their adjustment and resistance are full of knowledge on how to build and sustain resistance against socio-ecological catastrophes. The case study demonstrates the transnational impacts of climate change on the livelihoods of Indigenous people and highlights the significance of the subregional adaptation planning using traditional ecological knowledge in promoting the large-scale climate resilience (Helgesen et al., 2024; Næss et al., 2021). The insights play a pivotal role in the international policymaking endeavor to ensure the incorporation of Indigenous perspectives in the policies of environmental and sustainable development (Jääskeläinen, 2020; Oksanen, 2020). These insights will prove quite handy to international climate policymakers who desire to incorporate the Indigenous voices in climate resilience planning and sustainable development planning.

1.4.12 Case Study 5: Indigenous Communities of the Himalayas in India

People who have always lived in the Himalayas, like the Sherpa, Tamang, Lepcha, Bhutia, and Limbu, have been very important in protecting the environment for a long time. They were able to live together in the Himalayas' fragile ecosystem for hundreds of years because they had a spiritual connection to nature. They have had good ecological knowledge and stable ways to make a living. Climate change, cutting down trees, and building new roads and bridges are all things that could happen today. In response, these people are innovating with trail-blazing conservation innovations marrying traditional ecological knowledge with modern principles of sustainability (Folke et al., 2016). Mountains and rivers with forests are sacred for many Indigenous people of the Himalayas and are therefore reflected in religious and culture-based belief systems in conservation. In Sikkim, India, for example, the Lepcha people hold the landscape in the Khangchendzonga as sacred, and it was conserved with ensuing conservation of the biodiversity. Community Conservation Areas (CCAs) are delineated as forests where there is ownership and control over management of the forests and forests' products, crucial for their livelihoods, and therefore linking people and Indigenous lands with conservation of biodiversity. These are exemplified in the North Eastern Indian States with a variety of ownership, size, systems of management, and purposes of conservation. Regional and Autonomous District Councils of the Sixth Schedule of the Indian Constitution safeguard people's traditional and customary rights in North East India. There are as many as 16 district councils in North East India today, with three in Manipur, Mizoram, Assam and Tripura. In Arunachal Pradesh, India, Indigenous people have established Community Conservation Areas such as the Chug Valley CCA spanning over 92.5 km^2 and with bans against commercial logging and hunting. The initiatives there have been very rewarding with the conservation of an endangered fauna such as the red panda and clouded leopard among other faunas. Sustainable agriculture systems such as terrace farming, agroforestry, and crop rotations are against soil erosion and ensure preservation of soil fertility, major reasons for preserving crop and agricultural biodiversity with pressure from climate changes. Transhumance with the Sherpa and Tamang people against overgraze prevention leads to keeping the soil fit and healthy with agrobiodiversity in Himachal Pradesh. An ancient activity nicknamed Barah Anaaj as a polyculture system double-serves as food provision and with soil health and with maintenance of agrobiodiversity in Himachal Pradesh.

1.4.13 Contemporary Conservation Initiatives: Surya-Kunj Eco-Restoration Project, India

Surya-Kunj Eco-Restoration Project situated over an extent of about 71 acres and established in the year 1992 on the property of the Govind Ballabh Pant National Institute of Himalayan Environment (NIHE) at the city of Almora, India, is an exemplary case of community participation in forests with sustainable income and education extension for the promotion of recovery of biodiversity (Negi et al., 2025). It is an excellent example of community-led forests with sustainable income and recovery of biodiversity. Surya-Kunj encourages the recovery of degenerated ecosystems and biodiversity with participatory management of forests and community participation with income and education extension. Surya-Kunj demonstrated the wonder of restoration of forests for nearly three decades with better forest arrangement, recovery of biodiversity and better carbon sequestration, benefiting both the people and the environment. Important restoration challenges faced are forest fires, grazing along with livestock, and the non-availability of plenty of water needed for plant regeneration. Surya-Kunj is exemplified as a robust arboretum and an exemplary restoration case with immense impact on society and the environment. UN Decade on Ecosystem Restoration emphasizes large-scale long-duration restoration of ecosystems, and Surya-Kunj, being one such restoration, is an exemplary case. Restoration activity was complemented with supporting afforestation with native vegetation for fostering biodiversity, rainwater harvesting techniques with supply of soil moisture and better fertility, and eco-tourism and sustainable livelihood for the people. The project was an enormous achievement with the augmentation of forest coverage, the restoration of the fertility level of the soil and the minimisation of human-animal conflict.

In 2024, the Wildlife Institute of India (2020a, 2020b) reported that people converged at the Harela Festival where they planted native plants during the forest fire season. This was an indication of a powerful desire to replant trees and powerful intentions. The primary objectives of the project will be to engage individuals in the community, to enable them earn income in the long run, and educate them on the environment. In the year 2024, there was a Harela festival where people gathered to plant native plants to restore the forest following a fire. This demonstrated their extreme dedication to the issue of planting trees once more. It is significant that the local and global reasons are the ways indigenous people in the Himalayas care about nature; they are a good example of taking care of nature in the long run. They also show how traditional rights can help protect biodiversity (Dangwal et al., 2022). The case from the region is useful in teaching other international conservation initiatives that seek to balance modern practices with contemporary sustainability paradigms (Friends of GHNP; Wildlife Institute of India, 2020a, 2020b).

1.4.14 Great Himalayan National Park Community Partnership, India

The Great Himalayan National Park (GHNP) cooperates with the people efficiently as well by initiating livelihood development, eco-tourism, and awareness-building programs with the intention of maintaining a harmonious co-relation between the people and the natural values of the park. It created a sustainable balance between people's livelihood and natural resource protection. Park managers, along with institutions such as BTCA, work toward people empowerment, especially of women, through asset building, logical utilization of natural assets, as well as socio-economic development. All of the above initiatives have a propensity for less dependency of people on the assets of the park through initiating alternative livelihood development opportunities. The Sainj Wildlife Sanctuary, populated with 120 people, and the Tirthan Wildlife Sanctuary, without people but open for traditional grazing, are part of the GHNPCA, and hence, active grazing and human settlement management are required. Friends of GHNP is a volunteering society that plays an indispensable role in generating awareness about the conservation intervention of the park and supporting the development of the rural societies. Friends of GHNP have spread eco-tourism training among the men and initiated early development of street theater. The GHNPCA is an exemplary of community-based eco-tourism in the Himalayan domain. The GHNP cooperates with institutions such as the Wildlife Institute of India for research concerning the socio-economic and biological aspects. Region-wide intimate knowledge of the people about the landscapes governs efficient and culture-sensitive management regimes. The Great Himalayan National Park Conservation Area was inscribed as a UNESCO World Heritage Site in the year 2014. Inscription of the park deems that it is an exceptional site for the conservation of biodiversity with the protection of over 1,000 plant, 31 mammal, and 209 avifaunal species. The Great Himalayan National Park (GHNP) in Himachal Pradesh, a UNESCO World Heritage Site, actively incorporates Indigenous people in conservation. The management of the park cooperates with hunter-gatherer people and makes sustainable livelihood programs (beekeeping, eco-tourism, cultivation of medicinal plants) for the people. They train people concerning community training on monitoring of the biodiversity and introduce sacred forest sites, where no cutting and hunting is tolerated. This co-operation not only lessened the extent of the spread of poaching but also attained an increasing number of wild animals and livelihood promotion, and as a whole, saved biodiversity better with an increase in people-friendly stewardship of the natural assets.

1.4.15 Case Study 6: Amazonian Tribes in Brazil—Indigenous Resistance and Environmental Stewardship

The Amazon Rainforest, the "lungs of the planet," is home to over 300 Indigenous groups, including the Yanomami, Munduruku, Ashaninka, and the Kayapo, all of whom long coexisted with the forest. For hundreds of years, people lived in harmony with the forest, using traditional knowledge to manage the environment in a way that was good for the long term. But they are always in danger from deforestation, illegal mining, and climate change. They are still important to the work of stopping ecological decline because they are flexible and use smart environmental methods. Their Traditional Ecological Knowledge (TEK) and Sustainable Practices are both important parts of the answer.

Indigenous people in the Amazon use Traditional Ecological Knowledge (TEK) to keep the forest healthy and promote biodiversity. They do it through controlled burning, rotational farming, and agro forestry. They do not cut a great deal of trees, but attempt to add nutrients to the soil using controlled burning, rotational farming, and agro forestry (Fearnside, 2017). An example is the mosaic landscape adopted by the Kayapo where there is an amalgamation of forests and a large amount of farmland in a manner that does not affect the environment (Nepstad et al., 2006). This forms a system that is sustainable and with low ecological footprint. The Yanomami have been able to learn to grow medicinal plants productively and preserve the biodiversity, and this is depicted by their conservation skills. Whatever they do, the Indigenous lands remain accessible to illegal mining, gold mining, and large-scale farming. The Amazon deforestation has also increased in the recent years due to the change in policy to benefit agribusiness and extraction economies (Ferrante & Fearnside, 2019). This endangers the life of the Indigenous people and endangers the forest survival. During the recent years, the Amazon has been increasingly deforested due to the alterations in the policy that favored the agribusiness and extractive economies. This endangers the existence of this forest just as it endangers the lives of Indigenous people (Díaz et al., 2018). Unlawful mining polluted rivers with mercury that caused serious medical crises among Indigenous people and especially among Yanomami and Munduruku communities. Further, the customary fishing and agriculture practices have been adversely affected by the infrastructure development (hydropower dams) thereby affecting food security.

1.4.16 Community-Led Environmental Protection Initiatives

Amazonian people have responded to the necessity of the territory protection. The Kayapo have formed patrol teams to clear out and expel miners and loggers and applied custom knowledge along with technological knowledge to protect the territory. With the help of NGOs as varied as the Amazon Watch and Instituto Socioambiental (ISA), they are using satellite monitoring and planes for delineating deforestation in real time and reporting violation cases to the authorities. The Munduruku people won against the São Luiz do Tapajós dam on grounds of the law, stopping the project due to environmental and societal effects (Ferrante & Fearnside, 2019). In addition, Indigenous people are partakers in international climate negotiations and are demanding greater legal protection of lands. Amazon leaders, whom COIAB Indigenous people represent, in 2021 urged an affirmation of Indigenous rights over territory as one key climate solution. There are Restoration and Conservation Projects as well.

There are a lot of tribes that are working on reforestation and conservation projects. For example, there is an Ashaninka reforestation project that works with conservation groups. They mix traditional agroforestry with scientific research to restore the ecosystem and keep Indigenous people's jobs (Nepstad et al., 2006). The Xingu Seed Network is another successful project that helps Indigenous women collect and share native seeds to help with reforestation projects and promote forest regrowth and diversity. These projects show that Indigenous conservation can be a different way for the Amazon to grow in a way that lasts. The Amazonian Indian people of Brazil are leaders in environmental protection. They use traditional knowledge and new technologies to protect their lands. They keep fighting against deforestation, pollution, and land grabs, even when there are threats from outside. They do this by using legal challenges, surveillance campaigns, and international advocacy. The importance of Indigenous land rights for protecting the environment and the climate is what is most important. It is not only Amazon, but also the entire world, that cares what will preserve their knowledge and assist their projects.

Little (2008) assumes that the environmental preservation of the Indigenous inhabitants of Amazonia serves as a prime example of the primary relationship between the traditional ecological knowledge and the struggle with transnational deforestation and climate change (Fearnside, 2017; Ferrante & Fearnside, 2019). All people can find value in the way that they use new technology and Indigenous self-governance, and their models of sustainability and climatic justice, led by Indigenous people, can serve as examples that can be emulated by others (Nepstad et al., 2006; Coordination of Indigenous Organisations of the Brazilian Amazon, 2010).

1.5 Conclusion

The six case studies that have been generated through this study provide unique information about the Indigenous societies in different parts of the world in their efforts to solve ecological problems. The case studies and research papers reveal some of the common strategies and context-specific approach to methods of deep ecological understanding and cultural energy, through comparative analysis. Table 1.1, "Response of the Indigenous Community to Socio-Ecological Crises," shows examples of the Indigenous societies of various parts of the world, facing the ecological issue with the help of community-based, culturally-oriented, and sustainable approaches. It unveils the tendons of customary knowledge, environmental management and flexibility of knowledge as cultures respond. The following discussion highlights the themes that are replicated in the six Indigenous cases, namely: common environmental issues, dependence on traditional ecological knowledge, community-based reactions, and seeking cultural and sustainable ways of existence. As it can be observed with the Sami and the folks of the Amazon, the Maasai Mara and the Kayapo, the fact that Indigenous people possess a very valuable ecological knowledge, known as TEK, is vital in ensuring that ecosystems are healthy and diverse.

The social cultures all employ the traditional practices such as agro forestry, grazing on a rotational basis, as well as the controlled burning of plants to acquire the natural resources in a manner that is environmentally friendly. To prevent the destruction of the systems and to prevent climate change, we should defend native knowledge systems. Some of the issues that continue to take place among the Indigenous societies are deforestation, economic development (infrastructure construction and mineral exploration), land acquisitions, and climate change. Groups such as the Kayapo, the Maasai, the Sami, and the Himalaya people have consequently devised a successful strategy in countering this. These do much, whether it comes to direct involvement, such as protesting against the dam projects, and monitoring of individuals who partake in deforestation without authorisation, or it is political, judicial, and collaborations with the environmental non-governmental organizations.

Cultural and livelihood adaptations are the overriding theme in all the case studies. The native cultures have exceptional skills to survive the environmental changes and financial difficulties. The Sami rely on GPS technology to assist them in managing their reindeer, the Maasai earn money through handicraft and eco-tourism and the Amazonian societies participate in the mass reforestation programs. Such societies evolve with the existing cultural systems. Such flexibility demonstrates that community-based conservation can be permanent and lead to the positive effects on the surrounding and economy. The second most common thing is the Collaborations and Partnerships. Some of the societies in these case studies collaborate with external parties in an attempt to advance their interests as far as conservation is concerned. In the Sami case, the Sami collaborate with environmental organizations to conserve the reindeer grazing areas and tribes in the Amazon conserve cut down areas by collaborating with conservation organizations. Such unions emphasize collaboration

Table 1.1 Overview of case studies presenting the key elements of each Indigenous community's response to socio-ecological crises

Indigenous group	Environmental challenges	Sustainable solutions	Key successes
Zapatista Movement (Chiapas, Mexico)	Deforestation, exploitation of natural resources, marginalization	Autonomy, agroecology, seed-saving, community-led governance	Reforestation, Revival of traditional knowledge, sustainable livelihoods, local food security
Kayapo People (Xingu, Brazil)	Logging, mining, deforestation, threats from outsiders	Territorial surveillance, partnerships with NGOs, cultural revitalization	Legal recognition of land rights, effective territorial protection
Maasai (Mara, Kenya)	Climate change, loss of grazing land, biodiversity decline	Community conservancies, eco-tourism, wildlife protection, rotational grazing	Improved biodiversity, strengthened community governance, alternative income
Sámi Reindeer Herders (Northern Europe)	Climate change, habitat disruption, industrial encroachment	Traditional herding knowledge, participatory mapping, policy advocacy	Enhanced resilience, preservation of cultural heritage, regional collaboration
Himalayan Indigenous Communities (India)	Glacier retreat, soil erosion, deforestation, biodiversity loss	Surya-Kunj Eco-Restoration Project, sacred groves protection, sustainable agriculture, local stewardship	Restored habitats, community empowerment, cultural and ecological integration
Amazonian Tribes (Brazil)	Deforestation, illegal mining, environmental degradation, displacement	Forest guardianship, territorial monitoring, Indigenous political mobilization, traditional ecological knowledge	Slowing deforestation, global recognition of Indigenous rights, ecological preservation

across cultures in the pursuit of the long-term environmental objectives. The policy implications and protections of law are provided in each case study.

The reasons of indigenous people to seek legal regulations of territories is a valuable aspect in long-term conservation sustainability. Weak enforcement of the laws such as the one involving the fight of the Kayapo against illegal mining and logging in Brazil shows that Indigenous territory is not strongly safeguarded. On the other hand, the Sami right to wind power development and the campaign to protect the proclamation of the Himalayas Community Conserved Areas (CCAs) has examples of the power of law tools of proclaiming the authority of Indigenous territory. International Environmental Justice: All the above case studies have a basis for the contention that Indigenous peoples constitute important actors of global environmental justice and greenhouse gas mitigation. Their grassland, forest, and other land conservation constitute an international role in carbon sequestration. Attaining their rights, expanding their government role, and ensuring sustainable livelihoods constitute opportunities for enhancing environmental sustainability and social justice, according to Ostrom (2009).

Himalayan Indigenous communities, as one example, intertwine ecological value and reforestation, and Amazon Indigenous communities intertwine customary knowledge and new monitoring gear for the protection of large tracts of forests. NGO networks and international partners have facilitated so much of what would otherwise be a complex and time-consuming practice for communities like the Sámi and the Kayapo, whose cooperative work among other nations and institutions was provided legal ratification and wider global media exposure. These accomplishments coexist with institutional inequalities such as the lack of protection by the law and climatic stress that overwhelm traditional adaptation mechanisms. The case studies combined argue that Indigenous peoples do not present ecological catastrophe but are active participants of sustainable change. These responses of Indigenous people pose essential doubts of dominant paradigms of resource extraction in terms of reciprocity, in terms of diversity and terms of intergenerational care-taking. In the case of global environmental policy, the worth of Indigenous empowerment, of the incorporation of TEK in the formulation of climate policy, and fair investment in community-led conservation is what is documented by Indigenous responses.

As more and more emergencies on the planet threaten the Earth, indigenous responses promise not only to make it but to give the future of ecological justice and sustainability. It contributes to the world discussion on socio-ecological sustainability by writing about Indigenous-led actions in the presence of environmental destruction, displacement, and culture disruption in a series of ecological conditions. Building on comparative research on community-led approaches to the building of resilience among the case studies considered, the research sheds critical light on the possibility of traditional ecological knowledges, of order of governance adapted with an inclination of care, and of community-led solutions as distributive and scalable solutions for planetary environmental emergencies. These are the scientific and practical outcomes of the work.

In science, it fills the gap in the literature by centering Indigenous voices, and not on the margin, of mitigation of climate change, of conserving nature in the appreciation of biodiversity, and of sustainable development. It broadens the paradigm on resilience studies by bringing and encompassing sociocultural orderings and lived experiences under top-down and technocratic orderings of resilience. Globally, the research is enriching as it looks at a range of international agendas such as the UN Sustainable Development Goals (SDGs), Paris Agreement, and the post-2020 Global Biodiversity Framework, each of which looks at rising appreciation of the value of knowledge systems which are Indigenous and place-based. Lessons of the experience of policymakers, green activists, and developers' planners globally have value for building fairer, greener, and people-friendly development trajectories. Against the rising tide of climatic and ecological crises, it is demonstrated in the research here that the empowerment of peoples is not only morally imperative but ecologically possible for long-duration sustainability and planetary health.

Acknowledgements We would like to express our indebted gratitude to the altruistic assistance of the Manav Rachna International Institute of Research and Studies (MRIIRS), Department of English, School of Media Studies and Humanities (SMeH), Faridabad, for providing us with the much-needed support and facilities needed to complete this work. We are especially grateful to the professors for their guidance, encouragement, and unwavering belief in the value of researching emergent trends in literary and technological fields.

Authors Contribution All authors have contributed equally to this work.

Funding The research was not supported by any funding.

Potential Conflicts of Interest On behalf of all authors, the corresponding author states that there is no conflict of interest.

Informed Consent No individual participants were included in the study.

Data Availability Statement The data used in this study is derived from publicly available sources, including academic literature, environmental organization reports, and statistical databases on deforestation rates, climate change impacts, and biodiversity loss. No proprietary or restricted datasets were utilized. Any additional data supporting the findings of this study can be provided upon reasonable request.

AI Disclosure Statement The authors acknowledge the use of Grammarly [https://www.grammarly.com] to assist in the refinement of this manuscript, specifically to correct grammar, improve sentence structure, and enhance clarity in accordance with academic writing standards. The prompts used included: "Check grammar and punctuation across the full manuscript," "Enhance sentence clarity and eliminate redundancy," and "Improve coherence and flow in academic. The output from these prompts was used to enhance language accuracy, readability, and stylistic consistency, without modifying the content, interpretation, or originality of the research. While the authors acknowledge the usage of AI, they maintain that they, Ms Swati Mittal and Dr Shivani Vashist, are the sole authors of this article and take full responsibility for the content therein, as outlined in COPE recommendations.

References

Abas, A., Aziz, A., & Awang, A. (2022). A systematic review on the local wisdom of Indigenous people in nature conservation. *Sustainability, 14*(6), 3415. https://doi.org/10.3390/su14063415

Adger, W. N., Brooks, N., Bentham, G., Agnew, M., & Eriksen, S. (2005). Social-ecological resilience to climate change in coastal regions. *Tyndall centre for climate change research. Working paper 27.* http://www.tyndall.ac.uk/theme3/final_reports/it1_11.Pdf

Agrawal, A., & Ostrom, E. (2001). Collective action, property rights, and decentralization in resource use in India and Nepal. *Politics & Society, 29*(4), 485–514. https://doi.org/10.1177/0032329201029004002

Bedelian, C., & Ogutu, J. O. (2017). Trade-offs for climate-resilient pastoral livelihoods in wildlife conservancies in the Mara ecosystem, Kenya. *Pastoralism, 7*(1), 1–22. https://doi.org/10.1186/s13570-017-0085-1

BenYishay, A., Heuser, S., Runfola, D., & Trichler, R. (2017). Indigenous land rights and deforestation: Evidence from the Brazilian Amazon. *Journal of Environmental Economics and Management, 86(C)*, 29–47. https://doi.org/10.1016/j.jeem.2017.07.008

Berkes, F., Colding, J., & Folke, C. (2000). Rediscovery of traditional ecological knowledge as adaptive management. *Ecological Applications, 10*(5), 1251–1262. https://doi.org/10.2307/2641280

Berkes, F. (2018). *Sacred ecology* (4th ed.). Routledge.

Bobrow-Strain, A. (2007). *Intimate enemies: Landowners, power, and violence in Chiapas*. Duke University Press. https://doi.org/10.1215/9780822389521

Brockington, D., Duffy, R., & Igoe, J. (2008). *Nature unbound: Conservation, capitalism and the future of protected areas*. Earthscan. https://danbrockington.com/wp-content/uploads/2020/09/dan-brockington-rosaleen-duffy-jim-igoe-nature-unbound_-conservation-capitalism-and-the-future-of-protected-areas-2008-1.pdf

Brondizio, E. S., Ostrom, E., & Young, O. R. (2009). Connectivity and the governance of multilevel social-ecological systems: The role of social capital. *Annual Review of Environment and Resources, 34*(1), 253–278. https://doi.org/10.1146/annurev.environ.020708.100707

Brown, B. J., Hanson, M. E., Liverman, D. M., & Merideth, R. W. (1987). Global sustainability: Toward definition. *Environmental Management, 11*(6), 713–719. https://doi.org/10.1007/BF01867238

Buscher, B., & Fletcher, R. (2020). *The conservation revolution: Radical ideas for saving nature beyond the Anthropocene*. Verso Books. https://www.academia.edu/download/103471618/14888386.2020.181176920230617-1-1m55wa.pdf

Cerqueira, B. L. (2024). *Narrating nature: Zapatista encounters and the remaking of human–environmental relations*. Transnational Press London.

Chhatre, A., & Agrawal, A. (2009). Trade-offs and synergies between carbon storage and livelihood benefits from forest commons. *Proceedings of the National Academy of Sciences of the United States of America (PNAS), 106*(42), 17667–17670. https://doi.org/10.1073/pnas.0905308106

Cleaver Jr, H. M. (1998). The Zapatista effect: The Internet and the rise of an alternative political fabric. *Journal of International Affairs, 51*(2), 621–640. https://www.jstor.org/stable/24357524

Collier, G. A., & Quaratiello, E. R. (2005). *Basta! Land and the Zapatista Rebellion in Chiapas*. Food First Books.

Coordination of Indigenous Organizations of the Brazilian Amazon (COIAB). (2010). The nature conservancy in Brazil - COIAB Amazon Indigenous Federation. https://web.archive.org/web/20100317212302, http://www.nature.org/wherewework/southamerica/brazil/work/art18068.html

Cutter, S. L., Barnes, L., Berry, M., Burton, C., Evans, E., Tate, E., & Webb, J. (2008). A place-based model for understanding community resilience to natural disasters. *Global Environmental Change, 18*(4), 598–606. https://doi.org/10.1016/j.gloenvcha.2008.07.013

Dangwal, B., Rana, S. K., Negi, V. S., & Bhatt, I. D. (2022). Forest restoration enhances plant diversity and carbon stock in the sub-tropical forests of western Himalaya. *Trees, Forests and People, 7*, Article 100201. https://doi.org/10.1016/j.tfp.2022.100201

Díaz, S., Pascual, U., Stenseke, M., Martín-López, B., Watson, R. T., Molnár, Z., Hill, R., Chan, K. M. A., Baste, I. A., Brauman, K. A., Polasky, S., Church, A., Lonsdale, M., Larigauderie, A., Leadley, P. W., van Oudenhoven, A. P. E., van der Plaat, F., Schröter, M., Lavorel, S., & Shirayama, Y. (2018). Assessing nature's contributions to people. *Science, 359*(6373), 270–272. https://doi.org/10.1126/science.aap8826

Fa, J. E., Watson, J. E. M., Leiper, I., Potapov, P., Evans, T. D., Burgess, N. D., Molnár, Z., Fernández-Llamazares, Á., Duncan, T., Wang, S., Austin, B. J., Jonas, H., Robinson, C. J., Malmer, P., Zander, K. K., Jackson, M. V., Ellis, E., Brondizio, E. S., & Garnett, S. T. (2021). The role of Indigenous Peoples' lands in conservation. *Frontiers in Ecology and the Environment, 18*(3), 135–140. https://doi.org/10.3389/fevo.2020.578014

FAO. (2020). *Global forest resources assessment 2020*. Food and Agriculture Organization of the United Nations. https://doi.org/10.4060/ca8753en

Fearnside, P. M. (2017). Deforestation of the Brazilian Amazon. In H. Shugart (Ed.), *Oxford research encyclopedia of environmental science*. Oxford University Press. https://doi.org/10.1093/acrefore/9780199389414.013.102

Ferrante, L., & Fearnside, P. M. (2019). Brazil's new president and 'ruralists' threaten Amazonia's environment, traditional peoples, and the global climate. *Environmental Conservation, 46*(4), 261–263. https://doi.org/10.1017/S0376892919000213

Folke, C. (2006). Resilience: The emergence of a perspective for social–ecological systems analyses. *Global Environmental Change, 16*(3), 253–267. https://doi.org/10.1016/j.gloenvcha.2006.04.002

Folke, C., Biggs, R., Norström, A. V., Reyers, B., & Rockström, J. (2016). Social-ecological resilience and biosphere-based sustainability science. *Ecology and Society, 21*(3), 41. https://doi.org/10.5751/ES-08748-210341

Forbes, B. C., Stammler, F., Kumpula, T., Meschtyb, N., Pajunen, A., & Kaarlejärvi, E. (2016). Sea ice, rain-on-snow and tundra reindeer nomadism in Arctic Russia. *Biological Reviews, 91*(1), 123–145. https://doi.org/10.1098/rsbl.2016.0466

Friends of GHNP. (n.d.). *Overview: Community*. Great Himalayan National Park. Retrieved June 13, 2025, from https://greathimalayannationalpark.com/overview-community/

Galvin, K. A., Reid, R. S., Behnke, R. H., & Hobbs, N. T. (2008). Fragmentation in semi-arid and arid landscapes: Consequences for human and natural systems. *Springer Science & Business Media*. https://doi.org/10.1007/978-1-4020-4906-4

Garnett, S. T., Burgess, N. D., Fa, J. E., Fernández-Llamazares, Á., Molnár, Z., Robinson, C. J., Watson, J. E. M., Zander, K. K., Austin, B., Brondizio, E. S., Collier, N. F., Duncan, T., Ellis, E., Geyle, H., Jackson, M. V., Jonas, H., Malmer, P., McGowan, B., Sivongxay, A., & Leiper, I. (2018). A spatial overview of the global importance of Indigenous lands for conservation. *Nature Sustainability, 1*(7), 369–374. https://www.nature.com/articles/s41893-018-0100-6

Goldman, M. J., & Riosmena, F. (2013). Adaptive capacity in Tanzanian Maasailand: Changing strategies to cope with drought in fragmented landscapes. *Global Environmental Change, 23*(3), 588–597. https://doi.org/10.1016/j.gloenvcha.2013.02.010

Haddad, N. M., Brudvig, L. A., Clobert, J., Davies, K. F., Gonzalez, A., Holt, R. D., Lovejoy, T. E., et al. (2015). Habitat fragmentation and its lasting impact on Earth's ecosystems. *Science Advances, 1*(2), Article e1500052. https://doi.org/10.1126/sciadv.1500052

Harvey, N. (1998). *The Chiapas rebellion: The struggle for land and democracy*. Duke University Press. https://doi.org/10.1215/9780822398301

Helgesen, I. S., Johannesen, A. B., Bostedt, G., & Sandorf, E. D. (2024). Climate change and reindeer herding—A bioeconomic model on the impact of climate change on harvesting profits for Saami reindeer herders in Norway and Sweden. *Ecological Economics, 223*, Article 108227. https://doi.org/10.1016/j.ecolecon.2024.108227

Homewood, K., Trench, P., & Brockington, D. (2012). Pastoralist livelihoods and wildlife revenues in East Africa: A case for coexistence? *Pastoralism: Research Policy and Practice, 2*(1), 36–61. https://doi.org/10.1186/2041-7136-2-19

IPCC. (2022). *Climate change 2022: Impacts, adaptation, and vulnerability*. Contribution of working group II to the sixth assessment report of the intergovernmental panel on climate change. https://www.ipcc.ch/report/sixth-assessment-report-working-group-ii/

Jääskeläinen, T. (2020). The Sámi reindeer herders' conceptualizations of sustainability in the permitting of mineral extraction – Contradictions related to sustainability criteria. *Current Opinion in Environmental Sustainability, 43*, 49–57. https://doi.org/10.1016/j.cosust.2020.02.002

Johnsen, K. I., Benjaminsen, T. A., & Eira, I. M. G. (2015). Seeing like the state or like pastoralists? Conflicting narratives on the governance of Sámi reindeer husbandry in Finnmark, Norway. *Norsk Geografisk Tidsskrift - Norwegian Journal of Geography, 69*(4), 230–241. https://doi.org/10.1080/00291951.2015.1033747

Kimmerer, R. W. (2013). *Braiding sweetgrass: Indigenous wisdom, scientific knowledge, and the teachings of plants*. Milkweed Editions.

Lantto, P., & Mörkenstam, U. (2007). Sámi rights and Sámi challenges: The modernization process and the Swedish Sámi movement, 1886–2006. *Scandinavian Journal of History, 33*(1), 26–51. https://doi.org/10.1080/03468750701431222

Little, K. (2008). Lepcha narratives of their threatened sacred landscapes. *Transforming Cultures eJournal, 3*(1), 225–228. http://www.actsikkim.com/docs/Little_Lepcha_Narratives.pdf

Maison, L. M. (2023). Beyond western and Indigenous perspectives on sustainability: Politicizing sustainability with the Zapatista rebellious education. *Journal of Transformative Education, 21*(1), 41–58. https://doi.org/10.1177/15413446221079595

Maurya, P. K., Ali, S. A., Ahmad, A., Zhou, Q., da Silva Castro, J., Khan, E., & Ali, H. (2020). *An introduction to environmental degradation: Causes, consequence and mitigation.* Agriculture and Environmental Science Academy. https://doi.org/10.26832/aesa-2020-edcrs-01

McMichael, P. (2012). *Development and social change: A global perspective.* SAGE Publications.

Mishra, M., Sudarsan, D., Santos, C. A. G., Mishra, S. K., Kar, D., Baral, K., & Pattnaik, N. (2021). An overview of research on natural resources and indigenous communities: A bibliometric analysis based on Scopus database (1979–2020). *Environmental Monitoring and Assessment, 193*(2), 59. https://doi.org/10.1007/s10661-020-08793-2

Murphree, M. W. (2009). The strategic pillars of communal natural resource management: Benefit, empowerment and conservation. *Biodiversity and Conservation, 18*(10), 2551–2562. https://doi.org/10.1007/s10531-009-9644-0

Nadasdy, P. (1999). The politics of TEK: Power and the “Integration” of knowledge. *Arctic Anthropology, 36*(1–2), 1–18. https://www.jstor.org/stable/40316502

Næss, M. W., Fisktjønmo, G. L. H., & Bårdsen, B.-J. (2021). The Sami cooperative herding group: The siida system from past to present. *Acta Borealia, 38*(1), 81–103. https://doi.org/10.1080/08003831.2021.1972265

Negi, V. S., Pathak, R., Dangwal, B., & Joshi, R. K. (2025). Biodiversity conservation and management through forest landscape restoration in the Western Himalaya, India. *Frontiers in Conservation Science, 6*, 1663322. https://doi.org/10.3389/fcosc.2025.1663322

Nepstad, D. C., Stickler, C. M., & Almeida, O. T. (2006). Globalization of the Amazon soy and beef industries: Opportunities for conservation. *Conservation Biology, 20*(6), 1595–1603. https://doi.org/10.1111/j.1523-1739.2006.00510.x

Ogar, E., Pecl, G., & Mustonen, T. (2020). Science must embrace traditional and Indigenous knowledge to solve our biodiversity crisis. *One Earth, 3*(2), 162–165. https://doi.org/10.1016/j.oneear.2020.07.006

Ogutu, J. O., Owen-Smith, N., Piepho, H. P., & Said, M. Y. (2011). Continuing wildlife population declines and range contraction in the Mara region of Kenya during 1977–2009. *Journal of Zoology, 285*(2), 99–109. https://doi.org/10.1111/j.1469-7998.2011.00818.x

Oksanen, A.-A. (2020). The rise of Indigenous (pluri-)nationalism. *Sociology, 54*(6), 1141–1158. https://doi.org/10.1177/0038038520943105

Onchwati, J., Sommerville, H., & Brockway, N. (2010). Sustainable tourism development in the Masai Mara National Reserve, Kenya, East Africa. *WIT Transactions on Ecology and the Environment, 139*, 319–330. https://doi.org/10.2495/ST100281

Ostrom, E. (2009). A general framework for analyzing sustainability of social-ecological systems. *Science, 325*(5939), 419–422. https://doi.org/10.1126/science.1172133

Ostrom, E. (2012). Governing the commons: The evolution of institutions for collective action. *Cambridge University Press.* https://doi.org/10.1017/CBO9780511807763

Pretty, J., Adams, B., Berkes, F., de Athayde, S. F., Dudley, N., Hunn, E., Maffi, L., Milton, K., Rapport, D., Robbins, P., Sterling, E., Stolton, S., Tsing, A., Vintinner, E., & Pilgrim, S. (2009). The intersections of biological diversity and cultural diversity: Towards integration. *Conservation and Society, 7*(2), 100–112. https://doi.org/10.4103/0972-4923.58642

Redford, K. H., & Fearn, E. (Eds.). (2007). *Protected areas and human displacement: A conservation perspective.* Wildlife Conservation Society.

Reid, R., Fernandez-Gimenez, M. E., & Galvin, K. (2014). Dynamics and resilience of rangelands and pastoral peoples around the globe. *Annual Review of Environment and Resources, 39*(1), 217–242. https://doi.org/10.1146/annurev-environ-020713-163329

Reyes-García, V., Fernández-Llamazares, Á., Guèze, M., et al. (2019). The contributions of Indigenous Peoples and local communities to ecological restoration. *Ecological Economics, 160*, 111–119. https://doi.org/10.1111/rec.12894

Rimisho, E., & Matei, O. (2024). The role of Maasai culture in tourism industry development in Ngorongoro Conservation Area, Tanzania. *Pan African Journal of Business Management, 8*(1), 1–195. https://doi.org/10.61538/pajbm.v8i1.1492

Rist, S., & Dahdouh-Guebas, F. (2006). Ethnosciences—A step towards the integration of scientific and indigenous forms of knowledge in the management of natural resources for the future. *Environment, Development and Sustainability, 8*(4), 467–493. https://doi.org/10.1007/s10668-006-9050-7

Rus, J. L., Hernández Castillo, R. A., & Mattiace, S. L. (2003). *Mayan Lives, Mayan Utopias: The Indigenous Peoples of Chiapas and the Zapatista Rebellion.* Rowman & Littlefield.

Russell, A. (2005). Myth and the Zapatista movement: Exploring a network identity. *New Media & Society, 7*(4), 559–577. https://doi.org/10.1177/1461444805054119

Schwartzman, S., & Zimmerman, B. (2005). Conservation alliances with Indigenous peoples of the Amazon. *Conservation Biology, 19*(3), 721–727. https://doi.org/10.1111/j.1523-1739.2005.00695.x

Schwartzman, S., Moreira, A., & Nepstad, D. (2013). Rethinking tropical forest conservation: Perils in parks and promises in Indigenous lands. *Conservation Letters, 6*(1), 21–27. https://doi.org/10.1111/j.1755-263X.2012.00284.x

Sikor, T., & Lund, C. (2009). Access and property: A question of power and authority. *Development and Change, 40*(1), 1–22. https://doi.org/10.1111/j.1467-7660.2009.01503.x

Sobrevila, C. (2008). *The role of Indigenous peoples in biodiversity conservation: The natural but often forgotten partners.* World Bank.

Stahler-Sholk, R. (2010). The Zapatista social movement: Innovation and sustainability. *Alternatives, 35*(3), 269–290. https://doi.org/10.1177/030437541003500306

Steffen, W., Richardson, K., Rockström, J., Cornell, S. E., Fetzer, I., Bennett, E. M., & Sörlin, S. (2015). Planetary boundaries: Guiding human development on a changing planet. *Science, 347*(6223), 1259855. https://doi.org/10.1126/science.1259855

Stevens, S. (2014). *Indigenous peoples, national parks, and protected areas: A new paradigm linking conservation, culture, and rights.* University of Arizona Press.

Tengö, M., Brondizio, E. S., Malmer, P., et al. (2014). Connecting diverse knowledge systems for enhanced ecosystem governance: The multiple evidence base approach. *Ambio, 43*(5), 579–591. https://doi.org/10.1007/s13280-014-0501-3

Tiwari, K. N. (2010). Traditional knowledge systems and biodiversity conservation in the Indian Himalayas. *Indian Journal of Traditional Knowledge, 9*(2), 225–231.

Toledo, V. M., Ortiz-Espejel, B., Cortés, L., Moguel, P., & Ordoñez, M. D. J. (2003). The multiple use of tropical forests by indigenous peoples in Mexico: A case of adaptive management. *Conservation Ecology, 7*(3). Retrieved from http://www.consecol.org/vol7/iss3/art9/

Turner, T. (1993). The role of Indigenous peoples in environmental crisis: The example of the Kayapo of Brazil. *Perspectives in Biology and Medicine, 36*(3), 526–545. https://doi.org/10.1353/pbm.1993.0027

Turner, B. L., Kasperson, R. E., Matson, P. A., McCarthy, J. J., Corell, R. W., Christensen, L., & Schiller, A. (2003). A framework for vulnerability analysis in sustainability science. *Proceedings of the National Academy of Sciences, 100*(14), 8074–8079. https://doi.org/10.1073/pnas.1231335100

Tyler, N. J., Turi, J. M., Sundset, M. A., Strøm Bull, K., Sara, M. N., Reinert, E., & Oksanen, L. (2007). Sámi reindeer pastoralism under climate change: Applying a generalized framework for vulnerability studies to a sub-arctic social-ecological system. *Global Environmental Change, 17*(2), 191–206. https://doi.org/10.1016/j.gloenvcha.2006.06.001

UNEP. (2021). *Making peace with nature: A scientific blueprint to tackle the climate, biodiversity and pollution emergencies*. United Nations Environment Programme. https://www.unep.org/resources/making-peace-nature

United Nations. (2021). *State of the world's Indigenous peoples: Rights to lands, territories and resources*. United Nations Department of Economic and Social Affairs. https://www.un.org/development/desa/indigenouspeoples/publications

Van der Haar, G. (2004). The Zapatista uprising and the struggle for Indigenous autonomy. *European Review of Latin American and Caribbean Studies, 76*, 99–108. https://www.jstor.org/stable/25676074

Vásquez Coronado, M. H., Medina Valderrama, C. J., Valencia-Arias, A., Morales Huamán, H. I., Valencia, J., & Cardona-Acevedo, S. (2024). Analysis of research trends on economic growth and environmental degradation: A bibliometric study. *Sustainable Environment, 10*(1). https://doi.org/10.1080/27658511.2024.2345445

Walker, B., Holling, C. S., Carpenter, S. R., & Kinzig, A. (2004). Resilience, adaptability, and transformability in social–ecological systems. *Ecology and Society, 9*(2), 5. https://doi.org/10.5751/ES-00650-090205

Walker, W., Gorelik, S. R., & Baccini, A. (2020). The role of forest conversion, degradation, and disturbance in the carbon dynamics of Amazon Indigenous territories and protected areas. *Proceedings of the National Academy of Sciences, 117*(6), 3015–3025. https://doi.org/10.1073/pnas.1913321117

Wang, L., Buitenwerf, R., Munk, M., Amoke, I., Bøcher, P. K., & Svenning, J.-C. (2020). Accelerating savanna degradation threatens the Maasai Mara socio-ecological system. *Global Environmental Change, 60*, Article 102030. https://doi.org/10.1016/j.gloenvcha.2019.102030

Whyte, K. P. (2018). Indigenous science (fiction) for the Anthropocene: Ancestral dystopias and fantasies of climate change crises. *Environment and Planning E: Nature and Space, 1*(1–2), 224–242. https://doi.org/10.1177/2514848618777621

Wildlife Institute of India. (2020a). *Biodiversity and conservation in the Indian Himalayas: Annual report*. Wildlife Institute of India. https://wii.gov.in

Wildlife Institute of India. (2020b). *Great Himalayan National Park conservation initiatives*. WII Technical Report.

Zimmerman, B., Peres, C. A., Malcolm, J. R., & Turner, T. (2001). Conservation and development alliances with the Kayapo of southeastern Amazonia, a tropical forest Indigenous people. *Environmental Conservation, 28*(1), 10–22. https://doi.org/10.1017/S0376892901000029

Zipperer, W. C., Northrop, R., & Andreu, M. (2020). Urban development and environmental degradation. *Oxford Research Encyclopedia of Environmental Science*. https://doi.org/10.1093/acrefore/9780199389414.013.97

Chapter 2
Social and Environmental Development: Opportunities for Neo-Urbanization Through the Lens of Ecological Modernization

Amar M. Dhere and Sachin J. Deore

Abstract Global South Urbanization is a complex issue, with rapid urbanization, ecological and social inequities that interact and thus, sustainable urban development should be an urgent issue. In this setting, two theoretical solutions Ecological Modernization Theory (EMT) and emergent paradigm of Neo-Urbanization may prove useful in resolving the technologization advancement, ecological care, and social integration. Although the existing body of knowledge has proved the applicability of such models in developed economies, little is known about the applicability of the concept of these models in less developed economies, like India, which is resource-limited. This gap has been addressed in this chapter by talking about how the Indian urban and infrastructural projects can be used as experimental platforms to construct climate-resilient and socially equitable development in line with the Sustainable Development Goals (SDGs) SDG 11 (Sustainable Cities), SDG 7 (Clean Energy), and SDG 13 (Climate Action). The research methodology is the multi case analysis qualitative method, and a total of seven model projects were identified where the green infrastructure, forms of government participatory, and environmental novelty were experimented, including Magarpatta City SEZ, Cochin international airport, Palava Smart City, Auroville township, and Indira Paryavaran Bhawan. Types of data sources used are the planning materials, sustainability reports, peer-reviewed literature, and specific performance measures of the site. These results show that the projects incorporate renewable energy systems, water sensitive planning, circular waste management, and community-based governance concepts in their development models. By so doing, they confirm fundamental postulations of

A. M. Dhere (✉)
Department of Science, SNDT College of Home Science, Pune 411038, MS, India
e-mail: prof.amardhere@gmail.com

S. J. Deore
Department of Geography, Post Graduate Studies and Research, SNDT Women's University, Pune 411038, MS, India
e-mail: deoresachinj@gmail.com

G. K. Panda et al. (eds.), *Socio-Ecological Crisis of Development*, Sustainability Solutions, https://doi.org/10.1007/978-3-032-22171-1_2

EMT that it is possible to have environmental protection and economic development reinforcing each other, as well as illustrate the neo-urbanization reconciliation of regenerative and inclusive urban regeneration. Considerably important, the projects feature how grassroots involvement and the indigenous knowledge systems can complement the technological informativeness, confronting the supremacy of Global North-focused sustainability discourses. This study, by applying a combination of theoretical and empirical data, leads to a new integrative paradigm, according to which EMT and neo-urbanization are integrated so that Indian cities can be a force of sustainability instead of a source of environmental pressure. The chapter provides a lesson to be used not only regionally, but also globally as it provides a blue print that can be replicated to rethink urban development as a co-evolutionary process among human settlements, technology and ecosystems.

Keywords Neo-urbanization · Ecological modernization theory · Sustainable urban development · Smart cities · Green infrastructure · Community resilience · Renewable energy · SDG · Climate action · Case studies in India

2.1 Introduction

The modern global environment is characterized by urbanization, which has boosted growth of the economies, development of technology, and social change. Cities have become the drivers of innovation, cultural blurring, and financial prosperity that draws millions of people across continents to seek better living standards (Ghahremani et al., 2021). A fast and unchecked growth of urban cities has in the past been one of the reasons as to why environmental degradation, depletion of resources, and the increase of socio-economic inequities have increased. Urban sprawls which are not controlled through proper structural planning, poor infrastructure, and administration systems will worsen problems such as air and water pollution, loss of biodiversity, congestion, and informal settlements. This is why the classical paradigm of urbanization is becoming increasingly unsustainable in the framework of the exhaustive planetary resources (Kajiita & Kangethe, 2024).

The environmental footprint of cities is significant because about 70 percent of global greenhouse gases are produced in cities and over two out of three thirds of the world's energy is consumed by cities (Almihat et al., 2022). Since by 2050, it is estimated that urban populations will be over 68 percent of the world population (UN DESA, 2022), there has never been a greater imperative that funks the standard models of urban development. The recent policy agenda of Sustainable Development Goals (SDGs) and SDG 11, specifically, devoted to sustainable cities, and communities, highlights the necessity of sustainable transformation of cities in the inclusive, resilient and environmentally responsibility manner. In this context, a paradigm of neo-urbanization has been developed. It presents a reorganization of urban development by introducing the ecological sensibility, technological advancement, and participatory governance into the structure of planning and development.

Contrary to the more traditional linear models of urban expansion, partly based on the prevailing legitimacy of short-term economic gains, neo-urbanization is shifting toward the regeneration of societies that are more eco-social and more in line with the ideas of the Ecological Modernization Theory (EMT): technological advancement and environmental friendliness have not been the dominant attribution of each other up till now (Clark & Wu, 2021; Wang et al., 2023). Rather, they can develop together using cleaner technologies, institutional change, and joint forms of government (Du & Yang, 2025; Jorgenson et al., 2024).

In this sense, neo-urbanization is not merely a hypothetical change rather a rebirth of a new avenue of urban development that aims at achieving a blend between economic performance and ecological conservation and inclusivity. Ecological modernization also questions the classic idea that environmental protection is holding back development and instead argues that technological progress, an energy transition toward renewables, and a circular economy can help to promote sustainability without necessarily stopping modernization (Rodenbiker, 2021). The necessity to redefine urbanization is predetermined by urgent problems of the global agenda: growing effects of the climate change, increasing socio-economic disparities, and growing the susceptibility of urban residents to environmental hazards. The ecological impact of cities is so significant because of the emissions produced by cities, which nearly 70 percent of carbon emissions are produced by cities and two-thirds of all global energy is consumed in cities (Almihat et al., 2022). However, the very urban environments have the huge potential: it can spearhead the testing of sustainable technologies, promote the models of green governance, and jumpstart the ways toward the green economic transitions.

The chapter is a critical look at the interaction between the ecological modernization and the neo-urbanization and more importantly the practical prospects of developing sustainable and inclusive cities. It investigates how built in solutions, as far as smart grids and decentralized renewable energy systems to green mobility and waste-to-resource innovations can alleviate the ecological strain and at the same time enhance the quality of lives in urban areas. Moreover, the discussion is reinforced by how the involvement of capitalists, channeled through environmentally oriented approaches, could become an ordeal toward transformational urban change. The chapter provides an input on the prevailing discussions regarding sustainable urban development. It not only places the concept of neo-urbanization which is based on the principles of ecological modernization as a vision of the future, but as a dynamic model that can be utilized to find solutions to the complex social, economic, and environmental problems that characterize the present and future urbanization (Abujder et al., 2025). It provides specific interventions and reflexive criticism to policy-makers, urban planners, researchers, and urban communities who care about the way in which the cities become sustainable and equitable by the sheer grace of ecological wisdom through a hypothetical lens.

2.1.1 Theoretical Framework: Ecological Modernization Theory

Originally introduced in the 1980s, the Ecological Modernization Theory (EMT) is essentially the rejection of the conventional approach to zero-sum relation between economic growth and environmental protection (Jorgenson et al., 2024). Ecological Modernization Theory (EMT) follows on the notion that environmental issues could create the forces of technological change, institutional change, social change, as a response to previous environmental theories, which held economic growth and sustainability as opposing concepts. The earliest researchers, including Joseph Huber and Martin Janicke, argue that ecological pressures can trigger the implementation of greener technologies, better industrial operations and innovative administrative systems, and symbolize societies on a new path of environmental development. This view brings an alternative understanding of modernity as being ecologically refined and being able to achieve development without the need to forego the gains of the industrialization and modernization process (Foggin et al., 2021).

This is the core of EMT and technological innovation by underlining that breakthroughs in renewable energy, sustainable buildings, sustainable agriculture, and waste-to-resource systems would help decouple the growth of the economy with environmental destruction. Through the incorporation of eco-efficiency in the production and consumption cycle, the urban development initiatives can continue to improve the economic output as well as the protection of the natural systems. The next element is equally fundamental, as it is institutional restructuring that points out the importance of collaborative governance when undergoing a transition toward sustainability. EMT suggests that the governments, businesses, civil society groups, and community players will all be expected to embrace systemic change and beyond small action or sector-based actions (Nandy, 2024). This kind of reconfiguration demands institutions to be more adaptive, participatory and long-term ecologically focused.

The concept of reflexive modernization which entails the society evaluating and remodeling its industrial apparatus, consumption patterns, and policy structures is another distinguishing feature of EMT. Reflexivity stances make sustainability not an addition to any social and economic system but a component of the system. Such a continued review and re-evaluation of various procedures leads to the shift in the methods of reactive and problem-solving techniques to the proactive strategies which do not violate the ecological health. In the environment of the fast urbanization, EMT offers the most well-meaning framework. Cities as densely populated economic and technology innovation centers as well as social interaction form an optimal case in which ecological transformation can be introduced in trials (Liu & Lo, 2023). Not only can city centers set examples of sustainable living that combine compact planning, effective infrastructure and cutting-edge technologies to ease the environmental impact and enhance living standards, it can help urban centers become a hub of tourism and workplaces. This means that cities and mega projects will no longer be seen as solely a source of environmental pressure, but as spaces of solution as well where a connection between the three questions of innovation, governance, and

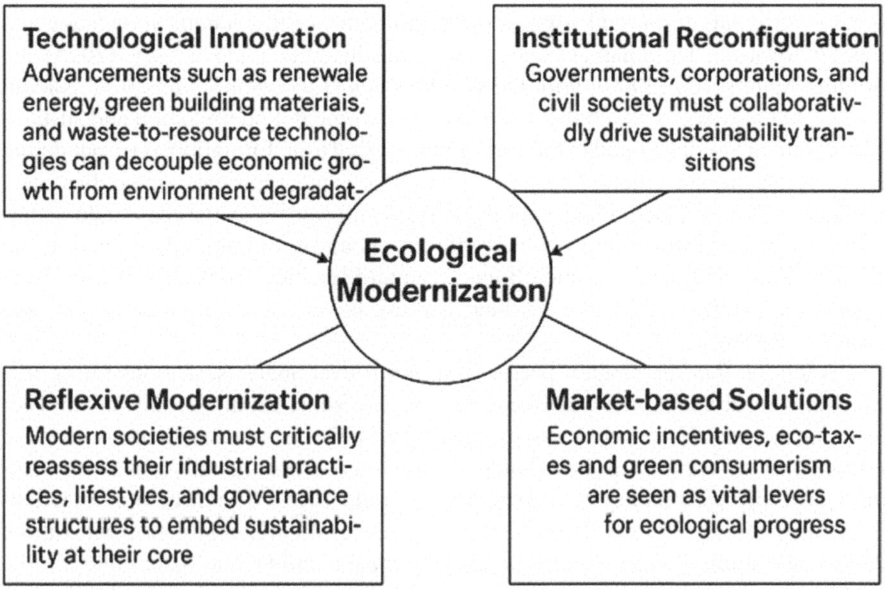

Fig. 2.1 Theoretical pragmatics of ecological modernization theory (EMT)

civic engagement is generated to create resilient, equitable, and sustainable futures. Through the ecological modernization principle, neo-urbanization has the power to change the conventional city growth models into dynamic and environmentally sensitive ones, which could address ecosystem demands of the twenty-first century in a complex and diverse manner (Fig. 2.1).

2.1.2 Understanding Neo-Urbanization

This chapter places the debate on neo-urbanization and Ecological Modernization Theory (EMT) on a global-regional continuum and makes them important in a sustainable urban development. Cities globally are confronted with accumulating challenges of high population growth, climate susceptibility, socio-economic disparities, and environmental pressure (Arku & Marais, 2021). Although global supercities like Amsterdam, Copenhagen, and Singapore exhibit the prospects of ecological modernization with its smart technologies, citizen-driven governing models, and strong policies (Bera et al., 2023; UN-Habitat, 2024), the emerging areas also have a lot to teach. The examples of cities within South Asia, Sub-Saharan Africa, and Latin America demonstrate how they adopted adaptive strategies in the state of resource scarcity, with Indian cities, including Pune, Surat, Bhubaneswar, and Kochi, leading the way in decentralized governance, digital inclusion, and nature-based solutions (Peter, 2021; Kaviya Darsini et al., 2026; Shirke et al., 2025). These practices shake

North-centric paradigms and suggest the significance of indigenous knowledge and frugal innovation. This chapter connects previous insights of Du & Yang, (2025) and Mol and Spaargaren (2000), with the current ones (Jorgenson et al., 2024; Katyal, 2024), and the ways in which the discourse of ecological modernization and sustainable urbanism has developed in the wake of the digital transformation, post-pandemic recovery, and climate emergencies. Based on theory and empirical case-studies, such as Magarpatta City, Cochin International Airport and Auroville, the chapter describes how eco-oriented infrastructure, renewable energy and community participation can create resilient and inclusive and climate compatible cities. These kinds of projects support the priorities of SDGs as they increase clean energy, climate action, and resource efficiency.

It is unique that the chapter performs a multi-level analysis, demonstrating that ecological modernization is not exclusive to high-resource settings, and can be redefined on the multi-level contexts in rapidly urbanizing and resource-constrained settings through approaches based on participation and ecologically conscious environments (Kaviya Darsini et al., 2026; Shirke et al., 2025). The threefold objectives of its study include (1) the theoretical convergence of EMT and neo-urbanization, (2) the case studies of eco-centric urban governance and technological integration, and (3) the sustainable urban transformation model that may be replicated following the SDGs. This way, the chapter offers not only the conceptual clarity but theory practical frameworks to scholars, policy-makers and the urban practitioners (Swales and Feak, 2012; Booth and others, 2016). Finally, it shows how the urban project can be transformed into its role in creating the forces of ecological degradation and promote resiliency and sustainability, which further advances the discussion of creating regenerative and equitable urban regeneration through the world.

2.2 Research Methodology

The chapter research design is the case chapter research design that will take an in-depth approach to research on the potentials of neo-urbanization based on the conceptual model of ecological modernization. Since urban environments are complex and social, technological and ecological systems are interconnected, the case chapter approach is a perfect approach to methodology to investigate the application in the real environment, contexts and transformative practices within modern urban development.

2.2.1 Case Study Approach and Rationale

With the help of the case study approach, such vivid, situation-based research of urban spaces that actively interact with the principles of ecological modernization is achievable (Peck, 2015). The chapter brings out the variety and multi-faceted nature of the process of neo-urbanization through the choice of cities or urban areas that can be identified as having made attempts at combining the green technologies, developing sustainable infrastructures, and pursuing a community-focused constructive vision in development. The reason why it was selected as a case study is that it has the potential to give a comprehensive picture of the phenomena in their natural environment and consequently identify trends, innovations, concerns and learnings that otherwise might be obscured when the results were analyzed on a grander scale, whether quantitatively or through theoretical means (Cleland et al, 2021). It is a multiple case study approach that allows the cross-case comparison and reveals similarities and individual patterns of ecological modernization in various urban settings.

2.2.2 Case Selection Criteria

The case is carefully chosen on the unique parameters defined to ascertain relevance in data and richness (Alam, 2021). The selection criteria

(1) signs of technological creation (e.g., the use of renewable energy, smart city mobility, the promotion of a cycle economy),
(2) active institutionalization of reconfiguring partnerships between governments, private sectors and the civil society with regard to sustainability transitions,
(3) reflexivity in modernization like urban policy changes, participatory governance, and behavioral changes toward sustainable living,
(4) administration of market-based policies such as eco-taxes, green certifications, or incentives of sustainable consumption and production.

The geographic location, social-economic situation, and governance structure diversity is also believed to attract in-depth insights that depict different urban realities.

2.2.3 Data Collection Methods

Information gathering will mainly be based on qualitative methods that ensure that the complexity and contextualism of the chosen cases are obtained. Analysis of urban development policies, policy documents, environmental reports, sustainability assessment and other secondary materials that describe the approaches and the results of the neo-urbanization efforts. Each and every data were collected systematically so as to maintain consistency between the chosen cases and as a result achieve the depth of analysis and fruitful comparability. After this, cross-case synthesis was performed to find common patterns, outstanding differences, and emerging themes in the cases. This two-level methodology increases the internal validity because in this way, both within case detailed insights and generalizations across cases can be provided (Fig. 2.2).

Fig. 2.2 Research flowchart. *Source* Authors

2.3 Results and Discussion

(1) **Magarpatta City SEZ**—Magarpatta City SEZ in Pune is an example of sustainable development of urban areas characterized by an eco-based vision built around the Vedic philosophy and five classical elements: Jal, Vayu, Pruthvi, Agni, and Akash. It has an inbuilt waste management system that treats approximately 15 tons of waste per day where organic matter is left to create composts and biogas as well as reusing non-biodegradable waste hence forming a closed-loop ecosystem. Its use of renewable energy is highlighted and the two-tonne biogas plant provides 270 units of electricity every day and the largest single-organization solar water heating system in Asia (7,000 collectors) should warm 900 kilolitres of water each day; emissions cutting 13,483 tons/year and saving 5 crore in energy expenses. To achieve reduction of carbon emission, construction practices use 1.5 lakh tons of fly ash, with the other activities of rainwater collection in 515 recharge wells with artificial lakes contribute to the recharge of groundwater and green supply throughout the year. The infrastructure planning incorporates four sewage treatment plants, passive design, and environment friendly architecture such as raised parking, insulated IT buildings, and traditional Wada designs to light and ventilate the buildings naturally. Urban greening further reduces dust and noise, earning the township carbon credits worth ₹3.8 crore annually and a 10% property tax rebate from Pune Municipal Corporation. These initiatives position Magarpatta as a benchmark for sustainable SEZs and urban development (Dhere, 2018) (Table 2.1).

(2) **Cochin International Airport: Solar Airport**—The case of Cochin International Airport's transition to full solar power provides a significant example of

Table 2.1 List of Indian projects showcasing ecological modernization and neo-urbanization

Project Name	Location	Highlights	Key ecological aspects	SDG component
Magarpatta city SEZ	Pune, Maharashtra	Privately developed integrated eco-city by farmers; green spaces, energy efficiency, waste management, community governance	Sustainable land use, energy-efficient infrastructure, integrated waste management, community-driven sustainability	SDG 11 (Sustainable Cities), SDG 7 (Clean Energy), SDG 12 (Responsible Consumption)
Cochin international airport—solar airport	Kochi, Kerala	World's first fully solar-powered airport; sustainable infrastructure, renewable energy leadership	100% solar energy utilization, carbon footprint reduction, renewable energy integration	SDG 7 (Affordable and Clean Energy), SDG 13 (Climate Action)
Palava Smar AND eco city	Near Mumbai, Maharashtra	Private smart city with sustainable infrastructure, walk-to-work concepts, smart grids, green spaces	Urban sustainability, energy-efficient smart grids, mixed land use planning	SDG 11 (Sustainable Cities), SDG 9 (Industry, Innovation)
Charanka Solar park	Gujarat	One of the largest solar parks globally, promoting large-scale renewable energy deployment	Massive renewable energy deployment, large-scale solar farming, reduction of fossil fuel dependency	SDG 7 (Affordable and Clean Energy), SDG 13 (Climate Action)
Sabarmati riverfront development	Ahmedabad, Gujarat	Eco-urban renewal project with flood protection, public spaces, waste water treatment integration	Integrated flood resilience, eco-restoration, sustainable urban public spaces	SDG 11 (Sustainable Cities), SDG 6 (Clean Water and Sanitation)
Net-zero energy buildings (Indira Paryavaran Bhawan)	New Delhi	India's first net-zero energy building using solar energy, green materials, water conservation techniques	Net-zero carbon emissions, green material usage, water and energy conservation	SDG 7 (Affordable and Clean Energy), SDG 12 (Responsible Consumption), SDG 13 (Climate Action)

(continued)

Table 2.1 (continued)

Project Name	Location	Highlights	Key ecological aspects	SDG component
Auroville experimental township	Auroville, Tamil Nadu, India	Living laboratory for sustainable development integrating nature and technology	Ecological modernization through land regeneration, water conservation, renewable energy, and reforestation	SDG 11 (Sustainable Cities), SDG 13 (Climate Action), SDG 15 (Life on Land)
Auroville green belt land use plan	Auroville, Tamil Nadu, India	Strategic plan balancing settlement growth with ecological preservation	Native forest restoration, soil rehabilitation, ecosystem-based zoning, agroforestry promotion	SDG 15 (Life on Land), SDG 6 (Clean Water), SDG 12 (Responsible Consumption)

Source Authors

how large-scale infrastructure can integrate renewable energy solutions effectively. The installation and operation of a 12 MWp solar photovoltaic plant enabled the airport to meet its complete electricity requirements while reducing carbon emissions substantially. Over a period of one year, the plant produced over 18 million units of clean energy, contributing meaningfully to the airport's environmental goals. The operational data indicated that renewable energy integration did not compromise the airport's functionality or performance. On the contrary, the project proved to be economically viable, demonstrating a favorable payback period and considerable savings on energy costs. The success of the initiative supports the broader understanding that clean energy technologies can be both efficient and financially sustainable when properly implemented in large infrastructure projects (Sreenath et al., 2022) (Table 2.2).

The example of Cochin International Airport can be discussed within the framework of the Ecological Modernization Theory since it shows that economic growth and environmental protection may be reinforcing factors. By massively implementing the solar photovoltaic systems, the airport has transformed its operations, with limited environmental effect meaning that environmental responsibility has become a part of strategic decision making in the airport. The project was not only able to achieve significant cost savings, but also to demonstrate the role that the innovation and institutional leadership can play in achieving sustainable results without the need of top-down regulation. Cochin Airport can serve as an effective example of ecological modernization in the context of critical infrastructure development by presenting the way to demonstrate how localized efforts can bring substantial ecological and financial results (Yousuf et al., 2024) (Table 2.3).

Table 2.2 Key feature of Magarpatta City SEZ for making it modern eco entity

Aspect	Key highlights
Vision and design philosophy	Inspired by five Vedic elements (Jal, Vayu, Pruthvi, Agni, Akash); symbolized by five windows on the main gate; focus on pollution-free, green, and clean environment
Waste management	Daily segregation and treatment of 15 tons of waste; organic waste used for composting and biogas; non-biodegradable waste safely recycled or sold, earning ₹20,000 weekly
Compost production	Produces 3 tons of vermicompost monthly; compost used in gardens and nurseries; over 400 tonnes/month of waste managed with 280 tons bio-composted and 120 tonnes recycled
Biogas plant	2-tonne capacity plant generates 80–100 m^3 biogas daily, converted to 270 units of electricity; powers garden pumps and lights; replaces 118 commercial gas cylinders/month
Solar water heating system	Largest in Asia by a single organization; 7,160 collectors producing 900 KLD hot water; saves ₹5 crore and 13,483 tons CO_2 annually; installed on all residential and commercial units
Fly ash utilization	1.5 lakh tons of fly ash used in construction; reduces 150,000 tons of CO_2; bricks are more durable, less water-absorbing, and have better finish than traditional bricks
Rainwater harvesting	Collects and recharges groundwater using 515 bore wells and artificial lakes; ensures year-round water availability; improves microclimate with Pune's largest mist fountain
Groundwater recharge	Landscaping with water-absorbent pavers; mounds and permeable blocks increase water retention; helps raise water table and prevent urban flooding
Water and sewage treatment	Water treatment plant supplies 20 lakh L/day; STPs treat 2 MLD with expansions to 4.5 MLD; treated water reused for gardening and flushing
Passive building design	Natural ventilation in parking; air insulation in IT buildings; "Wada" style homes ensure light and airflow; greenery buffers noise and dust
Carbon credits and incentives	Solar system generates ₹3.8 crore in carbon credits annually; PMC gives 10% property tax rebate for eco-initiatives; promotes replication by other SEZs and developers

Source Authors (Complied from Dhere, 2018)

(3) **Palava Smart & Eco City**

Palava Smart City demonstrates that highly developing technologies might assist in the sustainable urbanization process and raise the quality of life. The city is based on the postulates of Ecological Modernization and Neo-Urbanization and shows that the concepts of technological innovation and environmental protection can mutually support each other (Madakam, 2022). Palava facilitates resource real-time tracking using ICT and IoT systems, incorporating smart grids and renewable energy and encouraging energy-efficient development without slowing down the growth. The fact that it has been focusing on sustainable movement, smart traffic, and eco-friendly

Table 2.3 Key feature of Cochin International Airport one of the first renewable energy airports

Aspect	Key Highlights
Project type	Fully solar-powered international airport
Solar plant capacity	12 MWp
Annual energy generation	Over 18 million kWh
Carbon emissions reduced	Approximately 15,000 metric tons per year
Economic impact	Significant cost savings; payback period ~ 5–6 years
Technology used	Solar photovoltaic (PV) systems
Operational impact	Full energy self-sufficiency; no disruption in services
Environmental impact	Major reduction in carbon footprint
Institutional significance	Shift toward sustainable energy integrated in operations
Broader implication	Model for ecological modernization in infrastructure

Source Authors (Complied from Sreenath et al., 2022)

public transportation indicate the neo-urbanization pattern of compact and inhabitable, and technologically-oriented cities. In promoting a symbiotic approach toward urban growth and ecological accountability, facilitating public–private cooperation, and using data-driven governance, Palava can be seen as a working example on how smart cities can balance both approaches to urban growth and the environmental concern, providing a model of sustainable and inclusive development that could be applied to urban development in India and other countries (Madakam, 2022) (Table 2.4).

(4) **Charanka Solar Park, Gujarat—**A good example of the application of Ecological Modernization Theory (EMT) can be seen in the adoption of solar energy in India where it is stated that technological innovations can spearhead the sustainability and expansion of the environment. Photovoltaic (PV) panels, solar micro grids, and rooftop-based systems are examples of high-tech solar energy systems, which alleviate the reliance of fossil fuels and lower carbon footprint of electricity generation. The more sustainable energy systems make their solutions decentralized, which means these systems serve the remote regions to make their energy more reliable and efficient to distribute it. India is optimizing its solar infrastructure by means of state-of-the-art technologies such as smart inverters and energy storage systems to guarantee the flow of power at all times, even when the sun is not shining (Ullah, 2021).

India is technologically moving toward solar power due to innovations in improving efficiency and cost-effectiveness. The solar PV systems have become more effective in turning sunlight to electricity and smart grid technologies allow integrating the solar energy with the current power grids. Through these advancements, real time monitoring, energy forecasting, and optimization can be done, working around the issues with solar energy being intermittent. Through these kinds of technologies, India is improving its grid stability, minimizing wastes of energy, and a

Table 2.4 Key feature of Palava smart and eco city has PPP model of ecological modernization

Aspect	Key Highlights
Development objective	Creation of a sustainable, smart urban environment integrating advanced technologies
Key areas of focus	Smart governance, intelligent infrastructure, sustainable mobility, citizen engagement
Technological integration	ICT, IoT, data analytics, and smart grids for efficient management and monitoring
Public–private partnerships	Collaborative efforts between government and private sector for infrastructure development and innovation
Environmental impact	Reduction in carbon footprint, energy-efficient buildings, and sustainable urban planning
Quality of life indicators	Improved public services, enhanced mobility, reduced congestion, and better safety
Challenges addressed	Urban inefficiencies, resource management, environmental degradation, and urbanization pressures
Outcome focus	Long-term sustainability, improved urban resilience, and inclusivity through smart technologies

Source Authors (Complied from Madakam, 2022)

step toward a low-carbon economy, in accordance with the ecological principles of modernization. The importance of solar energy as the key to the ecological modernization explains the nature of the environmental sustainability. Solar energy is a clean and a renewable source of energy and thus decreases the use of polluting fuels in the form of fossil fuels, alleviating the environmental effects of the conventional energy production systems. The added application of solar technologies is useful in reducing carbon emissions, which would help in the work against climate change. Recycling of solar panels can be regarded as evidence of continuous development and improvement of the sustainable approach, which supports solar energy as an option characterized by the use of a limited number of resources. Such developments illustrate how solar conversion in India remains both beneficial to ideal environmental protection and to the development of technology, which makes it a central part of the modernization and sustainability agenda in the country (Ullah, 2021) (Table 2.5).

(5) **Sabarmati Riverfront Development**

Sabarmati Riverfront project is an example of how an urban transformation can be achieved using the concepts of Ecological modernization theory (EMT) and neo-urbanization. It will combine sustainable water management, green space, and an increase in public amenities by adopting technological innovation and recovering the ecological environment to create a livable urban ecosystem (Simons et al., 2024). The project transforms the classical city environments and places an emphasis on

Table 2.5 Key feature of Charanka Solar Park gives access to renewables for rural community

Aspect	Key highlights
Focus area	Solar energy's role in India's economic development
Economic contribution	Enhances energy security and reduces dependence on fossil fuels
Technological integration	Utilization of various solar-based technologies for electricity generation
Job creation	Generation of employment opportunities in the solar energy sector
Rural development	Improvement of energy access in rural areas through decentralized solar solutions
Industrial development	Promotion of industrial growth via solar energy adoption
Technology transfer	Facilitation of knowledge and technology transfer in the solar energy sector
Government initiatives	Implementation of policies and incentives to promote solar energy adoption
Environmental impact	Reduction in carbon emissions and mitigation of climate change effects

Source Authors (Complied from Ullah, 2021)

the ecological design, green technologies and the relationships between the city and the business as the core in balancing the city development and the environmentally sustainable levels. This contemporary fixity of modernization and inclusiveness as well as environmental sustainability illustrates how the urban areas can become more modernized and less ecological footprints are to be provided hence making the Sabarmati Riverfront an exemplar of green urbanization (Shah & Kunte, 2025) (Table 2.6).

Table 2.6 Key features of Sabarmati Riverfront project to show way for biodiversity preservation

Aspect	Key highlights
Water treatment systems	Advanced systems for treating and managing river water quality
Smart infrastructure	Integration of digital technologies for efficient city management
Green spaces	Creation of parks and recreational areas along the riverfront
Energy-efficient lighting	Use of solar-powered streetlights and energy-efficient lighting
Waste management systems	Eco-friendly waste management solutions for a cleaner environment
Biodiversity preservation	Efforts to restore and protect local flora and fauna along the river
Public transport connectivity	Improved access through sustainable public transport options

Source Authors (Complied from Simons et al., 2024)

(6) **Net-Zero Energy Buildings (Indira Paryavaran Bhawan)**

Indira paryavaran bhavan (IPB) is one such example of ecological modernization that combines ecologically friendly architecture with best technology. The characteristics of its passive design, such as natural ventilation, optimum orientation, and daylighting, minimize the use of artificial lighting and cooling, whereas solar panels and geothermal systems provide clean energy (Manjunath et al., 2023). Other innovations like smart energy management and regenerative lifts make things efficient and reduction of the emissions to a minimum. Being a neo-urbanization model, the building shows the way urbanization can be combined with ecology principles with a net-zero energy model. Having demonstrated that eco-efficient technologies can operate in massive infrastructure, IPB provides an example that can be copy-and-pasted on sustainable urban development in India and elsewhere (Chaturvedi et al., 2024) (Table 2.7).

(7) **Auroville Township & Green Belt Model**

The example of Austville demonstrates how the ecological modernization process and neo-urbanization may be combined by means of land-use plans like the Green Belt Plan by Stein (2015), utilizing scientific concepts, renewable technologies, and the ecosystem approach to planning. The evolution of nature and technology is expressed by such initiatives in the rain-harvesting, organic agriculture, soil regeneration, and renewable energy (Viswanathan & Sharpe, 2024). Auroville is also a low-density (unlike traditional urban development), decentralized suburban area with dispersed cluster projects, green corridors, and buffer areas that incorporate ecological integrity within urban development. The emphasis on regenerative urbanism, such as the recovery of biodiversity and restoration, afforestation, and soil restoration, demonstrates how technological and managerial innovations can actively restore nature. Being an ecological and social experiment, Auroville shows how citics can

Table 2.7 Key feature of Indira Bhawan an modern model net zero energy building

Aspect	Key highlights
Energy efficiency	Passive design features (daylighting, natural ventilation) reduce energy consumption
Renewable energy	Solar panels and geothermal systems meet the building's energy needs
Technological innovation	Regenerative lifts and smart energy management systems optimize energy use
Carbon emissions reduction	Energy-efficient HVAC and lighting systems contribute to reduced emissions
Sustainability goals	Net-zero energy design minimizes reliance on non-renewable energy sources
Urban design	Integration of sustainability with modern urban development principles

Source Authors (Complied from Manjunath et al, 2023)

change their exploitative modes of growth into site-specific, ecologically restorative settlement to place it as a practical manifestation of ecological modernization and neo-urban consciousness (Table 2.8).

Table 2.9 which is a comparative analysis indicates the strengths of Indian urban projects that are unique in terms of addressing socio-ecological crises but is presented in relation to the world benchmarks so as to conceptualize it in greater depth into the theoretical impacts of ecological modernization and neo-urbanization (Tajuddin and Dabrowski, 2021). Based on the elaboration of Indian case studies that has already been presented, this section aims at considering relevant models at the international level in order to elicit sensitive global–local learnings. Vauban, located in Freiburg, Germany, is a globally acclaimed example of urban growth and development prepared by its citizens (Fastenrath & Preller, 2018), just as Magarpatta City SEZ. Both programs place greater emphasis on the ideas of participatory governance and socially friendly design but in Magarpatta with its distinctive way of integrating the traditional Vedic ecology principles into the modern eco-urbanism paradigm, the culturally oriented paradigm of eco-urbanism that Vauban lacks in its highly technocratic manner remains a significant feature of the latter project. This difference justifies the importance of introducing local knowledge systems into the contemporary city planning approach to increase the context and community ownership in the Global South (Table 2.9).

Equally, in the Netherlands, Schiphol Airport has incorporated solar energy systems to facilitate decarbonisation of airports (Davydenko & Hilbers, 2024). However, its partial dependence on solar energy stands in comparison to the total self-sufficiency of Cochin International Airport through solar energy that makes the Cochin the first eco-modernization project in the sphere of infrastructure. The model created by Cochin can be of special help to countries of the Global South, where

Table 2.8 Key feature of Auroville project an model of eco-urbanism

Aspect	Key highlight
Philosophy and ideals	Focus on human unity, sustainable living, and transcending political divisions
Governance and power dynamics	Power struggles emerge between idealistic governance models and legal realities
Land use and environmental planning	Green Belt Plan emphasizes ecological protection, water conservation, afforestation
Urban development strategies	Auroville applies non-traditional urban planning principles based on organic growth
Community participation	Participatory frameworks are essential, though difficult to consistently achieve
Challenges	Balancing vision with implementation, managing external pressures, resource constraints
Case specific studies	Site-specific land-use strategies, socio-political negotiation records, community initiatives

Source Authors (Complied from Viswanathan & Sharpe, 2024)

Table 2.9 Comparative implications of Indian urban projects in addressing socio-ecological crisis through ecological modernization and neo-urbanization

Indian project	Core contribution to study objectives	Comparable global model	Comparative insights
Magarpatta city SEZ (Pune)	Demonstrates community-led, eco-centric urban development integrating traditional knowledge systems (vedic ecology) and modern sustainability technologies	Vauban (Germany)—Eco-district with citizen participation and green infrastructure	Magarpatta surpasses in integrating cultural-ecological principles; both show successful citizen-driven urbanism
Cochin international airport (Kerala)	Validates large-scale renewable energy transition in infrastructure without compromising performance or cost-effectiveness	Schiphol Airport Solar Installations (Netherlands)	Cochin achieves full solar dependence; stands out for energy self-sufficiency in Global South context
Palava smart and eco city (Mumbai region)	Embeds IoT and PPP governance in sustainable urban design; focuses on smart mobility and resource efficiency	Songdo Smart City (South Korea)	Palava prioritizes inclusivity and environmental outcomes over high-tech spectacle; more adaptable for Indian conditions
Charanka solar park (Gujarat)	Advances decentralized renewable energy deployment with rural impact; boosts energy access and climate action	Topaz Solar Farm (USA)	Charanka's scale is comparable, but its developmental role in rural transformation is more pronounced
Sabarmati riverfront development (Ahmedabad)	Reclaims urban riverfront for ecology and public use; integrates flood management and biodiversity restoration	Cheonggyecheon Stream Restoration (South Korea)	Both reclaim ecological corridors, but Sabarmati combines with flood resilience in monsoon-prone region

(continued)

Table 2.9 (continued)

Indian project	Core contribution to study objectives	Comparable global model	Comparative insights
Indira Paryavaran Bhawan (New Delhi)	Demonstrates role of public institutions in Modeling net-zero energy infrastructure and governance innovation	Council House 2, Melbourne (Australia)	Comparable performance; India's model showcases potential of sustainable architecture under tropical, high-use conditions
Auroville township and green belt (Tamil Nadu)	Implements regenerative urbanism, land use ethics, and socio-ecological zoning with deep community involvement	Findhorn Ecovillage (Scotland)	Auroville features more complex governance and context-specific ecological restoration rooted in postcolonial reality

Source Authors

energy equity and cost-effective transformations are an issue as urgent as climate action. The Songdo Smart City in South Korea acts as the quintessence of hi-tech technological urbanism and involves the implementation of sensors, big data and automation, which facilitate life in an urban environment. Although it is spectacular, Songdo has been criticized because it was designed in a top-down manner and has a poor integration of citizens (Arend, 2023). Contrarily, the Palava Smart City focus on the public–private alliance, involvement of communities, and low-cost sustainability renders it more relevant in culturally varied and economically stratified situations such as India. Therefore, Palava is an example of a more inclusive and socially sensitive variant of neo-urbanization. The analogy between the Charanka Solar Park in Gujarat and the topaz Solar farm in California also clearly shows the socio-developmental promise of renewable energy facilities. Although both are enormous solar projects, Charanka has left a more significant impression than just generating clean energy to include rural development, employment of people, and decentralized energy delivery- something that is not as significant in Topaz model, which focuses on utility. This opposition describes the result of ecological modernization in the context in which it is integrated into the larger development planning such that it yields multi-dimensional returns. Sabarmati Riverfront project in Ahmedabad and the Cheonggyecheon Restoration projects of Seoul all manifest the ability of a city to be changed through its ecological regeneration. The design of Sabarmati however incorporates flood control measures to suit the Indian climate of the monsoon season, and thus, it is a stronger and hydrologically sensitive development. This strategy reminds the necessity of climate-contextual planning in urban environmental projects. One of the most recognized examples of energy-efficient buildings used by the community is Council House 2 (CH2) in Melbourne, Australia (Ashmawy et al, 2024). CH2 has such features as passive cooling, solar panels, and greywater recycling which closely match the design philosophy of Indira Paryavaran Bhawan in New Delhi. However, the fact that IPB works under the Indian conditions of tropical ambient temperature and high population density, demonstrates that even under the circumstances of the rising economy, the green building innovations can be expanded and valuable, demonstrating that these innovations are not exclusive to high-income and temperate neighborhoods (Chaturvedi et al, 2024).

The Findhorn Ecovillage in Scotland is ultimately a strong example of purposeful living, spiritual ecology, and sustainable community planning (Copeland et al, 2023). But the size, variety of socio-political experiments, and combination of ecosystem restoration built into Auroville in a postcolonial context, make it even more complicated and context-specific. Auroville can therefore be considered as regenerative urbanism that does not just simply maintain ecological functions, but rather works to revitalize the destroyed landscapes and create new forms of social-environmental relationships. These comparisons, taken collectively, confirm once again that Indian urban innovations despite being often developed within resource and institutional limits—are not only up-to-date with best practices worldwide, but are in most instances working to set new and more outreach-based course toward socio-ecological change. These insights can highly confirm the focal theme of the book that sustainable development should be locally based, socially adaptable, and ecologically revolutionary by social Resilience and Sustainable Solution.

Theoretical Analysis of Ecological Modernization with SDGs

The model of ecological modernization provides a dynamic approach to the interpretation of how the cities can turn the problem of climate change, pollution, as well as inequality, into the possibility of sustainable change. At the bottom, the Ecological Modernization Theory (EMT) aligns the technological innovation, institutional reform, reflexive modernization, and market-oriented mechanisms as mutually linked means to promote the furthering of the environmental and socio-economic objectives. Compromising cleaner technologies, inclusive types of governance, adaptive policy, eco-taxes or green certifications, EMT restructures ecological issues as challenges to development instead as sources of innovation and advancement. These values are directly overlapping with such phenomenon as Neo-Urbanization that develops city-building models based on intelligence, sustainability, and inclusivity. According to this formulation, neo-urbanization would be acting as an intermediary between the processes of ecological modernization and the enhancement of the Resilient Urban Systems. Regenerative infrastructure, adaptive urban planning and integration of ecosystems are the main aspects of such systems to ensure the long-term sustainability. Finally, this model is closely related to the Sustainable Development Goals (SDGs)—SDG 11 (Sustainable Cities and Communities), SDG 7 (Clean Energy), SDG 13 (Climate Action), as well as SDG 6 (Clean Water and Sanitation). It provides an outline of a logical theoretical route along which specific innovations and governance changes can shift sustainability quadrants of ideals into operational plans and position cities as participants in the management of the environment and social justice (Fig. 2.3).

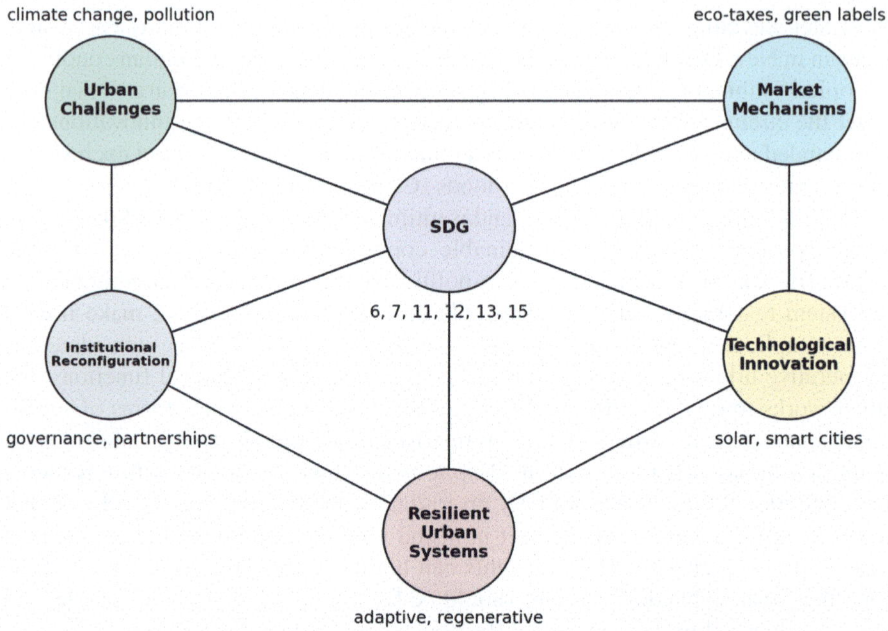

Fig. 2.3 Theoretical ecological modernization model in the present context. *Source* Authors

Analyzing Ecological Modernization Through the Lens of India's Socio-Ecological Development

A review of the key Indian urban and infrastructural projects shows an increasing trend toward sustainable development channels. These projects represent the plans of practical implementation of the Ecological Modernization Theory (EMT) and the neo-urbanization; they present solutions to the socio-ecological crises that conventional growth models have created. They demonstrate such localized propositions—a set of technological interventions, a community-based form of governance—that reinvent the ways of development in the Global South, especially in India (Vincent, 2023).

1. The Ecological Modernization of the Future of Resource-Intensive Growth—At the heart of socio-ecological issues in India, there is the reliance on fossil energy and centralized infrastructures. The Cochin International Airport, which is the first fully solar-powered airport in the world, and the Charanka Solar Park, one of the largest renewable clusters in India, are all the decisive steps toward abandoning the extractive systems. These examples can prove one of the key principles of EMT environmental protection and economic development could be achieved by technological innovation (Dias, 2021). They are also in line with SDG 7 (Affordable and Clean Energy) and SDG 13 (Climate Action) in that they minimize emissions of carbon and at the same time tackle energy poverty and prove to be economically viable through cost reductions and efficiencies. The

examples they provide demonstrate how the energy transition may be realistic, feasible, and transformative to sustainable development.

2. Neo-Urbanization and Re-Inventing the Urban Space—Traditional urbanization has caused ecological degradation and displacement of people. Contrastingly, other projects like Magarpatta City SEZ, Palava Smart City as well as the Sabarmati Riverfront development portray neo-urbanization which incorporates smart features, ecological planning and inclusive governance within the urban geography. Ecologically-oriented Magarpatta is based on the principles of community-centered planning and incorporates renewable energy sources, waste-to-compost facilities and water-conscious design (Dhere, 2017). These initiatives facilitate SDG 11 (Sustainable Cities and Communities) and SDG 12 (Responsible Consumption and Production) because of the establishment of socio-ecological resilience. On the same note, Palava Smart City will use ICT and IoT to manage resources efficiently, whereas the Sabarmati Riverfront can also apply regenerative urbanism by improving flood resistance, biodiversity, and open spaces.
3. Community Resilience as a Transition—The Catalyst Community-led initiatives are highlighted through the Auroville Green Belt Plan which engulfs the ecological premise, renewable technology and decentralized governance. It represents regenerative urbanism through its practices: organic farming, permaculture, rainwater gathering, and native afforestation, which develops the city in a proactive repairing way (Henfrey et al., 2023). Auroville, by putting participatory governance and ecological stewardship at the center of its motivational strategies, shows how bottom-up planning strategies can come up with more contextually meaningful and ethical solutions than top-down planning strategies.
4. Institutional Innovation and Governance Reform—This is another pillar of sustainable transformation, i.e., institutional leadership. Indira Paryavaran Bhawan (IPB) is the first net-zero energy building by the Indian Government, which combines passive architecture, solar energy delivery, and geothermal heating to obtain energy efficiency. This project shows that infrastructural leadership can be taken by the public institutions, which is transforming the environmental policy based on compliance instead of anticipatory governance (Muiderman et al., 2022). Being a prototype of the greening of institutions, IPB highlights the need to consider the environment at the design stage of planning.
5. Synthesizing Local Innovation to Sustainable Transformation—These projects are united by the emphasis on actions that are sufficiently geared toward practical outcomes that address the structural causes of issues such as socio-ecological ones in India. They show a multi-dimensional refocusing of development, based on decentralized solar at Cochin, community stewardship at Magarpatta, ecological restoration at the Sabarmati and regenerative land use at Auroville. Their results are evidence of a comprehensive treatment which is a combination of

   - adaptation at the technological level (smart grids, renewable energy, waste-to-energy),
   - social involvement (farmer- and community-based planning),

- policy co-arrangement (green building codes, public–private relationships),
- native afforestation (landscape restoration).

These initiatives prove that resilience and sustainability are not but can be attained because they connect local action with global sustainability objectives. The new patterns in India are active workshops of ecological change, which provide an opportunity to break the cycle of only extractive development and technological innovation, cultural continuity, and participatory government. All of them give important insights and empirical data towards formulating scalable solutions to the finances, environment and community wellness (Fuchs et al., 2025).

Aligning Gaps to Achieve **Sustainability Goal.**

There is a new trend of development that is appearing across India, it has shown that progress can be made without having to harm the environment. This change is observable in various innovations: Cochin International Airport that operates on pure solar energy indicates how renewable sources could be useful, and also in terms of economic feasibility. In Magarpatta City, the developers who are farmers have constructed communities that are environmentally friendly and include people rather than displacing them. Ecological restoration in the form of urban form, like the Sabarmati Riverfront Development, to net-zero urban infrastructure, like government programs like Indira Paryavaran Bhawan, are both on display. These instances together have been an indication of a sweeping shift in which technology, community involvement, and traditional knowledge are colliding together to create futures that are sustainable and that are equitable.

The case studies of Magarpatta City SEZ, Palava Smart City, Sabarmati Riverfront, Auroville Township, and Indira Paryavaran Bhawan portray the dynamic development of India in matters related to urbanization under the framework of neo-urbanization. They are not the regular types of city models but combine eco consciousness, technological advancements, and participation governing. The Magarpatta shows a distinctive combination of the traditional ecological values with the modern sustainability, which is used through waste segregation, solar energy systems, and rain water collection. The model of Palava Smart City develops the smart city model based on the IoT-based infrastructure and data-driven governance, and the recent Sabarmati Riverfront focuses on the restoration of the ecological setting by multiple functions. Auroville is a regenerative community based on organic agriculture, renewable energy and decentralized governance and Indira Paryavaran Bhawan is a good example of institutional leadership that is based on sustainability by integrating solar in its designs and passive architecture.

Combined, these initiatives highlight three main pillars of sustainable urban change, namely matching technology to ecology, enabling local communities, and progressive institutional frameworks. They show that the future of India cities lies in polycentric and context-altering strategies that can balance global sustainability goals and local demands (Iqbal, 2025). These models can be used as real-world solutions to climate-resilient and socially inclusive development through the integration of ecological modernization and community resilience. The results of the research prove that urbanization issues cannot be solved only by technological means but demand transformation of the urban structures with ecological awareness, participative actions and meaning making in the bureaucracies. The balanced approach has offered scaled approaches to balancing urban development, nature sustainability, and equity.

2.4 Conclusion

Scientifically and policy-wise, this chapter is a contribution to the debate of sustainable urban development since it criticizes ecological modernization and neo-urbanization as the means to reduce the socio-ecological crisis caused by traditional development patterns. Based on iconic Indian urban development projects, such as Magarpatta City SEZ, Cochin International Airport, Palava Smart City, Charanka Solar Park, Sabarmati Riverfront, Indira Paryavaran Bhawan and Auroville Township, the chapter grounds its position that ecological degradation and unsustainable urbanization are not only individual externalities, but are signs of dysfunction at the system level which is founded on stagnant growth paradigms. The considered projects present tangible evidence of the importance of Ecological Modernization Theory (EMT) of the necessity to state that environmental change can be implemented by using conventional methods of changing the institution, green technologies, and proactive regulation. Simply by connecting these efforts to Sustainable Development Goals (SDGs), the analysis provides a metric-based platform of assessing the sustainability of the urban environment, which confirms the SDGs as a valid instrument in sustainability studies.

The discussion also brings together the paradigm of neo-urbanization that no longer sees urban growth merely as a linear expansion, but rather as an aspect of regeneration and inclusive process that considers the limits of the environment, the promotion of social equity, and integration of technological adaptability. These Indian case studies show how cities can change their funding as environment sinks to being ecological resources through combined means: smart grids, decentralized waste and water-sensitive city design. This has a direct implication on scientific studies done in the areas of urban ecology, environmental sociology, infrastructure resilience, and environmental governance. More to the point, this study provides a Global South twist to what typically becomes a Global North literature regarding eco-cities and green urbanism. Although most Western patterns of sustainable cities are based on a high level of financial resources and the maturity of the institutions, the Indian examples discussed in this chapter prove that a significant ecological change can be achieved even in the conditions of limited resources and compact settlements. Ecological modernization in the Global South can meet (and possibly exceed) the international standards, shown by projects such as Cochin International Airport and Auroville, including even Schiphol Airport (Netherlands), Songdo Smart City (South Korea), or Findhorn Ecovillage (Scotland). The bigger picture of this chapter is that it re-invests urban planning as a socio-environmental engineering- at which local knowledge systems, public–private government models and technological advances are collaborative in terms of rising environmental challenges of the Anthropocene. Since global cities keep encountering more challenges caused by climate change, declining biodiversity, and socio-economic polarization, the urban paradigms arising in India are considered not only as the region-fixing models but also as the models of lessons in indomitable adaptation, inclusive growth, and planetary sustainability.

The conclusion of the research is that the study has provided a critical scientific impact by establishing the ecological modernization and neo-urbanization as globally adaptive frameworks of sustainable urban transformation, which are empirically based. It is based on Indian case studies demonstrating how cities can be changed into centers of environmental restoration, technological innovation and social equity using integrated and community-supported approaches. The chapter emphasizes the significance of interdisciplinary, culturally situated, and multi-scalar strategies that redefine development as a multi-faceted ecological development, but not a linear economic affair. By proving that sustainability models that integrate and support local visions of the Global South are feasible, it contributes to the discourse of urban sustainability globally and challenges scholars and policy-makers to perceive cities not as a liability to the environment, but as a source of inclusive and resilient futures.

AI Discloser—None of this chapter would be created using the help of artificial intelligence applications, such as the ChatGPT distro of OpenAI to aid in academic writing. While aids like language refinement, paraphrasing, content synthesis and formatting citations was done through AI. Where the authors takes care of the conceptual design, the analysis of the research, case selection, analysis of findings interpretation, and the final intellectual content. The AI content underwent a serious critique and editing to be accurate, original and meet the academic standards. No part of this work was solely or autonomously created by AI.

References

Abujder Ochoa, W. A., Iarozinski Neto, A., Vitorio Junior, P. C., Calabokis, O. P., & Ballesteros-Ballesteros, V. (2025). The Theory of complexity and sustainable urban development: A systematic literature review. *Sustainability, 17*(1), 3.

Alam, M. K. (2021). A systematic qualitative case study: Questions, data collection, NVivo analysis and saturation. *Qualitative Research in Organizations and Management: An International Journal, 16*(1), 1–31.

Almihat, M. G. M., Kahn, M. T. E., Aboalez, K., & Almaktoof, A. M. (2022). Energy and sustainable development in smart cities: An overview. *Smart Cities, 5*(4), 1389–1408.

Arend, A. (2023). Paved with good intentions? Urban planning and smart city policies in South Korea and Japan. *Temp. Int'l & Comp. LJ, 38*, 1.

Arku, G., & Marais, L. (2021). Global south urbanisms and urban sustainability—challenges and the way forward. *Frontiers in Sustainable Cities, 3*, Article 692799.

Ashmawy, R. A. E., Ragheb, A. A., Ragheb, G., & Marouf, O. (2024). Sustainable design principles for green office buildings: A comprehensive review. *International Journal of Sustainable Development & Planning, 19*(6).

Bera, B., Chinta, S., Mahajan, D. A., Sailaja, A., & Mahajan, R. (2023). Urbanization and its impact on environmental sustainability: A comprehensive review. *Journal of Harbin Engineering University, 44*(8), 1310–1318.

Chaturvedi, P. K., Kumar, N., & Lamba, R. (2024). Finding the gaps in design strategies and technological advancements for net-zero energy buildings development in India. *Energy & Environment, 35*(7), 3880–3920.

Clark, T. N., & Wu, C. (2021). Urbanization theorizing. In *Handbook of Classical Sociological Theory*, pp 421–438.

Cleland, J., MacLeod, A., & Ellaway, R. H. (2021). The curious case of case study research. *Medical Education, 55*(10), 1131–1141.

Copeland, C., MacKerron, G., & Foxon, T. J. (2023). Futures for Findhorn: Exploring challenges for achieving net zero in an ecological intentional community. *Futures, 149*, Article 103155.

Davydenko, I., & Hilbers, H. (2024). Decarbonization paths for the Dutch aviation sector. *Sustainability, 16*(3), 950.

Dhere Amar, M. (2017). Peasantry to industry: Innovative participatory model of development for policy implications-a case of Magarpatta City. *Int J Fam Busi Manag, 1*(2), 1–7.

Dhere Amar, M. (2018). Peasantry to industry: Innovative Participatory model of development for policy implications-a case of Magarpatta City. *Int J Fam Busi Manag, 1*(2), 1–7.

Dias, R. C., Seixas, P. C., & Lobner, N. (2021). Sustainable development and ecological modernization: Boundary discourses between "strong" and "weak" approaches. *Journal of Sustainable Development, 13*(4), 268–278.

Du, L., & Yang, K. (2025). Does institutional change during the green transition influence low-carbon innovation? Evidence from China's Low-Carbon City Project. *Technovation, 145*, 103274.

Fastenrath, S., & Preller, B. (2018). Freiburg: The emblematic green city. *Green building transitions: Regional trajectories of innovation in Europe, Canada and Australia* (pp. 69–98). Springer International Publishing.

Foggin, J. M., Brombal, D., & Razmkhah, A. (2021). Thinking like a mountain: Exploring the potential of relational approaches for transformative nature conservation. *Sustainability, 13*(22), 12884.

Fuchs, G., Kroos, F., Scherer, C., Seifert, M., & Stelljes, N. (2025). Exploring marine conservation and climate adaptation synergies and strategies in European seas as an emerging nexus: A review. *Frontiers in Marine Science, 12*, 1542705.

Ghahremani, H., Afsari Bajestani, S., McCarthy, L., & Jalalianhosseini, M. (2021). Transformation of urban spaces within cities in the context of globalization and urban competitiveness. *Journal of Urban Planning and Development, 147*(3), 05021026.

Henfrey, T., Feola, G., Penha-Lopes, G., Sekulova, F., & Esteves, A. M. (2023). Rethinking the sustainable development goals: Learning with and from community-led initiatives. *Sustainable Development, 31*(1), 211–222.

Iqbal, A., Nazir, H., & Qazi, A. W. (2025). Exploring the 15-minutes city concept: Global challenges and opportunities in diverse urban contexts. *Urban Science, 9*(7), 252.

Jorgenson, A., El Tinay, H., Fitzgerald, J., Givens, J., Goh, T., Huang, X., & Thombs, R. P. (2024). Advances in research on anthropogenic drivers of climate change. In *Routledge Handbook of Climate Change and Society*, pp. 60–76.

Kajiita, R. M., & Kangethe, S. M. (2024). Socio-Economic dynamics inhibiting inclusive urban economic development: implications for sustainable urban development in South African Cities. *Sustainability, 16*(7), 2803.

Katyal, N. (2024). Understanding about nature-based approaches for sustainable consumption, production and development. In *Nature-Based Solutions in Achieving Sustainable Development Goals: Harmonizing Nature and Progress* (pp. 317–367). Springer Nature Switzerland.

Kaviya Darsini, S., Madhivadhani, K., & Sudharsanamurthy, P. (2026). Urban sprawl: Methods and approaches in urban planning. In *Innovative and intelligent computing for global sustainability* (pp. 105–115). Springer Nature Switzerland.

Liu, M., & Lo, K. (2023). Eco-cities as urban laboratories of Chinese ecological modernization: Eco-experimentation and the modernization of urban governance. *Urban Geography, 44*(9), 2037–2061.

Madakam, S. (2022). An Exploratory Study on Internet of Things (IoT) in Indian Smart Cities: Lavasa, Palava, GIFT, Smartcity Kochi (SCK). *Palava, GIFT, Smartcity Kochi (SCK).* https://scholar.googleusercontent.com/scholar?q=cache:sKVwKHxvtwkJ:scholar.google.com/+Palava+Smar+%26+Eco+City&hl=en&as_sdt=0,5&as_ylo=2021

Manjunath, K. V., Patil, N., Aprameya, C. R., Sunil, S. K., Sirsgi, S., Sandeep, V., & Poojari, M. (2023, December). Zero-energy buildings—A new advancement on energy saving. In *International Conference on Smart and Sustainable Developments in Materials, Manufacturing and Energy Engineering* (pp. 271–276). Springer Nature Switzerland.

Muiderman, K., Zurek, M., Vervoort, J., Gupta, A., Hasnain, S., & Driessen, P. (2022). The anticipatory governance of sustainability transformations: Hybrid approaches and dominant perspectives. *Global Environmental Change, 73*, Article 102452.

Nandy, D. (2024). Global social movements (GSMs) and non-governmental organizations: Investigating the Impact on global social change. *Publica, 16*(1), 64–88.

Peck, J. (2015). Cities beyond compare? *Regional Studies, 49*(1), 160–182.

Peter, C. (2021). Social innovation for sustainable urban developmental transitions in Sub-Saharan Africa: Leveraging economic ecosystems and the entrepreneurial state. *Sustainability, 13*(13), 7360.

Rodenbiker, J. (2021). Making ecology developmental: China's environmental sciences and green modernization in global context. *Annals of the American Association of Geographers, 111*(7), 1931–1948.

Shah, K., & Kunte, K. (2025). Learning from Urban Riverfront Development: Revisiting the Sabarmati Model. In *Rivers, Cities and People* (pp. 91–111). Routledge India.

Shirke, V., Nabar, A., & Achari, A. (2025). Enhancing Urban Coastal Resilience Using Nature Based Solutions: A Study of Mumbai, India. *Environmentally Sustainable and Socially Resilient Urban Development and Management,* 789–820.

Simons, S., Kinjawadekar, A., & Kinjawadekar, T. A. (2024). Assessing the impacts of ecological framework of Indian riverfront revitalization projects. *Environment, Development and Sustainability, 26*(11), 27553–27583.

Sreenath, S., Sudhakar, K., & Yusop, A. F. (2022). Carbon mitigation potential of the airport-based solar PV plants in the Indian context. *International Journal of Ambient Energy, 43*(1), 1311–1319.

Tajuddin, N., & Dąbrowski, M. (2021). Enabling socio-ecological resilience in the global south: Insights from Chennai. *India. Sustainability, 13*(19), 10522.

Ullah, A. (2021). Public private partnership in solar energy development of India. *International Journal of Multidisciplinary Educational Research, 10*(12), 2. ISSN 2277-7881

UN-HABITAT. (2024). *World cities report 2024: Cities and climate action.* Stylus Publishing, LLC.

Vincent, K. (2023). Development geography II: Community-based adaptation and locally-led adaptation. *Progress in Human Geography, 47*(4), 604–612.

Viswanathan, P., & Sharpe, T. (2024). Ecourbanism: designing a 'green'city. *WIT Transactions on Ecology and the Environment, 84.*

Wang, Z., Lin, L., Zhang, B., Xu, H., Xue, J., Fu, Y., & Li, F. (2023). Sustainable urban development based on an adaptive cycle model: A coupled social and ecological land use development model. *Ecological Indicators, 154*, Article 110666.

Yousuf, M. U., Saleem, M. U., & Umair, M. (2024). Evaluating the 7E impact of solar photovoltaic power plants at airports: a case study. *Science and Technology for Energy Transition, 79*, 19.

Part II
Socio-ecological Crisis of Development and Community Resilience

Chapter 3
A Socio-ecological Issue that Increases Disaster Risk: Urbanization

Mehtap Kılıç Yildiz and Kader Demiröz Altintaş

Abstract Urbanization is defined as the process of population growth in urban areas, which is accompanied by significant changes in spatial, economic, and social structures. This phenomenon is influenced by various factors, including migration from rural to urban areas, natural population growth, and the intensification of economic activities in urban centers. Nevertheless, the absence of a deliberate and effective urbanization management strategy can engender a multitude of risks. These risks encompass security vulnerabilities, heightened disaster susceptibility, and concomitant social challenges, including poverty. In particular, unplanned and uncontrolled urbanization has been demonstrated to lead to environmental destruction and increase the vulnerability of human communities to disasters. Urbanization practices that are not conducive to health have been demonstrated to engender an increased probability of disaster occurrence. In the contemporary context, where the boundaries of nature are often disregarded and the consequences of climate change and associated disasters are recurrent, the significance of urbanization policies aimed at mitigating disaster risks is further accentuated. The objective of this study is to offer a comprehensive perspective on the aforementioned issue, in accordance with the findings derived from a systematic review of publications across diverse platforms. In this study, it is demonstrated how unplanned urbanization increases disaster risks through environmental degradation, social inequalities, and governance deficiencies, emphasizing that disaster risks stem not only from physical factors but also from the holistic weakening of socio-ecological systems. The model developed in the study systematically explains the mechanisms through which urbanization amplifies disaster risks, as well as holistic strategies for risk reduction, including nature-based solutions, inclusive governance, and the empowerment of vulnerable groups.

M. Kılıç Yildiz (✉)
Department of Emergency Aid and Disaster Management, Ardahan University, Ardahan Merkez, Türkiye
e-mail: mehtapkilic@ardahan.edu.tr

K. Demiröz Altintaş
Department of Emergency Aid and Disaster Management, Artvin Coruh University, Artvin Merkez, Türkiye
e-mail: kader.demiroz@artvin.edu.tr

G. K. Panda et al. (eds.), *Socio-Ecological Crisis of Development*, Sustainability Solutions, https://doi.org/10.1007/978-3-032-22171-1_3

Keywords Urbanization · Disaster · Socio-ecological · Climate change

3.1 Introduction

Urbanization is defined as the increase in the proportion of the population residing in urban areas, with individuals migrating from rural to urban regions (EEA, 2021). In the contemporary era, urbanization has become an imperative driven by a confluence of factors, including the escalating population, the advancement of industrialization, technological innovations, enhanced accessibility to services, infrastructure, and transportation networks, as well as employment opportunities (EEA, 2021; Satterthwaite, 2007; UN-Habitat, 2016; United Nations, 2018). Despite its growing imperative, unplanned urbanization introduces a multitude of risks. These risks may be interconnected and/or trigger each other. For instance, unplanned and unsustainable urbanization can result in increased poverty, public health and security risks, and damage to environmental factors. This, in turn, can increase the risk of disasters and lead to the development of disaster-resilient societies and communities (UN-Habitat, 2016; Wisner et al., 2004).

Unplanned and rapid urbanization, the realization of urbanization without considering regional requirements and characteristics, and the realization of urbanization in a way to accommodate more population than the area opened for urbanization can accommodate may increase disaster risks (Hooli, 2016; Mumuni et al., 2025; Peng & Li, 2021; Tajuddin et al., 2021; Woiwode et al., 2024). For instance, the establishment of settlements within river basins, the development of settlements at 0 points in sloping geography, the conversion of fertile agricultural lands into habitable zones, the implementation of impermeable asphalt systems, the undertaking of illicit and unauthorized construction in contravention of zoning plans, and the conversion of areas with a high disaster risk to construction zones have been identified as factors that escalate the risk and/or impact of disasters (Berke & Campanella, 2006; Satterthwaite, 2017; Solecki & Leichenko, 2006; United Nations, 2018).

There are various studies in the literature on urbanization and the risks it brings. This study is based on a systematic review of scientific articles on the subject. As a result of the review, it has been found that the realization of urbanization under inappropriate conditions and in the wrong way increases the risk of disasters in various aspects. It can be said that the realization of urbanization in accordance with the norms of the region and society in the right way is one of the most important factors in the creation of a resilient society for disasters. Especially in the execution of construction in urban areas, central and local governments should work in cooperation by considering environmental factors. Especially today, the increase in urban population, the concentration of industrial activities in these regions, the decrease in agricultural production, and climate change, drought and water stress due to various factors make permanent solutions for urbanization necessary (Cohen, 2006; IPCC, 2022; UN, 2019). In this context, negative situations caused by urbanization can be minimized with clearly defined international and national policies on urbanization and a cooperation process from individual to global (Bulkeley & Betsill, 2013; OECD, 2020; UN-Habitat, 2020). Unplanned and improper urbanization not only negatively affects the physical environment, but also social justice and inequality (Anguelovski et al., 2018; UNDRR, 2019b). Especially low-income groups may live in areas that are more vulnerable to disasters. This situation again increases the post-disaster recovery process and thus vulnerability. In solving all these problems, the use of nature-based solutions, integration of advanced technological developments into urbanization, and supporting community participation and local knowledge can be useful (Kabisch et al., 2017; World Bank, 2021). Among nature-based solutions, it can be said that green infrastructure, rain gardens, expanding the use of permeable surfaces, expanding the use of sponge cities, and providing permanent solutions intertwined with nature in urbanization are factors that reduce disaster risks in the urbanization process. In addition, it can be said that technological solutions such as smart cities, early warning systems, and disaster information systems, which are frequently used nationally and internationally, are also among the factors that reduce disaster risks in urbanization (EEA, 2021; UN-Habitat, 2020). There are many technologies that have become critical tools in the urban planning process. These are Geographic Information Systems, remote sensing systems, early warning systems, IoT technology, and a range of tools including smart city applications (UNDRR, 2019a; World Bank, 2021). Thanks to these technologies; risky area detection, infrastructure analysis, disaster scenarios, modeling, pre-detection of disaster risks, and warnings and increasing disaster awareness are provided (Burton, 2015; Zerger, 2002). In particular, supporting decision-making processes based on data before, during, and after disasters supports response and recovery efforts, and ensures that they gain speed and efficiency. This situation not only reduces disaster risks, but also contributes to the creation of more livable cities by ensuring the achievement of sustainability goals (IPCC, 2022; OECD, 2020). In order for all the above-mentioned issues to be carried out actively and effectively, social participation must be ensured. In this context, it will be beneficial to create disaster-resilient city models starting from the local level by utilizing the knowledge and experiences of the local people (Mercer et al., 2010; Wisner et al., 2004).

3.2 Methods

In this study, a systematic review of studies on socio-ecological risks caused by urbanization, factors that increase these risks, and risk mitigation strategies that can be developed against them has been conducted. The systematic review followed the steps of defining the study topic, reviewing the relevant literature around specific questions, data selection and data collection, and assessing and analyzing the quality of evidence. This systematic review was conducted in accordance with the PICOS framework and the Preferred Reporting Items for Systematic Reviews and Meta-Analyses (PRISMA) guidelines.

3.2.1 Information Source and Research Strategy

Defining the job: In order to explain the methodology used in the research, the work was defined within the framework of the proposed PICOS model. The four research questions (Problems) created in this direction are given below.

1. Through which basic dynamics and processes does the urbanization process influence the formation of socio-ecological risks?
2. Through which mechanisms do urbanization-related factors accelerate the increase in socio-ecological disaster risks?
3. What are the prominent strategies in the literature for reducing socio-ecological disaster risks due to urbanization and to what extent are the effectiveness of these strategies evaluated?
4. How can a holistic solution model be developed that takes into account the relationship between urbanization, socio-ecological risks, and disaster risk?

The articles to be included in the study were scanned as follows: The advanced search option was used on the Web of Science platform, the necessary scans were made by using the words "Disaster, Risk, Urbanization, Socio-Ecological" given in the inclusion criteria and selecting the "article title, abstract, keywords" option. Within the scope of this search, 3 articles that meet the criteria were listed.

The advanced search option was used on the Scopus platform, the necessary searches were made by using the words "Disaster, Risk, Urbanization, Socio-Ecological" given in the inclusion criteria and selecting the "article title, abstract, keywords" option. Within the scope of this search, 4 articles that meet the criteria were listed. Since one of these articles was also available on the "Web of Science" platform, it was evaluated through this platform.

The advanced search option was used on the Emerald Insight platform, the necessary searches were made by using the words "Disaster, Risk, Urbanization, Socio-Ecological" given in the inclusion criteria and selecting the "all fields" option. Within the scope of this search, 57 articles that meet the criteria were listed. The inclusion and exclusion criteria of the articles are presented in Table 3.1.

Table 3.1 Inclusion and exclusion criteria

Inclusion criteria	Exclusion criteria
• English language studies • Open access studies • Research and review articles • Disaster, risk, urbanization, socio-ecological • Emerald insight • Scopus • Web of science	• Other languages other than English • Closed access and paid journals • Publications other than articles • Publications outside the platforms included in the inclusion criteria

3.2.2 *Qualifying Criteria*

3.2.3 *Study Selection*

In data selection, studies were included in the study by reviewing the abstracts of the publications with two independent researchers. While one review author extracted the core data from the included studies, the second author checked the extracted data. Disagreements were resolved by discussing between the two review authors.

3.2.4 *Data Extraction*

In this study, a form was developed to extract data from publications on socio-ecological risks caused by urbanization, including factors that increase these risks and risk mitigation strategies by content analysis method. In the content analysis process, the researchers first produced meaningful codes by reading the data in detail, then created themes by combining these codes, and in the last stage, reviewed the themes and prepared a holistic report. A systematic, reliable and verifiable analysis method was adopted at every stage of the study, and transparency and traceability of the data obtained throughout the content analysis process were ensured. Content analysis is a method that emphasizes the active role of the researcher in the reporting process and offers a flexible approach to identifying, analyzing, and reporting themes (Braun & Clarke, 2006). In this context, the developed content analysis form enabled a systematic analysis of the mechanisms of socio-ecological risk generation of urbanization processes and strategies for risk mitigation.

3.2.5 *Data Synthesis and Analysis*

All the articles were systematically recorded and assessed for their suitability for the study through a four-stage process. These suitability checks included:

- Evaluating the relevance of the article titles to the study within the scope of the screening results,
- Evaluating and recording the suitability of the articles based on their abstracts and conclusions within the scope of the screening results,
- Recording and conducting a preliminary content check of the articles deemed potentially suitable for the study, and
- Systematically reviewing the full content of the articles that were preliminarily assessed as suitable.

Descriptive and content/thematic analysis methods were used to analyze the data. Two researchers systematically collected data from the included studies in line with the data collection form created in advance. In the first step, socio-ecological risks arising due to urbanization, factors that increase these risks, and recommended risk mitigation strategies were identified with the help of the data form. In the second step, these factors were elaborated in terms of their dimensions (e.g., risk types), antecedents (risk generating factors), and consequences (effects on disaster risk). In the third step, a model proposal conceptualizing the relationship between urbanization, socio-ecological risks, and disaster risk was developed. The content analysis process involves a detailed review of the articles identified as directly related to the topic of the study. The researchers analyzed the data in a systematic way, initially coding, then creating themes by establishing meaningful relationships between the codes, and completing all stages in a careful and timely manner. Each theme was supported by relevant pieces of data and checked for both intra-theme consistency and data set integrity. Assumptions and the specific approach taken during the analysis process were clearly articulated, ensuring consistency between the methodology used and the reporting process. Figure 3.1 visualizes the flow of information during the systematic review stages of the study.

3.3 Findings

As a result of the systematic review, a total of 13 articles were analyzed, and a table was created to systematically present and compare the findings.

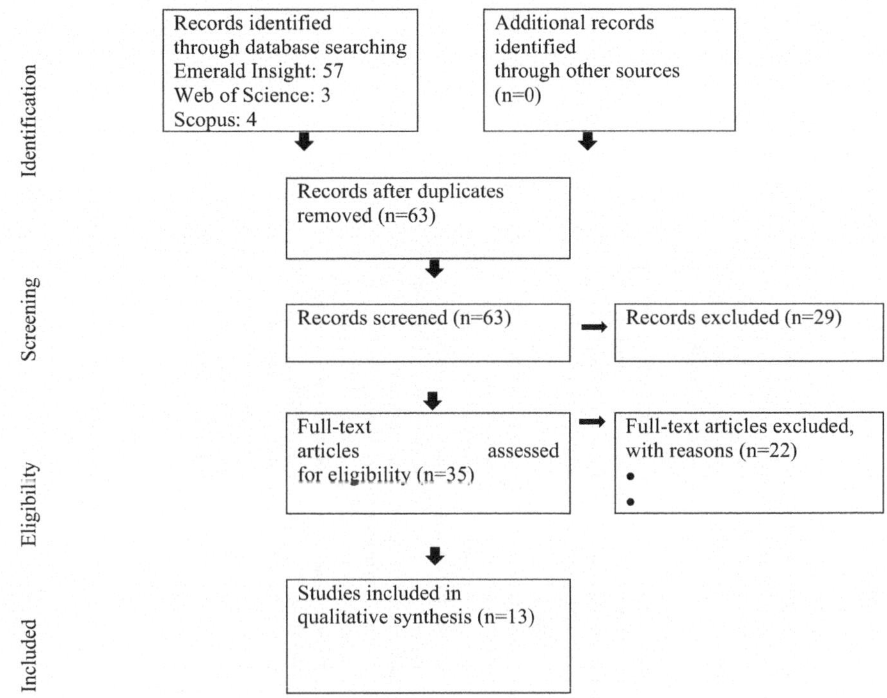

Fig. 3.1 Flow of information through the different phases of a systematic review

Title (Makale başlığı)	Author/year (Yazar/yıl)	Socio-ecological risks associated with urbanization	Factors of urbanization increasing disaster risk	Risk reduction strategies	Type of disaster	Strages of disaster
1. Resilience of the poorest: coping strategies and indigenous knowledge of living with the floods in Northern Namibia	Hooli (2016)	Flood risk Infrastructure deficiencies Housing insecurity Dysfunctionalization of traditional knowledge	Unplanned urbanization Inadequate infrastructure Construction without environmental impact assessment Rapid population growth and rural-urban migration Destruction of natural cover (degradation of waterways)	Improving spatial planning Strengthening infrastructure investments Integration of indigenous knowledge into planning processes Community participation in decision-making Fighting poverty and equitable distribution of resources	Drought Extreme Weather Floods	Risk Management Risk Mitigation Preparedness
2. Identifying entry points for adaptive governance in peri-urban Chennai (India): a multi-dimensional, multi-level, and multi-scalar approach	Woiwode et al. (2024)	Degradation of water systems Degradation of ecosystems Deepening social inequalities Social and economic conflicts in rural-urban transition areas Vulnerability to climate change	Unplanned and rapid peri-urban sprawl Destruction of water areas and development pressure Inadequate infrastructure and drainage systems Fragmented governance and institutional incompatibility Loss of ecosystem services	Strengthening multi-level and multi-actor governance structures Adaptive development pathways focused on water management and agro-ecology Assessing local knowledge and traditional practices Implementing nature-based solutions Overcoming governance syndromes and institutional reform	Extreme Weather Floods Water scarcity	Risk Management

(continued)

(continued)

Title (Makale başlığı)	Author/year (Yazar/yıl)	Socio-ecological risks associated with urbanization	Factors of urbanization increasing disaster risk	Risk reduction strategies	Type of disaster	Strages of disaster
3. Enabling socio-ecological resilience in the global south: insights from Chennai, India	Tajuddin et al. (2021)	Increased flood risk and inundation Water insecurity Heat stress and microclimate degradation Weakening of ecosystem services Deepening social inequalities Land conflicts and spatial injustice Lack of public awareness of the value of water assets	Degradation of natural water systems Disregard of physiographic structure Inadequate and improper planning of drainage infrastructure Decrease in green areas and open spaces Opening wetlands for construction and misclassification Spread of informal settlements in floodplains Institutional fragmentation and lack of coordination Uncertainties of ownership and management	Protection and restoration of natural water systems Integrated Urban Water Management Increasing Green and Open Spaces Participatory and Inclusive Planning Processes Strengthening Institutional Coordination Implementation of Disaster Risk Reduction Strategies	Floods	Risk Management Planning Preparedness Mitigation

(continued)

(continued)

Title (Makale başlığı)	Author/year (Yazar/yıl)	Socio-ecological risks associated with urbanization	Factors of urbanization increasing disaster risk	Risk reduction strategies	Type of disaster	Strages of disaster
4. Characteristics of urban flood resilience evolution and analysis of influencing factors: a case study of Yingtan City, China	Yuan et al. (2024)	Inadequacy and obsolescence of infrastructure systems Population aging Inefficiency of financial expenditures	Inadequate infrastructure maintenance Deficiencies in emergency management Distribution of vulnerable population	Improving fiscal spending • Increasing efficiency in public services • Reducing unnecessary spending • Priority investment in critical areas (infrastructure, education, health) • Flood control, disaster facilities, investment in technological innovation Adjusting population structure • Fertility policy optimization (young population growth) • Skilled population attraction policies • Contingency planning for vulnerable populations Infrastructure maintenance and emergency management • Regular maintenance of storm water pipe Systems • Regular inspection and repair • Optimized distribution of emergency resources (flood management)	Floods	Risk Management Planning Preparedness Mitigation

(continued)

(continued)

Title (Makale başlığı)	Author/year (Yazar/yıl)	Socio-ecological risks associated with urbanization	Factors of urbanization increasing disaster risk	Risk reduction strategies	Type of disaster	Strages of disaster
5. Spatial constraints or spatial dynamics? The spatial spillover effect of networks of flood regulation service flows on land-use degree	Wang and Li (2024)	Deforestation Decrease in ecosystem services Increase in land use intensity Disaster vulnerability in coastal areas	Increased demand for flood regulation services Increased risk in coastal areas Frequency and impact of disasters Imbalance in flood service flows	Increasing forest cover Reducing flood risk Maintaining the quality of ecosystem service flows Strengthening infrastructure Multi-actor governance Flexibilizing land use planning Ecological compensation and financial incentive systems	Floods	Risk management
6. Ensemble flood risk assessment in the Yangtze River Economic Belt under CMIP6 SSP-RCP Scenarios	Peng and Li (2021)	Large floodplains and environmental problems caused by rapid urbanization Increased vulnerability due to dense population growth Infrastructure deficiencies and inadequate drainage systems	Extreme precipitation and climate change Land use changes High economic and population density Intensive construction in coastal and low-lying areas	Improving urban infrastructure Using early warning systems and monitoring technologies Developing regional disaster management strategies Increasing urban resilience Developing climate change adaptation policies	Floods	Risk Management Preparedness

(continued)

(continued)

Title (Makale başlığı)	Author/year (Yazar/yıl)	Socio-ecological risks associated with urbanization	Factors of urbanization increasing disaster risk	Risk reduction strategies	Type of disaster	Strages of disaster
7. Implementation of a specific urban water management-Sponge City	Nguyen et al. (2019)	Urban floods Loss of natural infrastructures Reduction of groundwater resources Heat island effect Water pollution and ecosystem degradation	Increase in impervious surfaces Inappropriate drainage systems Climate change Urbanization in wrong areas Unplanned growth Poor urban planning	Implementing the Sponge City approach Rainwater harvesting and reuse Ecological water management Promoting public–private partnerships Developing data-based decision support systems Awareness raising and training	Flood Water scarcity	Risk Management Mitigation
8. A systematic review on the causes and effects of urbanization on wetlands in Sub-Saharan Africa	Mumuni et al. (2025)	Loss of biodiversity Flood risk and altered hydrology Degradation of water quality Loss of livelihoods	Population growth and migration Infrastructure projects Unplanned urbanization and slums Land use changes	Protection and restoration of wetlands Green infrastructure practices Sustainable land use planning Legal regulations and strict supervision Community engagement and awareness Multi-actor cooperation	Flood	Risk Management Policy development Prevention

(continued)

(continued)

Title (Makale başlığı)	Author/year (Yazar/yıl)	Socio-ecological risks associated with urbanization	Factors of urbanization increasing disaster risk	Risk reduction strategies	Type of disaster	Strages of disaster
9. Reconnecting with nature: developing urban spaces in the age of climate change	Lehmann (2019)	Decrease in biodiversity and degradation of ecosystems Increase in urban heat island effect Decrease in green areas Overuse and depletion of natural resources	Urbanization effect combined with climate change Loss of green areas and natural infrastructure Increase in energy and water consumption Infrastructure deficiencies	Implementing Nature-Based Solutions Re-greening cities Enhancing and protecting the value of natural capital Enhancing urban resilience Sustainable land use planning	Heatwaves Flooding / Urban Flooding	Risk Management Mitigation
10. We said, they said: the politics of conceptual frameworks in disasters and climate change in Colombia and Latin America	Lizarralde et al. (2020)	Increased risk of flooding Land use changes and environmental degradation Heat island effect and extreme temperatures Pollution and depletion of water resources	Unplanned urbanization Poor infrastructure policies Social inequalities and vulnerability Climate change impacts Insufficient participation and wrong prioritization	Policy development based on the needs of local people Contextual sensitivity in infrastructure projects Elimination of socio-economic inequalities Aspirational development" approach Strengthening governance and participation	Floods / Flash Floods) Hurricanes / Tropical Storms	Risk Management

(continued)

(continued)

Title (Makale başlığı)	Author/year (Yazar/yıl)	Socio-ecological risks associated with urbanization	Factors of urbanization increasing disaster risk	Risk reduction strategies	Type of disaster	Strages of disaster
11. Artificial neural networks for flood susceptibility analysis in Gangarampur sub-division of Dakshin Dinajpur, West Bengal, India	Paul (2025)	Land use changes Increase in impervious surfaces Disruption of the water cycle Erosion and sedimentation	Urbanization of low slope and low elevation areas High rainfall intensity Construction along streams and rivers High stream density Soil compaction in agricultural areas	Restricting construction in high-risk areas (Zoning Regulations) Infrastructure strengthening Creation of reservoirs and artificial wetlands Protection of existing natural wetlands Community participation and awareness raising Development of early warning systems	Floods	Risk Management Mitigation and Prevention
12. Disaster preparedness and well-being among Cambodian- and Laotian-Americans	DeYoung et al. (2020)	Degradation and erosion of coastal areas Collapse of physical infrastructure Increased dependence on marine resources Decline in ecosystem services	Rapid population growth and construction in coastal areas Infrastructure inadequacies Climate change impact (sea level rise and storms) Language barriers and difficulties in accessing information	Community-based disaster preparedness and resilience activities Infrastructure improvement activities Training and information programs Ecosystem restoration and protection Increasing financial resilience Developing climate change adaptation strategies	Hurricanes Flooding/coastal flooding	Risk Management Preparedness

(continued)

(continued)

Title (Makale başlığı)	Author/year (Yazar/yıl)	Socio-ecological risks associated with urbanization	Factors of urbanization increasing disaster risk	Risk reduction strategies	Type of disaster	Strages of disaster
13. Increasing vulnerability to floods in new development areas: evidence from Ho Chi Minh City	Duy et al. (2017)	Destruction of wetlands and natural waterways Increase in impervious surfaces Decrease in ecosystem services Decrease in groundwater levels and soil subsidence	Establishment of new development areas in low elevation and flood risk areas Increasing population density Inadequate and delayed infrastructure investments Precipitation and sea level rise due to climate change	Implementation of compact city approach Creation of resilient city structure in newly developing areas Making transportation infrastructure resilient to disasters Restoring old wetland functions Developing long-term spatial planning and climate change adaptation strategies	Floods	Risk Management Mitigation and Prevention

3.4 Socio-ecological Risks Associated with Urbanization

Natural disasters such as floods, combined with the negative effects of urbanization and population growth, exacerbate potential negative impacts, making it difficult for local communities, particularly those struggling with poverty, to cope. Inadequately constructed settlements bring about housing insecurity and, when combined with factors such as fragile infrastructure, create obstacles to development (Hooli, 2016). High population density is one of the factors that reduce urban resilience. Social vulnerability is one of the main factors that reduce resilience, particularly among individuals with low-income levels (Yuan et al., 2024). The damage caused to the environment by greenhouse gas emissions, which cause climate change, has been linked to the increase in urban population as well as the increased risk of flooding in cities (Peng & Li, 2021). Flood risk is particularly high in low-lying areas with dense drainage networks and is significantly affected by factors such as rainfall, topography, land use, and water availability (Paul, 2025). Furthermore, the scarcity of land for the design of wetlands and ponds due to industrialization in cities hinders environmental planning (Nguyen et al., 2019). The continuation of housing investments in new development areas despite their high flood risk both increases existing risks and deepens environmental impacts (Duy et al., 2017).

3.5 Factors of Urbanization Increasing Disaster Risk

Although seasonal rainfall is heavy, natural water systems have been disrupted due to settlements forming around water channels. Consequently, flooding has become an inevitable result of unplanned urbanization (Hooli, 2016). Among the main factors that increase the impact of disasters such as floods in cities, population density and structure, as well as the lack of adequate infrastructure to drain floodwater, are noteworthy (Woiwode et al., 2024). In addition, the decrease in forested areas with the potential to retain floodwaters is another significant factor that increases the risk (Wang & Li, 2024). Coping with urban flood events is largely related to water drainage (Nguyen et al., 2019). While current flood management relies heavily on hard engineering solutions, the insufficient use of green infrastructure and nature-based approaches weakens urban resilience (Duy et al., 2017). Migration and population mobility caused by urbanization lead to insufficient capacity in cities, consequently disrupting natural and ecological processes and increasing pressure on infrastructure (Mumuni et al., 2025). Communities are becoming increasingly vulnerable to disaster risks due to environmental changes, economic dependence, and social vulnerabilities (DeYoung et al., 2020).

3.6 Risk Reduction Strategies

Eliminating the socio-political conditions that increase the vulnerability of communities will enhance resilience to disasters. Preventing uncontrolled population growth resulting from unplanned urbanization, spatial planning, infrastructure improvement, land use, and environmental resource management are among the key measures that increase resilience (Hooli, 2016). Unplanned urbanization, a major obstacle to development, requires not only an environmentally focused approach to the process, but also an adaptable governance approach (Woiwode et al., 2024). In stakeholder collaboration, cooperation between the private sector and public institutions is also critical. Although collaboration is sought in the city's transformation processes, public participation in urban planning processes remains limited. Furthermore, the absence of feedback mechanisms and the differences between needs and policies also pose obstacles to risk mitigation strategies (Tajuddin et al., 2021).

Investments focused on risk reduction must prioritize urban resilience; basic public services such as infrastructure, health, and education must be prioritized (Yuan et al., 2024). Many cities face the threat of flooding due to climate change. As a risk mitigation method for these problems, sponge city approaches, which emphasize mechanisms that increase water permeability, are gaining prominence (Nguyen et al., 2019). Integrated water management strategies that consider structural, non-structural, and ecological approaches together are needed to reduce flood risk (Paul, 2025). In this context, understanding the long-term impacts of urbanization processes on natural ecosystems, especially wetlands, emerges as a critical requirement for sustainable urban planning (Mumuni et al., 2025). However, achieving urban and ecological sustainability is possible not only through technical solutions but also through the development of participatory and multi-level management approaches that strengthen cooperation among different actors (Lehmann, 2019). For more effective and equitable policies, new conceptual frameworks developed from the bottom up, based on local knowledge and experience, are proposed (Lizarralde et al., 2020). Building resilience requires community-based and participatory interventions that focus not only on environmental risks but also on the root causes of vulnerability, such as poverty, lack of infrastructure, and dependence on a single source of livelihood. Furthermore, psychosocial factors and historical traumas should be considered in this process, and changes in communities' resilience over time should be monitored (DeYoung et al., 2020).

3.7 A Model Proposal

A proposed model incorporating all elements of socio-ecological disaster risks caused by urbanization is presented in Fig. 3.2. The model links socio-ecological risks associated with urbanization and urbanization factors that increase disaster risk, proposing strategies to reduce disaster risks.

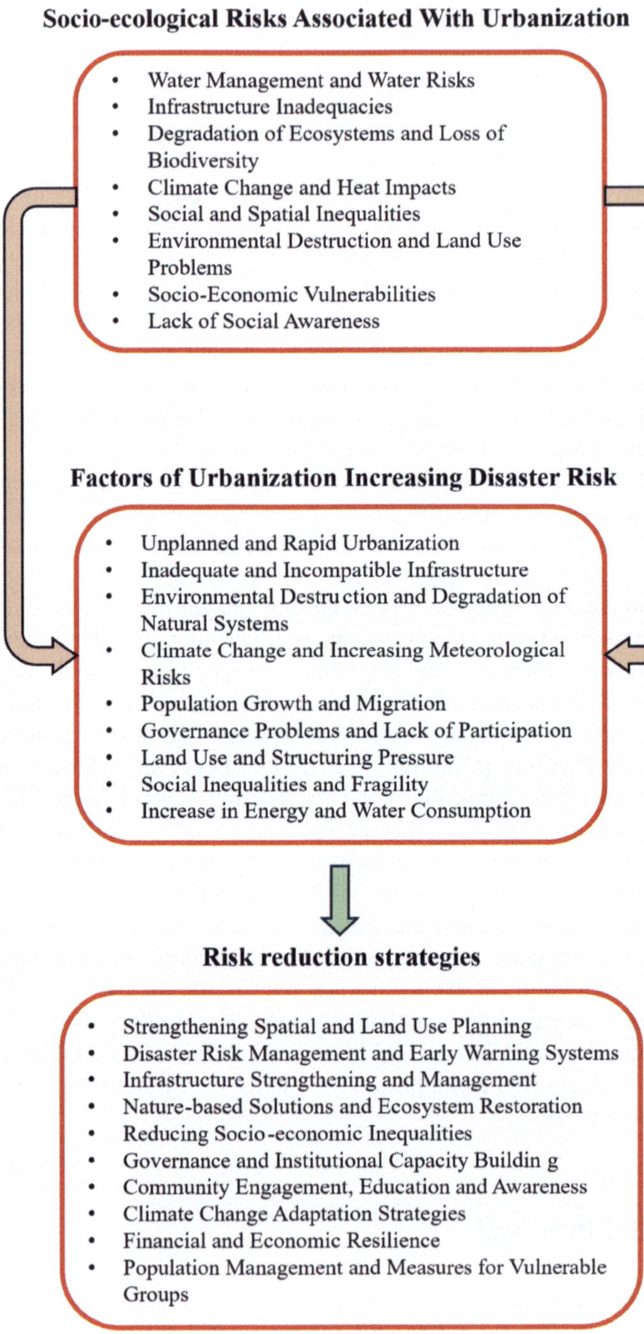

Fig. 3.2 A conceptual model of socio-ecological disaster risks driven by urbanization

It is emphasized that socio-ecological risks associated with urbanization, combined with urban factors that increase disaster risks, create obstacles to development. For this reason, the implementation of risk reduction strategies included in the model has been defined as fundamental strategies for overcoming these obstacles. Risk reduction strategies, primarily improving spatial planning and implementing land use planning in urban development processes, will determine the city's development focus and direction, enabling measures to be taken according to disaster types and contributing to risk reduction. The adoption of ecosystem-based solutions in planning processes contributes to sustainable development by making cities resilient to climate change risks. All of the aforementioned processes must be carried out collaboratively. For this reason, all institutions and organizations must cooperate in urban planning processes, and the needs of the community must be taken into account and incorporated into these processes. Furthermore, institutional capacity is extremely important in combating disaster risks. Adopting economically resilient and sustainable policies in the solutions to be developed will ensure financial sustainability.

Today, technological developments such as early warning systems are frequently used, particularly for meteorological disasters, ensuring that the public is informed before disasters occur. It is crucial to involve the public in all processes related to disaster risk reduction, especially because they are aware of local risks. In addition to rapid population growth, a number of factors are emerging that increase vulnerability. Poverty reduction is at the forefront of these factors. Social risks, such as the exclusion of disadvantaged groups from the process, are key factors that increase vulnerability in urban planning.

3.8 Conclusion

Various social and ecological risks increase disaster risks in cities. Poverty, inequality, rapid population growth, and uneven population distribution are social risks, while unplanned urbanization, destruction of green spaces, and settlements built without land-use planning are ecological risks that reduce resilience. To prevent the aforementioned risks, ecosystem-based spatial planning should focus on eliminating social risks.

3.9 AI Disclosure Statement

Where AI tools (e.g., language models, grammar checkers) were used to support language refinement or formatting, the authors supervised all usage, and the intellectual content remains solely the product of human authorship. No AI-generated content was accepted without critical human oversight and revision.

References

Anguelovski, I., Connolly, J. J. T., Masip, L., & Pearsall, H. (2018). Assessing green gentrification in historically disenfranchised neighborhoods: A longitudinal and spatial analysis of Barcelona. *Urban Geography, 39*(3), 458–491. https://doi.org/10.1080/02723638.2017.1349987;REQUESTEDJOURNAL:JOURNAL:RURB20;WGROUP:STRING:PUBLICATION

Berke, P. R., & Campanella, T. J. (2006). Planning for Postdisaster Resiliency. *The ANNALS of the American Academy of Political and Social Science, 604*(1), 192–207. https://doi.org/10.1177/0002716205285533

Braun, V., & Clarke, V. (2006). Using thematic analysis in psychology. *Qualitative Research in Psychology, 3*(2), 77–101. https://doi.org/10.1191/1478088706qp063oa

Bulkeley, H., & Betsill, M. M. (2013). Revisiting the urban politics of climate change. *Environmental Politics*. https://doi.org/10.1080/09644016.2013.755797

Burton, C. G. (2015). A validation of metrics for community resilience to natural hazards and disasters using the recovery from Hurricane Katrina as a Case Study. *Annals of the Association of American Geographers, 105*(1), 67–86. https://doi.org/10.1080/00045608.2014.960039;PAGE:STRING:ARTICLE/CHAPTER

Cohen, B. (2006). Urbanization in developing countries: Current trends, future projections, and key challenges for sustainability. *Technology in Society, 28*(1–2), 63–80. https://doi.org/10.1016/J.TECHSOC.2005.10.005

DeYoung, S. E., Lewis, D. C., Seponski, D. M., Augustine, D. A., & Phal, M. (2020). Disaster preparedness and well-being among Cambodian- and Laotian-Americans. *Disaster Prevention and Management: An International Journal, 29*(4), 425–443. https://doi.org/10.1108/DPM-01-2019-0034/FULL/PDF

Duy, P. N., Chapman, L., Tight, M., Linh, P. N., & Thuong, L. V. (2017). Increasing vulnerability to floods in new development areas: Evidence from Ho Chi Minh City. *International Journal of Climate Change Strategies and Management, 10*(1), 197–212. https://doi.org/10.1108/IJCCSM-12-2016-0169

EEA. (2021). Nature-based solutions in Europe : policy, knowledge and practice for climate change adaptation and disaster risk reduction (p. 159). https://www.eea.europa.eu/en/analysis/publications/nature-based-solutions-in-europe

Hooli, L. J. (2016). Resilience of the poorest: Coping strategies and indigenous knowledge of living with the floods in Northern Namibia. *Regional Environmental Change, 16*(3), 695–707. https://doi.org/10.1007/S10113-015-0782-5/TABLES/1

IPCC. (2022). *Climate change 2022: Impacts, adaptation and vulnerability. contribution of working Group II to the sixth assessment report of the intergovernmental panel on climate change*. https://www.ipcc.ch/report/ar6/wg2/

Kabisch, N., Korn, H., Stadler, J., Bonn, A., Kabisch, N., Korn, H., Stadler, J., & Bonn, A. (2017). Nature-based solutions to climate change adaptation in urban areas—Linkages between science, policy and practice. *Theory and Practice of Urban Sustainability Transitions*, 1–11. https://doi.org/10.1007/978-3-319-56091-5_1

Lehmann, S. (2019). Reconnecting with nature: Developing urban spaces in the age of climate change. *Emerald Open Research, 1*(5). https://doi.org/10.1108/EOR-05-2023-0001

Lizarralde, G., Páez, H., Lopez, A., Lopez, O., Bornstein, L., Gould, K., Herazo, B., & Muñoz, L. (2020). We said, they said the politics of conceptual frameworks in disasters and climate change in Colombia and Latin America. *Disaster Prevention and Management: An International Journal, 29*(6), 909–928. https://doi.org/10.1108/DPM-01-2020-0011/FULL/PDF

Mercer, J., Kelman, I., Taranis, L., & Suchet-Pearson, S. (2010). Framework for integrating indigenous and scientific knowledge for disaster risk reduction. *Disasters, 34*(1), 214–239. https://doi.org/10.1111/J.1467-7717.2009.01126.X

Mumuni, A. S., Mensah, H., Asamoah, S., & Simpeh, E. K. (2025). A systematic review on the causes and effects of urbanization on wetlands in Sub-Saharan Africa. *Urbanization, Sustainability and Society, 2*(1), 153–177. https://doi.org/10.1108/USS-10-2024-0063

Nguyen, T. T., Ngo, H. H., Guo, W., Wang, X. C., Ren, N., Li, G., Ding, J., & Liang, H. (2019). Implementation of a specific urban water management—Sponge City. *Science of the Total Environment, 652*, 147–162. https://doi.org/10.1016/j.scitotenv.2018.10.168

OECD. (2020). *The territorial impact of COVID-19: Managing the crisis across levels of government*. https://www.oecd.org/coronavirus/policy-responses/the-territorial-impact-of-covid-19-managing-the-crisis-across-levels-of-government-d3e314e1/

Özer, Y. E. (2017). İKLİM DEĞİŞİKLİĞİ YÖNETİŞİMİNDEKİ AKTÖRLERİN ANALİZİ VE TÜRKİYE. *Uluslararası Yönetim İktisat ve İşletme Dergisi, 13*(4), 833–851. https://doi.org/10.17130/IJMEB.2017433411

Paul, A. (2025). Artificial neural networks for flood susceptibility analysis in Gangarampur sub-division of Dakshin Dinajpur, West Bengal, India. *Frontiers in Engineering and Built Environment, 5*(1), 1–21. https://doi.org/10.1108/FEBE-09-2024-0061/FULL/PDF

Peng, L., & Li, Z. (2021). Ensemble Flood Risk Assessment in the Yangtze River Economic Belt under CMIP6 SSP-RCP scenarios. *Sustainability, 13*(21), 12097. https://doi.org/10.3390/SU132112097

Satterthwaite, D. (2007). *The transition to a predominantly urban world and its underpinnings. Human settlements discussion*. www.iied.org/pubs/display.php?o=10550IIED

Satterthwaite, D. (2017). The impact of urban development on risk in sub-Saharan Africa's cities with a focus on small and intermediate urban centres. *International Journal of Disaster Risk Reduction, 26*, 16–23. https://doi.org/10.1016/J.IJDRR.2017.09.025

Solecki, W., & Leichenko, R. M. (2006). Assessing the vulnerability of urban areas to the impacts of climate change: A case study of New York City. *Environmental Hazards*, 10–23.

Tajuddin, N., Abrowski, M. D., Musco, F., & Morello, E. (2021). Enabling socio-ecological resilience in the Global South: Insights from Chennai, India. *Sustainability, 13*(19), 10522. https://doi.org/10.3390/SU131910522

UN. (2019). World urbanization prospects: The 2018 revision. United Nations, Department of Economic and Social Affairs, Population Division. https://population.un.org/wup/

UNDRR. (2019a). Global assessment report on disaster risk reduction 2019.

UNDRR. (2019b). Global assessment report on disaster risk reduction 2019. United Nations. https://www.undrr.org/gar2019

UN-Habitat. (2016). World Cities Report 2016: Urbanization and Development—Emerging Futures.

UN-Habitat. (2020). World Cities Report 2020: The value of sustainable urbanization. https://unhabitat.org/sites/default/files/2020/10/wcr_2020_report.pdf

United Nations. (2018). World urbanization prospects: The 2018 revision.

Wang, T., & Li, H. (2024). Spatial constraints or spatial dynamics? The spatial spillover effect of networks of flood regulation service flows on land-use degree. *Water Science and Technology, 89*(3), 682–713. https://doi.org/10.2166/WST.2024.009/1366613/WST089030682.PDF

Wisner, B., Blaikie, P., Cannon, T., & Davis, I. (2004). *At risk: Natural hazards, people's vulnerability and disasters*.

Woiwode, C., Ramachandran, A., Philip, T., Rishika, D., & Rajan, S. C. (2024). Identifying entry points for adaptive governance in peri-urban Chennai (India): A multi-dimensional, multi-level, and multi-scalar approach. *Frontiers in Sustainable Cities, 6*, 1368240. https://doi.org/10.3389/FRSC.2024.1368240/BIBTEX

World Bank. (2021). *The evolving role of resilience in urban development*. https://openknowledge.worldbank.org/handle/10986/35284

Yuan, D., Wang, H., Wang, C., Yan, C., Xu, L., Zhang, C., Wang, J., & Kou, Y. (2024). Characteristics of urban flood resilience evolution and analysis of influencing factors: A case study of Yingtan City, China. *Water (Switzerland), 16*(6), 834. https://doi.org/10.3390/W16060834/S1

Zerger, A. (2002). Examining GIS decision utility for natural hazard risk modelling. *Environmental Modelling and Software, 17*(3), 287–294. https://doi.org/10.1016/S1364-8152(01)00071-8

Chapter 4
Kahramanmaraş Earthquakes: Vulnerability and Coping Capacity of Community

Sevda Demiröz Yıldırım

Abstract The 6 February 2023 Kahramanmaraş (Türkiye) earthquakes are recorded as the world's most destructive disasters in 2023. These devastating earthquakes in Türkiye and Syria account for two-thirds of the total number of disaster fatalities in 2023. A total of 11 provinces in Türkiye were severely affected by the epicentre of these earthquakes. These earthquakes, which were declared a level 4 disaster requiring an international appeal for assistance for Türkiye, caused devastation far beyond the country's capacity to cope. In addition to the response activities of local and central government, the processes by which people coped with the disaster have become one of the world's exemplary studies in the field of disaster management. Although different impacts are observed in each society, differences in the vulnerability and coping capacities of disaster victims have been identified according to disaster resilience. In this study, the disaster vulnerability and coping capacity of the disaster-stricken community are evaluated with examples from the field experience of the author's disaster field studies conducted in the region immediately after the Kahramanmaraş earthquakes. As a result of the study, policy, legislation and resilience recommendations will be presented in integrated disaster management processes for the creation of disaster-resilient communities worldwide.

Keywords Coping capacity · Disaster management · Disaster resilience · Kahramanmaraş earthquakes · Public administration · Vulnerability

4.1 Introduction

The Kahramanmaraş earthquakes of 6 February 2023 rank among the most devastating disasters in recent global history. These two catastrophic earthquakes, with magnitudes of 7.7 Mw (04:17—Türkiye/Kahramanmaraş Pazarcık) and 7.6 Mw

S. Demiröz Yıldırım (✉)
Department of Emergency Aid and Disaster Management, Burdur Mehmet Akif Ersoy University Bucak School of Health, Bucak, Burdur, Türkiye
e-mail: sdemirozyildirim@mehmetakif.edu.tr

G. K. Panda et al. (eds.), *Socio-Ecological Crisis of Development*, Sustainability Solutions, https://doi.org/10.1007/978-3-032-22171-1_4

(13:24—Türkiye/Kahramanmaraş Elbistan), which occurred approximately nine hours apart, represent the most destructive seismic events ever experienced in Türkiye to date (Fig. 4.1) (AFAD, 2023a, 2023c). These two devastating earthquakes affected 11 provinces in Türkiye: Kahramanmaraş, Gaziantep, Şanlıurfa, Hatay, Adana, Adıyaman, Elâzığ, Diyarbakır, Kilis, Malatya and Osmaniye. According to official figures, 50,783 people died and 107,204 were injured in Türkiye. The earthquakes affected 9.2 million people in Türkiye. In Syria, 5,900 deaths were recorded, and 8.8 million people were affected by the earthquakes. According to official data, half a million buildings were damaged in Türkiye, with 58,039 buildings classified as collapsed/requiring immediate demolition and a further 205,534 buildings severely damaged (AFAD, 2023a; Cumhurbaşkanlığı, 2023). In the immediate aftermath of the earthquakes in Türkiye, 3.3 million people migrated from the affected regions (Sağıroğlu et al., 2023). Financial losses due to the earthquakes in Türkiye are currently estimated at $34 billion (USD), with an additional $8.9 billion (USD) estimated for Syria (AFAD, 2023a; CRED, 2024). Based on previous local and global experience, the economic losses are expected to increase over time.

Earthquakes were classified as a Level 4 (S4) disaster in the Turkish disaster management system, indicating a level of impact that exceeded local, regional and national response capacities. Turkish disaster plans classify disasters that require national capacity and international support due to their destructive nature as level 4 disasters, together with the Disaster and Emergency Management Presidency of the affected province (Provincial AFAD center) and its first and second group of supporting provinces according to the Disaster Response Plan (AFAD, 2022a). As a result, a significant amount of national and international assistance was sent to the provinces affected by this level 4 disaster. However, during the first 72 h, considered the critical 'golden hour' before the arrival of aid, the disaster victims displayed a vulnerable tendency, exacerbated by the inadequacy of their prior disaster preparedness in the face of the catastrophic event. People experienced an abrupt shock from collapsed buildings, the loss of loved ones and the disruption of their social environment. The immense devastation caused by the earthquake hampered society's ability to cope with the disaster. In the immediate aftermath, disaster victims were expected to cope on their own, until basic needs such as search and rescue, medical care, food, water, basic shelter, heating (due to the harsh winter conditions) and the needs of people with special needs could be met by incoming aid. However, experience and observations from the Kahramanmaraş earthquake suggest that resilience and coping capacities were not fully adequate.

This study examines the vulnerability and coping capacity of the community in the epicentre of the disaster, Kahramanmaraş province, within the debris fields of the destruction caused by the earthquake and in temporary shelter areas, based on the researchers' experiences and observations. Within this framework, both vulnerability and coping capacities are analysed in detail in the debris fields and in a tent city in the city centre of Kahramanmaraş, which serves as a temporary accommodation area for disaster victims whose houses were damaged or destroyed.

This section examines how communities in regions affected by large-scale disasters respond when local, regional and national resources are limited, with a focus

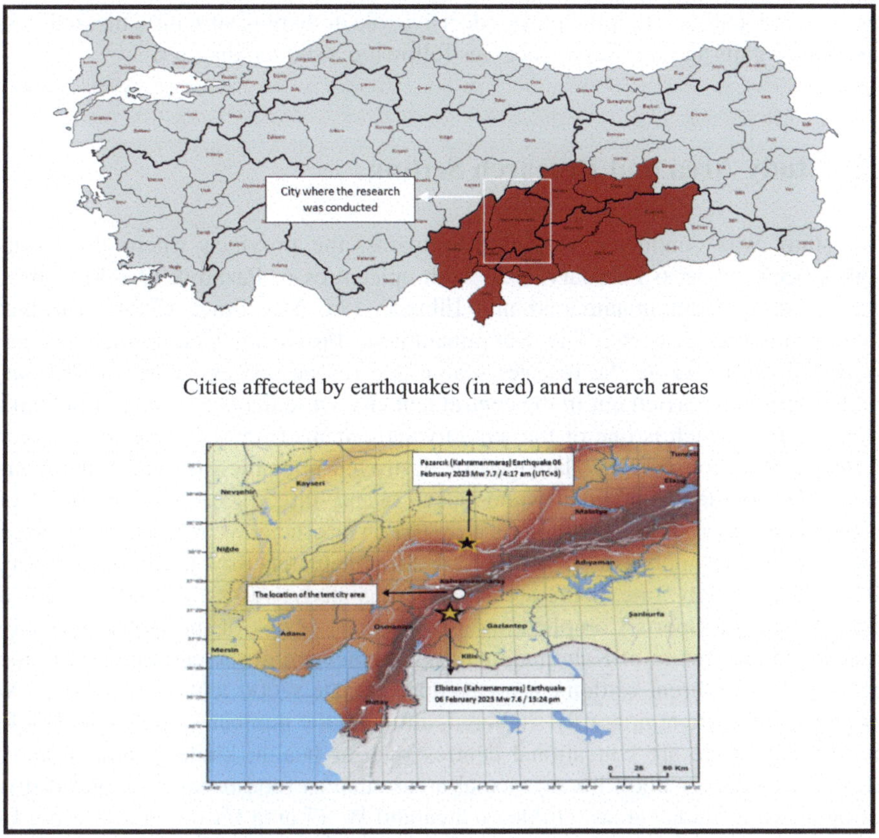

Map of the epicenter city of earthquakes and Tent city area

Fig. 4.1 Location of Kahramanmaraş city, origin of earthquakes and location of research area. *Source* AFAD (2023a)

on their lived experiences and adaptation strategies. Drawing on field observations conducted immediately after the disaster, the study offers contextual insights into grassroots coping mechanisms that require urgent post-disaster intervention. These mechanisms are often overlooked in large-scale statistical disaster research. In this respect, the study makes an innovative contribution. Furthermore, by situating local community responses within a broader socio-ecological framework, this section highlights the interconnected dimensions of vulnerability and social resilience in disaster-prone areas. The findings of this study offer two key insights. Firstly, they provide region-specific knowledge, and secondly, they offer valuable implications for international readers seeking to understand disaster resilience at a global scale.

This study is congruent with the overarching theme of the book, Socio-Ecological Crisis of Development: Community Resilience and Sustainable Solutions, which reveals how local social capital plays a decisive role in shaping structural inequalities, governance gaps and recovery processes following a sudden disaster.

4.2 Study Area and Research Setting

This study was conducted in Kahramanmaraş, the province where the earthquakes occurred on 6 February 2023 with epicentres in Pazarcık (7.7 Mw time: 04:17—Türkiye/Kahramanmaraş) and Elbistan (7.6 Mw time: 13:24—Türkiye/Kahramanmaraş) districts. The Kahramanmaraş Provincial Centre includes the provincial centre where the disaster search and rescue processes are carried out, and the activities carried out in the central tent city settlements established in Oniki Şubat district, which is one of the tent city settlements (Fig. 4.1). For the disaster victims whose houses were damaged or destroyed after the disaster, Kahramanmaraş Metropolitan Municipality announced twenty-three (23) safe assembly, tent and container areas in the centre of the province on its official website and announced that citizens can apply for these areas (Kahramanmaraş Büyükşehir Belediyesi, 2023). The tent city, one of the areas where this study was conducted, is located in the Provincial Sports Complex, and this tent city is one of the largest tent city areas in the city centre of Kahramanmaraş, the epicentre of the earthquakes. In the tent city, five different settlement areas, approximately 600 tents and 6000–9000 disaster victims are temporarily sheltered, although the number varies daily. Inside the tent city, there are educational centres such as health, kindergarten, primary school, secondary school, library, social areas such as food production and distribution areas, religious areas, childcare area and Wi-Fi area for communication. In addition, the other area within the study area is the 12 February District Centre of Kahramanmaraş, where the devastating effects of the disaster were observed. The 12 February District Centre is one of the regions that suffered severe destruction due to the Kahramanmaraş earthquakes (Fig. 4.2).

4.3 Material and Methods

This research is based on fieldwork conducted in Kahramanmaraş, the province most severely impacted by the 6 February 2023 earthquakes. Field research allows for observing events and human responses in their natural contexts, and this study applied both direct observation and participant observation methods (Çev: Dikeçligil & HP, 2011; Kaptan, 1995). In direct observation, the researcher remains external to events and records them without interference, whereas participant observation involves the researcher's active engagement in events. The researcher arrived at the disaster zone approximately 40 h after the first earthquake and stayed in the field for 62 days. Drawing on her professional background in disaster management, the researcher participated in both search and rescue operations in debris zones and humanitarian

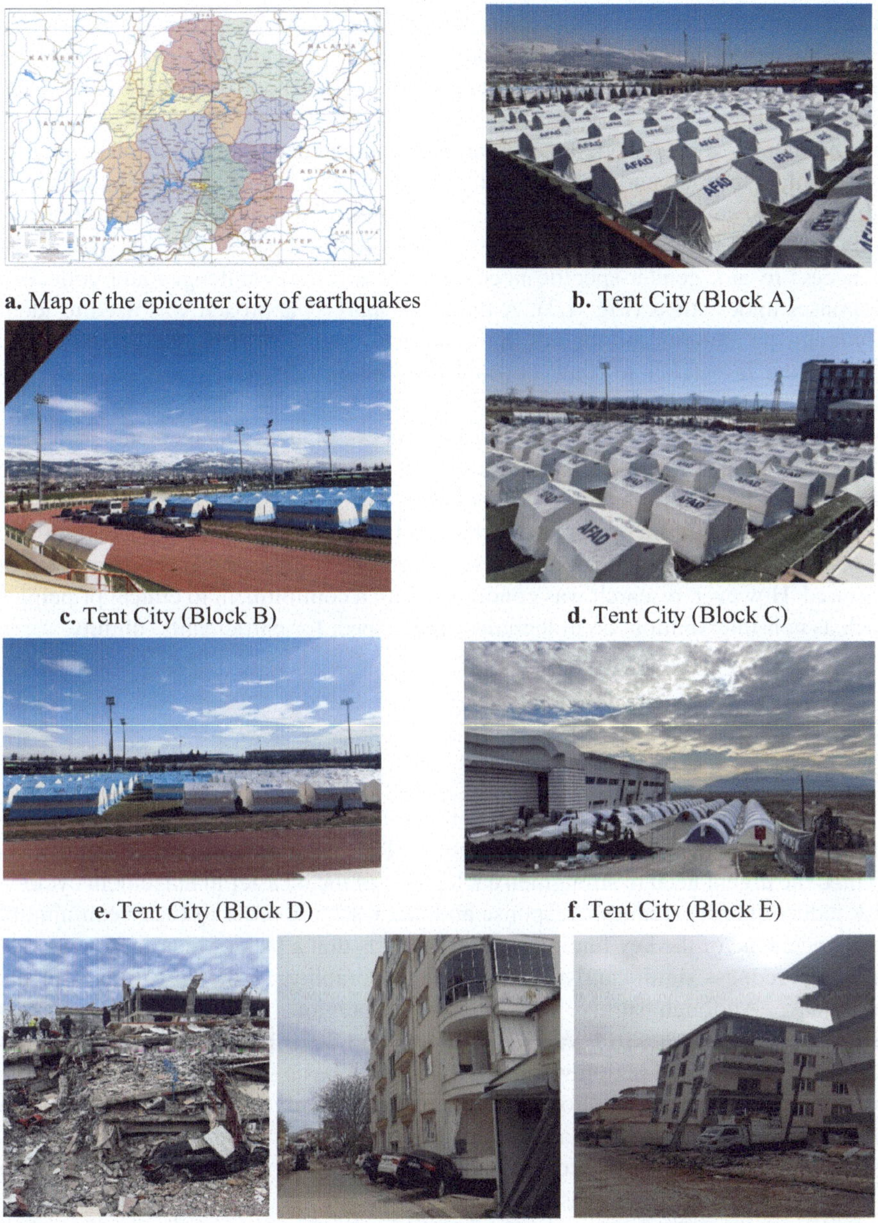

a. Map of the epicenter city of earthquakes
b. Tent City (Block A)
c. Tent City (Block B)
d. Tent City (Block C)
e. Tent City (Block D)
f. Tent City (Block E)
g. Collapsed structures
h. Damaged structures
i. Damaged structures

Fig. 4.2 Study area. *Source* **a** HGM (2025). **b–i** Author

aid activities in temporary shelter areas. Observations were carried out in two primary locations: debris fields in the 12 February District Centre, and the central tent city in the Oniki Şubat district of Kahramanmaraş, which hosted approximately 6,000–9,000 disaster victims.

Throughout this period, the researcher documented the conditions, interactions, vulnerabilities and coping strategies of disaster-affected populations. Field notes were systematically recorded in a research diary and categorised thematically following the fieldwork. The observations focused particularly on issues such as access to aid, gender-specific needs, communication challenges and collective responses under stress (Fig. 4.3). A thematic analysis approach was used to identify recurring patterns and themes in the observational data. This analytical method enabled the researcher to draw out key insights on community vulnerability and resilience in post-disaster settings.

This figure illustrates the step-by-step process of field engagement, including observation types, data documentation and thematic analysis conducted in the aftermath of the Kahramanmaraş earthquakes.

No personal identifiers were collected, and all observations took place in public or institutional settings with open access. Therefore, formal ethical approval was not required. However, research was conducted with a commitment to ethical principles such as non-intervention, confidentiality and respect for participants' dignity.

4.4 Results and Discussion

The Kahramanmaraş earthquakes of 6 February 2023 are recorded as the most devastating disaster in the history of the Republic of Türkiye. This catastrophic event highlighted the urgent need to strengthen each phase of the disaster management cycle—risk reduction, preparedness, response and recovery—in order to foster community resilience. One of the key findings of this study is that a lack of disaster knowledge and preparedness significantly exacerbated vulnerability in the early stages of the earthquake. Although Türkiye had experienced previous major earthquakes—such as the 1999 Marmara, 2011 Van, 2020 Elazığ and 2020 İzmir events—many individuals in the affected region were unfamiliar with basic self-help strategies for the critical initial 72-h period following a disaster, commonly referred to as the 'golden hour'. Although it is globally accepted that survivors must rely on their own capacities within the first 72 h before external aid arrives, observations from this research indicate that this principle was not effectively internalised by the affected community (İstanbul AFAD, 2009). The first 72 h are vital for ensuring access to food, water and shelter, and delays in these areas often result in increased mortality and suffering. Survival statistics from global earthquake events (Huang & Lien, 2012; Ochoa & Santos, 2015; Ochoa et al., 2007) support the critical role of community preparedness in reducing fatality rates.

The findings from this fieldwork suggest that the lack of knowledge, training and resources among local populations critically impaired their ability to respond

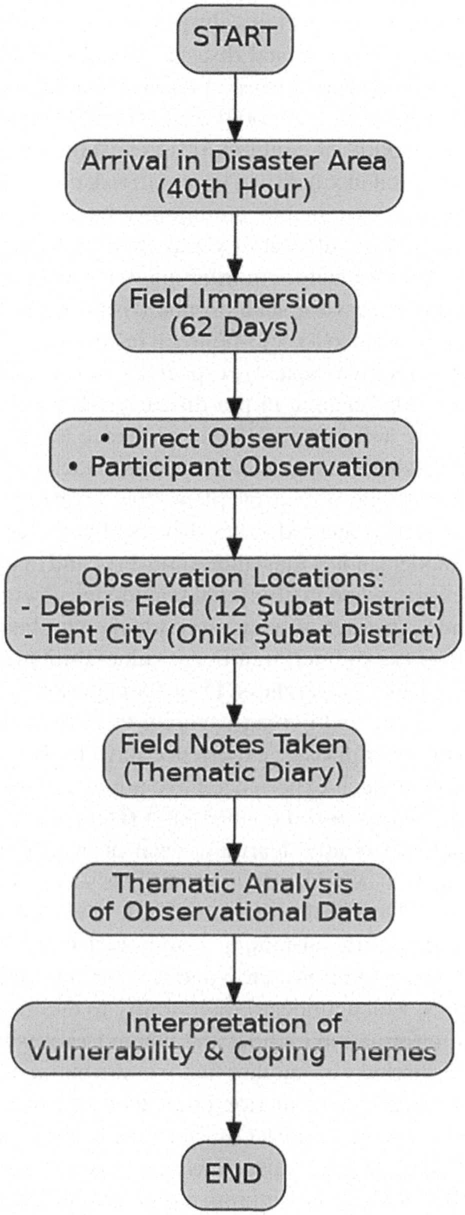

Fig. 4.3 Methodological flowchart of the field study. *Source* Figure created by the author

effectively, especially during this self-reliance window. This insight aligns with broader international disaster literature, which emphasises that resilience begins at the community level and requires locally appropriate knowledge and preparedness strategies (Twigg, 2007; UNDRR, 2015). From a socio-ecological crisis perspective, the inability of communities to sustain themselves in the early hours reflects a deep structural inequality in access to risk education and preparedness. This reinforces the need for sustainable, community-based disaster education programmes that address both social vulnerability and environmental exposure.

Another factor contributing to the vulnerability of disaster victims is their lack of preparedness for such events. Disaster preparedness is not only a technical requirement, but also a socio-political necessity that directly shapes communities' resilience and adaptive capacity, especially in the context of socio-ecological crises. Individuals who engage in pre-disaster risk management processes experience less harm during a disaster and their resilience is also enhanced (UNISDR, 2014). Following the Kahramanmaraş earthquakes, a lack of previous disaster experience, insufficient knowledge of disaster plans, low levels of preparedness and economic constraints (poverty) emerged as key drivers of vulnerability (Fig. 4.4). These findings align with broader studies that suggest poverty and marginalisation limit access to preparedness resources and institutional knowledge (Twigg, 2007; UNDRR, 2015). First and foremost, disaster-affected individuals who had no previous experience of any type of disaster exhibited significant vulnerability in the earthquake due to their inexperience and lack of awareness. Disaster experience is a critical factor in increasing both the resilience and coping capacity of individuals. Studies have shown that individuals with prior disaster experience have higher levels of disaster resilience, while those without such experience have a lower capacity to respond to disasters compared to their experienced counterparts (Demiröz Yıldırım, 2023). This supports the notion that experiential learning, even in the absence of formal education, builds critical adaptive behaviours necessary for survival in socio-ecological emergencies. In the first 72 h after the disaster, disaster victims lacked the knowledge and skills to secure essential life-sustaining resources (water, food, blankets, appropriate clothing for winter conditions, etc.) and to carry out initial response activities (rescue, first aid, etc.), which hindered their ability to demonstrate resilience. Furthermore, their lack of information on how and where to access the aid that arrived in the region during this period was another factor contributing to their vulnerability. The failure to utilise and access information embedded in national preparedness policies, such as those found in the Turkish Disaster Risk Reduction and Management Plans (AFAD, 2022a, 2022b), reveals a disconnection between institutional planning and community-level communication. Unprepared communities that lack disaster knowledge and experience have a reduced capacity to cope with disasters (Genç, 2023). This case study illustrates how systemic inequities—particularly those related to access to knowledge, infrastructure and economic resources—undermine community resilience, thereby reflecting the broader socio-ecological crisis of development discussed in this book.

One of the societal vulnerability factors observed during the devastating Kahramanmaraş earthquakes is related to the income level of the affected community

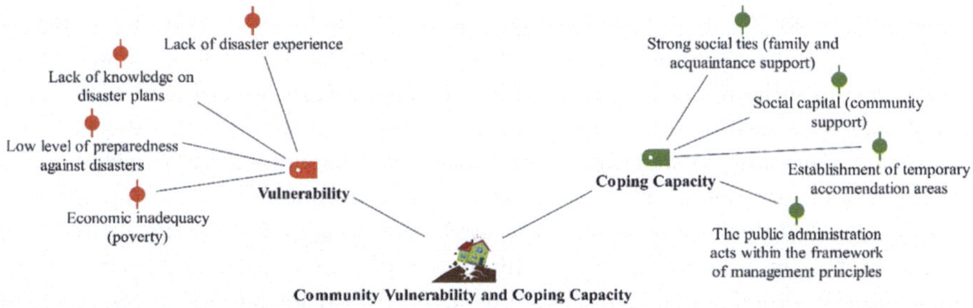

Fig. 4.4 Kahramanmaraş earthquakes: vulnerability and coping capacity of community. *Source* Author

(Genç, 2023). In particular, the presence of impoverished populations in the region was an important element in triggering vulnerability. Low-income communities often live in poorly constructed, unregulated housing, lack access to early warning systems and face systemic barriers in reaching preparedness resources, which collectively intensify the socio-ecological impacts of disasters (UNDRR, 2015). Disasters disproportionately affect countries with high poverty rates. Poverty-stricken societies are more vulnerable to disaster damage due to the absence or inadequacy of early warning systems, the lack of resilience of existing buildings and critical infrastructure and insufficient resources for disaster risk reduction and mitigation, response and recovery.

The different mortality rates in earthquakes of similar magnitude, such as the Iran Bam earthquake (2003/6.6 Mw/25,000 deaths) and the Japan Kobe earthquake (1995/6.9 Mw/5,000 deaths), which are directly correlated with the poverty levels of the affected populations, serve as compelling examples of this phenomenon. Despite the greater magnitude of the Japanese earthquake, the death toll in Iran was five times higher than in Kobe. The weaker disaster infrastructure and lack of adequate disaster response resources in Bam resulted in five times more fatalities than in the Kobe earthquake (Kett et al., 2005). Another example of the link between poverty and disaster vulnerability is the earthquake and subsequent tsunami that struck Haiti and Chile in 2010. In Haiti, the earthquake and tsunami caused 222,570 deaths and 300,000 injuries, while in Chile, a much larger earthquake in the same year (2010) resulted in 521 deaths and 12,000 injuries from the earthquake and subsequent tsunami.

The reason for the much higher death toll in Haiti, despite the smaller magnitude of the earthquake compared to Chile, is that 80% of Haiti's population lived below the poverty line and in rural areas with disaster-prone structures. Despite Haiti's frequent exposure to various disasters throughout its history (storms, earthquakes, tsunamis, etc.), the necessary legislation for disaster-resilient structures to protect its citizens has reportedly not been enacted or effectively implemented. Governments often take a high level of risk in disaster management because of inadequate pre-disaster information and limited resources due to economic factors, resulting in inadequate structural design efforts. Conversely, in Chile, despite the magnitude of the earthquake and the population and area affected, the lower rates of death, injury and structural damage compared to Haiti can be attributed to Chile's much more developed and affluent economy. In addition, the Chilean government recognises earthquakes as a serious threat and requires its building stock to be designed and constructed according to specific disaster-resistant rules and standards (Nguyen & Corotis, 2011). Similarly, in the context of the Kahramanmaraş earthquake, the research revealed that the most severely affected populations were those living in structurally fragile, low-cost housing without compliance to earthquake safety standards. These findings clearly illustrate how poverty-induced settlement patterns and infrastructural neglect not only exacerbate exposure and vulnerability, but also reflect deep-rooted socio-economic inequalities embedded in development processes.

Thus, poverty should not be seen merely as a background condition but as a dynamic force shaping the nature of vulnerability in disaster contexts. This case aligns with the focal theme of this book, showing how the socio-ecological crisis of development manifests in physical risks, institutional weaknesses and disproportionate harm to the most marginalised communities. The findings from Kahramanmaraş contribute to the growing international discourse on how disaster risk is socially constructed and demand urgent, equity-based, sustainable solutions.

In addition to the various factors that contributed to the community's vulnerability during the disaster, several elements were also identified that strengthened its coping capacity. These resilience-enabling factors are crucial for understanding how communities manage to survive and recover despite structural disadvantages. While these elements may vary in different regions, in the context of Kahramanmaraş city centre, which is the focus of this study, the following factors played a role in strengthening societal resilience: strong social ties (family and acquaintance support); the timely delivery of critical disaster relief to the region, facilitated by the community's previous disaster experiences (reflecting a form of experiential learning and localised disaster culture); the establishment of temporary shelter areas; the public administration's ethical, fair and just approach to disaster victims, as well as its efforts to conduct disaster management operations within the framework of public administration principles (Fig. 4.4).

This finding illustrates the dual nature of socio-ecological systems, where both vulnerability and resilience can coexist in the same context, shaped by dynamic social interactions, governance capacity and local knowledge. The prominence of social capital in post-disaster recovery in Kahramanmaraş echoes global findings that trust-based networks often outperform formal institutions in the critical hours following a disaster (Aldrich, 2012). Furthermore, the transparent and inclusive role of local public administration in the aftermath enhanced community trust and facilitated collective action—a key dimension of sustainable resilience. By highlighting community-driven coping mechanisms and the importance of ethical governance, this research contributes to the overarching theme of this book. It shows how community resilience emerges not only from infrastructure or technical preparedness, but also from social solidarity, local knowledge systems and governance practices rooted in equity. In doing so, the study bridges environmental, social and administrative dimensions, offering practical insights for sustainable development in disaster-prone settings.

Scientific research conducted after disasters has shown that strong social networks provide both psychological protection and contribute to coping mechanisms during such events. Robust social networks enable disaster survivors to share their experiences and facilitate the use of these experiences. However, social networks are vulnerable in disaster situations due to factors such as the disruption of social settlements and displacement. Studies have shown that individuals with weak social support are negatively affected after a disaster, while those with strong social support networks have better psychological well-being in the aftermath (Demiröz Yıldırım, 2023; Stough et al., 2017). In this context, social capital plays a critical role in strengthening the coping capacities of individuals and communities, particularly among vulnerable populations such as people with disabilities and the elderly. For example, following the Canterbury earthquake, people with orthopaedic disabilities, a vulnerable group, reported that social networks were vital for them. This study, which focused on the Canterbury earthquake, found that social networks enabled people with orthopaedic disabilities to feel included in the community rather than isolated or excluded, thereby improving their well-being (Bourke et al., 2017). These findings parallel the observations made in this study improve academic tone and readability, where social cohesion among survivors emerged as a key component of post-disaster resilience. Another study conducted in Türkiye found that strong social networks contribute significantly to the coping capacity of people with disabilities who have experienced disasters (Demiröz Yıldırım, 2023). While the presence of strong social networks facilitates coping processes by fostering a sense of community integration, the weakening or loss of these networks due to various factors negatively affects the coping mechanisms of vulnerable individuals. The Kahramanmaraş case shows that, despite structural weaknesses in the built environment, the high level of social capital among residents contributed meaningfully to community resilience. Disaster victims formed strong social bonds with each other, providing mutual support in various aspects, both material and emotional. Particularly in the immediate aftermath of a disaster, social solidarity in search and rescue efforts, ensuring safety and meeting shelter-related needs emerged as a crucial element of disaster resilience. In Türkiye, social capital—defined as the trust-based networks and norms that enhance cooperation among citizens and between institutions—remains a powerful buffer in disaster contexts (Aldrich, 2012; Tüysüz, 2011). Citizens outside the disaster zone often mobilised rapidly based on prior disaster experiences, providing targeted aid and logistic support. As in previous disasters, these mutual aid and solidarity processes ensured the rapid delivery of aid to the region. As mentioned above, another element in increasing disaster coping capacity is the timely delivery of vital disaster assistance to the region, facilitated by the community's previous disaster experience, which is a factor directly influencing survival rates. The public administration, including the accredited AFAD, worked with other civil society organisations, public agencies and the private sector to transparently share information on all needs in the affected region, such as search and rescue equipment, personnel data, medical needs and disaster relief supplies, thereby speeding up the delivery of aid to the area.

Search and rescue teams arriving in the region have made extraordinary efforts to ensure the survival of disaster victims. The delivery of adequate food and clean water to the region, particularly to prevent malnutrition among newborns and children, was of paramount importance in the disaster situation. In addition, given the harsh winter conditions that prevailed on the days of the earthquakes, the provision of aid to address this critical factor was vital. Delivering winter clothing, blankets and fuel/stoves to the region to prevent malnutrition in infants and children and to mitigate the potential increase in mortality due to hypothermia (a drop in body temperature below 35 °C) proved to be a life-saving element. The findings of this study contribute to the focal theme of the book by highlighting how community resilience is not solely a function of state response or infrastructure, but also of social solidarity, lived disaster experience and inclusive governance. These elements are essential to achieving sustainable and equitable disaster recovery. By focusing on locally embedded social capital as a sustainable solution in crisis contexts, this research also addresses a gap in existing disaster studies literature that often overlooks informal coping systems.

For the millions of disaster victims whose homes were damaged or destroyed by the earthquake, post-disaster shelter became one of the most critical issues. Severe winter conditions, including sub-zero temperatures and heavy snowfall, intensified the urgency of accessing safe and adequate shelter in the immediate aftermath. In Kahramanmaraş, the epicentre of the earthquake and the primary site of this study, the Metropolitan Municipality promptly announced on its official website the locations of 23 designated safe assembly points and tent and container settlements across the city centre, inviting citizens to seek refuge (Kahramanmaraş Büyükşehir Belediyesi, 2023). Despite the differences in the characteristics and social facilities available in each tent city, the critical needs for shelter and food were generally met in all of them in the initial phase. In subsequent phases, social facilities such as health centres, education centres (including kindergartens, primary and secondary schools, libraries and reading rooms), psychological counselling centres, food preparation and distribution areas, religious worship areas, childcare facilities and Wi-Fi access points for communication were established within the tent cities (Fig. 4.5). These facilities contributed positively to the community's ability to cope with the aftermath of the earthquake (Aslan et al., 2025). These facilities not only responded to immediate survival needs, but also contributed significantly to the psychosocial rehabilitation and resilience of the affected population. In particular, ensuring continuity in children's education and providing designated play spaces helped support emotional healing and a sense of normalcy. For adults, communal religious and social gathering spaces became sources of emotional support and spiritual resilience. Moreover, the establishment of age-inclusive psychological counselling areas served a preventive role in mitigating long-term mental health risks, such as Post-Traumatic Stress Disorder (PTSD), depression and anxiety. This integrated and staged sheltering approach exemplifies the interconnectedness of environmental, social and developmental dimensions in post-disaster recovery, aligning directly with the focal theme of this book. By highlighting how multi-dimensional shelter responses strengthen community resilience and well-being, this study contributes to a deeper understanding of sustainable and inclusive disaster recovery frameworks. In doing so, it also fills a gap in the literature by documenting how localised, participatory shelter practices can serve as scalable models for building socio-ecological resilience in future crises.

In the Kahramanmaraş earthquakes, in addition to strong social networks, support from family and acquaintances, and strong social capital, the rapid delivery of critical disaster relief to the region facilitated by the community's previous disaster experience, and the establishment of temporary shelters, the public administration's ethical, fair and just approach to disaster victims also stands out as a factor supporting the community's coping capacity. Efforts to provide initial post-disaster relief equally and equitably to all accessible areas provided critical support to communities in their coping processes. The public administration's transparent sharing of information received from the region about the disaster and clear articulation of needs facilitated the acceleration of bureaucratic processes, thereby speeding up the delivery of aid to disaster victims. Throughout the response to these earthquakes, the public administration demonstrated principles of public management such as transparency, equity, equality, effectiveness and efficiency. These principles not only ensure successful management during the disaster response process, but also enhance societal resilience. Given that transparency, effectiveness, efficiency, justice and equity are also components of coping capacities analysed in this study, it is important to briefly elaborate on these principles and their significance in disaster management.

The principle of transparency, a fundamental tenet of public administration, underscores the need for public information to be accessible and readily available to the public, thereby strengthening the legitimacy of administrative processes by increasing public confidence in government institutions. In the context of disasters, this principle promotes greater public confidence in government bodies, leading to wider acceptance of administrative decisions and providing the basis for informed public action. Informing citizens about the activities of public institutions encourages their participation in decision-making processes (Eken, 2005). In democratic societies, transparency is a cornerstone of public accountability and ethical governance, ultimately enhancing the credibility and legitimacy of public administration (Keleş, 2017). Transparency in disaster management processes aims to ensure the correct use of resources and public confidence (Hallwright & Handmer, 2019; Kadıoğlu, 2011; Liswanty & Prabowo, 2021). During a disaster, accurate public information, reliable communication channels and community involvement are critical to effective crisis management (AFAD, 2022a; Seetharam Sridhar et al., 2020). Transparency in the management of disaster relief and donations is fundamental to maintaining public trust (AFAD, 2023b). Sharing information publicly during the disaster management process prevents misinformation and supports effective response (AFAD, 2022b). The incomplete implementation of the principle of transparency has a negative impact on accountability (Liswanty & Prabowo, 2021). In the case of the Kahramanmaraş earthquakes, the initial approach of the public administration to share all available information, even if incomplete, with the public through an authorised source is an

a. Temporary toilets and baths

b. Laundry washing and drying truck

c. AFAD cooking container

d. Social Market Tent

e. Guidance and Research Center

f. Children's playground

Fig. 4.5 Tent city areas where the study was conducted. *Source* Author (The images contained herein are the intellectual property of the author. Unauthorised reproduction or distribution of these images is strictly prohibited)

example of the principle of transparency. Furthermore, the frequent public updates on ongoing and upcoming efforts, challenges and benefits observed in the region serve as a globally significant example in terms of fostering public trust and enhancing coping capacity.

Efficiency, a principle of public administration that emphasises the achievement of defined objectives in the best possible way and in the shortest possible time, is a principle that prioritises responding to the needs of society, the accurate allocation of public resources and the continuous improvement of the quality of services (Keleş, 2017; OECD, 2020, 2023). The principle of efficiency, which is directly related to the principle of effectiveness, involves using limited resources in such a way as to achieve the greatest benefit at the lowest cost (Akyel & Köse, 2010; Australian Government Productivity Commission, 2013; Keleş, 2017; OECD, 2020, 2023). The principles of effectiveness and efficiency aim to maximise societal benefits and ensure the sustainability of the public budget by promoting the effective use of natural and financial resources (Keleş, 2017; OECD, 2023; Sayıştay Kanunu, 2010). Effectiveness and efficiency are particularly paramount to the seamless and rapid execution of disaster management processes. In disaster crisis management, rapid and effective decision-making, coupled with the optimal use of limited resources, provides a crucial opportunity to reduce loss of life and property. Adherence to these principles is essential for the sound management of scarce human resources, machinery, equipment, vehicles and other elements during disaster crisis management. In the post-disaster recovery phase, these principles facilitate rapid and high quality reconstruction efforts. In the context of the Kahramanmaraş earthquake examined in this section, the public administration's allocation of resources based on the scale of devastation across all affected provinces and its coordination of personnel, equipment and supplies for the most rapid and effective response can be cited as an example.

The principles of justice and equity are fundamental tenets of public administration that ensure the provision of equitable services to all segments of society during disaster management processes. Within the framework of these principles, many provisions are necessary, including the implementation of special measures to protect vulnerable groups such as women, children, the elderly and persons with disabilities from the effects of disasters. For example, the organisation of post-disaster shelter and health services to prioritise the needs of these vulnerable groups is an example of these principles (AFAD, 2021, 2022b). In the Kahramanmaraş earthquake examined in this section, the public administration's efforts to deliver aid to each affected province and the initiatives taken for disadvantaged groups can be analysed through the lens of these principles. Although not all public administration principles were fully implemented during the Kahramanmaraş earthquakes, the aforementioned principles played a crucial role in increasing community and disaster victim trust in the administration and promoting individual involvement in the disaster management process. Thus, they are considered important components of coping capacity.

In conclusion, similar to other disaster experiences, the 2023 Kahramanmaraş earthquakes were characterised by factors influencing both the vulnerability and coping capacity of the affected population. This study identified the following elements as increasing societal vulnerability: lack of previous disaster experience, insufficient knowledge of disaster plans, low level of disaster preparedness and economic constraints (poverty). Conversely, the following factors were identified as enhancing the community's coping capacity and resilience: strong social ties (family and kinship support); the rapid delivery of critical disaster relief to the region facilitated by the community's prior disaster experience (strong social capital); the establishment of temporary shelters; the public administration's ethical, fair and equitable approach to disaster victims and its efforts to conduct disaster management operations within the framework of public administration principles. These findings align with existing international literature emphasising the dual role of vulnerability and resilience factors in shaping disaster outcomes (Cutter et al., 2010; Wisner et al., 2007). However, this study contributes uniquely by contextualising these dynamics within the socio-political and administrative framework of the Kahramanmaraş earthquakes, highlighting the critical role of transparent and equitable governance in enhancing coping capacities. While these findings are based on the researcher's field experiences and observations, there remains a need for further empirical research employing comparative analyses across different regions and disaster types to validate and expand upon these conclusions. Moreover, integrating quantitative measures of coping capacity alongside qualitative insights could deepen understanding of the mechanisms underlying community resilience.

4.5 Limitations of the Study

This study has several limitations. The first limitation is that the study covers the province of Kahramanmaraş (Türkiye), the epicentre of the earthquake. Social vulnerability and coping capacity in the other ten disaster-affected provinces of Hatay, Gaziantep, Kilis, Adana, Elâzığ, Adıyaman, Diyarbakır, Malatya, Osmaniye, Şanlıurfa and Syria are not included in this study. Another limitation is that the study only analyses crisis situations that involve disaster response and recovery processes. Risk management processes and other factors affecting the crisis are not included in the evaluation. The final limitation is that the study focused only on the vulnerability and coping capacity of the disaster victims that the author was able to observe. Therefore, the findings and results of the study cannot be generalised to the approximately 15 million disaster victims in Türkiye. I would like to thank all the employees of public institutions who supported the implementation of this study and shared their disaster experiences, all the volunteers who participated in the disaster and all the scholars who shared their experiences. I would also like to thank the disaster victims who, despite their painful experiences, shared all their thoughts and experiences in order to contribute to science.

4.6 Conclusions

In order to reduce vulnerability and increase resilience to disasters, it is necessary to demonstrate successful management that encompasses all stages—from risk reduction to preparedness, response and recovery. For disaster management processes to be effective and sustainable, public administrations must adopt a transparent, inclusive and participatory approach (Keleş, 2017; Liswanty & Prabowo, 2021; Seetharam Sridhar et al., 2020). This includes not only sharing disaster legislation and plans with citizens and stakeholders, but also actively involving vulnerable groups such as women, the elderly, children and persons with disabilities, along with the civil society organisations that represent them, in the design of disaster policies (Aslan et al., 2025; Demiröz Yıldırım, 2023). Furthermore, comprehensive and sustainable public education initiatives are required to inform society about the diverse types of disasters—earthquakes, floods, fires, landslides, avalanches—that frequently affect the country. These initiatives must ensure equitable access to disaster education and preparedness resources across all segments of society (Demiröz Yıldırım, 2023).

This study highlights that poverty, lack of support networks and disability are key structural barriers to disaster preparedness at the community level. Therefore, public institutions, local authorities and Non-Governmental Organisations (NGOs) must provide targeted support to overcome these inequalities within a good governance framework. By grounding these recommendations in direct field observations following the 2023 Kahramanmaraş earthquakes, this research contributes to international disaster literature by linking local social vulnerabilities to broader issues of governance, equity and sustainable resilience, as evidenced by the grounding of these recommendations in direct field observations following the 2023 Kahramanmaraş earthquakes. It offers a community-level perspective on how disaster management can be both inclusive and responsive, thereby advancing global knowledge on socio-ecological crisis and sustainable development.

Acknowledgements I would like to thank all the employees of public institutions who supported the realisation of this study and shared their disaster experiences, all the volunteers who participated in the disaster and all the scientists who shared their experiences. I would also like to thank the disaster victims who, despite their painful experiences, shared all their thoughts and experiences to contribute to science.

Funding The research was not supported by any funding.

Potential Conflicts of Interest The corresponding author states that there is no conflict of interest.

Informed Consent Informed consent was obtained from all individual participants included in the study.

Authors Contribution All authors have contributed equally to this work.

Data Availability Statement The datasets during and/or analysed during the current study available from the corresponding author on reasonable request.

AI Disclosure Statement In preparing this chapter, the author used the AI-assisted language tool ChatGPT (OpenAI, https://chat.openai.com; accessed between Apr and May 2025) solely for the purposes of improving linguistic clarity, sentence structure and English academic expression. The content, field observations, analysis and scientific interpretations presented in this chapter are entirely the author's original work. The AI tool was not used for generating ideas, arguments or content, but only as a language editing assistant. All intellectual responsibility for the chapter remains with the author.

References

AFAD. (2021). *T.C. İçişleri Bakanlığı Afet ve Acil Durum Yönetimi Başkanlığı—İl Afet Risk Azaltma Planları*. https://www.afad.gov.tr/il-planlari

AFAD. (2022a). *TAMP Türkiye Afet Müdahale Planı*. https://www.afad.gov.tr/kurumlar/afad.gov.tr/e_Kutuphane/Planlar/TAMP.pdf

AFAD. (2022b). Türkiye Afet Risk Azaltma Planı TARAP 2022–2030. In *2022*. https://www.afad.gov.tr/kurumlar/afad.gov.tr/e_Kutuphane/Planlar/28032022-TARAP-kitap_V6.pdf

AFAD. (2023a). *06 Şubat 2023 Pazarcık-Elbistan Kahramanmaraş (Mw: 7.7–Mw: 7.6) Depremleri Raporu*.

AFAD. (2023b). *Dünyada ve Türkiye'de Afet Finansmanı*. Deprem ve Risk Azaltma Genel Müdürlüğü.

AFAD. (2023c). *Kahramanmaraş Depremleri Ön Değerlendirme Raporu*.

Akyel, R., & Köse, H. Ö. (2010). Kamu Yönetiminde Etkinlik Arayışı: Etkin Kamu Yönetimi için Etkin Denetimin Gerekliliği. *Türk İdare Dergisi, 466*, 9–45.

Aldrich, D. P. (2012). Building resilience: Social capital in post-disaster recovery. In *Journal of Comparative Policy Analysis* (2) (University of Chicago Press).

Aslan, Z., Demiröz, K., & Demiröz Yıldırım, S. (2025). An examination of temporary shelter units in terms of minimum standards: The case of the Kahramanmaraş earthquake. *Sustainability, 17*(8 (3587)). https://doi.org/10.3390/SU17083587

Australian Government Productivity Commission. (2013). On efficiency and effectiveness: Some definitions Productivity Commission Staff Research Note. In *Productivity commission, commonwealth of Australia*. Commonwealth of Australia.

Bourke, J. A., Hay-Smith, E. J. C., Snell, D. L., & Schluter, P. J. (2017). Community inclusion of wheelchair users during the long-term recovery phase following the 2010/2011 Canterbury earthquakes. *International Journal of Disaster Risk Reduction, 23*. https://doi.org/10.1016/j.ijdrr.2017.05.004

Çev: Dikeçligil, B., & HP. (2011). Saha Çalışması. *Istanbul Journal of Sociological Studies, 24*, 35. https://dergipark.org.tr/tr/pub/iusoskon/issue/9526/119050

CRED. (2024). 2023 disasters in numbers. In *Nature medicine* (No. 7). Nature Research.

Cutter, S. L., Burton, C. G., & Emrich, C. T. (2010). Disaster resilience indicators for benchmarking baseline conditions. *Journal of Homeland Security and Emergency Management, 7*(1). https://doi.org/10.2202/1547-7355.1732

Cumhurbaşkanlığı, T. C. (2023). *2023 Kahramanmaraş ve Hatay Depremleri Raporu*.

Demiröz Yıldırım, S. (2023). Integrated disaster management experience of people with disabilities: A phenomenological research on the experience of people with orthopedic disabilities in Türkiye. *International Journal of Disaster Risk Reduction, 88*. https://doi.org/10.1016/j.ijdrr.2023.103611

Eken, M. (2005). *Yönetimde Şeffaflık* (1. Baskı). Seçkin Yayıncılık. https://www.seckin.com.tr/kitap/yonetimde-seffaflik-teori-uygulama-musa-eken-s-p-787865882?srsltid=AfmBOoqYPMSN9xBWew3O07z40P6QpDL1MeB7TMBAqCO3kE5JAQZWHQoB

Genç, F. N. (2023). 6 Şubat Kahramanmaraş Depremleri ve Türkiye'de Afet Yönetim Sistemi—Google'da Ara. *TESAM Strateji Dergisi*. https://www.google.com/search?q=6+%C5%9Eubat+Kahramanmara%C5%9F+Depremleri+ve+T%C3%BCrkiye%27de+Afet+Y%C3%B6netim+Sistemi&oq=6+%C5%9Eubat+Kahramanmara%C5%9F+Depremleri+ve+T%C3%BCrkiye%27de+Afet+Y%C3%B6netim+Sistemi&gs_lcrp=EgZjaHJvbWUqCAgAEEUYJxg7MggIABBFGCcYO9IBBzQ2OWowajeoAgCwAgA&sourceid=chrome&ie=UTF-8

Hallwright, J., & Handmer, J. (2019). Accountability and transparency in disaster aid: Cyclone Pam in Vanuatu. *International Journal of Disaster Risk Reduction, 36*, 101104. https://doi.org/10.1016/J.IJDRR.2019.101104

HGM. (2025). *Kahramanmaraş Mülki İdare İl Haritası*. https://www.harita.gov.tr/urun/kahramanmaras-mulk-idare-il-haritasi/389

Huang, J. S., & Lien, Y. N. (2012). Challenges of emergency communication network for disaster response. *2012 IEEE International Conference on Communication Systems, ICCS 2012*. https://doi.org/10.1109/ICCS.2012.6406204

İstanbul AFAD. (2009). *First 72 h for the individual and family in an earthquake*. https://www.ipkb.gov.tr/wp-content/uploads/2019/07/First-72-hours-for-the-individual-and-family-in-an-earthquake.pdf

Kadıoğlu, M. (2011). *Afet Yonetimi: Beklenilmeyeni Beklemek, En Kötüsünü Yönetmek*. Marmara Belediyeler Birliği. https://www.academia.edu/22920333/Afet_Yonetimi_Beklenilmeyeni_Beklemek_En_K%C3%B6t%C3%BCs%C3%BCn%C3%BC_Y%C3%B6netmek

Kahramanmaraş Büyükşehir Belediyesi. (2023). *Şehirde Güvenli Toplanma Alanları*. https://kahramanmaras.bel.tr/haber/2023/02/08/sehirde-guvenli-toplanma-alanlari

Kaptan, S. (1995). *Bilimsel Araştırma ve İstatistik Teknikleri*. Tekışık.

Keleş, R. (2017). *Kamu Yönetimi ve Demokratik Değerler*. İmge Kitapevi.

Kett, M., Stubbs, S., & Yeo, R. (2005). *Disability in conflict and emergency situations: Focus on Tsunami affected areas*.

Liswanty, I., & Prabowo, H. Y. (2021). Transparency and accountability mechanisms for disaster management funds: A case study. *Journal of Contemporary Accounting, 3*(1). https://doi.org/10.20885/jca.vol3.iss1.art2

Nguyen, L., & Corotis, R. B. (2011). The relationship between hazards vulnerability and stage of economic development in Haiti and Chile. In *Vulnerability, Uncertainty, and Risk: Analysis, Modeling, and Management—Proceedings of the ICVRAM 2011 and ISUMA 2011 Conferences*. https://doi.org/10.1061/41170(400)59

Ochoa, S. F., Neyem, A., Pino, J. A., & Borges, M. R. S. (2007). Supporting group decision making and coordination in urban disasters relief. *Journal of Decision Systems, 16*(2). https://doi.org/10.3166/jds.16.143-172

Ochoa, S. F., & Santos, R. (2015). Human-centric wireless sensor networks to improve information availability during urban search and rescue activities. *Information Fusion, 22*. https://doi.org/10.1016/j.inffus.2013.05.009

OECD. (2020). *OECD Public Governance Reviews*.

OECD. (2023). *The principles of public administration*. https://www.sigmaweb.org/publications/Principles-of-Public-Administration-2023.pdf

Sağıroğlu, Z. A., Ünsal, R., & Özenci, F. (2023). *6 Şubat 2023 Deprem Sonrası Göç ve İnsan Hareketlilikleri: Durum Değerlendirme Raporu*. www.aybu.edu.tr/gpm

Sayıştay Kanunu, Pub. L. No. 6085 (2010).

Seetharam Sridhar, K., Gadgil, R., & Dhingra, C. (2020). Good governance in the transparency, accountability, public participation and capacity (TAP-C) framework. *Paving the Way for Better Governance in Urban Transport, 1–18*. https://doi.org/10.1007/978-981-13-9620-5_1

Stough, L. M., Ducy, E. M. A., & Holt, J. M. (2017). Changes in the social relationships of individuals with disabilities displaced by disaster. *International Journal of Disaster Risk Reduction, 24*. https://doi.org/10.1016/j.ijdrr.2017.06.020

Tüysüz, N. (2011). *Sosyal Sermayenin Ekonomik Gelişme Açısından Önemi ve Sosyal Sermaye Endeksinin Hesaplanması*. T.C. Kalkınma Bakanlığı.

Twigg, J. (2007). Characteristics of a disaster-resilient community: A guidance note. *Natural Hazards Review, 1*(August).

UNDRR. (2015). *Sendai framework for disaster risk reduction 2015–2030*. https://www.undrr.org/publication/sendai-framework-disaster-risk-reduction-2015-2030

UNISDR. (2014). *Living with disability and disasters*. https://www.unisdr.org/2014/iddr/documents/2013DisabilitySurveryReport_030714.pdf

Wisner, B., Blaikie, P., Cannon, T., & Davis, I. (2007). *At risk: Natural hazards, people's vulnerability and disasters* (2nd ed.). Routledge.

Chapter 5
Soil Contamination with Metal(loid)s: Exploring the Complex Interconnections with Human and Livestock Health

Rajashekhar Rao Bangady Killur

Abstract Soil contamination by metal(loid)s stems from numerous anthropogenic sources, including agriculture, industrial activities, waste disposal, accidental spills, and mining. These contaminants threaten soil biota and ecosystem health. Metal(loid) pollution is harmful because these elements persist in the environment for longer periods and can bioaccumulate in organisms. Toxic metal(loid)s can enter the food webs, intensifying their harmful effects on crop plants, animals, and humans. This chapter reviews the highly toxic metal(loid) contaminants—cadmium, chromium, lead, nickel, and arsenic-based on studies from 2014 to 2025. By examining their origins, pathways, and impacts, we aim to support mitigation efforts and promote ecological and public health protection. Literature reveals widespread soil contamination, leading to bioconcentration of hazardous metal(loid)s in staple grains, tubers, forages, feed ingredients, and vegetable, and contamination of freshwater and marine ecosystems. These accumulations have caused dangerously high metal(loid) levels in terrestrial and aquatic food chains. Many studies report toxic metal(loid) levels in food staples, fish, meat, poultry, and dairy. Consuming these contaminated products exposes humans—especially children and vulnerable groups—to health risks. In response, a seven-pronged strategy was proposed to address the threats of this developmental crisis comprising of effective regulatory control, policy reforms and monitoring, soil rehabilitation or remediation, holistic agroecological management, integrated soil, water, and waste management, community capacity building through engagement, and research, innovation, and trans-national cooperation. Fostering community resilience depends on empowering local actors with knowledge, building adaptive governance, and inculcating agro-ecological solutions that restore soil health and protect food systems.

Keywords Bioaccumulation factor · Contaminants · Staples · Remediation · Oral ingestion · Metal(loid) mobility · Animal products

R. Rao Bangady Killur (✉)
School of Agriculture, Faculty of Natural Resources, Papua New Guinea University of Technology, Lae, Papua New Guinea
e-mail: rsraobk@rediffmail.com; rajashekhar.rao@pnguot.ac.pg

G. K. Panda et al. (eds.), *Socio-Ecological Crisis of Development*, Sustainability Solutions, https://doi.org/10.1007/978-3-032-22171-1_5

5.1 Introduction

Economic activities such as mining, industry, agriculture, and services are at the center of national development strategies. Of these, extractive activities such as mining, especially, are at the forefront of resource-rich countries in terms of generating employment, income, and Gross Domestic Product (GDP) (Dehkordi et al., 2024; United Nations, 2007). Agriculture remains important, especially in the least developed countries where it contributes over 25% of GDP (Statista, 2025; World Bank, 2024), and livestock alone contributes to about 40% of the total agricultural GDP. Despite these contributions, these sectors typically follow unsustainable development trends, with economic gains at the expense of public health and environmental degradation. Expanded agriculture, industrialization, and unrestricted mining have contributed to resource depletion, degradation of ecosystems, and widespread pollution. One of the key consequences is the contamination and pollution of soil with metals and metalloids which are commonly termed as heavy metals. Among metalloids, arsenic (As), and among metals, chromium (Cr), lead (Pb), cadmium (Cd), and nickel (Ni) are of specific concern because of their persistence and high toxicity in the environment (Martín-Pozo et al., 2019). Technological advancement, that enhances productivity, has also accelerated the release of pollutants from industrial, agricultural, medical, and domestic processes (Li et al., 2024; Wang et al., 2024). Besides metal(loid)s, other pollutants of major concern include pesticides, particulate matter ($PM_{2.5}$, PM_{10}), microplastics, volatile organic compounds, greenhouse gases, and non-degradable organic contaminants such as dioxins (Wang et al., 2024).

Soil is a substantial environmental sink for the wide array of contaminants, particularly under high-input agricultural systems or in industrial settings. Soil pollution remains neglected in environmental policy and monitoring frameworks (Puri & Rajashekhar Rao, 2023). Contamination arises from processes including injudicious use of agrochemicals (fungicides, insecticides, weedicides, and fertilizers etc.,), unscientific use or disposal of untreated sewage sludge, mining and smelting, e-waste processing, and irrigation with contaminated water. In urban areas and peri-urban areas, exhausts from vehicles and construction waste contribute to metal(loid) load, especially Pb and Zn (Divuniwaqa & Rajashekhar Rao, 2021). Atmospheric deposition by industrial air emission and traffic, and improper waste management, result in chronic exposure risk. Certain metal(loid)s (Cd, Pb, and Hg) are non-biodegradable and are noxious even at ultra-low concentrations, having dangers on soil fauna, plant growth, livestock, and human populations.

Contamination by metalloids has serious and wide-ranging health effects on both humans and animals. By releasing toxins into food chains, contaminated soils reduce crop yields and quality and cause diseases related to the nervous system, reproduction, and immunity in both humans and animals. Metalloid is accumulated by livestock from tainted feed. Animal products and livestock become secondary sources of human exposure. Public health, income stability, and food security are all jeopardized by these integration effects, primarily in marginalized or resource-constrained settings. This chapter investigates the degree of metalloid contamination of soil and the unseen links between it and the more significant socio-ecological disasters brought on by unsustainable growth goals. Based on recent findings (2014–2025) of the contamination of terrestrial and aquatic food webs by specific extremely toxic metal(loid)s, this synthesis proposes a comprehensive seven-faceted approach based on agroecological management, governance, and community resilience—a combination that has rarely been used in prior research. This chapter examines at how metal(loid) pollution affects food systems, ecological health, and the health of people and animals. By doing this, the chapter maps policy, technological, and research options for resilience and advances our understanding of the interactions between environmental degradation and community vulnerability. In order to guide sustainable management and policy interventions, the study aimed to: (i) examine recent evidence (2014–2025) regarding the causes, pathways, ecological effects, and health impacts on humans and livestock of metal(loid) pollution, particularly that of Cd, Cr, Pb, Ni, and As, across terrestrial and aquatic food chains; and (ii) present an integrated strategy for regulatory control, agroecological practice, and community-based resilience.

5.2 Methodology

This review takes a systematic approach to evaluate peer-reviewed studies from 2014 to 2025 on toxic metal(loid) contamination in soils for As, Cd, Pb, Cr, and Ni. The methodological flow of the study is presented in Fig. 5.1. Articles relevant to six major staples (rice, wheat, maize, cassava, potato, and soybean) were sourced from the key science databases Web of Science, Scopus, and Google Scholar using search words "heavy metals," "oil metal(loid) contamination," "toxic metal(loid)s in staples," "cadmium in crops," "metal(loid)s in feedstuff," "metal(loid)s in animal products," and "metal(loid) bioaccumulation." Articles were eligible if they also addressed metal(loid)s in animal products including milk, sheep, goat, poultry, fish, pork, and beef. The studies from geographical regions where potential food chain movement of metal(loid)s possible were prioritized. Precedence was accorded to studies that reported quantitative contamination levels, human and ecological exposure routes, and mitigation measures. Policy reports, regional and environment reports, were also considered. Information was extracted regarding the geographical coverage of the contamination, contaminant concentration ranges, affected crops or organisms, and consequences for human health. Trends were synthesized to show emerging patterns, vulnerable areas, and at-risk segments in the population. The methodology ensures an integrative and interdisciplinary review to inform evidence-based policy, research, and public health practice recommendations.

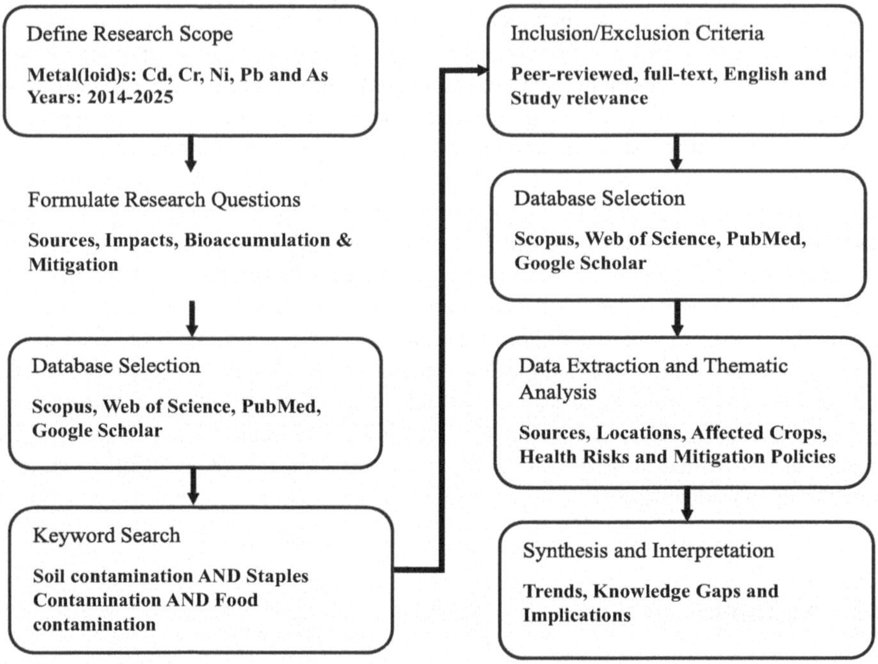

Fig. 5.1 Study methodology for the review of literature

5.3 Metal(loid)s and Soil Pollution

Low concentrations of metal(loid)s occur naturally in soils through geological processes like rock weathering, volcanism, and atmospheric deposition. The background concentrations of metal(loid)s such as Pb, Cd, Ni, Cr, and As typically range between 10–40, 0.1–1.0, 3–1000, 5–1100, and 0.1–40 mg kg^{-1}, respectively. However, these values can vary significantly depending on local conditions, including soil type, climatic factors, geographic location, and the nature of the parent rock material (Puri & Rajashekhar Rao, 2023, 2024).

Table 5.1 The concentration range of metal(loid)s in soils

Symbol of metal(loid)	Concentration (mg kg^{-1})	Source
Pb	10–40	Oorts et al. (2021)
Cd	0.1–1.0	Kubier et al. (2019)
Ni	3–1000	El-Naggar et al. (2021)
Cr	5–1100	Iyaka (2009)
As	0.1–40	Morosini et al. (2023)

However, human activities like agriculture, industry, and mining have raised metal(loid) concentrations much higher than in nature, with significant environmental and health consequences. Soils worldwide are now commonly contaminated, especially around industrial and mining sites, where pollution that was decades ago is still present. Toxic metal(loid)s like Cd, As, Pb, Cr, and Ni are especially dangerous if above safe thresholds. There are an estimated 5 million contaminated sites globally with around 20 million hectares (Li et al., 2022). The rising threat underscores the value of the One Health approach, advocating the unified animal, human, and environmental well-being intercessions to mitigate soil contamination, enhance soil security, and promote sustainable development (Queenan et al., 2017; Swan et al., 2024) (Table 5.1).

Metal(loid)s in soil undergoes varying physico-chemical and biological transformations that influence their mobility, bioavailability, and toxicity (Divuniwaqa & Rajashekhar Rao, 2021). Soil processes like adsorption and desorption of the metal(loid) on to materials of the soil—like clay minerals, organic matter, and oxides of Fe, Mn, and Al—influence the behavior of metal(loid)s. Adsorption limits mobility and plant uptake, while desorption increases metal(loid) availability in the soil solution. Metal(loid)s also precipitates as insoluble precipitates (e.g., sulphides, hydroxides, carbonates) that reduce toxicity, but environmental changes such as reduced pH or the addition of chelating agents will dissolve these compounds, increasing leaching and uptake hazards. Redox reactions also significantly influence metal(loid) speciation and distribution; in oxidizing conditions, Fe and Mn oxides can precipitate and further affect metal(loid) dynamics by adsorbing other metal(loid) ions.

5.4 Metal(loid)s in Livestock and Human Food Chains

Metal(loid)s is entering legumes, cereals, and forages through roots and translocated to above-ground plant portions, typically without obvious symptoms. Uptake is controlled by soil chemistry, metal(loid), plant species, and environmental factors. Agricultural management practices such as contaminated compost, sewage sludge, or municipal waste use can increase long-term risks to crops, water, animals, and human beings. In non-hyperaccumulators, metal(loid)s are generally sequestered in vacuoles or roots to limit injury. These metal(loid)s may hamper crop development, metabolism, and yield (Rashid et al., 2023). Hyperaccumulator crops, approximately 400 plants of mainly Brassicaceae, Asteraceae, Cyperaceae, and Fabaceae, can uptake 100–1000 times more metal(loid)s than ordinary crops and transport them to leaf vacuoles and are hence beneficial in phytoremediation of contaminated soils (Sharma et al., 2016).

Metal(loid)-hyperaccumulating plants are significantly important in the remediation of the environment by reducing metal(loid) concentration in the contaminated environment. Such plants are hazardous for consumption as feed or food because metal(loid)s can be transported through the food chain. Animals fed with polluted plants or water can concentrate metal(loid)s in the kidney, liver, and bones and suffer from growth deficiencies, reproductive impairment, and organ toxicity. Human beings are also impacted through the ingestion of polluted food, milk, or meat. Metal(loid)s is not metabolized or eliminated from the body and hence can build up in tissues via bioaccumulation. While inadvertent consumption of soil may lead to contamination of humans with metal(loid)s (Wieczorek et al., 2023), the most significant risk is the ingestion of metal(loid)-contaminated drinking water (Rahman et al., 2023; Singh et al., 2024) and contaminated food (Rahman et al., 2023). The Bioconcentration Factor (BCF) is routinely employed in the calculation of metal(loid)s uptake in crops. This index provides a measurable indication of extent of metal(loid)s movement from water, soil, or air, into terrestrial or aquatic organism, including food crops. BCF values can be computed using the Eq. (5.1):

$$\mathrm{BCF} = \frac{\text{Concentration of a metal in crop/forage}}{\text{Concentration of that metal in soil/water}} \tag{5.1}$$

The Bioconcentration Factor (BCF) is one of the most important environmental science metrics that quantify the bioavailability, ecotoxicological risk, and metal(loid) transfer ability. BCF measures the potential of a metal(loid) to accumulate within an organism. A higher BCF value indicates a bigger potential for toxic accumulation in living tissues. BCF plays a unique role in the determination of plant efficacy in eliminating soil pollutants, a critical factor in bioremediation and contaminant mobility through various food chains ultimately entering livestock and human systems.

5.5 Metal(loid)s Exposure to Humans and Health Risk Assessment

Chronic or acute metal(loid) exposure are associated with human health risks. Here exposure refers to ingestion of metal(loid) through contaminated food or water. Cancerous and non-cancerous health effects are triggered about by exposure to toxicants. The Target Hazard Quotient (THQ), which takes into account consuming metal(loid) contaminated food, is typically used to evaluate non-carcinogenic health risk. By comparing the estimated daily intake to a reference (acceptable dose), it assesses the risk. Equation (5.2), proposed by Taiwo et al. (2019), can be used to calculate THQ values.

$$\mathrm{THQ} = \frac{EDI}{RfD} \tag{5.2}$$

where *RfD* is the oral reference dose that corresponds to the highest allowable risk-free exposure level, and *EDI* is the population's estimated daily intake of a metal(loid) ($\mathrm{mg\ day^{-1}\ kg^{-1}}$ body weight). Further, Hazard Index (HI) is widely used as a measure of potential health risk arising from exposure to multiple metal(loid)s. According to

Taiwo et al. (2019), HI can be calculated using Eq. (5.3) and represents an aggregate measure that combines risks from various metals:

$$HI = \sum THQ_{As} + THQ_{Cd} + THQ_{Cr} + THQ_{Ni} + THQ_{Pb}. \tag{5.3}$$

The life-time Cancer Risk (CR) from a metal(loid) oral ingestion can be estimated by multiplying Cancer Slope Factor (CSFing) of a given metal(loid) with the estimated daily intake value as shown in Eq. (5.4):

$$CR = EDI \times CSFing. \tag{5.4}$$

Target Cancer Risk (TCR) is the sum of Cancer Risk from all metals contaminating the food and is usually calculated from Eq. (5.5) as suggested by Taiwo et al. (2019):

$$TCR = \sum CR_{As} + CR_{Cd} + CR_{Cr} + CR_{Ni} + CR_{Pb}. \tag{5.5}$$

5.6 Bioconcentration of Metal(loid)s in Staples

Current reports indicate widespread metal(loid) contamination of edible staples (Table 5.2), and vegetables, fruits, spices, and herbs are also affected. Staple foods such as rice, wheat, maize, legumes, and tubers form the foundation of diets globally and, since they are eaten in large quantities, represent a major source of potential metal(loid) intake. Allowable As levels in cereals are 0.5 mg kg^{-1}, while for Cd, the values range from 0.01 to 3 mg kg^{-1} depending on the product, with no limits for Cr and Ni, according to the European Commission (2023). The data in Table 5.2 demonstrate widespread contamination of grains—rice, lentils, beans, oats, wheat, and peas—with poisonous metal(loid)s namely Cd, As, and Pb in the U.S., India, China, and Australia (Dressler et al., 2023; TatahMentan et al., 2020). The outcomes reflect varied Bioconcentration Factor (BCF) among metal(loid)s and crops, with some of the crops having selective uptake of some metal(loid)s to a significant extent. BCF values between less than 1 reflect minimal accumulation and exclusion of metal(loid) by the plant. BCF values 1 to 100 reflect moderate accumulation, 100–1000 reflects high accumulation, and BCF $\geq$ 1000 reflects extremely high accumulation that might pose risk of toxicity through edible plant parts (Pachura et al., 2015). Increased metal(loid) exposure due to consumption of grains, tubers, and fodder contaminated with them is on the rise.

BDL-Below Detection Limit.

Table 5.2 Bioconcentration Factor (BCF) of major staple crops. The table provides range or mean values of metal(loid)s

Crops	No of samples (n)	As	Pb	Cd	Cr	Ni	References
Rice grain	30	0.010–0.023	BDL-0.039	0.087–0.095	0.283–0.678	BDL-0.004	Hasan et al. (2022)
Rice grain	20	0.08–1.04	0.15–0.86	0.03–0.48	0.28–2.84	0.10–1.10	Kormoker et al. (2020)
Rice grain	55		0.460–0.543	0.359–0.387	0.147–0.224	0.089–0.096	Setia et al. (2021)
Rice grain	56		BDL-0.102	BDL-0.203	BDL-0.055	0.001–0.156	Daulta et al., (2020)
Rice grain	159	0.01	0.002	0.62	0.003		Zhou et al. (2024)
Wheat grain	48	0.028–0.14	0.001–0.005	0.12–1.40	0.006–0.014	0.006–0.014	Zhang et al. (2018)
Wheat grain	55		0.483–0.526	0.542–0.614	0.211–0.363	0.045–0.050	Setia et al. (2021)
Wheat grain	25					0.000–0.058	Xu et al. (2024)
Maize grain	10		0.005–0.011	0.41–0.92		0.009–0.025	Rezapour et al. (2022)
Maize grain	6	0.00037–0.0005	0.0019–0.0033	0.016–0.019	0.013–0.018		Xie et al. (2022)
Cassava tuber	7		1.16	1.2			Omeka and Igwe (2021)
Cassava tuber	12	0.01	0.667	1.25	0.037		Samlafo et al. (2022)
Potato tubers	56	<0.001	<0.003–0.009		<0.001–0.032	0.001–0.024	Shi et al. (2022)
Potato tubers	9	0.00018–0.00024	0.0006–0.0007	0.026–0.036	0.011–0.012		Xie et al. (2022)
Soybean	4	0.00037–0.00038	0.0007–0.0008	0.028–0.033	0.014–0.016		Xie et al. (2022)
Soybean	12		0.03–0.87				Blanco et al. (2017)

5.7 Metal(loid)s in Forages, Feed Ingredients, and Animal Feeds

Farm animals are usually feeding on the polluted pastures and consuming polluted water, thereby exposed to metal(loid)s. Table 5.3 shows widespread contamination of forages, feeds, and feed ingredients with many samples exceeding the Maximum Limits (MLs) of the European Union. For example, the ML for Pb in general feed stuffs is 10 mg kg^{-1}, although higher levels are permissible for some materials like forages (30 mg kg^{-1}) and phosphates (15 mg kg^{-1}). Arsenic is restricted at 2 mg kg^{-1} in complete diets, and Cd at 1 mg kg^{-1} (European Commission, 2023). Chronic exposure to the two metal(loid)s affects animal health by distorting growth, reproduction, immunity, and product quality. Lactating animals produce less milk, poultry produce fewer eggs, and meat quality decreases in exposed livestock. Metal(loid)s like As, Pb, and Cd get accumulated in organs like the kidneys, bones, and liver, finally entering the human food chain through milk, meat, and eggs. Food safety is now a constraint and challenge for trade. Toxicity effects are metal(loid) and exposure duration dependent. Pb intoxicates and affects the nervous system and causes anemia, weakness, and reproductive damage. Cd targets the kidneys and bones and depresses immunity. Mercury affects enzymes and behavior, while as causes gastrointestinal distress and long-term toxicity. Exposure to Cr—typically caused by industrial pollution—damages organs and reduces fertility and feed intake (Afzal & Mahreen, 2024).

5.8 Metal(loid) Exposure to Animals

Metal(loid)s, once in animals, get concentrated in individual organs—Pb and Cd in liver, kidney, and bones, and Hg in brain, muscles, and fat, and As in muscle and organs. These, in turn, can concentrate in animal products for instance eggs, meat, milk, and fish (Table 5.4). People consuming offal (liver and kidney) from cattle and sheep around mines have significant metal(loid) intake. Milk obtained from cows, buffaloes, goats, and sheep that ingest contaminated feed, poultry eggs on contaminated feed, and fish on contaminated water are health-hazardous. Pb, which is commonly present in animal sources, is of special concern due to its toxic and bio-accumulative nature. Chronic exposure of Pb can be observed in fish up to a concentration of 44.3 mg kg^{-1} (Panda et al., 2023). Near industrial areas, consumption of metal(loid)s causes metabolic, structural, and functional organ malformations in animals. Chickens store As, Hg, Cd, and Pb in organs like lungs and reproductive tissue and are impacted by Cd in growth and egg production through feed conversion (Aljohani, 2023). Chronic exposure to metal(loid)s even in low doses reorganizes histology in tissues of kidneys, reproductive organs, brain, and liver reducing the activity of enzymes and altering serum biochemistry. Dose, duration, and route of exposure determine severity of toxicity. While trace amounts of some

Table 5.3 Metal(loid)s in livestock forages, feed, and feed ingredients (mg kg^{-1})

Feed/feed ingredient/ forage	Number of samples (n)	Cd	Cr	Ni	Pb	References
Feed ingredients						
Rice bran	13	0.001–0.02				Tao et al. (2020)
Rape seed cake	10	0.023–0.056				Tao et al. (2020)
Fish meal	5		0.27–1.49			Tao et al. (2020)
Micro-minerals	12	BDL	0.097–0.281		0.188–0.260	Kamal et al. (2022)
Forages						
Barseem	21			6.53–81.2	ND-5.73	Kanwal et al. (2022)
Various fodders	75			5.34–10.43	6.69–11.24	Hussain et al. (2024)
Green grass	111	BDL-0.35	BLD	BLD-6.03	BDL-2.97	Fechete et al. (2024)
Green grass	10	0.03–0.079	3.6–5.10		32.0–41.2	Liu et al. (2020)
Feeds						
Poultry feed	16	ND-1.64	ND-12.90	ND-2.64	ND-4.21	Shahariar et al. (2025)
Duck feed	20	0.10–0.52	ND-95.69		1.08–4.42	Aendo et al. (2018)
Pig formula feed	87	0.002–0.052	0.22–1.00			Tao et al. (2020)
Beef cattle	9	BDL	0.251–0.457		0.256–0.343	Kamal et al. (2022)

BDL-Below Detection Limit, ND-Not Detected

metal(loid)s like Se, Co, Cu, Zn, and Mn are necessary for the body, others—Cd, Ni, Hg, As, and Pb—are toxic. Exposure to animals occurs primarily through polluted water and food from industrial, agricultural, and geogenic origins, although dermal and inhalation exposures have been noted as well. Once ingested, metal(loid)s are retained in the body and acquire acute or chronic impacts based on exposure trends (Afzal & Mahreen, 2024). They cause disruption to cellular processes by creating free radicals and oxidative stress. Excessive content of metal(loid)s in animal feedstuff poses severe public health risks, but their early detection, antidotes, and correct management help reduce impacts on livestock productivity and well-being (Afzal & Mahreen, 2024).

5.9 Metal(loid)s Exposure to Humans Through Staples and Animal Products

Major pathway of metal(loid)s exposure to humans is through the ingestion of contaminated food and water. Studies have reported exposure to As and Hg through fish and livestock, while paddy rice (*Oryza sativa* L.), a significant staple in Asia, is a significant dietary source of As and Cd. Leafy greens, tubers, and cereal grains contaminated with Cd, Pb, and Ni also contribute to exposure risk. Drinking water consumption of As, Cr, and Pb, and incidental ingestion of soils by children, is of significant health concern in areas where the water has become contaminated. Inhaling suspended Pb, Cd, and Cr due to industrial processes is especially dangerous. Dermal exposure to the contaminated water, dust, or material may lead to Ni, Hg, Pb, and Cr intake. Of note, Pb and Hg are able to cross the placental barrier and damage fetal development. Among all the routes of exposure, intake via contaminated food continues to be the most significant pathway. Regulatory agencies such as WHO, FAO, EU, and U.S. EPA have established tolerances for foodborne metal(loid)s in order to protect public health as well as to control international trade. For instance, for Cd in rice, FAO/WHO has a 0.2 mg kg^{-1} limit, Codex Alimentarius allows 0.2 mg kg^{-1} for inorganic As, and the EU limits Pb in most crops to 0.1 mg kg^{-1}. Despite all these regulations, studies often report exceedances in contaminated areas. Wei et al. (2023) found that chronic rice intake, even within tolerable amounts, may pose severe risks—primarily to infants and children.

Enduring exposure to metal(loid)s affects millions of people globally and has been linked to cancers and other severe health outcomes. Heavy contamination of rice, fish, and animal products by epidemiological and toxicological research evidence is apparent, and non-carcinogenic risks are of concern (Table 5.5). The Hazard Index (HI) that assesses the non-cancer health hazard is generally greater than 1 for metal(loid)s like Cd, Cr, Pb, Ni, and As in rice, vegetables, tubers (taro, sweet potato, Singapore taro), fish, and meat. HI values in children are generally 2–4 times higher than in adults, indicating increased susceptibility. These exposures are associated with kidney and liver damage, skin irritation, disturbed blood function, nervous

Table 5.4 Metal(loid)s concentration and contamination in animal products (mg/kg on wet weight basis)

	Cd	Cr	Pb	Ni	Number of samples (N)	References
Milk						
Bovine milk	<0.001	<0.070	0.001–0.004	<0.050	5	Di Bella et al. (2020)
Goat and sheep milk	<0.001	<0.070	0.001–0.007	<0.050	11	Di Bella et al. (2020)
Cow and buffalo	0.006–0.032	0.007–0.040	0.040–0.204	0.078–0.295	7	Sharma et al. (2025)
Buffalo milk	0.0012–0.058		0.0023–0.029		90	Shahzad et al. (2025)
Poultry meat						
Muscle	ND-0.035	0.869–2.27	0.289–0.915	BDL-0.42		Shahariar et al. (2025)
Duck eggs	ND-2.55	ND-4.05	ND-9.70		20	Aendo et al. (2018)
Muscle	0.02–0.08		0.30–1.07			Zhuang et al. (2014)
Fish						
Fish muscle	0.00–509		<0.007–599			Ali and Khan (2018)
Fresh water fish		0.07–27	0.0016–44.3			Panda et al. (2023)
Marine fish	0.002–0.04	0.016–0.173	0.021–0.082		652	Han et al. (2021)
Marine fish	0.001–0.290		0.001–16.68		324	Prabakaran et al. (2025)
Pork						
Pork meat	0.01–0.03	0.15–0.23	0.22–0.38		100	Barone et al. (2020)

(continued)

Table 5.4 (continued)

	Cd	Cr	Pb	Ni	Number of samples (N)	References
Pork products	ND-0.51	0.012–0.26	0.006–0.121		80	Pei et al. (2020)
Pork products	0.0007–0.223	0.0056–0.258	0.0001–0.196	0.0064–0.212	70	Bilandzic et al. (2020)
Red meat						
Beef products	0.0007–0.54	0.007–0.903	0.0001–0.196	0.0064–0.212	84	Bilandzic et al. (2020)
Beef	BDL	BDL	0.002–0.098	BDL	11	Di Bella et al. (2020)
Beef parts	BDL-0.41	BDL-1.05	0.41–1.50		30	Akele et al. (2022)
Sheep	0.009–0.018	0.342–0.921	0.023–0.123	0.001–0.665	17	Zeinali et al. (2019)
Cow muscle	0.010–0.019	0.438–0.995	0.015–0.072	0.017–0.130	17	Zeinali et al. (2019)

BDL-Below Limit of Detection, ND-Not Detected

system damage, and reproductive dysfunction (Puri & Rajashekhar Rao, 2024). It is long proven that metal(loid)s (As, Cd, Cr, and Ni) are Group 1 carcinogens and cancer-causing agents as per International Agency for Research on Cancer. Risk is high in tubers like sweet potato and taro due to high Ni accumulation (Puri & Rajashekhar Rao, 2024). Of the animal product consumers, fish consumers are more susceptible to Pb and Cr, and marine and freshwater fish have risks from As and Hg contamination (Han et al., 2021; Prabakaran et al., 2025). Pork products with high Pb and As concentrations are unsafe for children (Barone et al. 2021; Zhang et al., 2022), and consumption of beef is hazardous via Group 1 metal(loid) contaminants (Emurotu et al., 2024; Zeinali et al., 2019). Cow kidney and liver contain maximum Cd level and related cancer hazard. In polluted areas consumption of these organs should be avoided. Although buffalo milk risks are quite low, there is still a need to monitor inorganic contaminants (Shahzad et al., 2025). Metal(loid)s induces oxidative stress and genotoxicity resulting in DNA damage even at low exposures. They affect DNA repair and induce epigenetic changes. Some metal(loid)s disrupt cell signaling and gene expression, causing mutations. Arsenic exposure results in lung, skin, bladder, and liver cancers. Cadmium induces lung, prostate, and kidney cancers, and Cr is associated with nasal and sinus cancers and skin ulcers. Nickel causes lung and nasal cancers, whereas Pb has been linked with stomach, brain, and kidney cancers, particularly in the workplace (IARC, 2012).

Heavy metal(loid) pollution is detrimental to livestock, humans, and society at large. Exposure to metal(loid)s over long durations leads to cancer, organ damage, neurological problems, and developmental problems, which consequently increases healthcare expenditure and reducing workforce productivity. Polluted livestock commodities—meat, milk, and eggs—threatens food safety and also may put nutritional security at stake. Farmers economically lose livestock productivity, are subjected to trade barriers, and bear high costs of remediation, offsetting agricultural profitability. Socially, rural livelihoods rooted in livestock decline, poverty deepens, and the affected communities experience stigma and reduced resilience. Poor management and governance can lead to public loss of trust in food systems, industries, and government and aggravates social challenges (Tables 5.5 and 5.6).

Table 5.5 Non-carcinogenic risk as measured by Hazard Index (HI) values in adults and children ingesting metal(loid)s contaminated cereal grains, edible tubers, and some animal products

Metal(loid)s studied	Most important contributor to HI	HI in children	HI in adults	Crops/animal products	References
As, Pb, Cd, Cr, Cu, Zn, Mn, and Ni	Cu and As	–	2.87–10.71	Rice	Shaheen et al. (2024)
Cd, Pb, Hg, Cr, Ni, As, Zn, and Cu	As and Cd	6.014	3.558	Rice	Guo et al. (2022)
Cd, Hg, As, Pb, Cr, Ni, Zn, and Cu	As and Cd	6.014	3.558	Rice	Guo et al. (2022)
Cu, Cd, Zn, Pb, As, and Hg	Cd and Cr		>1.0	Wheat	Meng et al. (2025)
Cd, Pb, Cr, and Ni	Cr, Cd, Pb, Ni	7.11–9.82	4.03–7.32	Tuberous staples	Puri and Rajashekhar Rao (2024)
Pb, Cr, and Cd	Cr, Pb, and Cd	<0.1	<0.1	Fish	Ramish et al. (2024)
Cd, Cr, Pb, and Ni	Cd, Cr, Ni, Pb, and Al	5.19	1.48	Cow	Emurotu et al. (2024)
Cd, Pb, Cr, and Ni		5.21	1.49	Goat	
Cr, Cd, Ni, Cu, and Pb	Pb	<1.0	<1.0	Sheep and cow	Zeinali et al. (2019)
Cd, Pb, Hg, Zn, Cu, and Cr	Pb, Cd, Cr, Zn, Cu, Hg	0.21–0.30	0.10–0.13	Pork products	Barone et al. (2021)
Cd, Pb, Cu, Fe, and Mn	Cd, Pb, Cu, Mn, Fe	20.6	14.72	Pork roasted	Nambowa et al. (2024)
		64.98	46.45	Goat roasted	
As, Mn, Fe, Hg, Cd, and Pb		0.52–0.68	0.07–0.09	Buffalo milk	Shahzad et al. (2025)

Table 5.6 Cancer risk values in adults and children consuming cereal grains, staple tubers (sweet potato, yams, and taro), and animal products contaminated with metal(loid)s

Metal(loid)s studied	Most contributor to cancer risk	Children	Adults	Crops/ Animal products	References
Cd, As, Hg, Pb, Cr, Ni, Cu, and Zn	As and Cd	8.4E−03	5.00E–03	Rice	Guo et al. (2022)
Cu, Cd, Zn, Pb, Hg, and As	Cd and Cr		>1.00E–04	Wheat	Meng et al. (2025)
Pb, Cd, Cr, As, and Hg	As, Cd, and Cr		>1.00E–04	Wheat	Wang et al. (2024)
Cr, Cd, Ni, and Pb	Ni	3.61E−03 to 1.54E−02	1.16E–02 to 2.88E–02	Tuberous staples	Puri and Rajashekhar Rao (2024)
Cr, Pb, and Cd	Pb, Cr, and Cd	2.70E–01	2.10E–01	Fish	Ramish et al. (2024)
Cd, Pb, Ni, and Cr	Cd, Cr, Ni, and Pb	9.55E + 00	2.73E + 00	Cow	Emurotu et al. (2024)
Cd, Cr, Ni, and Pb	Cd, Cr, Ni, and Pb	4.62E + 00	2.54E + 00	Goat	Emurotu et al. (2024)
Pb, Cr, and Cd	Pb	2.25E–04–4.42E–04	0.97E–04–1.97E–04	Pork products	Barone et al. (2021)
Pb, Hg, Cd, and As	As	>1.00E–04		Pork	Zhang et al. (2022)
Cd, Ni, Cr, Cu, and Pb	Pb and Cd	>1.00E–04	>1.00E–04	Cow and sheep muscles	Zeinali et al. (2019)

5.10 Sustainable Solutions to the Developmental Crisis

The world is facing a deepening environmental crisis due to unsustainable practices in industries, agriculture, and waste management. This has led to widespread pollution, contaminating ecosystems and food chains, posing health risks, and disproportionately impacting marginalized communities. In response, a framework of community-based and interdisciplinary methods has been proposed to address the root causes of environmental degradation and promote resilience, justice, and sustainability (Fig. 5.2). At the center of this approach is preventive resilience, which focuses on controlling and regulating pollution at the source. This systems-level change in development management demands that all communities, especially the historically disadvantaged, are protected from the threat of ecological degradation through strict implementation of pollution controls. By tackling the problem at its core, this strategy

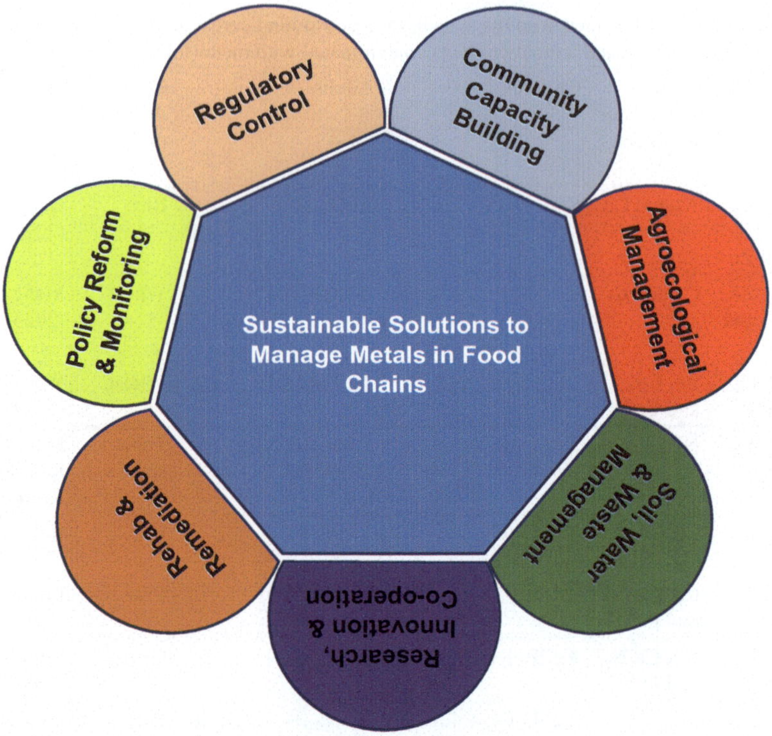

Fig. 5.2 Proposed seven-integrated strategies as robust and long-term developmental solutions to control metal(loid)s in food chains

aims to build a more sustainable and equitable future for all. Preventing environmental harm before it happens is possible through proactive policies and monitoring.

Robust regulations, transparent data, and community involvement are key to effective environmental governance. These resilient institutions build trust and shared responsibility among governments, citizens, and stakeholders. Moreover, nature-based solutions like phytoremediation and composting can restore degraded ecosystems in a cost-effective and locally adaptable way. These practices encourage self-sufficiency and cultivate a culture of ecological stewardship rooted in local knowledge. By taking a proactive and community-driven approach, environmental health can be safeguarded and this in turn creates a sustainable future. Environmental disasters from metal(loid)s pollution of food chains can be achieved through proactive pollution regulation. By implementing robust policies and monitoring systems, a more resilient and transparent environmental governance framework can be instituted. Ensuring stakeholder communities participating in this process, and responsible and accountable authorities can become true stewards of vulnerable local ecosystems. Restoration of degraded landscapes requires policies and procedures that build trust, fostering collective responsibility, and embracing nature-based solutions.

Simple and cost-effective tools such as phytoremediation techniques, application of organic amendments, including composts, farm yard manure, crop residues, or waste biomass-generated biochar can be an effective means to revitalize soils and boost safety of ecosystems. These environment-friendly techniques attenuate or inactivate metal(loid)s in soils thus reducing mobility of toxicants into plant-livestock-human continuum. These strategies are particularly well-suited for underprivileged rural and urban areas, providing accessible and adaptable solutions to tackle environmental challenges. Embracing this engaging approach to environmental protection through decreased load of contaminants in the feed, animal products and reduced human exposure to metal(loid)s thus demonstrating harmony with nature. Their use promotes local self-sufficiency and the establishment of a culture of ecological stewardship based on indigenous knowledge. Agroecological land management practices are the basis for making the food system stronger and more independent. Agroecology is based on how biodiversity, soil health, and the health of the community all affect each other. These methods lower the amount of metal(loid) in crops and stop more pollution by using fewer chemicals and improving the natural cycle of nutrients. To lessen the overall severity of environmental problems, soil, water, and waste must be managed together (Li et al., 2024). Restoring the environment and protecting public health are achieved through nutrient loop closure, wastewater purification, and proper solid waste disposal. Integrated management restores and preserves essential ecological services for future generations, whether in rural villages or growing urban settlements.

Community capacity building and engagement is pivotal in resilience building. With meaningful involvement of communities in issue identification, co-solutioning, and management of resources, the outcomes are more just and sustainable. Local and indigenous knowledge systems uniquely offer insights into ecosystem processes and sustainable practices lacking in mainstream scientific discourse. Education, training, and participatory governance enable communities to support adaptive learning, long-term ownership, and collective action in addressing complex environmental issues. Often, simple food preparation and processing techniques can also effectively reduce exposure to metal(loid)s. For instance, modified parboiling or soaking rice overnight before cooking can reduce inorganic arsenic levels by up to 25% (Rahman et al., 2019). Research, innovation, and cooperation provide the knowledge infrastructure for driving progress. Innovation and science must be made relevant, local, and co-produced with impacted communities; they are no longer the exclusive domain of prestigious institutions. In order to provide rural or underserved communities with access to international best practices, global cooperation can help leverage knowledge exchange, capital flows, and policy coordination.

These seven approaches collectively provide a logical and cohesive response to the socio-ecological crisis of metal(loid) development in food chains. They address the institutional and societal structures that lead to marginalization and inequality in addition to the environmental effects of pollution from metal(loid)s. Such a response aims to achieve sustainability by providing social justice, economic viability, and environmental integrity. Above all, resilience cannot be thought of as a passive state. Such a response aims to achieve sustainability by providing social justice, economic viability, and environmental integrity. Most importantly, resilience is a dynamic, growing, and transformative process that turns ecological crisis into a catalyst for renewal and justice rather than a passive state of overcoming adversity.

5.11 Conclusions

This chapter draws on interconnections between metal(loid)s contamination in human and livestock systems to the animal welfare and human social well-being in the context of rapid technological growth. It informs adherence to best practices in livestock and agricultural management and offers evidence to establish strict regulatory limits for metal(loid)s in foods, animal feeds, and soils. A framework for developing and managing the developmental crisis of metal(loid) contamination in vulnerable populations is also provided by the findings. Metal(loid) accumulation in food webs has been made worse by unsustainable development, and this poses a serious and long-term health risk to ecosystem resilience and community well-being. The chapter goes into detail about how harmful metals (loids) such as Cd, Cr, Pb, Ni, and As continue to accumulate in crops and pasture, persist in the soil matrix, and eventually enter the bodies of people and animals, leading to severe chronic disease and weakened food security. One of the main obstacles to controlling these metalloids in any ecosystem is a lack of knowledge about how they move and behave in those environments. Upcoming technologies, like enhanced detection methods, artificial intelligence, and machine learning, have the potential to forecast contamination risk and direct mitigation choices. But technological advancements alone won't suffice for sustainable solutions. Participatory policy processes, ecologically sustainable crop and livestock production systems, enhanced environmental regulation, and community-based strategies will all be necessary. Developing adaptive governance, empowering local stakeholders with knowledge, and preserving nature-based solutions to restore soil health and guarantee food security are the foundations of community resilience.

Authors Contribution All contributions by the single author.

Funding The research was not supported by any funding.

Potential Conflicts of Interest The author declares that there is no conflict of interest.

Informed Consent Not applicable.

Data Availability Statement All data collected is presented in the paper.

AI Disclosure No AI tool was used in writing this chapter.

References

Aendo, P., Netvichian, R., Viriyarampa, S., Songserm, T., & Tulayakul, P. (2018). Comparison of zinc, lead, cadmium, cobalt, manganese, iron, chromium and copper in duck eggs from three duck farm systems in Central and Western, Thailand. *Ecotoxicology and Environmental Safety, 161*, 691–698. https://doi.org/10.1016/j.ecoenv.2018.06.052

Afzal, A., & Mahreen, N. (2024). Emerging insights into the impacts of heavy metal(loid)s exposure on health, reproductive and productive performance of livestock. *Frontiers in Pharmacology, 15*, 1375137. https://doi.org/10.3389/fphar.2024.1375137

Akele, M. L., Desalegn S. K., Asfaw, T. B., Assefa, A. G., Alemu, A. K., de Oliveira, R. R. (2022). Heavy metal contents in bovine tissues (kidney, liver and muscle) from Central Gondar Zone, Ethiopia. *Heliyon 8,*e12416. https://doi.org/10.1016/j.heliyon.2022.e12416

Ali, H., & Khan, E. (2018). Bioaccumulation of non-essential hazardous heavy metal(loid)s and metal(loid)loids in freshwater fish. *Risk to Human Health, Environmental Chemistry Letters, 16*(3), 903–917. https://doi.org/10.1007/s10311-018-0734-7

Aljohani, A. S. M. (2023). Heavy metal(loid) toxicity in poultry: A comprehensive review. *Frontiers in Veterinary Science, 29*(10), 1161354. https://doi.org/10.3389/fvets.2023.1161354

Barone, G., Storelli, A., Quaglia, N. C., Garofalo, R., Meleleo, D., Busco, A., & Storelli, M. M. (2020). Trace metal(loid)s in pork meat products marketed in Italy: Occurrence and health risk characterization. *Biological Trace Element Research, 199*, 2826–2836. https://doi.org/10.1007/s12011-020-02417-z

Barone, G., Storelli, A., Quaglia, N.C., Garofalo, R., Meleleo, D., Busco, A., Storelli, M.M. (2021). Trace metals in pork meat products marketed in Italy: Occurrence and health risk characterization. *Biological Trace Element Research 199,* 2826–2836. https://doi.org/10.1007/s12011-020-02417-z

Bilandžić, N., Sedak, M., Čalopek, B., Đokić, M., Varenina, I., Solomun Kolanović, B., Božić Luburić, Đ., Varga, I., Roncarati, A. (2020). Evaluation of element concentrations in beef and pork meat cuts available to the population in the Croatian capital. *Foods 9,* 1861. https://doi.org/10.3390/foods9121861

Blanco, A., Salazar, M. J., Cid, V. C., Pignata, M. L., & Rodriguez, J. H. (2017). Accumulation of lead and associated metal(loid)s (Cu and Zn) at different growth stages of soybean crops in lead-contaminated soils: Food security and crop quality implications. *Environmental Earth Sciences, 76*, 182. https://doi.org/10.1007/s12665-017-6508-x

European Commission. (2023). Commission Regulation (EU) 2023/915 of 25 April 2023 on maximum levels for certain contaminants in food and repealing Regulation (EC) No 1881/2006. *Official Journal of the European Union, L 119*, 103–157. Retrieved from https://eur-lex.europa.eu/legal-content/EN/TXT/?uri=CELEX%3A32023R0915

Daulta, R., Sridevi, T., & Garg, V. K. (2020). Spatial distribution of heavy metal(loid)s in rice grains, rice husk, and arable soil, their bioaccumulation and associated health risks in Haryana India. *Toxin Reviews, 40*(4), 859–871. https://doi.org/10.1080/15569543.2020.1793781

Dehkordi, M. M., Nodeh, Z. P., Dehkordi, K. S., Salmanvandi, H., Khorjestan, R. R., & Ghaffarzadeh, M. (2024). Soil, air, and water pollution from mining and industrial activities: Sources of pollution, environmental impacts, and prevention and control methods. *Results in Engineering, 23*, Article 102729. https://doi.org/10.1016/j.rineng.2024.102729

Di Bella, C., Traina, A., Giosuè, C., Carpintieri, D., Lo Dico, G.M., Bellante, A., Del Core, M., Falco, F., Gherardi, S., Uccello, M.M., Ferrantelli, V. (2020). Heavy metals and PAHs in meat, milk, and seafood from Augusta area (Southern Italy): Contamination levels, dietary intake, and human exposure assessment. *Frontiers in Public Health 8,* 273. https://doi.org/10.3389/fpubh.2020.00273

Divuniwaqa, C., & Rajashekhar Rao, B. K. (2021). Biochar application alters soil Ni fractions and phytotoxicity of Ni to pakchoi plants. *Environmental Technology and Innovation, 23*, Article 101751. https://doi.org/10.1016/j.eti.2021.101751

Dressler, J. M., Raab, A., Wehmeier, S., & Feldmann, J. (2023). Arsenic, cadmium, and lead in rice and rice products on the Austrian market. *Food Additives and Contaminants: Part B Surveillence, 16*(2), 185–195. https://doi.org/10.1080/19393210.2023.2194061
El-Naggar, A., Ahmed, N., Mosa, A., Niazi, N. K., Yousaf, B., Sharma, A., Sarkar, B., Cai, Y., & Chang, S. X. (2021). Nickel in soil and water: Sources, biogeochemistry, and remediation using biochar. *Journal of Hazardous Materials, 419*, Article 126421. https://doi.org/10.1016/j.jhazmat.2021.126421
Emurotu, J. E., Olawale, O., Dallatu, E. M., Abubakar, T. A., Umudi, Q. E., Eneogwe, G. O., & Atumeyi, A. (2024). Carcinogenic and non-carcinogenic health risk assessment of heavy metal(loid)s in the offal of animals from Felele Abattoir, Lokoja, Nigeria. *Toxicology Reports, 13*, Article 101701. https://doi.org/10.1016/j.toxrep.2024.101701
Fechete, F. I., Popescu, M., Mârza, S. M., Olar, L. E., Papuc, I., Beteg, F. I., Purdoiu, R. C., Codea, A. R., Lăcătuş, C. M., Matei, I. R., Lăcătuş, R., Hoble, A., Petrescu-Mag, I. V., & Bora, F. D. (2024). Spatial and bioaccumulation of heavy metal(loid)s in a sheep-based food system: Implications for human health. *Toxics, 12*(10), 752. https://doi.org/10.3390/toxics12100752
Guo, Y., Huang, M., You, W., Cai, L., Hong, Y., Xiao, Q., Zheng, X., & Lin, R. (2022). Spatial analysis and risk assessment of heavy metal(loid) pollution in rice in Fujian Province, China. *Frontiers in Environmental Science, 10*, 1082340. https://doi.org/10.3389/fenvs.2022.1082340
Han, J.L., Pan, X.D., Chen, Q., & Huang, B.F. (2021). Health risk assessment of heavy metal(loid)s in marine fish to the population in Zhejiang, China. *Scientific Reports, 11*, 11079. https://doi.org/10.1038/s41598-021-90665-x
Hasan, G.M.M.A., Das, A.K., & Satter, M.A. (2022). Accumulation of heavy metal(loid)s in rice (Oryza sativa. L) grains cultivated in three major industrial areas of Bangladesh. *Journal of Environment and Public Health*, 1836597. https://doi.org/10.1155/2022/1836597
Hussain, M. I., Khan, Z. I., Ahmad, K., Naeem, M., Ali, A., Elshikh, M. S., Zaman, Q., Iqbal, K., Muscolo, A., & Yang, H. H. (2024). Toxicity and bioassimilation of lead and nickel in farm ruminants fed on diversified forage crops grown on contaminated soil. *Ecotoxicology and Environmental Safety, 283*, Article 116812. https://doi.org/10.1016/j.ecoenv.2024.116812
IARC. (2012). Arsenic, metal(loid)s, fibres, and dusts. In *IARC Monographs on the Evaluation of Carcinogenic Risks to Humans Volume 100C, International Agency for Research on Cancer*. WHO.
Iyaka, Y. A. (2009). Chromium in soils: A review of its distribution and impacts. *Continental Journal of Environmental Sciences, 3*, 13–18.
Kamal, M.T., Al-Mamun, M., Hossain, M.M., Razzaque, M.A., & Hashe, M.A. (2022). Assessment of heavy metal(loid)s in feed and beef in Bangladesh: A safety issues. *Meat Research, 2*(1). https://doi.org/10.55002/mr.2.1.10
Kanwal, H., Raza, A., Zaheer, M. S., Nadeem, M., Ali, H. H., Manoharadas, S., Rizwan, M., Kashif, M. S., Ahmad, U., Ikram, K., Riaz, M. W., & Rasool, F. (2022). Transformation of heavy metal(loid)s from contaminated water to soil, fodder and animals. *Scientific Reports, 14*, 11705. https://doi.org/10.1038/s41598-024-62038-7
Kormoker, T., Proshad, R., Islam, M. S., Shamsuzzoha, M., Akter, A., & Tusher, T. R. (2020). Concentrations, source apportionment and potential health risk of toxic metal(loid)s in foodstuffs of Bangladesh. *Toxin Reviews*. https://doi.org/10.1080/15569543.2020.1731551
Kubier, A., Wilkin, R. T., & Pichler, T. (2019). Cadmium in soils and groundwater: A review. *Applied Geochemistry, 108*, Article 104388. https://doi.org/10.1016/j.apgeochem.2019.104388
Li, X., Shen, X., Jiang, W., Xi, Y., & Li, S. (2024). Comprehensive review of emerging contaminants: Detection technologies, environmental impact, and management strategies. *Ecotoxicology and Environmental Safety, 278*, Article 116420. https://doi.org/10.1016/j.ecoenv.2024.116420
Li, C, Zhou, K, Qin, W., Tian, C., Qi, M., Yan, X., Han, W. (2022). Recent advances in soil remediation technology for heavy metal contaminated sites: A critical review. *Journal of Environmental Management 312,* 114925. https://doi.org/10.1016/j.jenvman.2022.114925
Liu, W., Ahmad, K., Khan, Z. I., Mughal, N., Wajid, K., Munir, M., Malik, I. S., Bashir, H., Mehmood, N., Muqadas, H., & Noorka, I. R. (2020). Comparative study of forage toxic

metal(loid)s of conventional versus non-conventional pastures in relation to animal mineral nutrient allowance. *Environmental Science and Pollution Research*. https://doi.org/10.1007/s11356-020-09599-y

Martín-Pozo, L., de Alarcón-Gómez, B., Rodríguez-Gómez, R., García-Córcoles, M.T., Çipa, M., Zafra-Gómez, A. (2019). Analytical methods for the determination of emerging contaminants in sewage sludge samples: A review. *Talanta 192,* 508–533. https://doi.org/10.1016/j.talanta.2018.09.056

Meng, F., Liu, D., Bu, T., Zhang, M., Peng, J., & Ma, J. (2025). Assessment of pollution and health risks from exposure to heavy metal(loid)s in soil, wheat grains, drinking water, and atmospheric particulate matter. *Journal of Environmental Management, 376*, Article 124448. https://doi.org/10.1016/j.jenvman.2025.124448

Morosini, C., Terzaghi, E., Raspa, G., Grotti, M., Armiraglio, S., Anelli, S., & Di Guardo, A. (2023). Arsenic movement and fractionation in agricultural soils which received wastewater from an adjacent industrial site for 50 years. *Science of the Total Environment, 898*, Article 165422. https://doi.org/10.1016/j.scitotenv.2023.165422

Nambowa, B., Aruwa, J.O., Adepoju, A.A., Asanga, O.P., Sanusi, I.O. (2024). Health risk assessment of exposure to some heavy metals from roasted pork and goat meats sold in Ishaka-Bushenyi Municipality, Uganda. *Journal of Agricultural Chemistry and Environment 13,* 4022. https://doi.org/10.4236/jacen.2024.134022

Omeka, M. E., & Igwe, O. (2021). Heavy metal(loid)s concentration in soils and crop plants within the vicinity of abandoned mine sites in Nigeria: An integrated indexical and chemometric approach. *International Journal of Environmental Analytical Chemistry*. https://doi.org/10.1080/03067319.2021.1922683

Oorts, K., Smolders, E., Lanno, R., & Chowdhury, M. J. (2021). Bioavailability and ecotoxicity of lead in soil: Implications for setting ecological soil quality standards. *Environmental Toxicology and Chemistry, 40*, 1948–1961. https://doi.org/10.1002/etc.5051

Pachura, P., Ociepa-kubicka, A., & Skowron-Grabowska, B. (2015). Assessment of the availability of heavy metal(loid)s to plants based on the translocation index and the bioaccumulation factor Assessment of the availability of heavy metal(loid)s to plants based on the translocation index and the bioaccumulation factor. *Desalination and Water Treatment, 57*, 1469–1477. https://doi.org/10.1080/19443994.2015.1017330

Panda, B. P., Mohanta, Y. K., Parida, S. P., Pradhan, A., Mohanta, T. K., Patowary, K., Mahari, W. A. W., Lam, S. S., Ghfar, A. A., Guerriero, G., Verma, M., & Sarma, H. (2023). Metal(loid) pollution in freshwater fish: A key indicator of contamination and carcinogenic risk to public health. *Environmental Pollution, 330*, Article 121796. https://doi.org/10.1016/j.envpol.2023.121796

Pei, F., Wang, Y., Fang, Y., Li, P., Yang, W., Ma, N., Ma, G., & Hu, Q. (2020). Concentrations of heavy metal(loid)s in muscle and edible offal of pork in Nanjing city of China and related health risks. *Journal of Food Science, 85*(2), 493–499. https://doi.org/10.1111/1750-3841.15014

Prabakaran, K., Charoenpong, C., Bureekul, S., Wang, X., & Sompongchaiyakul, P. (2025). Heavy metal(loid) contamination in marine fish from the Andaman sea: Influence of habitat and health risk assessment. *Marine Pollution Bulletin, 210*, Article 117299. https://doi.org/10.1016/j.marpolbul.2024.117299

Puri, S. B., & Rajashekhar Rao, B. K. (2023). Ecotoxicological risks of metal(loid)s in the subsistence food garden soils of Watut river floodplains, Papua New Guinea. *Environmental Geochemistry and Health, 45*(11), 8403–8415. https://doi.org/10.1007/s10653-023-01735-0

Puri, S. B., & Rajashekhar Rao, B. K. (2024). Health risks of metal(loid)s in soils and staple foods of the subsistence food gardens in the floodplains of Watut river, Papua New Guinea. *Environmental Monitoring and Assessment, 196*, 12765. https://doi.org/10.1007/s10661-024-12765-1

Queenan, K., Garnier, J., Nielsen, L., Buttigieg, S., Meneghi, D., Holmberg, M., Zinsstag, J., Rüegg, S., Häsler, B., & Kock, R. (2017). Roadmap to a one health agenda 2030. *CABI Reviews, 24*, 1–17. https://doi.org/10.1079/PAVSNNR201712014

Rahman, H., Carey, M., Hossain, M., Savage, L., Islam, M. R., & Meharg, A. A. (2019). Modifying the parboiling of rice to remove inorganic arsenic while fortifying with calcium. *Environmental Science and Technology, 53*(9), 5249–5255. https://doi.org/10.1021/acs.est.8b06548

Rahman, M. S., Reza, A. H. M. S., Siddique, M. A. B., Akbor, M. A., & Hasan, M. (2023). Accumulation of arsenic and other metal(loid)s in soil and human consumable foods of Meherpur district, Southwestern Bangladesh, and associated health risk assessment. *Environ Sciences Europe, 35*, 47. https://doi.org/10.1186/s12302-023-00751-2

Ramish, Irshad, M.A., Nawaz, R., Nasim, I., Irfan, A., Hussain, A., Al-Mutairi, A.A., Rizwan, M., Al-Hussain, S.A., & Zaki, M.E.A. (2024). Bioaccumulation of carcinogenic metal(loid)s in river fish: A quantitative investigation of public health risk. *Ecological Indicators, 162*, 112057. https://doi.org/10.1016/j.ecolind.2024.112057

Rashid, A., Schutte, B. J., Ulery, A., Deyholos, M. K., Sanogo, S., Lehnhoff, E. A., & Beck, L. (2023). Heavy metal(loid) contamination in agricultural soil: Environmental pollutants affecting crop health. *Agronomy, 13*(6), 1521. https://doi.org/10.3390/agronomy13061521

Rezapour, S., Moghaddam, S., Nouri, A., & Aqdam, K. K. (2022). Urbanization influences the distribution, enrichment, and ecological health risk of heavy metal(loid)s in croplands. *Scientific Reports, 12*, 3868. https://doi.org/10.1038/s41598-022-07789-x

Samlafo, V. B., Essel, K., & Ankudze, B. (2022). Health risk assessment of heavy metal(loid)s via consumption of cassava, cultivated on reclaimed mining land sites in Prestea-Huni Valley District. *Ghana Food and Public Health, 12*(1), 7–13. https://doi.org/10.5923/j.fph.20221201.02

Setia, R., Dhaliwal, S. S., Singh, R., Kumar, V., Taneja, S., Kukal, S. S., & Pateriya, B. (2021). Phytoavailability and human risk assessment of heavy metal(loid)s in soils and food crops around Sutlej river, India. *Chemosphere, 263*, Article 128321. https://doi.org/10.1016/j.chemosphere.2020.128321

Shaheen, N., Hasan, T., Sultana, M., Akhter, K. T., Khan, I. N., Irfan, N. M., & Ahmed, M. K. (2024). Carcinogenic and non-carcinogenic health hazards of potentially toxic elements in commonly consumed rice cultivars in Dhaka city, Bangladesh. *PLoS ONE, 19*(5), Article e0303305. https://doi.org/10.1371/journal.pone.0303305

Shahzad, S., Ashraf, K., Ehsan, N., Sultan, K., Abbasi, A., Tabassum, S., Muzamil, A., Rebouh, N. Y., & Zaman, Q. U. (2025). Assessment of hazardous trace metal(loid)s and associated health risk as affected by feed intake in buffalo milk. *Scientific Reports, 15*, 9841. https://doi.org/10.1038/s41598-025-92256-6

Shahriar, S.M.S., Haque, N., Hasan, T., Sufal, M.T.A., Hassan, M.T., Hasan, M., Salam, S.M.A. (2025). Heavy metal pollution in poultry feeds and broiler chickens in Bangladesh. *Toxicology Reports 14,* 101932. https://doi.org/10.1016/j.toxrep.2025.101932

Sharma, S. S., Dietz, K. J., & Mimura, T. (2016). Vacuolar compartmentalization as indispensable component of heavy metal(loid) detoxification in plants. *Plant, Cell and Environment, 39*(5), 1112–1126. https://doi.org/10.1111/pce.12706

Sharma, A., Gupta, S., Shrivas, K., Chakradhari, S., Pervez, S., & Deb, M. K. (2025). Heavy metal(loid) contamination in cow and buffalo milk from industrial and residential areas of raipur, India: A health risk assessment. *Food and Chemical Toxicology, 196*, Article 115178. https://doi.org/10.1016/j.fct.2024.115178

Shi, X., Lin, Q., Deng, P., Feng, T., & Zhang, Y. (2022). Assessment of heavy metal(loid) Uptake in potatoes cultivated in a typical Karst Landform, Weining County China. *Foods, 11*(15), 2379. https://doi.org/10.3390/foods11152379

Singh, V., Ahmed, G., Vedika, S., Kumar, P., Chaturvedi, S. K., Rai, S. N., Vamanu, E., & Kumar, A. (2024). Toxic heavy metal(loid) ions contamination in water and their sustainable reduction by eco-friendly methods: Isotherms, thermodynamics and kinetics study. *Scientific Reports, 14*, 7595. https://doi.org/10.1038/s41598-024-58061-3

Statista. (2025). Global gross domestic product (GDP) at current prices from 1985 to 2029 https://www.statista.com/statistics/268750/global-gross-domestic-product-gdp/. Reterieved March 30, 2025

Swan, T., McBratney, A., & Field, D. (2024). Linkages between soil security and one health: Implications for the 2030 sustainable development goals. *Frontiers in Public Health, 12*, 1447663. https://doi.org/10.3389/fpubh.2024.1447663

Taiwo, A.M., Oyeleye, O.F., Majekodunmi, B.J., Anuobi, V.E., Afolabi, S.A., Idowu, O.E., Ojekunle, Z.O., Taiwo, O.T. (2019). Evaluation of health risk of metals (Zn, Cr, Cd, Ni, Pb) in staple foods from Lagos and Ogun States, Southwestern Nigeria. *Environmental Monitoring and Assessment 191,* 167. https://doi.org/10.1007/s10661-019-7307-5

Tao, C., Wei, X., Zhang, B., Zhao, M., Wang, S., Sun, Z., Qi, D., Sun, L., Rajput, S. A., & Zhang, N. (2020). Heavy Metal(loid) content in feedstuffs and feeds in Hubei Province China. *Journal of Food Protection, 83*(5), 762–766. https://doi.org/10.4315/0362-028X.JFP-18-539

TatahMentan, M., Nyachoti, S., Scott, L., Phan, N., Okwori, F. O., Felemban, N., & Godebo, T. R. (2020). Toxic and essential elements in rice and other grains from the United States and other countries. *International Journal of Environmental Research and Public Health, 17*(21), 8128. https://doi.org/10.3390/ijerph17218128

United Nations. (2007). Industrial Development for the 21st Century (p. 432). Sustainable Development Perspectives, United Nations.

Wang, P., Wang, X., Wang, H., & Zhang, G. (2024). Health risk assessment of heavy metal(loid)s from grains and grain products in Gansu Province, China, 2012–2020. *Journal of Food Composition and Analysis, 125*, Article 105851. https://doi.org/10.1016/j.jfca.2023.105851

Wei, R., Chen, C., Kou, M., Liu, Z., Wang, Z., Cai, J., & Tan, W. (2023). Heavy metal(loid) concentrations in rice that meet safety standards can still pose a risk to human health. *Communications Earth and Environment, 4*, 84. https://doi.org/10.1038/s43247-023-00723-7

Wieczorek, J., Baran, A., & Bubak, A. (2023). Mobility, bioaccumulation in plants, and risk assessment of metal(loid)s in soils. *Science of the Total Environment, 882*, Article 163574. https://doi.org/10.1016/j.scitotenv.2023.163574

World Bank (2024). Agriculture, forestry, and fishing, value added (% of GDP) https://data.worldbank.org/indicator/NV.AGR.TOTL.ZS. Retrieved September 30, 2024, Retrieve March 30, 2025.

Xie, N., Kang, C., Ren, D., & Zhang, L. (2022). Assessment of the variation of heavy metal(loid) pollutants in soil and crop plants through field and laboratory tests. *Science of the Total Environment, 811*, Article 152343. https://doi.org/10.1016/j.scitotenv.2021.152343

Xu, S., Li, C., Wang, Y., Wu, A., Gao, G., Zang, F. (2024). Characteristics and evaluation of heavy metal pollution in a soil-wheat system of an arid oasis city in northwest China. *Ecotoxicology and Environmental Safety 271,* 115958. https://doi.org/10.1016/j.ecoenv.2024.115958

Zeinali, T., Salmani, F., & Naseri, K. (2019). Dietary intake of cadmium, chromium, copper, nickel, and lead through the consumption of meat, liver, and kidney and assessment of human health risk in Birjand, Southeast of Iran. *Biological Trace Element Research, 191*, 338–347. https://doi.org/10.1007/s12011-019-1637-6

Zhang, Y., Yin, C., Cao, S., Cheng, L., Wu, G., & Guo, J. (2018). Heavy metal(loid) accumulation and health risk assessment in soil-wheat system under different nitrogen levels. *Science of the Total Environment, 622–623*, 1499–1508. https://doi.org/10.1016/j.scitotenv.2017.09.317

Zhang, H., Zhou, G., Zhang, S., Yang, Y., Dev, S., Su, Q., Deng, X., Chen, Q., & Niu, B. (2022). Risk assessment of heavy metal(loid)s contamination in pork. *Food Control, 135*, Article 108793. https://doi.org/10.1016/j.foodcont.2021.108793

Zhou, M., Liu, H., Liu, B., Zhang, M., Tang, Z., Zong, L., Wang, S., Niu, X., & Tian, F. (2024). Geochemical characterization of heavy metal(loid)s in the soil rice system and risk assessment for human health. *Environmental Pollutants and Bioavailability, 36*(1), 2368592. https://doi.org/10.1080/26395940.2024.2368592

Zhuang, P., Zou, B., Lu, H., & Li, Z. (2014). Heavy metal(loid) concentrations in five tissues of chickens from a mining area. *Polish Journal of Environmental Studies, 23*, 2375–2379.

Chapter 6
Spatial Analysis of Social Vulnerability to Flood Hazards in Sylhet Division, Bangladesh: A PCA-Based Index Approach

Rifat Bin Hossain, Md. Iqbal Hasan, Tasneem Sharmin, Nafisa Nandini, and Md. Juhayr Akbar

Abstract Flood vulnerability assessment is critical for effective disaster risk management in flood-prone regions like Sylhet Division, Bangladesh. This study applied a multi-stage, index-based approach grounded in Principal Component Analysis (PCA) to evaluate social vulnerability across the division at the upazila level. The framework integrated three key dimensions: socio-economic characteristics, infrastructure, and exposure, with indicators selected based on local relevance, data availability, and literature review. Indicators included demographic variables (age, gender, employment), urbanization metrics, population density, and access to safe drinking water, sourced from Bangladesh Bureau of Statistics, satellite remote sensing, and national hazard reports. Data normalization and PCA were applied, with three principal components explaining 76% of the variance: socio-economic (46.7%), infrastructure (17.3%), and exposure (12.1%). Weights derived from factor loadings were used to construct a composite Social Vulnerability Index (SoVI), mapped at the upazila level using ArcGIS. Results reveal significant spatial heterogeneity in vulnerability. Socio-economic factors were the most influential, with 26.83% of upazilas classified as highly vulnerable due to aging populations, unemployment, and gender disparities. Infrastructure vulnerability (7.32% extreme vulnerability) was linked to unplanned urbanization in floodplains, while exposure vulnerability

R. B. Hossain (✉) · Md. I. Hasan · T. Sharmin · N. Nandini · Md. J. Akbar
Department of Geography and Environment, University of Dhaka, Dhaka, Bangladesh
e-mail: rifat-63-2017616638@geoenv.du.ac.bd

Md. I. Hasan
e-mail: mdiqbal-63-2017916671@geoenv.du.ac.bd

T. Sharmin
e-mail: tasneem-65-2019717634@geoenv.du.ac.bd

N. Nandini
e-mail: nafisa-66-2020517436@geoenv.du.ac.bd

Md. J. Akbar
e-mail: mdjuhayr-66-2020017404@geoenv.du.ac.bd

G. K. Panda et al. (eds.), *Socio-Ecological Crisis of Development*, Sustainability Solutions, https://doi.org/10.1007/978-3-032-22171-1_6

(14.63% extreme) correlated with high population density and limited water access. Composite vulnerability mapping identified hotspots, with over 51% of the region falling into high or moderate vulnerability categories. The study recommends the need for targeted interventions, prioritizing socio-economic resilience alongside infrastructure and exposure mitigation. The upazila-level assessment provides granular insights for policymakers to allocate resources effectively and enhance flood risk management strategies in vulnerable communities.

Keywords Flood vulnerability · Social vulnerability index (SoVI) · Principal component analysis (PCA) · Sylhet division · Geospatial risk mapping

6.1 Introduction

Floods are considered the most frequent natural disaster worldwide, affecting approximately 3.2 billion people and resulting in more than 218,000 deaths and $1.3 trillion in economic losses between 1990 and 2022 (Dewa et al., 2022). The greatest repercussions of floods are found in Asia and Africa, with low-income countries facing increased mortality and vulnerability (Liu et al., 2024). Global projections indicate that climate change, combined with rapid urbanization and population growth, will disproportionately increase flood risk exposure in low-income and densely populated countries (Xie & Meng, 2023) and are anticipated to raise no matter how much efforts have been put in to implement flood protection measures (Ajtai et al., 2023). Although earlier researches focused mainly on hazard analysis, flood reduction, and control recent flood hazard assessment especially after the inauguration of the Flood Directive of the European Parliament and Council, adopts a broader perspective that address all key components including hazard, exposure, and vulnerability (Lund et al., 2021; Pittore et al., 2020).

Flood vulnerability assessment is critical for gaining a better understanding of potential flood risks and developing mitigation methods (Ajtai et al., 2023). Throughout the years, the physical vulnerability to flooding has received significant attention, with most studies focusing on building location, infrastructure resilience, drainage, and land use analysis through vulnerability curves, indices, matrices, and modeling tools (Nguyen et al., 2023). Alternatively, studies focusing on the ecological susceptibility to flood emphasized on the consequences of environmental degradations and loss of natural floodplain to increase the floods risk (Abdrabo et al., 2023; Sarker et al., 2025; Wu, 2021). Besides, various research has explored the damage that future flood catastrophes might cause in the background of climate and socio-economic changes (Arnell & Gosling, 2016; Kam et al., 2021; Pariartha et al., 2023; Vousdoukas et al., 2018). However, over the years, the dimension of social vulnerability has received less attention. The concept of "social vulnerability to flood" describes how individuals, communities, or groups may be vulnerable to the devastating effects of floods. This vulnerability is influenced by institutional, demographic, and socio-economic factors and includes aspects of exposure, sensitivity, and adaptability (Bucherie et al., 2022; Spielman et al., 2020).

The problem of flooding in Bangladesh is multi-layered as it is affected by natural and human causes. It leads to extensive human, agricultural, and economic destruction, and the most vulnerable populations are rural populations. Sylhet division of Bangladesh is especially vulnerable to the extreme occurrence of flood events on a regular basis due to the topography and hydroclimatic characteristics. Over the last number of years, the Sylhet division experienced historic floods that have had a devastating impact on the people, economy, and the environment of the region (Shafiq, 2023). Some of the effects are massive displacement, crop destruction, poverty, food insecurity, and multiple mental health challenges (Howlader et al., 2024). Inundation caused by floods, especially the 2017 and 2022 ones, covered up to 90% of Sylhet Division displacing millions of people and causing enormous losses of crops (Saunders et al., 2023). Although the study of flood vulnerability in Bangladesh is developing, there are still major gaps in knowledge and the analysis of social vulnerabilities that curtail the performance of disaster management and resilience-building strategies (Orru et al., 2022) especially in the Sylhet haor area. The majority of the studies followed a theoretical focus and tend to reduce social vulnerability to only a few elements, including physical exposure and simple coping capacity, instead of the existing variety of social, economic, and institutional impacts (Aktar et al., 2021; Dhawale et al., 2024; Younus, 2014). In this connection, multidimensional and interconnecting factors in social vulnerability studies may turn into significant milestone to predict the ransom of flood disasters, and act and implement mitigating and sustainable measures and policy.

The relevance of formulating detailed methodologies of measuring social vulnerability has also been emphasized and vulnerability indices have been used as the most common analytical measure. They consist of a wide range of indicators that present social, demographic, economic, and built-environment characteristics and, hence, impact community resilience (Bronfman et al., 2024; Bucherie et al., 2022; Derakhshan et al., 2022). The vulnerability assessment is highly dependent on the choice, weighting, and aggregation of variables (Azizi et al., 2023). As indicators do not equate the contribution to vulnerability, a variety of studies have embraced a heterogeneous approach to weighting, whereby the Principal Component Analysis (PCA) remains the most commonly used approach (Abdrabo et al., 2023; Islam & Meng, 2024; Wu, 2021). To illustrate, Torok (2018) estimated the social vulnerability to floods in Romania by generating a PCA-based index of 27 variables and enhanced the evaluation with Geographically Weighted Regression (GWR) to query the spatial correlations between highly vulnerable areas and flood impacted areas.

Similarly, by comparing the aggregation of the component scores, the first score of the components and weighted sums, Machado and Ratick (2018) illustrated how sensitive vulnerability indices are toward the aggregation processes.

Recognizing the spatial–temporal variety of variables, it is difficult to construct a universally used strategy for analyzing social vulnerability. The research in the field is still characterized by various uncertainties that are less addressed, such as: subjective selection of indicators, exclusion of the hazard dimension, equal or subjective weighting methods, and a decreased number of variables due to unavailability of data. Although various studies addressed these difficulties separately, there are a few complicated ways that aimed to eliminate all the gaps stated above. Besides, the majority of the flood risks assessments carried out in Bangladesh are at national/ district level with a limited scope of assessing the physical aspects of the hazards (Haque et al., 2022; Hossain, 2025; Islam & Chowdhury, 2024; Rahman & Salehin, 2013). Such a narrow lens overlooks major intra-regional differences in vulnerability, and it cannot take into consideration the development of upazila-level disaster management (Aroca-Jiménez et al., 2020). Moreover, the full triad of vulnerability—exposure, sensitivity, and coping capacity has yet to be analyzed within one analytical framework.

Therefore, the goal of this study is to build a comprehensive index-based methodology approach grounded in Principal Component Analysis (PCA) for analyzing and mapping the social flood vulnerability, solving the main gaps in the field. Its novelty lies in the fine-scale, upazila-level analysis that integrates socio-economic, infrastructural, and exposure-related factors. As such, the study aims at (a) developing a composite SoVI through the use of PCA to assess the upazila-level vulnerability to floods, (b) identifying the important socio-economic and infrastructural and exposure variables at the upazila level, and (c) mapping the spatial distribution of flood vulnerability to help identify the high-risk areas. The research helps in developing the local resilience strategies, which is stronger as the most vulnerable segments are established. Its findings provide a scientifically robust basis on data-driven policy, resource movement, and adaptive planning—major actions on community resilience improvement and a sustainable inclusive development of flood-prone areas spatially.

6.2 Study Area

The study was conducted in the Sylhet Division, located in the northeastern region of Bangladesh, encompassing four districts: Sylhet, Moulvibazar, Habiganj, and Sunamganj (Fig. 6.1). The division spans an area of approximately 12,635 square kilometers and is characterized by a diverse topography that includes low-lying haor wetlands, floodplains, river valleys, and hillocks. Hydroclimatic conditions and the topography of the Sylhet Division are unique and hence this area is highly susceptible to flooding. In 2017, 2022, and 2024, flash floods resulted in the widespread relocation, loss of crops, and loss of socio-economic activities (Rahman et al., 2024; Shafiq, 2023). The northeast region (Sunamganj, Sylhet, Kishoreganj, Netrokona, Brahmannbaria, and Moulovibazar) was inundated (55.76%) with the 2022 flood. It affected millions of people, ruined infrastructures, and farms (Rahman et al., 2024).

Vulnerability is aggravated by the fact that socio-economic problems prevent a community to manage and adapt to flood disasters. These problems are extreme poverty, poor educational levels, poor health care, and poor infrastructure. Rural

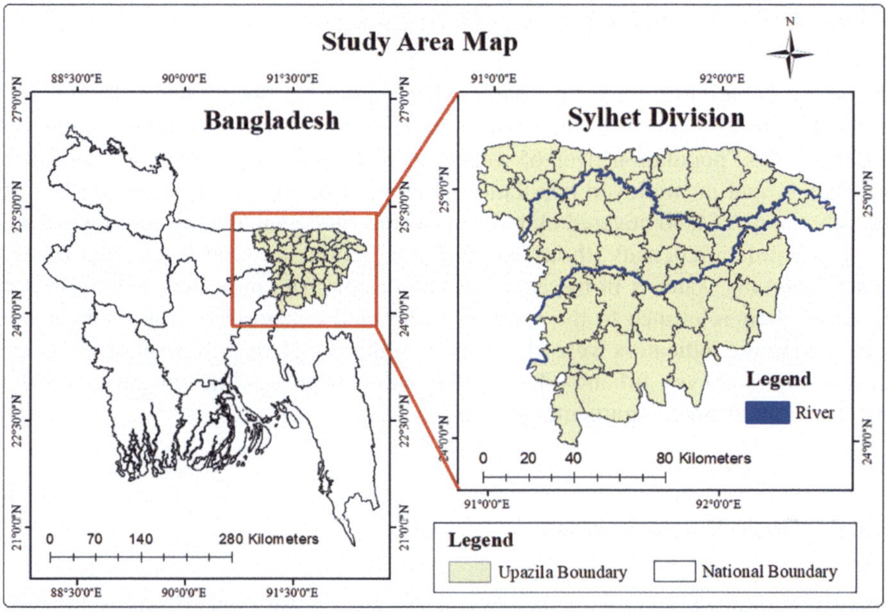

Fig. 6.1 Study area map of the Sylhet Division, Bangladesh

communities and vulnerable population are affected disproportionately after the flood, food insecurity, mental health, and limited access to official support networks. Flood susceptibility mapping in Sylhet by means of advanced geospatial/machine learning methods shows that there is an essential part of the terrain, particularly the northwestern haor areas, which are very susceptible to floods. Moreover, Rudra and Sarkar (2023) stated that over 78% of the area is of this type. The risk reduction and adaptation plans should prioritize locally focused and integrated vulnerability assessments, as both social and physical conditions significantly influence overall vulnerability.

6.3 Material and Methods

The social vulnerability to flood assessment for Sylhet Division was conducted using a multi-step, PCA-based index architecture similar to that of Ajtai et al. (2023). The aspects of vulnerability determination include socio-economic parameters, infrastructural, and exposure-related parameters. Data normalization, PCA analysis, index aggregation, and spatial mapping were carried out after the indicator selection process. To obtain a composite vulnerability score, the conventional index-based methodology was followed to combine multiple indicators. The analyses were conducted at the upazila level in Sylhet Division.

6.3.1 Indicator Selection and Data Sources

The study is based on three dimensions that capture the local factor of the study area. The first dimension is the socio-economic dimension, including population under 5 years, population over 65 years, share of women population, employed population, and unemployed population. The infrastructure dimension consists of the percentage of land covered by built-up surface and annual built-up growth rate. Finally, population density (thousand per km^2), population growth rate, and access to safe drinking water is in the exposure dimension. The indicators were selected based on their relevance to the state of Sylhet and categorized. The data for the socio-economic indicators were obtained from the Bangladesh Bureau of Statistics (census and surveys), urban/built-up information from satellite imagery, and the exposure data from governmental reports.

6.3.2 Data Normalization

Initially, normalization of the indicators on a 0–1 scale was done. It was established whether each variable indicated a positive (high value = low vulnerability) or negative (low value = high vulnerability) relationship.

For positively correlated indicators, the normalized score was calculated as

$$X_n = \frac{X - X_{\text{min}}}{X_{\text{max}} - X_{\text{min}}} \tag{6.1}$$

and for negatively correlated indicators as

$$X_n = \frac{X_{\text{max-}}X}{X_{\text{max}} - X_{\text{min}}} \tag{6.2}$$

where X is the original indicator value, and $X_{\text{min}}, X_{\text{max}}$ are its minimum and maximum observed values. The calculation ensuring the range of vulnerability value is from 0 (least vulnerable) to 1 (highest vulnerable). It also keeps the direction of influence of the vulnerability constant for all indicators (Fig. 6.2).

6.3.3 Principal Component Analysis (PCA)

Principal Component Analysis (PCA) was applied to reduce the size of the dataset and obtain statistically significant weights of each indicator. Before PCA was implemented, the appropriateness of the data belonging to factor analysis was determined through Kaiser–Meyer–Olkin (KMO) measure of sampling adequacy and Bartlett test of sphericity. KMO of more than 0.80 and statistically significant test Bartlett (p

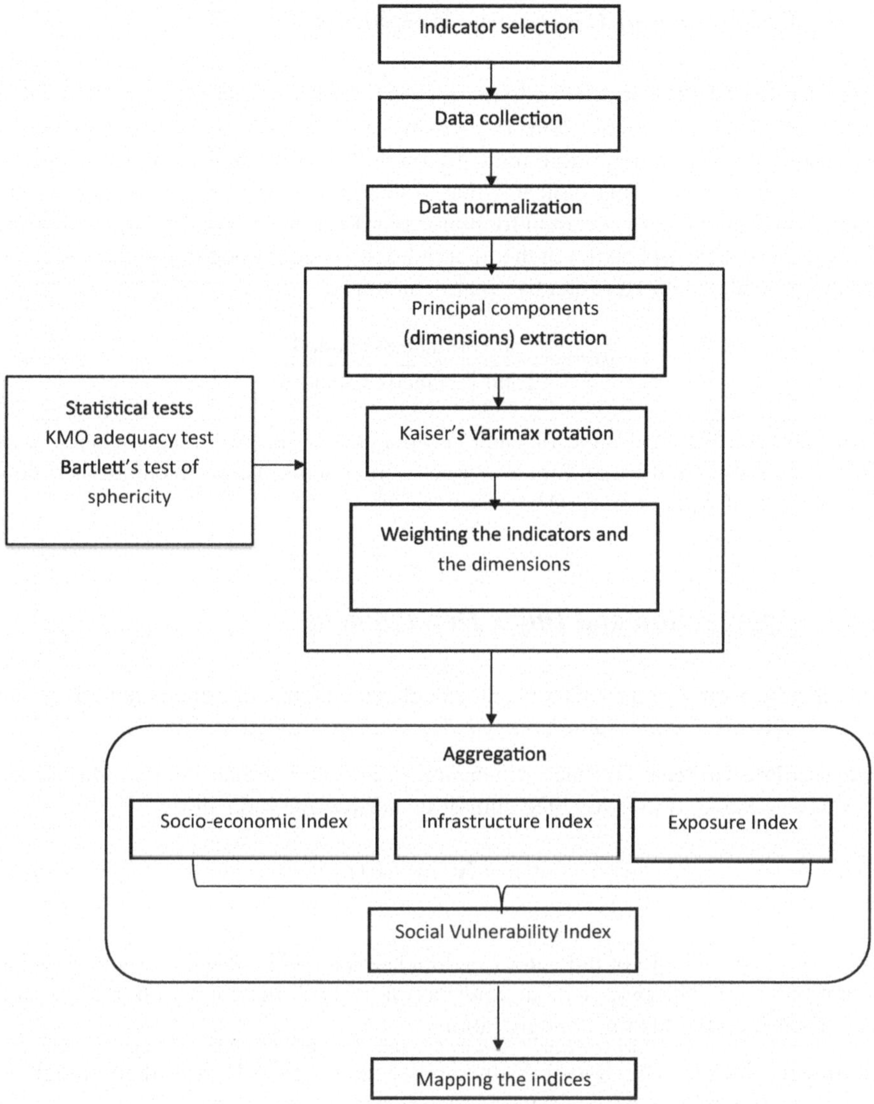

Fig. 6.2 Methodological workflow of the study

< 0.05) showed that the data was suitable in PCA (Kaiser, 1959). Components whose eigenvalues were greater than 1, according to the Kaiser criterion, were extracted to analyze it. Varimax (orthogonal) rotation was used in order to improve interpretability. Factors that had factor loadings of 0.50 or above were included in the further analysis (Ajtai et al., 2023). The retained principal components were then placed against one of three dimensions, which was socio-economic, infrastructure, or exposure, according to the strength and nature of the variable loading.

6.3.4 Indicator and Dimension Weighting

From the rotated PCA results, weights for each indicator and for each dimension were derived. In each dimension, the weight of each indicator was calculated as the squared factor loading of that indicator on the component, normalized to ensure that the weights total. This indicates that indications with a greater burden on the component (i.e., those that account for more volatility) are assigned to larger weights. The weight of each dimension in the final index was determined in proportion to the variance elucidated by its respective main component.

$$Wd = \frac{\%\ \text{variance explained}}{\text{Total variance explained.}} \tag{6.3}$$

In practice, the eigenvalue of each component, or its explained variance, was utilized in relation to the total for all retained components. These calculations follow the approach of Ajtai et al. (2023).

6.3.5 Aggregation and Index Development

Utilizing the normalized indicators and calculated weights, intermediate indices and the comprehensive flood vulnerability index were developed as follows:

Intermediate Indices: For each dimension d, an intermediate vulnerability index V_d was calculated as the weighted sum of its normalized indicators:

$$V_d - \sum_i W_{di} * I_{di} \tag{6.4}$$

where W_{di} is the weight of indicator i in dimension d and I_{di} is the normalized value of that indicator. The proportional contribution of each indicator to the dimension score is ensured by this linear aggregation.

Composite Index: The Social Vulnerability Index (SoVI) was determined by weighing and adding the three intermediate indices.

$$SoVI = W_1V_1 + W_2V_2 + W_3V_3 \tag{6.5}$$

where W_d is the weight of dimension d and V_d is the index. Each upazila receives a vulnerability score. The higher the vulnerability score, the higher the flood vulnerability. The continuous vulnerability scores were classified into qualitative categories (e.g., very low to very high). The Jenks natural breaks approach was used for the classification, which optimizes between-class variation and ensures that each upazila is allocated to one of five hierarchical vulnerability groups.

6.3.6 Spatial Mapping

The spatial distribution of the vulnerability indices was visualized using GIS techniques. ArcGIS 10.7 was used to map upazila-level SoVI scores after they were combined with administrative boundary shapefiles. Each upazila was symbolized according to its vulnerability class. This geospatial representation facilitates the identification of spatial patterns and priority zones for flood risk management and adaptation strategies (Xie & Meng, 2023).

6.4 Results

6.4.1 The Weighting Results Using PCA

The Kaiser–Meyer–Olkin (KMO) measure of sampling adequacy produced a value of 0.727, indicating meritorious suitability for factor analysis, according to the findings of the statistical analyses carried out in SPSS. Furthermore, Bartlett's test of sphericity was statistically significant ($p < 0.001$), confirming that the correlation matrix was appropriate for Principal Component Analysis (PCA). Three principal components were extracted using PCA in this study, keeping only those with eigenvalues greater than 1 (Table 6.1 and Fig. 6.3). Together, these factors accounted for 76% of the dataset's total variance, with the first principal component explaining 46.7%, the second for 17.3%, and the third for 12.1% (Table 6.1).

To enhance interpretability, a rotated component matrix (using Varimax rotation) was applied to identify the key indicators associated with each component, with significant factor loadings highlighted (Table. 6.2). Different dimensions of vulnerability were defined using indicators that showed the highest loadings (>0.5) within each component. These indicators' clustering led to the following conceptualization of the principal components: (1) the socio-economic dimension (PC1), encompassing demographic indicators such as population composition, female population, and employment status; (2) the infrastructure dimension (PC2), comprising metrics related to built-up area expansion; and (3) the exposure dimension (PC3), which included indicators pertaining to social development and resource accessibility (Table. 6.3).

For each dimension, by normalizing the squared factor loadings to a unity sum, the indicator weights were derived. Additionally, the relative contribution of each dimension was calculated using Eq. (6.4) (Table. 6.3). Within the socio-economic dimension, the highest weights were assigned to age structure and female population indicators, reflecting their substantial influence on vulnerability. These factors critically shape a community's adaptive capacity and resilience. On the contrary, in the exposure dimension, the indicator for access to drinking water was given the lowest weight, likely due to its comparatively indirect association with vulnerability. In the

Table 6.1 The principal components extracted

Total variance explained							Total variance explained			
Component	Initial eigenvalues			Extraction sums of squared loadings			Component	Rotation sums of squared loadings		
	Total	% of Variance	Cumulative %	Total	% of Variance	Cumulative %		Total	% of Variance	Cumulative %
1	4.667	46.666	46.666	4.667	46.666	46.666	1	4.664	46.64	46.64
2	1.727	17.272	63.938	1.727	17.272	63.938	2	1.642	16.417	63.056
3	1.211	12.106	76.045	1.211	12.106	76.045	3	1.299	12.988	76.045
4	0.898	8.978	85.023				4			
5	0.78	7.803	92.826				5			
6	0.395	3.947	96.772				6			
7	0.148	1.484	98.256				7			
8	0.118	1.181	99.437				8			
9	0.047	0.471	99.908				9			
10	0.009	0.092	100				10			

Kaiser–Meyer–Olkin Measure of Sampling Adequacy (KMO): 0.727
Bartlett's Test of Sphericity, Sig. (0.0001)

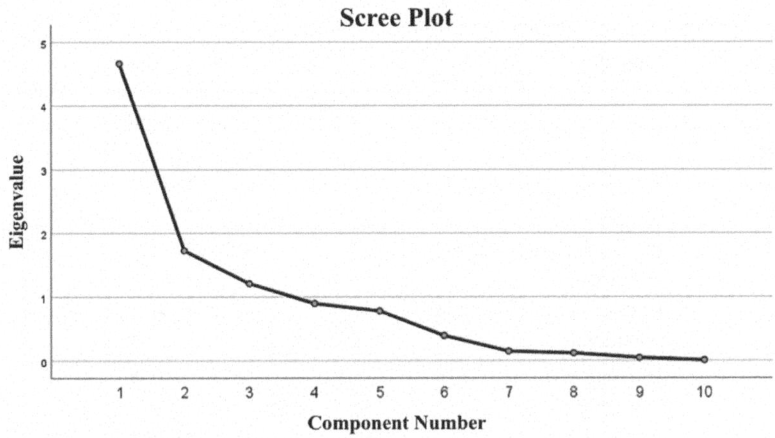

Fig. 6.3 Scree plot of eigenvalues explained by the resulting components

Table 6.2 Rotated component matrix from PCA and factor loadings after rotation

Rotated component matrix			
	Component		
	PC1	PC2	PC3
Population density	0.238	−0.196	0.675
Population under 5	0.939	−0.073	−0.088
Population over 65	0.954	0.117	0.095
Share of women	0.992	−0.008	0.058
Employed population	0.935	0.067	0.179
Unemployed population	0.918	0.067	0.123
Access to drinking water	−0.304	0.447	0.536
Population growth rate	−0.021	−0.144	0.666
Percentage of build-up	−0.025	0.791	−0.109
Built-up growth rate	−0.144	0.854	0.182

Extraction Method: Principal Component Analysis
Rotation Method: Varimax with Kaiser Normalization

infrastructure dimension, the built-up growth rate was the most heavily weighted indicator due to its significance in shaping spatial vulnerability.

Within the exposure dimension (PC3), the population growth rate (factor loading = 0.666) and population density (loading = 0.675) were the most influential indicators, underscoring their critical role in capturing the pressures exerted by increasing demographic concentrations in flood-prone regions. Access to drinking water (loading – 0.536) also contributes moderately to this dimension, implying that during flood events, vulnerability increases for a deficit in water accessibility.

Table 6.3 Weight calculation based on the rotated component matrix

Dimensions	Indicators	PC1	PC2	PC3	Indicator Weight	Dimension Weight
Socio-economic dimension	Population under 5 years	0.939			0.198184888	0.47
	Population over 65 years	0.954			0.201350781	
	Share of women population	0.992			0.209371043	
	Employed population	0.935			0.19734065	
	Unemployed population	0.918			0.193752638	
Infrastructure dimension	Percentage of built-up		0.791		0.480851064	0.16
	Built-up growth rate		0.854		0.519148936	
Exposure dimension	Population density			0.675	0.359616409	0.13
	Population growth rate			0.666	0.354821524	
	Access to drinking water			0.536	0.285562067	

The infrastructure dimension (PC2) is strongly associated with built-up area expansion, as evidenced by high factor loadings for the percentage of built-up area (0.791) and built-up area growth rate (0.854). These indicators show a unique aspect of exposure associated with urbanization patterns, where flood risk is increased by rapid and unplanned physical development. With the highest percentage of explained variance, the socio-economic dimension (PC1) is the most significant factor contributing to vulnerability at the community level. This component shows exceptionally high factor loadings for female population (0.992), population under 5 years (0.939), population over 65 years (0.954), employed population (0.935), and unemployed population (0.918). Together, these indicators demonstrate the increased socio-economic sensitivity of communities with higher percentages of economically marginalized (such as unemployed people) and demographically dependent groups (such as young and elderly populations). The size of these loadings indicates that these populations have limited resilience and recovery potential, making them disproportionately vulnerable to the effects of flooding. Consequently, the socio-economic dimension is assigned the highest weight in the composite vulnerability index. This result highlights the crucial interaction between flood exposure and socioeconomic fragility, as economic and demographic vulnerabilities greatly increase overall susceptibility throughout the upazilas under study.

6.4.2 Flood Vulnerability Index Calculation and Spatial Analysis

In order to assess flood vulnerability, this study integrated three intermediate indices that were derived from specific indicators, each of which was weighted according to its relative importance. These indices were then combined to construct the overall Social Vulnerability Index (SoVI), with Fig. 6.4 illustrating the contribution of individual indicators to their respective dimensions. While overall variability among indicators appears moderate, several demonstrate significant spatial heterogeneity in their influence. For example, in upazilas such as Madhyanagar, Chunarughat and Shayestaganj, the proportion of built-up land appeared as a key driver of vulnerability, underscoring the role of urbanization and infrastructure density in shaping flood risk. The spatial distribution of the SoVI, shown in Fig. 6.8, enables a separated analysis of vulnerability patterns across upazilas, highlighting localized risk factors. These insights are essential for evidence-based flood risk management and the prioritization of targeted adaptation strategies.

6.4.3 Socio-Economic Vulnerability Index Analysis

The analysis of the socio-economic vulnerability index (Fig. 6.4) reveals spatial disparities in vulnerability levels across the studied upazilas, with 21.95% of Local Administrative Units (LAUs) exhibiting moderate vulnerability and a substantial 26.83% categorized as highly vulnerable. In contrast, 24.4% demonstrate low vulnerability, while 14.63% fall into the very low category. It is interesting to note that, despite having higher population densities, urban upazilas typically exhibit moderate levels of vulnerability. This is mainly because of two mitigating factors, the presence of strong water supply infrastructure and higher employment rates, both of which contribute to enhanced adaptive capacity by guaranteeing access to potable water during floods and maintaining household economic stability. However, the analysis also identifies 11 upazilas facing extreme vulnerability, largely driven by a high proportion of elderly residents and elevated unemployment rates. These conditions are closely linked to selective outmigration patterns where younger, working-age individuals have moved to urban centers or emigrated internationally (notably to the United Kingdom and the Middle East) for employment opportunities. Rural areas now have aging populations and lower economic productivity as a result of this demographic shift. Moreover, spatial patterns indicate a correlation between high vulnerability and geographic location, particularly within the ecologically sensitive Haor basin, where flood risks are exacerbated by seasonal inaccessibility during monsoons and chronic infrastructure deficits. The combination of aging populations, economic stagnation, and inadequate infrastructure creates a multifaceted vulnerability scenario that greatly increases exposure to and sensitivity to hydrological hazards in these areas.

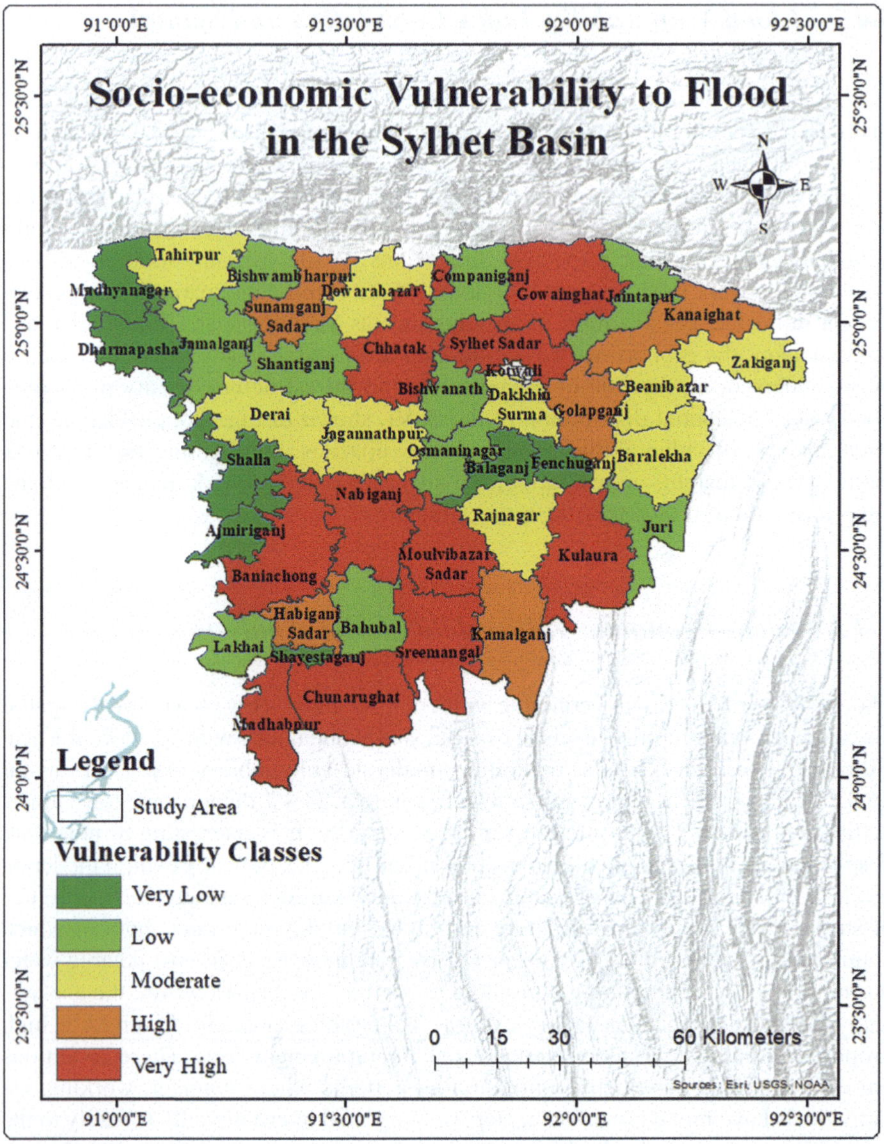

Fig. 6.4 Socio-economic vulnerability map

6.4.4 Infrastructure Vulnerability Index Analysis

The Sylhet Division exhibits significant spatial heterogeneity in flood susceptibility, according to the infrastructure vulnerability assessment (Fig. 6.5), which incorporates indicators like the percentage of built-up areas and urban expansion rates. According to the findings, 7.32% of upazilas are considered to be extremely vulnerable. This is mainly because of the fast urbanization that is taking place in hydrologically sensitive areas, which greatly raises the risk of flooding. The overall distribution follows a clear gradient: 36.6% of upazilas are highly vulnerable, 24.4% moderately vulnerable, 12.2% exhibit low vulnerability, and 19.51% are classified as minimally vulnerable. This pattern emphasizes a crucial connection between uncontrolled urban growth and increased flood risk, particularly in areas like alluvial floodplains where significant anthropogenic landscape modification has interfered with natural hydrological processes. The most severe vulnerabilities are concentrated in downstream, low-lying upazilas within the Haor basin, where geomorphological features—such as inadequate topographic relief and poor natural drainage capacity—combine with urban expansion to intensify floodwater retention and escalate the magnitude of flood events.

6.4.5 *Exposure Vulnerability Index Analysis*

The exposure vulnerability assessment (Fig. 6.6) identifies specific spatial patterns of flood susceptibility in Sylhet Division, emphasizing the relationship between demographic pressure and infrastructural limitations. Six upazilas, representing 14.63% of the study area, demonstrate significant vulnerability due to concurrent high population growth rates (mean annual increase of 2.3%) and population densities surpassing 800 persons per square kilometer.

In rapidly urbanizing areas, where uncontrolled development in flood-prone areas has greatly increased infrastructure and population exposure to hydrological hazards, the vulnerabilities are especially noticeable. A secondary cluster of exposure is identified in 6.5% of upazilas classified as highly vulnerable. These areas exhibit constant population growth rates of 1.5–2.0% annually and intermediate population densities ranging from 500 to 800 persons per square kilometer. Demographic pressures are exacerbated by deficiencies in access to safe drinking water, with only 62% of households enjoying reliable year-round access. The risks to public health during flood events are greatly increased by this circumstance. Conversely, most upazilas exhibit reduced exposure levels, with 24.4, 31.7, and 17.1% classified as moderate, low, and very low vulnerability, respectively. The regions exhibit stable demographic characteristics, with growth rates under 1.2%, low settlement densities of fewer than 400 persons per square kilometer, and balanced ratios of population to infrastructure. These factors collectively alleviate pressure on essential services and improve resilience against flood risks. According to the assessment, protective infrastructure has a moderating effect in a number of upstream upazilas where effective flood defense systems, like redundant water supply networks and engineered embankments that shield 65–80% of settlements, reduce significant theoretical exposure. This emphasizes the need for thorough assessments of flood vulnerability to include both exposure indicators and adaptive capacity measuresFig. 6.7).

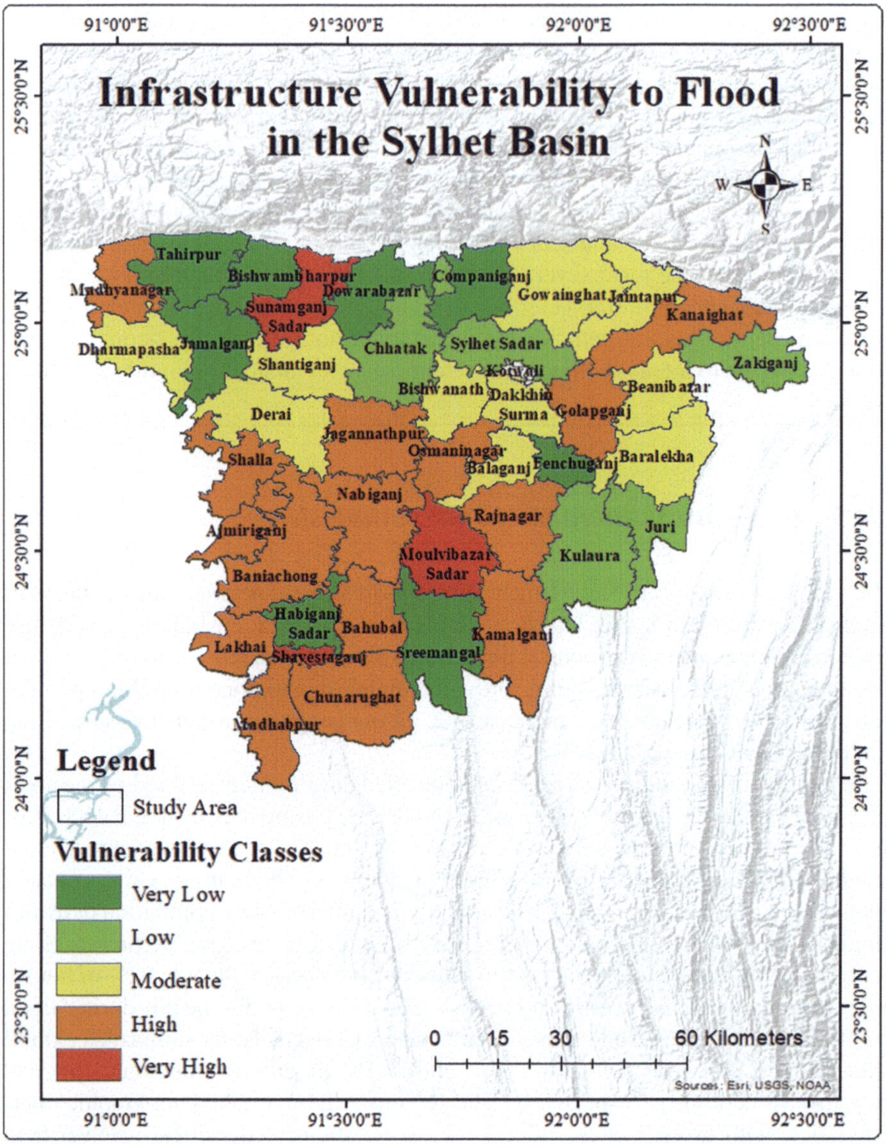

Fig. 6.5 Infrastructure vulnerability map

6.4.6 *Composite Social Vulnerability Assessment*

The social vulnerability map of the Sylhet Basin (Fig. 6.8) was developed by integrating socio-economic vulnerability with factors associated with infrastructure and

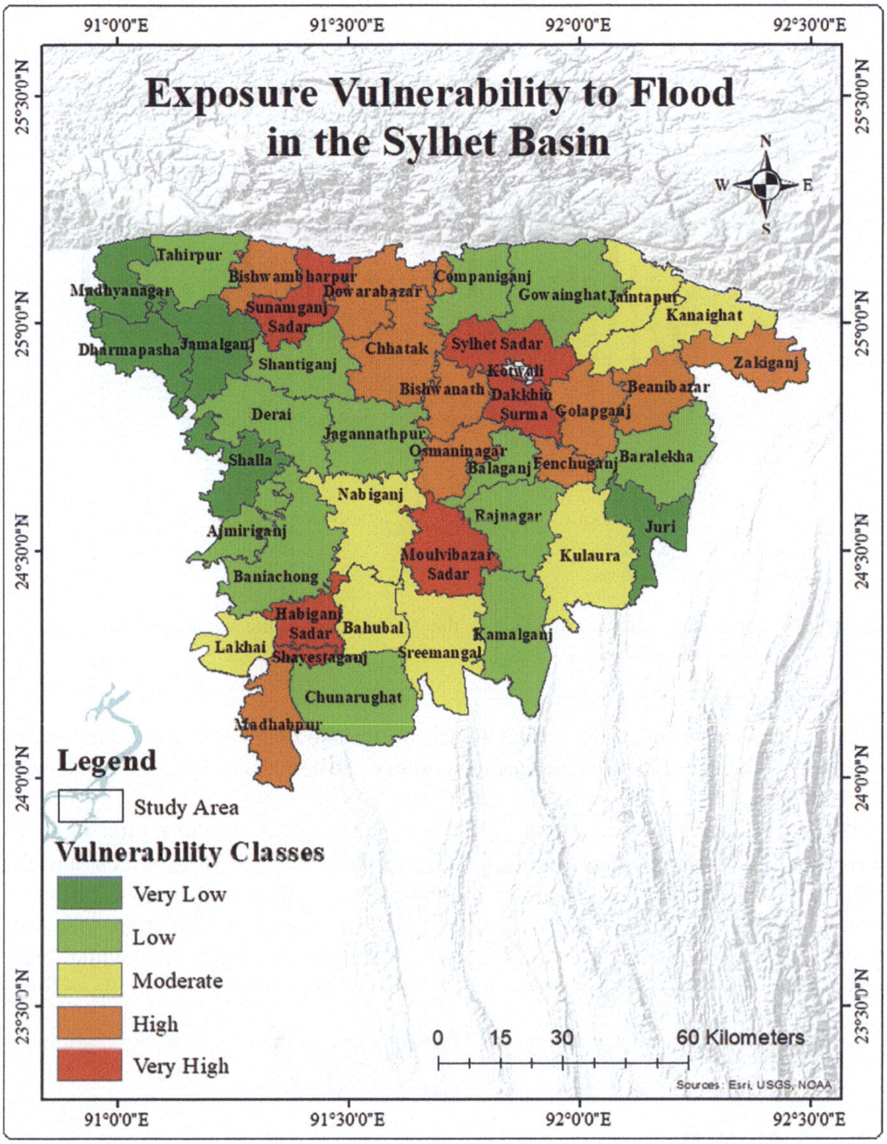

Fig. 6.6 Exposure vulnerability map

exposure, following the methodology outlined in Eq. 6.5. Broadly, infrastructure indicators are reflective of the social environment, while exposure-related indicators align with the physical environment. As these elements are often challenging to modify

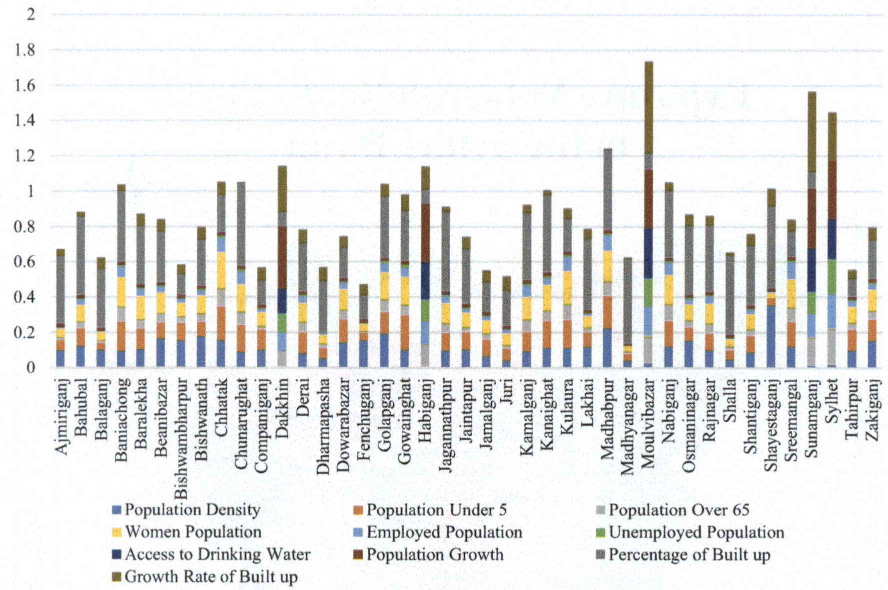

Fig. 6.7 The contribution of each indicator to the development of the intermediate vulnerability indices in the analyzed upazilas

directly to reduce vulnerability, policy interventions should prioritize reducing socio-economic vulnerability by enhancing community resilience and adaptive capacity to flood hazards.

The spatial patterns reveal noticeable heterogeneity across the region in terms of population distribution, community characteristics, access to resources, and the extent of flood-affected areas. This variation results in a non-uniform distribution of vulnerability levels across the upazilas. A significant portion of the study area falls into the high (26.83%) and moderate (24.4%) vulnerability categories, while fewer upazilas are classified under very high (12.2%) and very low (17.1%) vulnerability classes.

6.5 Discussion

The research problem of the study was to measure the spatial extent of social vulnerability to flood hazards in the Sylhet Division of Bangladesh using a composite index based on Principal Component Analysis (PCA). Social vulnerability, particularly in flood-prone areas like Sylhet, is influenced by specific hydro-meteorological and topographical conditions, making it a complex and localized phenomenon. This

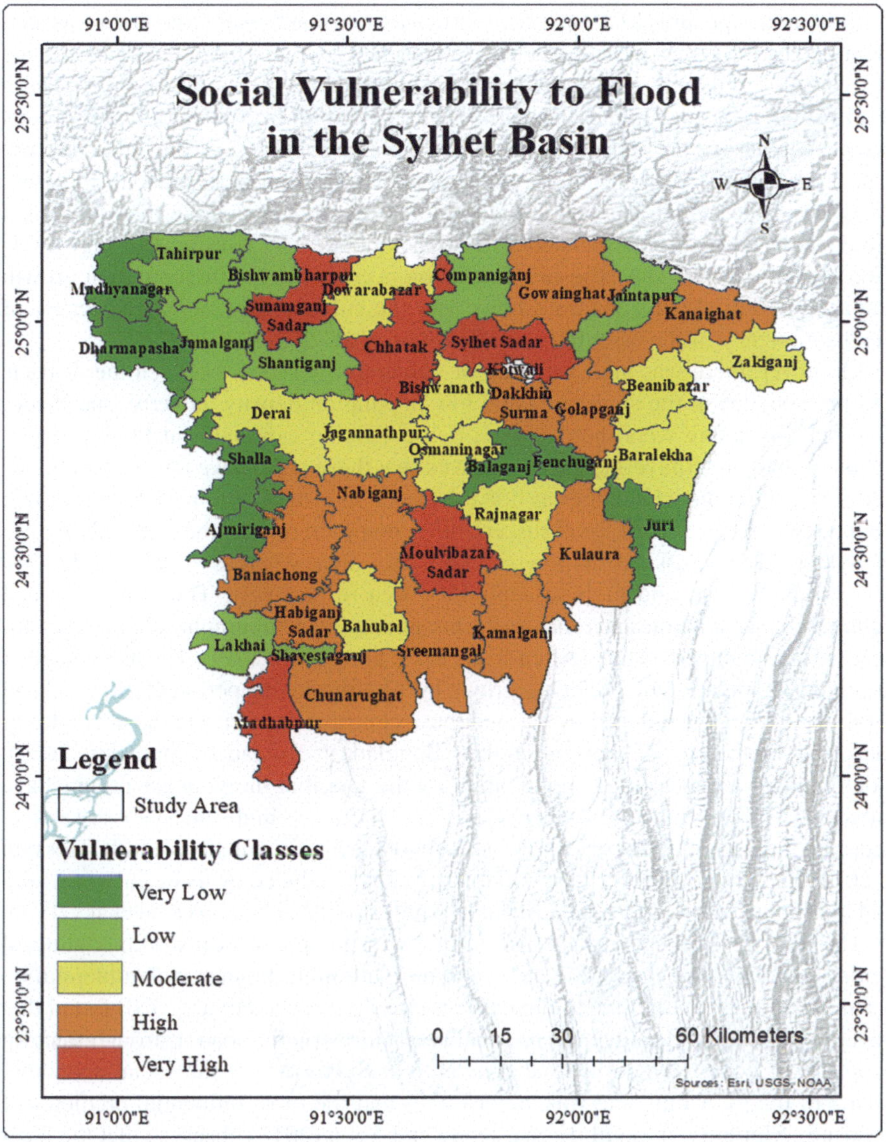

Fig. 6.8 Social vulnerability to flood hazards in the Sylhet Division

knowledge highlights the focused Disaster Risk Reduction (DRR) initiatives as identifying the most vulnerable groups and determining the policy actions to reduce the negative effects (Mizutori, 2020; Ray-Bennett et al., 2022).

The PCA indicated that there are three major components that interpret 76% of the total variance in social vulnerability: socio-economic factors (46.7%), infrastructure

(17.3%), and exposure (12.1%). It is important to note that the socio-economic aspect, identified by the proportion of the elderly inhabitants, women, children under the age of five years, and employed people, was the strongest factor of vulnerability. These results support the past studies that highlight the susceptibility of demographic groups that have lower coping abilities (Clar et al., 2023; Xu et al., 2024). The infrastructure aspect put a spotlight on the danger of the rapid urbanization process and the growing urban coverage, which has been backed by the research indicating that unplanned urbanization growth is associated with higher flood risks (Beshir & Song, 2021; Tierolf et al., 2021). The exposure dimension is concerned with population growth and access to drinking water and shows how the vulnerability increases because of the insufficiency of the essential services (Shen et al., 2023).

Moreover, there was high variation of vulnerability scores between the districts and the upazilas in the study area and low population density, poverty, and inadequate infrastructure were there as high-risk clustering especially in Madhyanagar, Chunarughat, and Shayestaganj. It is proved that these areas have always had floods and socio-economic challenges as it is evidenced in history (Rudra & Sarker, 2023; Shafiq, 2023). The fear of flood susceptibility mapping reveals that it is only 6.52% of Sylhet that is classified as low susceptibility area, whereas 40.98 and 37.43% are extremely high and high susceptibility area respectively. The issue of social vulnerability is dynamic and relies on the circumstance, demographic changes, socio-economic, and infrastructure (Allen et al., 2020; Song et al., 2019). The percentage of the elderly population in Sylhet has grown by 12% during the period of 2011–22 and rural population has reduced by 9% and these contribute to a rise in the dependency rates and weakening of the households (Bangladesh Bureau of Statistics, 2022). Communities are weakened significantly by the fact that the younger groups have moved. This means that areas, where the level of literacy is high and NGOs are active, decrease the scores in the susceptibility category by approximately a fifth (Hossain et al., 2023). The readiness of the community can be reduced by using early warning and traditional coping mechanisms, but it requires social cohesion (Kim et al., 2021).

The findings of the study are also in line with the previous research conducted among regions using composite indices to measure multidimensional vulnerability. As an illustration, in the Indian Sundarbans, Biswas and Nautiyal (2020) found that poverty, population density, and the quality of infrastructure had a strong effect on exposure to floods, which were also notable in Sylhet where the social-economic aspect of the PCA outcome was major (46.7) and the most influential predictor of exposure. Similarly, in South Korea, Jeong and Yoon (2018) observed that the local socio-economic conditions are significant predictors, and therefore are more effective as compared to hazard-only models. More to the point, the analysis at the upazila level indicated that eleven upazilas are undergoing severe socio-economic danger, particularly the Haor basin. This is mostly because the population is getting older and there are a lot of people out of work. This kind of separate study can help us understand how different places are vulnerable and help us make better and fairer plans for how to adapt.

Community involvement is important for managing flood risk, but our research shows that the most vulnerable upazilas in Sylhet are often not allowed to take part.

The socio-economic examination has been found to have 11 upazilas subject of extreme vulnerability which is due to the aging of populations, high unemployment, and selective outmigration—something that restricts the action of households in response to the early warning or to take part in community projects. Moreover, there is a significant shortage of infrastructure, with 38% of highly vulnerable regions lacking reliable access to safe drinking water. These results are consistent with those of Bouwer et al. (2021), who emphasize that socio-economic frailty compromises adaptation effort. The advantages of the participation can also be seen in comparative evidence: Dewa et al. (2022) established that the inclusion of women cooperatives in the process of disseminating early warnings led to the reduction of delays in evacuation by 30%, and Rasheed (2021) also found that community-based adaptation efforts were more effective in Vietnam than top-down interventions. The fact that demographic and employment factors (47% of variance in PCA) have high effects on the Sylhet area highlights the necessity to incorporate vulnerable groups into decision-making. Enhancement of local ownership by use of participatory strategies will help in making adaptation strategies fair and relevant to the vulnerability profile of the area.

The flood risk is a multi-level issue, and our results indicate why it is highly important in Sylhet. Extreme socio-economic and infrastructure hot spots including Madhyanagar, Chunarughat, and Shayestaganj do not have sufficient flood shelters, road access, and dependable water supplies (38% of highly vulnerable upazilas) indicating a poor local government dimension. On the same note, the percentage of upazilas categorized as extreme infrastructure vulnerability has been noted to be 7.32 with the Haor basin recording the highest levels of governance gaps that contribute to the vulnerability. The trends show what is happening in the country as a whole, which shows that local government is still weak because of poor coordination and lack of resources (Poudel & Shaw, 2024). It is, therefore, necessary to strengthen platforms that connect Union Parishads, district administrations, NGOs, and communities as stipulated in the Standing Orders on Disaster (MoDMR, 2019) and DRR strategies. However, such issues as funding, accountability, and integration of knowledge (Mandal et al., 2024) need to be considered. The coordination with upstream authorities is quite eminent in the case of Sylhet as it has transboundary rivers and therefore needs real-time prediction. Incorporating spatially explicit vulnerability data, encompassing the 11 upazilas facing significant socio-economic vulnerability, into governance frameworks would facilitate more targeted and equitable resilience-building initiatives.

The results of the SoVI clearly demonstrate the correspondence between the social vulnerability in Sylhet and the international priorities of resilience following the Sustainable Development Goals (SDGs). It is believed to be estimated that approximately 28% of the population in high-risk upazilas is forced to live below the poverty line, which implies SDG 1 (No Poverty) relationship and confirms the role of the socioeconomic status in disaster risk prediction (Naeimiasl et al., 2025). The weakness in health-related aspects is also apparent. The greatest SoVI scores were in the upazilas, which registered the greatest proportion of post-flood illness, which is directly linked to SDG 3 (Health and Well-being). The lack of secure housing and infrastructures is also indicated by the fact that 42% of the households in the high-risk areas can access flood-resilient housing, which is a problem as far as SDG 11 (Sustainable Cities and Communities) is concerned. The focus on early warning systems and the localization of the high-SoVI upazilas facilitates the achievement of SDG 13 (Climate Action), whereas the emphasis on the collaboration between the communities, non-governmental organizations, and governmental institutions is associated with SDG 17 (Partnerships for the Goals) (Pienaah et al., 2023). However, despite the solid spatial and statistical findings of the existing study, there are weaknesses of the SDG mapping, as the data of the direct income, health outcome, and longitudinal adaptation is unavailable. Future research should therefore integrate the participatory indicator design and the longitudinal tracking so as to advance the contribution of the SoVI-based methods in the SDG tracking.

Combining spatial vulnerability mapping with professional experience and experience of previous floods can make the policy more effective and build the resilience of the community (Ramadan et al., 2025). Of particular significance in this case are non-structural measures such as awareness campaigns, risk communication, and preparedness programs because of the resource constraints and location-specific approaches (Bodas, 2023). These priorities are reflected in our SoVI scores of high-vulnerabilities upazilas, the literacy rates are 15 points less than the regional ones, and only 42% of households have flood-resistant structures, which implies that there is a barrier to effective communication and preparedness. The strengthening of multi-level governance and stakeholders collaboration must also be intensified since studies have established that the greatest impacts on the undermining of the process of disaster risk reduction are lack of trust and inadequate coordination (Djalante, 2012). Not only does the active inclusion of communities in the decision-making process instill trust, but it is also used to ensure that the knowledge of the locals is not the only part of the adaptation strategies. Lastly, the spatial representation of the vulnerable regions as hotspots would help the decision-makers have a realistic approach of intervention on the areas that have the worst deficiencies in resiliency.

6.6 Conclusions

This study presents a scientifically sound and spatially explicit analytic methodology of gauging social vulnerability to flood risks in the Sylhet Division of Bangladesh using a PCA-based index methodology, offering a holistic, data-driven view of the patterns of spatial vulnerability by synthesizing multidimensional indicators into three main components, namely, socio-economic factors, infrastructural conditions, and exposure. The results demonstrate sharp differences between the upazilas, especially in Haor basin, where weak infrastructure, population pressure, and ecological sensitivity are combined to increase the risk of flood. These lessons demonstrate that it is relevant to plan according to the local needs and facts to address the weaknesses in the short term and the causes in the long term.

The research has a major implication on the disaster risk management and sustainable development since it provides a practical implementation of its procedures including vulnerability maps and composite indexes that can be implemented to conduct certain responses, equitable distribution of resources, and climate-resilient development of developmental structures. The study target of the fringe groups of the aging population, women, and the unemployed is compliant with the tenets of social equity and resiliency planning that is inclusive. The research presents to the scientific community a transferable, scalable methodology to the Global South with which is largely applicable in other flood-prone regions in the lack of data and the complexity of the socio-economic situations do not permit such a holistic risk assessment. This paper, integrating spatial analytics with a composite index model, assists the globe to play its part in the Sendai Framework of Disaster Risk Reduction, and other Sustainable Development Goals (SDGs), like poverty reduction, climate action, sustainable cities, and gender equality. The work creates a positive relationship between the local vulnerability processes and the international discourse on the resilience to disasters and climate change, which adds to the scientific discourse.

However, the research also has a number of limitations. This dependency on secondary data might restrict the time resolution and also suppress the ability to include context-specific variables like social capital, governance structures, and qualitative perceptions of risk which are in many cases important to represent the entire range of vulnerability. Also, the data being static restricts the possibility to capture recent changes in the demographics or development interventions, and the upazila-level resolution might hide the significant intra-regional differences. Future studies need to focus on incorporating longitudinal and high-resolution spatial information into the study to enable the capture of the temporal dynamics and micro-scale vulnerability. Moreover, the use of participatory and qualitative methods, like community-based mapping or stakeholder interviews, could help to better understand how well people in a community can adapt and how strong their social networks are. Extending the analytical framework to governance, institutional quality, and climate projections would also be more comprehensive and predictive. More than just enhancing the practice of vulnerability assessment, these developments would enhance the policy frameworks with a view to producing equitable and climate-resilient national and global development.

Acknowledgements The authors would like to express sincere gratitude to the Department of Geography and Environment, University of Dhaka, for providing academic support and access to essential data and tools necessary for this research. Special thanks are extended to the Bangladesh Bureau of Statistics (BBS) and the Flood Forecasting and Warning Centre (FFWC) for their valuable datasets. The authors also acknowledge the support of local administrative offices in the Sylhet Division for their cooperation during data collection.

Authors Contribution: **Rifat Bin Hossain:** Conceptualization, Methodology, Writing-original draft, Writing-review & editing, Supervision. **Md. Iqbal Hasan:** Methodology, Supervision. **Tasneem Sharmin:** Methodology, Software, Visualization. **Nafisa Nandini:** Writing-review & editing, Visualization. **Md. Juhayr Akbar:** Writing-original draft.

Funding: The research was not supported by any funding.

Potential Conflicts of Interest: On behalf of all authors, the corresponding author states that there is no conflict of interest.

Informed Consent: Informed consent was obtained from all individual participants included in the study.

DataAvailability Data Availability Statement: The datasets during and/or analyzed during the current study available from the corresponding author on reasonable request.

AI Disclosure Statement
This research benefited from the use of Artificial Intelligence (AI) tools to enhance the readability and language clarity of the manuscript. Specifically, Google's Gemini AI was used on July 15, 2025 via the interface available at https://gemini.google.com.

References

Abdrabo, K., Kantoush, S., Esmaiel, A., Saber, M., Sumi, T., Almamari, M., Elboshy, B., & Ghoniem, S. (2023). An integrated indicator-based approach for constructing an urban flood vulnerability index as an urban decision-making tool using the PCA and AHP techniques: A case study of Alexandria, Egypt. *Urban Climate* https://doi.org/10.1016/j.uclim.2023.101426

Ajtai, I., Ştefănie, H., Maloş, C., Botezan, C., Radovici, A., Bizău-Cârstea, M., & Baciu, C. (2023). Mapping social vulnerability to floods. A comprehensive framework using a vulnerability index approach and PCA analysis. *Ecological Indicators, 154*, 110838. https://doi.org/10.1016/j.ecolind.2023.110838

Aktar, M. A., Shohani, K., Hasan, M. N., & Hasan, M. K. (2021). Flood vulnerability assessment by flood vulnerability index (FVI) method: A study on Sirajganj Sadar Upazila. *International Journal of Disaster Risk Management, 3*(1), 1–14.

Allen, M., Gillespie-Marthaler, L., Abkowitz, M., & Camp, J. (2020). Evaluating flood resilience in rural communities: A case-based assessment of Dyer County, Tennessee. *Natural Hazards, 101*, 173–194. https://doi.org/10.1007/s11069-020-03868-2

Arnell, N., & Gosling, S. (2016). The impacts of climate change on river flood risk at the global scale. *Climatic Change, 134*, 387–401. https://doi.org/10.1007/s10584-014-1084-5

Aroca-Jiménez, E., Bodoque, J. M., & García, J. A. (2020). How to construct and validate an integrated socio-economic vulnerability index: Implementation at regional scale in urban areas prone to flash flooding. *Science of the Total Environment, 746*, Article 140905. https://doi.org/10.1016/j.scitotenv.2020.140905

Azizi, E., Nikoo, M., Mostafazadeh, R., & Hazbavi, Z. (2023). Flood vulnerability analysis using different aggregation frameworks across watersheds of Ardabil province, northwestern Iran. *International Journal of Disaster Risk Reduction.* https://doi.org/10.1016/j.ijdrr.2023.103680

Bangladesh Bureau of Statistics. (2022). *Statistical yearbook of Bangladesh 2022.* Statistics and Informatics Division, Ministry of Planning, Government of the People's Republic of Bangladesh.

Beshir, A., & Song, J. (2021). Urbanization and its impact on flood hazard: The case of Addis Ababa, Ethiopia. *Natural Hazards, 109*, 1167–1190. https://doi.org/10.1007/s11069-021-04873-9

Biswas, S., & Nautiyal, S. (2020). An assessment of socio-economic vulnerability at the household level: A study on villages of the Indian Sundarbans. *Environment, Development and Sustainability, 23*, 11120–11137. https://doi.org/10.1007/s10668-020-01085-2

Bodas, M. (2023). The dark side of the (preparedness) moon: Why promoting public preparedness remains illusive. *Prehospital and Disaster Medicine, 38*, s96–s97. https://doi.org/10.1017/S1049023X23002698

Bouwer, R., Pasquini, L., & Baudoin, M. (2021). Breaking down the silos: Building resilience through cohesive and collaborative social networks. *Environmental Development, 100646*. https://doi.org/10.1016/J.ENVDEV.2021.100646

Bronfman, N., Nikole, G., Castañeda, J., Cisternas, P., & Repetto, P. (2024). Relationship between social vulnerability and community resilience: A geospatial study in the context of natural disasters. *International Journal of Disaster Risk Reduction*. https://doi.org/10.1016/j.ijdrr.2024.104774

Brouwer, R., Akter, S., Brander, L., & Haque, E. (2007). Socioeconomic vulnerability and adaptation to environmental risk: A case study of climate change and flooding in Bangladesh. *Risk Analysis: An International Journal, 27*(2), 313–326.

Bucherie, A., Hultquist, C., Adamo, S., Neely, C., Ayala, F., Bazo, J., & Kruczkiewicz, A. (2022). A comparison of social vulnerability indices specific to flooding in Ecuador: Principal component analysis (PCA) and expert knowledge. *International Journal of Disaster Risk Reduction, 73*, Article 102897.

Clar, C., Junger, L., Nordbeck, R., & Thaler, T. (2023). The impact of demographic developments on flood risk management systems in rural regions in the Alpine Arc. *International Journal of Disaster Risk Reduction*. https://doi.org/10.1016/j.ijdrr.2023.103648

Derakhshan, S., Emrich, C., & Cutter, S. (2022). Degree and direction of overlap between social vulnerability and community resilience measurements. *PLoS ONE, 17*. https://doi.org/10.1371/journal.pone.0275975

Dewa, O., Makoka, D., & Ayo-Yusuf, O. (2022). A deliberative rural community consultation to assess support for flood risk management policies to strengthen resilience in Malawi. *Water*. https://doi.org/10.3390/w14060874

Dhawale, R., Schuster-Wallace, C., & Pietroniro, A. (2024). Assessing the multidimensional nature of flood and drought vulnerability index: A systematic review of literature. *International Journal of Disaster Risk Reduction*. https://doi.org/10.1016/j.ijdrr.2024.104764

Djalante, R. (2012). Review Article: "Adaptive governance and resilience: The role of multi-stakeholder platforms in disaster risk reduction." *Natural Hazards and Earth System Sciences, 12*, 2923–2942. https://doi.org/10.5194/NHESS-12-2923-2012

Haque, M. M., Islam, S., Sikder, M. B., & Islam, M. S. (2022). Community flood resilience assessment in Jamuna floodplain: A case study in Jamalpur District Bangladesh. *International Journal of Disaster Risk Reduction, 72*, Article 102861.

Hossain, M. S. (2025). Assessing smallholder farmers' flood risk behavior and their willingness to pay for crop insurance as a risk coping strategy in northern Bangladesh. *Natural Hazards, 121*(4), 4191–4217.

Hossain, B., Shi, G., Ajiang, C., Sohel, M. S., & Yijun, L. (2023). Social vulnerability, impacts and adaptations strategies in the face of natural hazards: Insight from riverine islands of Bangladesh. *BMC Public Health, 23*(1), 1737.

Howlader, R., Hossain, M., Jahan, C., Rahaman, M., & Chowdhury, M. (2024). Risk assessment and zonation of flash flood in Sylhet basin, Northeast Bangladesh using GIS-MCDM tool. *Safety in Extreme Environments*. https://doi.org/10.1007/s42797-024-00106-x

Islam, R., & Chowdhury, P. (2024). Local-scale flash flood susceptibility assessment in northeastern Bangladesh using machine learning algorithms. *Environmental Challenges, 14*, Article 100833.

Islam, M., & Meng, Q. (2024). Spatial analysis of socio-economic and demographic factors influencing urban flood vulnerability. *Journal of Urban Management*. https://doi.org/10.1016/j.jum.2024.06.001

Jeong, S., & Yoon, D. (2018). Examining vulnerability factors to natural disasters with a spatial autoregressive model: The case of South Korea. *Sustainability, 10*, 1651. https://doi.org/10.3390/SU10051651

Kaiser, H. F. (1959). Computer program for varimax rotation in factor analysis. *Educational and Psychological Measurement, 19*(3), 413–420.

Kam, P., Aznar-Siguán, G., Schewe, J., Milano, L., Ginnetti, J., Willner, S., McCaughey, J., & Bresch, D. (2021). Global warming and population change both heighten future risk of human displacement due to river floods. *Environmental Research Letters, 16*. https://doi.org/10.1088/1748-9326/abd26c

Kim, H., Woosnam, K., Marcouiller, D., & Kim, H. (2021). Seeking anticipatory adaptation: Adaptive capacity and resilience to flood risk. *Environmental Hazards, 21*, 36–57. https://doi.org/10.1080/17477891.2021.1902783

Liu, Q., Du, M., Wang, Y., Deng, J., Yan, W., Qin, C., Liu, M., & Liu, J. (2024). Global, regional and national trends and impacts of natural floods, 1990–2022. *Bulletin of the World Health Organization, 102*, 410–420. https://doi.org/10.2471/BLT.23.290243

Lund, A., Sokolow, S., Jones, I., Wood, C., Ali, S., Chamberlin, A., Sy, A., Sam, M., Jouanard, N., Schacht, A., Senghor, S., Fall, A., Ndione, R., Riveau, G., De Leo, G., & López-Carr, D. (2021). Exposure, hazard, and vulnerability all contribute to Schistosoma haematobium re-infection in northern Senegal. *PLoS Neglected Tropical Diseases, 15*. https://doi.org/10.1371/journal.pntd.0009806

Machado, E., & Ratick, S. (2018). Implications of indicator aggregation methods for global change vulnerability reduction efforts. *Mitigation and Adaptation Strategies for Global Change, 23*, 1109–1141. https://doi.org/10.1007/s11027-017-9775-7

Mandal, K. K., Dharanirajan, K., Meena, M. L., Jaman, T., & Rana, S. (2024). Application of geospatial tools in the assessment of Flood hazard impact on social vulnerability of Malda district, West Bengal. *India. Natural Hazards Research, 4*(3), 470–485.

Mizutori, M. (2020). Reflections on the Sendai framework for disaster risk reduction: Five years since its adoption. *International Journal of Disaster Risk Science, 11*, 147–151. https://doi.org/10.1007/s13753-020-00261-2

Naeimiasl, F., Vahidi, H., & Soheili, N. (2025). Leveraging principal component analysis for data-driven and objective weight assignment in spatial decision-making framework for qanat-induced subsidence susceptibility assessment in railway networks. *ISPRS International Journal of Geo-Information, 14*(5), 195. https://doi.org/10.3390/ijgi14050195

Nguyen, T., Asahi, C., & Tran, T. (2023). A systematic review with bibliometric analysis of different approaches and methodologies for undertaking flood vulnerability research. *Sustainable Water Resources Management, 9*, 1–20. https://doi.org/10.1007/s40899-023-00865-8

Orru, K., Klaos, M., Nero, K., Gabel, F., Hansson, S., & Nævestad, T. (2022). Imagining and assessing future risks: A dynamic scenario-base social vulnerability analysis framework for disaster planning and response. *Journal of Contingencies and Crisis Management*. https://doi.org/10.1111/1468-5973.12436

Pariartha, I., Aggarwal, S., Srinivas, R., Egodawatta, P., McGree, J., & Goonetilleke, A. (2023). Compounding effects of urbanization, climate change and sea-level rise on monetary projections of flood damage. *Journal of Hydrology*. https://doi.org/10.1016/j.jhydrol.2023.129535

Pienaah, C., Batung, E., Saaka, S., Mohammed, K., & Luginaah, I. (2023). Early warnings and perceived climate preparedness among smallholder farmers in the Upper West Region of Ghana. *Land*. https://doi.org/10.3390/land12101944

Pittore, M., Haas, M., & Silva, V. (2020). Variable resolution probabilistic modeling of residential exposure and vulnerability for risk applications. *Earthquake Spectra, 36*, 321–344. https://doi.org/10.1177/8755293020951582

Poudel, N., & Shaw, R. (2024). Significance of urban-rural programs for disaster-resilient and sustainable water systems in the Kathmandu Valley. *Modern Environmental Science and Engineering*. https://doi.org/10.15341/mese(2333-2581)/5-6.10.2024/02

Rahman, R., & Salehin, M. (2013). Flood risks and reduction approaches in Bangladesh. In *Disaster risk reduction approaches in Bangladesh* (pp. 65–90).

Rahman, M., Kamruzzaman, M., Deb, L., & Islam, H. T. (2024). Flood mapping, damage assessment, and susceptibility zonation in northeastern Bangladesh in 2022 using geospatial datasets. *Progress in Disaster Science, 100402*. https://doi.org/10.1016/j.pdisas.2024.100402

Ramadan, M., Almurshidi, A., Razali, S., Ramadan, E., Tariq, A., Bridi, R., Rahman, M., Albedwawi, S., Alshamsi, M., Alshamisi, M., Alrashdi, S., Alnaqbi, S., Alhammadi, H., & Hosani, N.

(2025). Spatial decision-making for urban flood vulnerability: A geomatics approach applied to Al-Ain City, UAE. *Urban Climate*. https://doi.org/10.1016/j.uclim.2025.102297

Rasheed, F. (2021). Devising a socioeconomic vulnerability assessment framework and ensuring community participation for disaster risk reduction: A Case-Study Post Kerala Floods of 2018. pp. 451–467. https://doi.org/10.1007/978-3-030-57332-4_32

Ray-Bennett, N., Clarke, K., & Mendez, D. (2022). Sendai framework's global targets A and B: Opinions from the global platform for disaster risk reduction's ignite stage 2019. *International Journal of Disaster Risk Science, 13*, 651–663. https://doi.org/10.1007/s13753-022-00432-3

Rudra, R., & Sarkar, S. (2023). Artificial neural network for flood susceptibility mapping in Bangladesh. *Heliyon, 9*. https://doi.org/10.1016/j.heliyon.2023.e16459

Sarker, S., Jahan, I., Wang, X., & Azad, A. (2025). Geospatial approach to assess flash flood vulnerability in a coastal District of Bangladesh: Integrating the multifaceted dimension of vulnerabilities. *ISPRS International Journal of Geo-Information*. https://doi.org/10.3390/ijgi14050194

Saunders, A., Giezendanner, J., Tellman, B., Islam, A., Bhuyan, A., & Islam, A. (2023). A comparison of remote sensing approaches to assess the devastating may-june 2022 flooding in Sylhet, Bangladesh. In *IGARSS 2023–2023 IEEE International geoscience and remote sensing symposium* (pp. 452–455). https://doi.org/10.1109/IGARSS52108.2023.10283378

Shafiq, S. (2023). The frequency and impact of flooding in the sylhet division of Bangladesh: An Investigation. *International Journal of Business, Social and Scientific Research*. https://doi.org/10.55706/ijbssr11103

Shen, S., Kim, K., & Liu, D. (2023). Aging in place or moving to higher ground: Older adults' adaptation to sea level Rise in Honolulu, Hawaii. *Sustainability*. https://doi.org/10.3390/su15129535

Song, J., Chang, Z., Li, W., Feng, Z., Wu, J., Cao, Q., & Liu, J. (2019). Resilience-vulnerability balance to urban flooding: A case study in a densely populated coastal city in China. *Cities, 95*, Article 102381. https://doi.org/10.1016/J.CITIES.2019.06.012

Spielman, S., Tuccillo, J., Folch, D., Schweikert, A., Davies, R., Wood, N., & Tate, E. (2020). Evaluating social vulnerability indicators: Criteria and their application to the Social Vulnerability Index. *Natural Hazards, 100*, 417–436. https://doi.org/10.1007/s11069-019-03820-z

Tierolf, L., Moel, H., & Vliet, J. (2021). Modeling urban development and its exposure to river flood risk in Southeast Asia. *Computers, Environment and Urban Systems, 87*, Article 101620. https://doi.org/10.1016/J.COMPENVURBSYS.2021.101620

Vousdoukas, M., Mentaschi, L., Voukouvalas, E., Bianchi, A., Dottori, F., & Feyen, L. (2018). Climatic and socioeconomic controls of future coastal flood risk in Europe. *Nature Climate Change, 8*, 776–780. https://doi.org/10.1038/s41558-018-0260-4

Wu, T. (2021). Quantifying coastal flood vulnerability for climate adaptation policy using principal component analysis. *Ecological Indicators, 129*, Article 108006. https://doi.org/10.1016/J.ECOLIND.2021.108006

Xie, W., & Meng, Q. (2023). An integrated PCA–AHP method to assess urban social vulnerability to sea level rise risks in Tampa, Florida. *Sustainability, 15*(3), 2400. https://doi.org/10.3390/su15032400

Xu, J., Takahashi, M., & Li, W. (2024). Identifying vulnerable populations in urban society: A case study in a flood-prone district of Wuhan, China. *Natural Hazards and Earth System Sciences*. https://doi.org/10.5194/nhess-24-179-2024

Younus, M. A. F. (2014). Flood vulnerability and adaptation to climate change in Bangladesh: A review. *Journal of Environmental Assessment Policy and Management, 16*(03), 1450024.

Chapter 7
Resilience Amidst Adversity: Navigating Environmental Shifts in the Gosaba Block of the Sundarbans

Lina Bose and Shayantani Banerjee

Abstract At this point, it is important to study the resilience of coastal communities to the mounting impacts of climate change like rising sea levels and severe weather. As is most evident in the Sundarbans case, a UNESCO World Heritage Site, the complex correlation between human vulnerability and environmental change in these regions can be simply explained by the intricate connexion between the environment and human vulnerability in these regions. Even though it is often highlighted in the literature that these communities are vulnerable, there is very little knowledge about the daily, highly engrained cultural practises that allow them to cope and cultivate resilience. The ordinariness of culture has been overlooked as the basis of adaptive strength in most studies in favour of the large scale interventions of the government or non-governmental organisations. The indigenous residents of the Gosaba Block of Sundarban who bear the brunt of the impacts of the climate change are grossly confronted with the environmental concerns that cut across the socio-economic and cultural lines as discussed in this chapter. Raymond (Keywords: A vocabulary of culture and society. Oxford University Press, 1985) says that resilience is embedded in the mundane of the culture and is shown in daily lives of the residents of the Gosaba and their methods of adaptation. The particular environmental conditions the Gosaba Block has to deal with, i.e. sea level rise, erosion, salinity intrusion, extreme weather events are discussed and established in the paper. Through elaborate ethnographical studies, we determine the issues that need further research to understand the resilience complexities in the Sundarbans. This paper will fill this gap by exploring the special challenges facing this Gosaba Block in the Sundarbans through an in-depth ethnography approach. In addition to simple measures of resilience, we analyse how resilience can be observed in the daily practises of the community and in the adaptive strategies used by the community. The outcomes of

L. Bose (✉)
Department of History, Amity University Jharkhand, Ranchi, India
e-mail: linabose08@gmail.com

S. Banerjee
Amity Institute of English Studies and Research (AIESR), Amity University Jharkhand, Ranchi, India
e-mail: sbanerjee@rnc.amity.edu

G. K. Panda et al. (eds.), *Socio-Ecological Crisis of Development*, Sustainability Solutions, https://doi.org/10.1007/978-3-032-22171-1_7

our work are going to indicate the specific spheres that need further research and provide the policymakers with ground-based information that is informative.

Keywords Gosaba block · Sundarbans · Resilience · Cyclone · Vulnerable communities · Social crisis

7.1 Introduction

The Sundarbans is a large collection of mangrove forests found in the delta area of the Ganges, Brahmaputra and Meghna rivers in Bangladesh and India. It demonstrates the strong relationship between the ecological richness and the human development (Blasco et al., 1998). Over time, the region has had prosperous communities that depend on the natural resources. It is considered a UNESCO world heritage as a result of its own biodiversity that comprises the famous Royal Bengal tiger and a complicated web of tidal waterways (Mallick & Rahman, 2010). The effects of climate change are however increasingly challenging this balance. Among these effects are increased sea levels, increased intensity of cyclones, increased intrusion of salinity as well as coastal erosion (IPCC, 2022).

The Gosaba Block (Fig. 7.1), which is located in the Indian Sundarbans in the state of West Bengal is a critical microcosm of these more general problems identified with the environment. One of the frontline regions i.e. Bay of Bengal face-on, Gosaba in particular is affected by these environmental changes and more vulnerable to these changes in terms of infrastructure, social fabric and subsistence (Dasgupta et al., 2014). As a consequence of the increased intensity and frequency, cyclones caused by Cyclone Aila in 2009 and Cyclone Amphan in 2020 have caused extensive damage, displacement of people, destruction of agricultural land and contamination of freshwater sources (Hazra et al., 2021). Also, the already unstable agricultural activities are affecting food security and conventional livelihoods such as fishing and aquaculture due to the unobtrusive rise in the sea level and subsequent salinity intrusion (Mukherjee et al., 2018). Conceptualisation of resilience has emerged to be a primary prism through which the capacity of the ecosystem and human societies in Sundarbans to withstand, adapt and bounce back to shocks and stresses in the face of such increasingly challenging circumstances can be viewed (Folke et al., 2010; Holling, 1973). Despite the abundance of literature on ecological resilience as it applies to mangrove ecosystems (Alongi, 2008), a thorough understanding of the social characteristics of the ecological resilience, particularly, the traditions and adaptational strategies that people in the Gosaba Block reside in apply (Adger, 2000; Berkes and Folke, 1998). This paper has indicated that the traditional knowledge and adaptive mechanisms that the local communities in Gosaba have are rich and have been developed over the decades as a result of interacting with such a dynamic environment. This informed by the concept put forward by Raymond Williams (1985) about the fact that resilience exists in the ordinariness of culture.

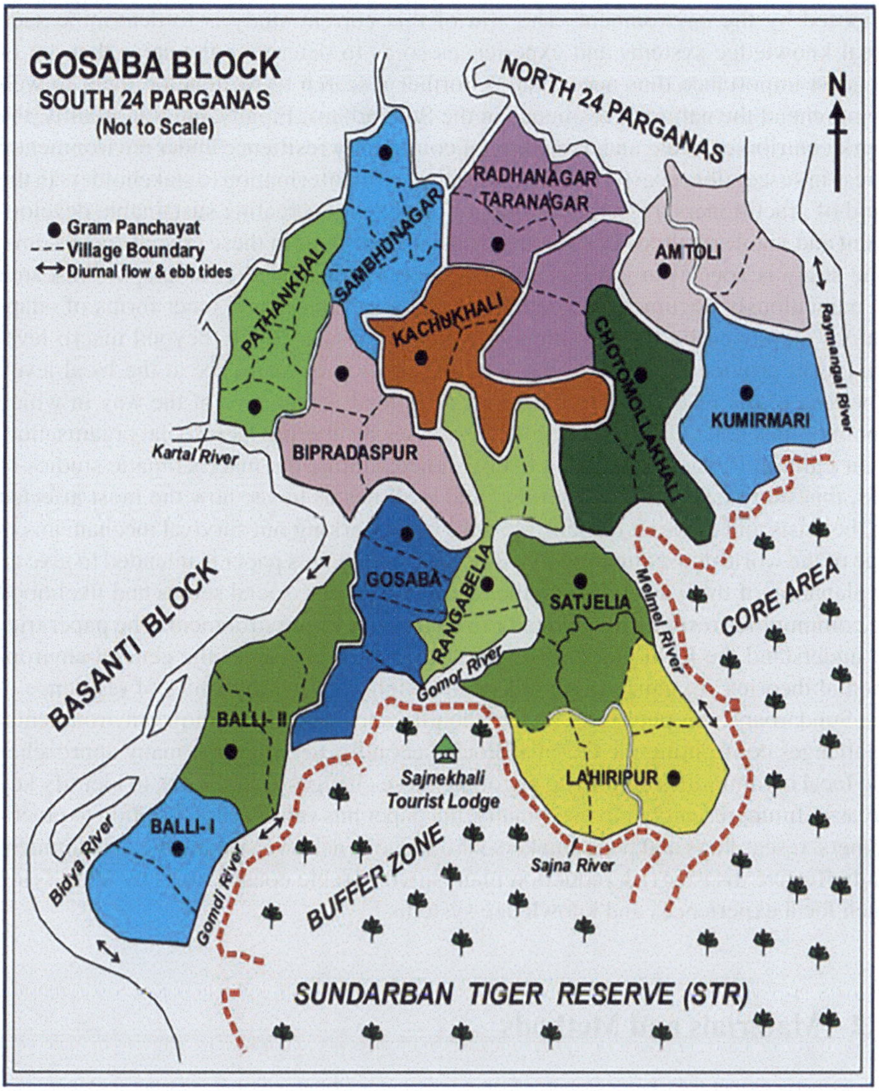

Fig. 7.1 The map shows Gosaba Block is bordered by Basanti Block to the west and North 24 Parganas to the north. The eastern and southern boundaries are defined by the Sundarban Tiger Reserve (STR). 2016 *Source* Chowdhury et al. ()

Besides the common environmental hypotheses, within the paper an effort is made to trace a comprehension of the local practical realities of resilience. The Gosaba Block is where most ethnographic research has been done to shed illumination on the complex way in which these communities are changing their social pattern, cultural patterns and livelihood mechanism as a response to the stresses imposed by the environment. The aim of this current study is to determine such local knowledge systems and experiences so as to determine the areas that are of greatest importance thus necessitating further research to be done on them to well comprehend the nature of resilience in the Sundarbans. Finally, but importantly, the work contributes to the understanding of community resilience under environmental stress in susceptible coastal regions, offering useful information to stakeholders in the field of practitioners, researchers and policymakers in creating sustainable development and viable methods of reducing disasters risks within these crucial ecosystems. The study is special in the sense that it adopts a dedicated ethnographic research to meticulously document the detailed, lived experiences and other forms of adaptations employed by local communities in the Gosaba Block, beyond macro level studies to provide insight into the actual utilisation of resilience at the local level. The idea of the research is to have a more refined perspective of the way in which communities react to the environmental stresses by altering their social organisation, their cultural life and the mode of subsistence. Unlike the macroclimatic studies of old, this study takes us to the ground and it allows us to see how the most affected in the crisis and the least responsible have been working out survival mechanisms in one of the world-most climatically vulnerable areas. This paper is intended to give an explanation of the ways in which the cultural practises, social setups and livelihood of communities respond dynamically to changes in the environment. The paper tries to understand the local practical realities of resilience besides the general environmental theories. To gain a more thorough insight into the dynamics of resilience in the Sundarbans, this paper will seek to identify and discuss the unique environmental challenges confronting the Gosaba Block; secondly, to record the many approaches the local communities employ to negotiate these changes; and thirdly, to identify key areas of future research interest. Finally, the paper has valuable information to practitioners, researchers and policymakers who strive to achieve sustainable development and effective disaster risk reduction plans in vulnerable coastal areas by identifying such local experiences and knowledge systems.

7.2 Materials and Methods

This paper embarks on a detailed exploratory and qualitative expedition into the Gosaba Block of Sundarbans. The paper predominately applies the ethnographic methods, as it puts itself in the perspective and world of the communities involved, to understand the nuances of life in the face of environmental changes. The data collection is carried out through a complex method that involves: focus group, semi-structured interviews, participant observation and key informant interviews (Table 7.1). To be able to record the multifaceted nuances of the life and the vision of people in the community, the main approach of the study is ethnographic, that is, it involves the focus on the comprehensive immersion into the local setting. The

Table 7.1 The methods and methodology pursued

Methods	Methodologies	
Qualitative research methods	Participant observation	Field visits and firsthand engagement and the capture of contextual understanding within the communities in the Gosaba block
	Semi-structured interviews	Conducted to explore individual experiences and perspectives in depth while maintaining a focus on key research questions
	Focus group discussions (FGDs)	FGDs conducted in general facilitated the rich discussions and uncovering shared understandings, challenges and coping mechanisms within community groups
	Key informant interviews (KIIs)	KIIs with the local leaders and stakeholders helped to understand the varied dynamics of the region

complexity of data collecting will guarantee the in-depth understanding of the information through triangulation. The exact qualitative research methods involve those listed in Table 7.1.

Methodological Flowchart

7.3 Review of Literature

Resilience is the concept of environmental science, psychology, sociology and disaster management. It defines the ability of individuals, groups or systems to cope, adjust to and recuperate from the demanding situations and sudden change. There have been many studies based on the basic principles of resilience, the aspects that influence resilience and its application in practise. Resilience is necessary to address such global challenges as social inequality, health crisis, economic instability and climate change. It is a process which evolves in reaction to different challenges and not a definite feature. Companies such as the World Bank and the United Nations are stressing its importance in socio-economic plans, sustainable development and disaster plans. Discussions about its definition, measurement and its ethical issues have not ended although it has a wide area of application. To illuminate the role of resilience in modern research and practise, this review explores several resilience models, how it has been used in different sectors and those who criticise it.

Resilience has been theorised variously based on the field. In psychology, resilience is generally perceived as an individual character that helps individuals to survive adversity (Masten, 2001). Bonanno (2004), on the other hand, differentiates reconfiguration, resistance and recovery following traumatic events. Psychological models of resilience emphasise protective factors such as emotional regulation, social support and cognitive adaptability as determining factors with regard to ability of an individual to overcome adversity. The concepts related to resilience in sociology are institutional support, social capital and community cohesion. Norris et al. (2008) argue that community resilience is a process, which implies social capital, economic development, communication and information and community competence.

In these views, resilience is not a single characteristic as it is a dynamic and systemic process that is influenced by several variables. Other social-ecological resilience theories, such as Folke (2006) place stress on cross-scale interactions, adaptation and transformation in ecological human system interactions. Economic definition of resilience is the ability of a system to absorb a shock and continue functioning (Rose, 2004). Economic resilience models difference between static resilience, or short-term absorptive capacity and dynamic resilience, or long-term adaptability and innovation to disruptions. The models are especially useful when describing the interruptions in the supply chain, financial crisis and job recovery. Due to the increasing attention to the issue of sustainability and disaster preparedness, the role of environmental and climate change resilience models has also increased. The adaptive cycle model of ecological system resilience is the capacity of the ecological system to accommodate the shock without losing the important functionalities of that system (Holling, 1973). The Intergovernmental Panel on Climate Change (IPCC, 2014) has included resilience in its climate adaptation policy which stipulates the importance of governance, knowledge sharing and adaptive learning to reduce climate risks. In indigenous and cultural contexts, resilience is closely associated with Traditional Ecological Knowledge (TEK) and community practice. Models such as Berkes et al. (2000) explore how indigenous institutions, oral culture and localised adaptation methods build resilience in socio-ecological systems. Such models diverge from Western-centric models of resilience because they focus on relationality, intergenerational transmission and integrated adaptations. The Intergovernmental Panel on Climate Change (IPCC) says that it is possible to respond to climate variability and extreme weather with the help of such a tactic as resilience (IPCC, 2014). Also noted by the study conducted by Adger (2000) is the significance of socio-ecological resilience in adaptation as a way of managing environmental risk in which the traditional governance frameworks and knowledge can assist communities to adapt (Adger, 2000). Another dimension of climate resilience is the ability of the ecosystems and the societies to adjust to the new environmental conditions without losing their basic functions.

As a way of enhancing environmental sustainability, Folke et al. (2010) argue that adaptive management techniques, participatory decision-making procedures and multi-scalar coordination should be prioritised in resilience-based governance. An example of this is that climate-resistant varieties of crops, soil protection measures and agro ecological diversifications all lead to agricultural system resilience. Due to the increased number of climate-related problems, including heatwaves, water shortages and extreme weather events, urban resilience has become popular in cities. The first priority in urban resilience models is given to green areas, intelligent technologies that enhance adaptive capacity and sustainable infrastructure. The Rockefeller foundation 100 Resilient Cities project highlights the importance of social equity, governance and coordinated planning in developing urban environments that are resilient to climate change. The local and indigenous peoples are resilient through their traditional land-use systems, water resource conservation and habitat protection. Empirical studies of the indigenous resilience to the Amazon, Sundarbans and the Arctic tell us about the indigenous systems of knowledge that enable environmental adaptation and sustainability. So as to reduce the risk of climate, Mukherjee and Nayak (2021) discuss the ways Sundarbans community employs the conventional fishing approaches and mangroves conservation strategies. Although possibility of climate change and environmental studies has laid more emphasis on resilience, there remain challenges in the implementation of resilience policies. Fragmented policies, lack of resources and socio-economic inequality tend to undermine the achievement of resilience. Also, resilience scholarship is attacked on the grounds of normalising the exposure to risks and shifting responsibility onto vulnerable populations instead of the governments (Joseph, 2013). This implies that in future studies, resilience must be combined with climate justice and that attention be given to provision of fair adaptation measures whereby needs of vulnerable populations should be given the priority.

Social resilience is a very similar concept, and the research focuses on how a society responds and alleviates social shocks such as pandemics, political turmoil and economic crises. Aldrich (2012) also brings the importance of social networks to recovery activities following the disaster and points out how collaboration, trust and collective actions help to improve the resilience of society. To reduce the level of social and economic suffering, the COVID-19 pandemic also revealed the importance of strong health systems, digital infrastructure and policy interventions (Vogel et al., 2020). Rose (2004) asserts that economic resilience is dynamic (adaptive and transformative capacity) and is also static (absorptive capacity). Studies on rural living have demonstrated that the resilience of the marginalised groups is enhanced through institutional support, access to resources and diversification of the economy (Ellis, 2000). Similarly, the articles of Mukherjee et al. present the argument of a new approach to social resilience to climate change. It focuses on the Indian Sundarbans, a region that is frequently hit by cyclones among other natural calamities. The authors propose transformative social resilience as opposed to the conventional views on resilience that often focus on top-down conservation and technical options. This approach puts a heavy stress on the reformation of the problem of social resiliency by considering the local community stories and real-life experiences.

Also, indigenous and cultural worldviews are highly resilient, where adaptability and sustainable utilisation of resources is closely linked to Traditional Ecological Knowledge (TEK) (Berkes et al., 2000). To respond to environmental uncertainty, the indigenous societies tend to resort to shared governance, oral tradition and customary law. At Sundarbans, local land use and state-supported disaster management, including the resilience of communities to coastal and forest-based community make communities better-managed in managing disasters (Mukherjee & Nayak, 2021). The significance of this is to investigate the resilience of Sundarbans or its ability to withstand shock and long-term change and bounce back as a result. The Sundarbans, the sole mangrove forest that is present in Bangladesh and India is highly vulnerable to extreme environmental hazards because of augmented sea level, frequent cyclones and saltwater intrusion. This is a witness to the will of the people to adapt to these dangers in this prone area of Gosaba Block. The case studies of the regions show levels of resilience that are founded on regional institutions, customary knowledge and flexible livelihoods. Restorative mangrove programmes on a community level have been recorded to be effective in addressing storm surge and coastal erosion (Ghosh, 2018). In the conditions of the environmental stress, now the local communities are able to secure food supply because of the saline-tolerant aquaculture such as saline-resistant fish farming (Chowdhury & Maiti, 2020). One of the government initiatives, which is aimed at integrating scientific studies with the traditional approaches to adaptation, is the Sundarbans Resilience Project. Cyclone shelters, early warning and alternative livelihood plans have improved community resilience (Mukhopadhyay et al., 2019). The issues of long-term sustainability and equal distribution of resources are still challenging. The resilience of the communities in Gosaba underscores the need to have holistic resilience models that incorporate ecological conservation, socio-economic development and participatory governance. The future should focus on unifying the local adaptation processes and should aim at the systemic vulnerability by the intervention of policy measures. Although resilience is considered to be a positive quality in most instances, scholars have criticised the use of resilience in the policy and governance level. According to Cote and Nightingale (2012), the discourse of resilience is prone to masking issues of power and asymmetry hence likely to place the adoption load on already vulnerable populations. Moreover, the discussion is occasionally dragged into the politics of neoliberalism that focuses on individualism and neglects the combination of structural factors like poverty, endemic marginalisation and destruction of the natural environment (Joseph, 2013). Another difficulty is measurement, since its multidimensional nature renders quantification difficult. Researchers have utilised various indicators, from economic stability to social capital, institutional resilience and environmental sustainability, but a single framework is not yet established (Carpenter et al., 2001). The resilience literature always has the tendency to be multidimensional, covering the psychological, ecological, social, economic and cultural areas. The limitations that are intrinsic to the concept of resilience should be understood, but at the same time, irrespective of the fact that the concept is effective in providing a framework for adaptation and recovery, it is still necessary to understand the intrinsic limitations that lie with it especially with regard to the power structures and injustices of systems. Future scholarship should also give priority to interdisciplinary collaboration in order to develop more equitable and sustainable resilience building efforts; this would combine resilience theory with justice-focused approaches. Considering the fragile nature of the ecosystem, Sundarbans are highly susceptible to effects of environmental change and as such the inhabitants of Sundarbans are at high risk. In this connection, the main weaknesses of this area may be listed as follows. The intrusion of seawater resulting from the rising sea levels has resulted in the high levels in both the soil and freshwater sources, a situation that has affected agriculture in a dire way. Young women who were in the vicinity of the formidable tsunami that struck the coastal area at Parigi were traumatised as fearsome waves sound was emitted and buildings and trees were destroyed. Leaving coastal zones and relocating them to higher places may be a solution in the short term, but on the other hand may be highly expensive. The area has been witnessing an increased number of natural calamities such as Cyclone Amphan and Cyclone Yaas which led to the damage of homes, croplands and employment (Mukherjee et al., 2022). The vulnerable or weak embankments guarding the low-lying areas are frequently breached making the event of flooding a normal scene. The reduction of mangrove forests which act as the first line of defence against storm surges has led to a loss of fish stocks and honey-extracting zones, thus affecting the local people working in traditional jobs directly (Rahman et al., 2020). The depletion of forest resources is another cause of ecological unrest and consequently the region is also at a higher risk. Also, due to the spikes in salinity, the sources of freshwater are getting reduced, thus the drinking water availability and the irrigation process are both affected. Poor access to clean water raises the probability of numerous health conditions, especially among women and children (Chowdhury et al., 2018).

The environmental vulnerabilities of Sundarbans are related to both environmental and social dimensions, as the latter, such as controversial land ownership and socio-economic differences, greatly impact the former, leading to greater vulnerabilities. In instances where there exist climate disasters, there are members of the local population who cannot be compensated by the government because they lack formal ownership of land. Secondly, gender-segmentation of political spaces and unequal resource distribution is the reason for weak access by women (Ahmed et al., 2021).

7.4 Case Study: An Overview of Coping Mechanisms in the Gosaba Block of the Sundarbans

The case study of resilience in people in the backdrop of increasing environmental stresses is the Gosaba Block in the complex estuarine network of the Indian Sundarbans. The area most at the frontline of the effects of climate change and other environmental issues is the low-lying deltaic region whose dynamic nature is characterised by tidal rivers, mangrove forest and human settlements. Throughout the generations, the Gosaba residents have been walking a thin rope life, with their lives being inseparably connected to the beating and ups and downs of this unique ecosystem. As the frequency and intensity of environmental shocks in the form of cyclones and storm surges increase with the rise of sea levels and the rise in salinity, however, their traditional adaptive strategies are being stretched to their limits and new adaptive strategies are necessary. The most immediate and devastating danger to the Gosaba Block is caused by cyclonic storms and storm surge. The region lies in the entrance of the Bay of Bengal and so, the region is mostly affected by such severe weather patterns. Over the years, people have come up with many coping strategies of dealing with the immediate effects. The traditional houses which are typically built on high plinths and composed of materials which are easy to find like mud and bamboo are somewhat resistant to frequent floods. Early warning systems, which are usually dependent on networks of communities and local expertise, are also important to enable the people to relocate to safer areas, which could be the cyclone shelters or the multi purpose community halls, which are constructed on elevated grounds. Nevertheless, these conventional defences are usually crushed by the growing severity of cyclones, as in the case observed over the past decades resulting in massive destruction of residential homes, infrastructure and farmlands. During an interview it has been highlighted by the locals.

> *We've seen cyclones before, but now... now it feels like the whole world is angry with us.*
> Samaddar Majhi
> (A villager)

The above statement experience of climate change in a vulnerable region. The feeling that 'the whole world is angry' speaks volumes about the perceived scale of the threat, the potential connection to human-induced environmental changes and the deep sense of vulnerability felt by the community in the Gosaba Block. Moving beyond traditional coping mechanisms, these communities increasingly leverage indigenous ecological knowledge—such as the granular interpretation of local weather patterns—and rely on informal social safety nets to navigate environmental volatility. They are increasingly using mobile early warning systems, cyclone shelters, aquaculture and salt-resistant agriculture to diversify their sources of income. As a testament to how communities adapt while working to defend their ecological rights, mangrove restoration is also a crucial endeavour to fortify natural defences. They are now building the earthen bunds to stop the cyclone water flow.

Today the degradation of the mangrove ecosystem further exacerbates the vulnerability of the Gosaba Block. Mangroves act as natural bio-shields, buffering the impact of cyclones and storm surges, preventing coastal erosion and providing crucial habitats for a variety of flora and fauna. However, decades of deforestation for agriculture, aquaculture and fuelwood have significantly reduced mangrove cover. Recognising the critical role of mangroves in building resilience, community-based mangrove restoration initiatives are gaining momentum. Local communities, often with the support of NGOs and forest departments, are actively involved in planting and protecting mangrove saplings. These efforts not only help in coastal protection, but also contribute to biodiversity conservation and the sustainable harvesting of non-timber forest products. The Bay of Bengal is prone to cyclones, over many years *bunds* and embankments have been built over the water to save the village from getting washed away. During cyclone these embankments often gets washed away, as seen in the picture, local communities often engage in the manual reinforcement and repair of earthen embankments using locally available materials like soil-filled sacks, bamboo and wooden stakes. This is a common immediate response to breaches or weaknesses in the flood defences. The image clearly shows collective action, highlighting the importance of community solidarity and cooperation in managing flood risks. Faced with increasing salinity due to flooding, communities are experimenting with salt-tolerant crop varieties and aquaculture practices can withstand higher salinity levels.

Beyond these direct environmental impacts, the communities of Gosaba also grapple with socio-economic vulnerabilities that compound their challenges. Social marginalization, poverty and restricted access to healthcare and education can all make it more difficult for them to adjust to and bounce back from environmental shocks. As a result, addressing these underlying vulnerabilities holistically is necessary to build resilience in Gosaba. Nowadays, a large number of SHGs (Self-Help Groups) are essential for encouraging group action, making resources and information more accessible and empowering underrepresented groups, especially women's communities. The Tiger Widow Organization, whose diverse initiatives are essential to increasing awareness and advancing tiger conservation in the area, is a striking example. The communities that live in the Sundarban block are the primary stakeholders in conservation because they face numerous difficulties on a daily basis. The communities of the Gosaba Block are exhibiting incredible resilience in the face of complex and escalating environmental challenges, battling nature on a daily basis. Traditional knowledge, cutting-edge technologies, community-based projects and a growing understanding of the need for sustainable practices are just a few of their varied and dynamic coping strategies. But because of the size and complexity of the environmental threats, local communities, governmental organisations, non-governmental organisations and researchers must continue to work together. Gosaba needs to come up with long-term resilience plan that is founded on wide range of strategies. This programme cannot be solely based on how to react to the immediate environmental ramifications; it must also consider intrinsic socio-economic vulnerabilities, encourage ecological regeneration and encourage the establishment of a framework of collaboration and knowledge based in fighting the uncertain and changing future.

Driven by a profound connection to their environment, the communities are taking significant risks to protect it, ultimately for their own well-being. The case study of Gosaba offer valuable lessons for other coastal communities facing similar environmental adversities, highlighting the crucial role of human agency, adaptation and community-led initiatives in building resilience in a rapidly changing world. The people of the Sundarbans, facing severe environmental problems, have made some positive changes in resource management to support their livelihoods and protect their rights and, at the same time, accommodate the environmental changes. Acknowledging that the protection of the mangroves is an essential task for most of the local communities, some of the NGOs have started tree planting activities since these areas have been so badly disturbed that the good forests have disappeared. These projects also provide ecotourism and conservation-based jobs as the alternative sources of income besides acting as organic barriers against floods and storms (Ghosh et al., 2019). Two of the adaptation measures employed by farmers in the Sundarbans are planting salt-tolerant rice varieties and practicing mixed-crop farming that in the long run will reduce the risks of losses in production. Hydroponics or the use of the major aqueous plant roots to form the floating beds are mainly formed using the aqueous vegetation, namely, bamboo rafts and water hyacinth. They can be a useful tool for dealing with unfavourable weather during harvest season, which invariably follows floods. This represents a profound ecological interaction between scientific understanding and hands-on learning, making it more than just a farming innovation. Due to environmental degradation with regard to fish and catching methods, local (male and female) fishermen have resorted to adopting crab farming and green/agriculture practices such as fish-rice cultivation and have been successful in this. The aforementioned techniques, apart from the economic reasons, also contribute to the conservation of the mismatched environments (Dasgupta et al., 2021). Efforts are being made to construct climate-resilient embankments using geotextiles and local vegetation. Additionally, traditional mud houses are being reinforced with stronger materials to withstand extreme weather conditions (Sen & Roy, 2018). To avoid depending too much on occupations that are affected by climate, people are creating various sources of income. Woman Handicrafts (Honey Processing) and Small-Scale Entrepreneurship are SHGs (Self-Help Groups) that are producing crafts, honey processing and small-scale entrepreneurship. Training seminars not only empower the uneducated young individuals with skill set, but also, in turn, create job opportunities in tourism and weaving of cloth and solar technology. Thus, they become economically self-sufficient and provide developmental support for the poor communities (Jalais, 2020). A great start has been made with rainwater harvesting systems that are in place in several villages to relieve the problem of water scarcity. These efforts are supported by some self-help activities such as sludge removal from plywood tanks and restoration of underwater springs (Bhattacharya et al., 2022). Local governance bodies, in collaboration with NGOs and government agencies, have strengthened disaster preparedness by establishing early warning systems and cyclone shelters. Community training programmes on disaster response and evacuation strategies have significantly reduced casualties during extreme events (Roy et al., 2021).

7.5 Limitations of the Study

Though this research on resilience in the Gosaba Block had some valuable insights, there were some challenges and limitations that were faced. First of all, the qualitative approach, though offering deep, profound data which is made possible by the ethnographic observation and thorough interviews, is inherently restricted in the way it can be applied to other geographical or socio-economic frameworks. In certain cases, due to the peculiarities of the culture and ecology of the Sundarbans, it is hard to reproduce or extrapolate some of the strategies of adaptation. Moreover, though the study records existing adaptive abilities, it has failed to determine numerically how effective or scalable such community-based programmes would be under the future and the most adverse impacts of climate change. The emphasis of the study in the inner processes of resilience of the community, though vital, may have disproportionately controlled the comprehensive investigation into the external structural barriers that could have influenced the vulnerability of the community and adaptive ability, including the governmental policies in general, market forces or the ineffective support of the community. The constraints of the field-based qualitative research in terms of resources and time also implied that more thorough historical examination of all the environmental changes or more profound evaluation of various socio-economic effects of the changes that occurred in the diversity of people groups in the Gosaba Block could not have been made with the balance of envisioned depth, thus suggesting areas for future, more longitudinal or quantitative research.

7.6 Conclusion and Recommendations

In a world of constant environmental upheaval, residents of the Gosaba Block represent what Raymond Williams used to refer to as structures of feeling the lived, experienced experience of community in the face of change not merely on a material level, but also, emotionally and culturally (Williams, 1977). Based on qualitative methodologies, such as ethnographic observation, in-depth interviews and examinations of community narratives, this study has carefully captured these lived experiences that has offered a balanced insight into how these people coped with their circumstances. Their strength is not as reactive as generative, made on the basis of indigenous knowledge, collective memory and a kind of symbiosis with the dynamic ecologies of the Sundarbans. Nor are they innocent of climate catastrophe. Rather, such communities are currently engaged in a process of improvisation, adaptation and innovation. It is not necessary to think that bureaucrats and politicians, however well-intentioned they are, are better placed to resolve problems than the people on the ground who, as Elinor Ostrom (1990) successfully argued, have the greatest interest in seeing the solution work (p. 24). The daily challenges and inventions such as restoring mangroves to salt farming, repairing embankments in communities to female entrepreneurship are forms of resilience that bring together tradition and pragmatism. This practise has been resonant with the concept of theorists like Dipesh Chakrabarty (2009) to think climate history in expanse of subaltern voices within the planetary history of ecological crisis. Indeed, the Gosaba communities are not merely surviving, but they are reasserting control over uncertainty by building new ecological futures via local governance and grassroots efforts.

The study leaves a major impression in the scientific research as it questions traditional and top-down methods of examining climate adaptation and points out the importance of local agency and indigenous ecological knowledge. The chapter keenly explains how a given community of the Sundarbans essentially experiences and addresses the pervasive environmental warfare, thereby providing an insight on the ways to go towards sustainable solutions. The study points out the socio-ecological crisis with real-life impacts of environmental changes and socio-economic displacements on the Gosaba Block, not just discourses on the subject, but on the vulnerability itself. It then fairly supposes that community resilience as the primary process to survive this crisis is responsive because indigenous knowledge and memory and an in-depth symbiosis with their ecosystem make improvisation, adaptation and innovation possible, in defiance of the orthodoxies of top-down development which usually overlooks local capabilities. Moreover, the chapter integrates the critically environment, development and social crisis orientations through mapping the interrelationship of the ecological issues, the need for decentralised and just development processes and the social conflicts and activism of 'subaltern voices' in charting new ecological futures. This fine-grained, on-the-ground view addresses a major gap in the literature, which tends to miss empirical assessments of community-based adaptation that comprehensively engage these aspects. By framing the Gosaba communities as neither passive sufferers but active actors 'reclaiming agency over uncertainty', your research presents new insights on 'sustainable solutions' and 'development', reiterating that efficient climate action and sustainable futures are grounded in empowering local people and acknowledging their self-organising potential, thereby presenting a strong case for the universal applicability of localized resilience in combating complex environmental and social problems. The value of this research is far-reaching. In proving its efficiency, in terms of community-based adaptation, it presents a good case of local empowerment in climatic policy-making and policy implementation. It has been demonstrated that the same communities that are suffering the most due to climate change tend to be the same communities that have been providing their own solutions to these problems in a sustainable manner. The experience of Gosaba is of serious world importance. With climate change being experienced in all parts of the world, many of the vulnerable and low-lying communities are faced with the same ecological risks and socio-economic discontinuities. The lessons learned in Gosaba as pertaining to adaptive capacity, harmonisation of the traditional and contemporary practises and the necessity of collective action are important messages to all policymakers, practitioners and researchers universally. The study is part of a crucial international discourse of climate justice that argues that commendable climate action must engage the recognition as well as encouragement of the self-organising capacity of the most impacted people. It highlights the applicability of localised resilience as a paradigm to negotiating the challenges faced by a shifting planet as universal, thus a curious and vital read to the global audience.

Finally, the history of Gosaba is a witness to the persistence of cultural and ecological connections in a time of disjuncture. It not only calls scholars, policymakers and the civil society to take note of such initiatives but to learn as well. Only uniformity and control should not be rated as the only valid sources of productivity and knowledge since this stifles the creativity of nature and people, as correctly observed by Vandana Shiva (1988) (p. 12). In its salty fields and muddy embankments, Gosaba tells another story, that of resiliency being created through group nurturing, collective memory and an implacable determination to survive in the face of destiny. It would be wonderful one day to find a research that develops climate adaptation models which are based on the ecological wisdom of the past and future technology. We need to examine such problems too as the migration of climates, the special susceptibilities of various genders and the long-term effects on society and economy due to the change in the environment. Such knowledge would assist in setting up sustainable development plans of the region. To ensure this occurs, it is necessary to have a concerted effort by local communities, governments and international organisations. By doing so, the vulnerable groups will be in a position to continue enjoying a sustainable relationship with nature besides securing their future against the uncertainties posed by climate changes.

AI Discloser

The authors ensure that all created material had been revised, edited and checked against factual truth and suitability. The original research findings and the interpretation of qualitative data were not created on an AI tool without human control and verification. The main arguments of the paper were not built on an AI tool without human supervision.

Acknowledgements This research would not have been possible without the invaluable contributions of many individuals and institutions. Foremost, we extend our deepest gratitude to the resilient communities of the Gosaba Block in the Sundarbans, whose openness, generosity and unwavering spirit in sharing their lived experiences, insights and traditional knowledge formed the very bedrock of this study. Their trust and active participation were truly inspiring. We are immensely thankful to the local community leaders, facilitators and interpreters whose guidance and support were instrumental in navigating the intricacies of the field and fostering meaningful connections with the residents. Their intimate understanding of the region and its people was indispensable. We would like to express our sincere gratitude to J. Mukherjee, A. Sen, K. Lahiri-Dutt and A. Ghosh for their foundational chapter, 'Towards Transformative Social Resilience: Charting a Path with Climate-Vulnerable Communities in the Indian Sundarbans', in the publication *Planetary Justice: Stories and Studies of Action, Resistance and Solidarity*. Their work provides an invaluable framework for understanding the complex interplay between climate vulnerability and social resilience, significantly shaping our perspective on community-led adaptation strategies in fragile ecosystems. Dr. Jenia Mukherjee is my supervisor who always guide us in every path. Her insightful analysis served as a crucial source of inspiration and guidance throughout this project. We are also immensely grateful to our mentor and Hon'ble Vice Chancellor Sir Dr. Ashok K Srivastava, our research director Dr. Jayeeta Chattopadhay for their continuous encouragement. Finally, we acknowledge the groundbreaking work of scholars like Raymond Williams, Elinor Ostrom, Dipesh Chakrabarty and Vandana Shiva, whose theoretical frameworks provided profound inspiration and intellectual grounding for understanding the complex dynamics of socio-ecological resilience.

Authors Contribution There has been equal contribution of both authors to this work.

Funding The research was not funded.

Potential Conflicts of Interest

On behalf of the authors, the corresponding author states that there is no conflict of interest.

Informed Consent

Informed consent was taken on all individual participants involved in the study.

References

Adger, W. N. (2000). Social and ecological resilience: Are they related? *Progress in Human Geography, 24*(3), 347–364.

Ahmed, S., et al. (2021). Climate change and human rights in the Sundarbans. *Climate Policy Journal, 15*(3), 245–260.

Aldrich, D. P. (2012). *Building resilience: Social capital in post-disaster recovery.* University of Chicago Press.

Alongi, D. M. (2008). Mangrove forests: Resilience, carbon cycles and climate change. *Progress in Oceanography, 78*(1–4), 1–13.

Berkes, F., Colding, J., & Folke, C. (2000). Rediscovery of traditional ecological knowledge as adaptive management. *Ecological Applications, 10*(5), 1251–1262.

Berkes, F., & Folke, C. (Eds.). (1998). *Linking social and ecological systems: management practices and social mechanisms for building resilience*. Cambridge University Press.
Bhattacharya, R., et al. (2022). Water management in the Sundarbans: A case study on rainwater harvesting. *Journal of Water Resources, 29*(4), 112–129.
Blasco, F., Saenger, P., & Janodet, E. (1998). Global coverage of mangroves: present status and future perspectives. In *Mangrove ecosystems: technical guidelines for assessment, monitoring and management* (pp. 3–27). International Society for Mangrove Ecosystems.
Bonanno, G. A. (2004). Loss, trauma, and human resilience: Have we underestimated the human capacity to thrive after extremely aversive events? *American Psychologist, 59*(1), 20–28.
Carpenter, S. R., Walker, B., Anderies, J. M., & Abel, N. (2001). From metaphor to measurement: Resilience of what to what? *Ecosystems, 4*(8), 765–781.
Chakrabarty, D. (2009). The climate of history in a planetary age. *Critical Inquiry, 35*(2), 197–222.
Chowdhury, M. R., & Maiti, S. K. (2020). Climate change and resilience strategies in the Indian Sundarbans: A socio-ecological perspective. *Environmental Science and Policy, 108*, 81–90.
Chowdhury, M. A., et al. (2018). Salinity intrusion and water scarcity in the Sundarbans. *Environmental Research Letters, 13*(6), 320–339.
Chowdhury, A., et.al (2016). Ecopsychosocial Aspects of Human–Tiger Conflict: An Ethnographic Study of Tiger Widows of Sundarban Delta, India. *Environmental Health Insights, 10*(10).
Cote, M., & Nightingale, A. J. (2012). Resilience thinking meets social theory: Situating social change in socio-ecological systems (SES) research. *Progress in Human Geography, 36*(4), 475–489.
Dasgupta, R., Shaw, R., & Takeuchi, Y. (2014). Climate change and community resilience in the Indian Sundarbans. *Environmental Hazards, 13*(3), 217–237.
Dasgupta, P., et al. (2021). Sustainable aquaculture practices in the Sundarbans. *Fisheries Journal, 27*(2), 188–205.
Ellis, F. (2000). *Rural livelihoods and diversity in developing countries*. Oxford University Press.
Folke, C. (2006). Resilience: The emergence of a perspective for social–ecological systems analyses. *Global Environmental Change, 16*(3), 253–267.
Folke, C., Carpenter, S. R., Walker, B., et al. (2010). Resilience thinking: Integrating resilience, adaptability, and transformability. *Ecology and Society, 15*(4), 20.
Ghosh, A. (2018). Mangrove restoration and resilience building in the Indian Sundarbans. *Journal of Coastal Conservation, 22*(4), 731–743.
Ghosh, T., et al. (2019). Community-based mangrove restoration in the Sundarbans: Challenges and success stories. *Ecology and Society, 24*(2), 58–73.
Hazra, S., Ghosh, T., DasGupta, S., & Sengupta, S. (2021). Impact of Cyclone Amphan on the coastal environment and livelihoods of West Bengal, India: A remote sensing and field-based assessment. *Science of the Total Environment, 769*, Article 144847.
Holling, C. S. (1973). Resilience and stability of ecological systems. *Annual Review of Ecology and Systematics, 4*(1), 1–23.
Intergovernmental Panel on Climate Change (IPCC). (2014). *Climate Change 2014: Impacts, adaptation, and vulnerability*. Cambridge University Press.
IPCC. (2022). *Climate Change 2022: Impacts, Adaptation, and Vulnerability. Contribution of Working Group II to the Sixth Assessment Report of the Intergovernmental Panel on Climate Change*. In H.-O. Pörtner, D. C. Roberts, M. Tignor, E. S. Poloczanska, K. Mintenbeck, A. Alegría, M. Craig, S. Langsdorf, S. Löschke, V. Möller, A. Okem, & B. Rama (Eds.) Cambridge University Press.
Jalais, A. (2020). *The Sundarbans: People, place, and livelihoods*. Oxford University Press.
Joseph, J. (2013). Resilience as embedded neoliberalism: A governmentality approach. *Resilience, 1*(1), 38–52.
Mallick, A. U., & Rahman, A. F. M. (2010). Environmental degradation and human vulnerability in the Bangladesh Sundarbans. *Journal of Environmental Management, 91*(9), 1921–1929.
Masten, A. S. (2001). Ordinary magic: Resilience processes in development. *American Psychologist, 56*(3), 227–238.

Mukherjee, S., & Nayak, P. K. (2021). Resilience of coastal communities in the Sundarbans: Traditional knowledge and adaptation. *Sustainability, 13*(6), 3402.
Mukherjee, N., Sutherland, W. J., Dicks, L. V., Fellman, N., et al. (2018). Local ecological knowledge and its role in adapting to environmental change: A case study from the Indian Sundarbans. *Journal of Environmental Management, 218*, 486–495.
Mukherjee, S., et al. (2022). Impact of cyclones on the Sundarbans and adaptive strategies. *Climate Risk Management, 40*(5), 305–321.
Mukhopadhyay, A., Danda, A. A., & Ghosh, A. (2019). Building climate resilience in the Sundarbans: Emerging lessons from field interventions. *Climate and Development, 11*(7), 608–620.
Norris, F. H., Stevens, S. P., Pfefferbaum, B., et al. (2008). Community resilience as a metaphor, theory, set of capacities, and strategy for disaster readiness. *American Journal of Community Psychology, 41*(1–2), 127–150.
Ostrom, E. (1990). *Governing the commons: The evolution of institutions for collective action.* Cambridge University Press.
Rahman, A., et al. (2020). Biodiversity loss in the Sundarbans: Causes and consequences. *Conservation Biology, 23*(3), 201–217.
Rose, A. (2004). Defining and measuring economic resilience to disasters. *Disaster Prevention and Management, 13*(4), 307–314.
Roy, K., et al. (2021). Disaster preparedness in the Sundarbans: A case study of early warning systems. *Natural Hazards, 55*(4), 312–329.
Sen, D., & Roy, B. (2018). Climate-resilient infrastructure in the Sundarbans. *Engineering Journal, 21*(2), 78–96.
Shiva, V. (1988). *Staying alive: Women, ecology, and development*. Zed Books.
Vogel, B., Hötte, N., Theodoridou, K., et al. (2020). Resilience and the COVID-19 pandemic: Implications for public policy and governance. *Journal of Risk Research, 23*(7–8), 1030–1051.
Williams, R. (1977). *Marxism and literature*. Oxford University Press.
Williams, R. (1985). *Keywords: A vocabulary of culture and society*. Revised edition. Oxford University Press.

Chapter 8
COVID-19 Vulnerability: A Societal Implication and Community Response in India

Chandrani Tiwari and Gouri Sankar Bhunia

Abstract The COVID-19 pandemic has revealed and heightened liabilities within Indian society, requiring the deficit of correspondence in health, economic permanency, and access to necessary services quite proclaimed. Present chapter examines the vulnerability of COVID-19 from a social perspective by evaluating varied demographic classes such as the elderly, people with comorbid environments, poor people, and marginalized communities. We adopt a mixed-methods approach to explore the systemic inequities and socioeconomic features contributed to the susceptibility of demographic clusters both to the virus and its impacts. Present research work discusses community responses to the pandemic, aiming on grassroots initiatives, reciprocal aid networks, and government interferences that were portrayed to alleviate the waves of the catastrophe. Outcome of the study highlights the consequence of inclusive public health strategies that concentrate root causes and help build community resilience. Present work is accomplished by putting forward references for policymakers and community leaders to progress preparedness for any future public health emergencies, captivating a collaborative approach that lays the health and well-being of the most vulnerable units of society at the forefront.

Keywords COVID-19 · Vulnerability · Societal implications · Community response · Public health · Socioeconomic disparities · Marginalized communities

8.1 Introduction

The COVID-19 pandemic has been a worldwide calamity that has publicized and deteriorated societal exposures. India's wide range of residents and disparities in socioeconomic status have made the epidemic a significant social impact. The

C. Tiwari
Siddharth College, Bilaspur, Chhattisgarh, India

G. S. Bhunia (✉)
Independent Researcher, Kharagpur, Paschim Medinipur, West Bengal, India
e-mail: rsgis_gouri@rediff.com

G. K. Panda et al. (eds.), *Socio-Ecological Crisis of Development*, Sustainability Solutions, https://doi.org/10.1007/978-3-032-22171-1_8

objective of COVID-19 vulnerability is to examine the numerous characteristics of susceptibility that emerged during this unusual crisis and the reactions of the community (Chatterjee & Purohit, 2021). India awaits a comprehensive approach to completely comprehend COVID-19 vulnerability by interchanging socioeconomic, health, geographical, gender, and organizational variables. Planning authorities as well as health professionals can build a more resilient civilization that is robust to potential tragedies by knowing and exchanging with these vulnerabilities (Sanghvi, 2020).

COVID-19 aroused attention to the susceptibility of societal frameworks, economic stability, and health systems as it propagated throughout the entire country. The availability of COVID-19 examination, therapy, and vaccination has been affected by geographical variations. Underreporting of cases has been caused by the dearth of laboratory space and health care facilities in several villages (Sahni & Chaudhary, 2020). While there were challenges in rural areas, urban areas had overcrowding as the result of COVID-19 and higher transmission rates which lead to inequities associated with resource distribution, testing, and treatment access (Kumar & Kumar, 2021). Rate of COVID-19 spread varied widely across states and union territories in India. The states of Maharashtra and Delhi reported high cases due to density, and rural and less populated regions had low reporting initially (Sinha, 2020). Communities who are marginalized including low-income groups who largely work in the informal economy, and especially preconditioned people, found themselves at the highest risk for morbidity and complications accessing healthcare, employment, and basic human needs (Rathi & Prakash, 2020). Additionally, there were implications beyond physical health due to the pandemic. Mental health was affected, along with education, and social ties and cohesion, demonstrating the interwoven fabric of various society constructs.

Communities across India found inventive ways to respond to the issues and challenges posed by the pandemic in a show of optimism and resilience. Local authorities, community volunteers, and small-scale groups worked collectively to ensure that the most vulnerable populations had possession of amenities, support, and information (Mohan, 2021). Several demographic groups throughout India have been identified to be more susceptible during the COVID-19 pandemic because of things like age, occupation, socioeconomic level, and underlying medical issues. Because of their frequent lack of access to healthcare, a healthy diet, and comfortable environments, people who belong to poverty or low-income households are more predisposed to COVID-19 and may have worsening conditions (Nanda & Choudhury, 2021). Pregnant women have been considered as a fragile category because of possible pregnancy and parenting difficulties, even though research on their risk for serious medical conditions was still ongoing. Laborers in the informal economy, such as street hawkers and building workers, are more susceptible to the global outbreak as they frequently have no health coverage and secure jobs. This research enhances our understanding of how disasters can reshape social structures and highlights the importance of building inclusive and resilient communities by examining the impacts of COVID-19 through the lenses of vulnerability and community functioning. Outcome of this study will expand policy suggestions and approaches for grabbing vulnerabilities in evolving public health consequences and advocating a more just and resilient Indian society.

Considering that the literature currently in publication has focused on the health and financial implications of COVID-19, a few research studies have comprehensively investigated the interconnections of different vulnerability aspects (socioeconomic circumstance, geographical region, gender, occupation, and institutional remedy) alongside grassroots and community-based programs in the Indian context. This study is unique because it implements an extensive approach that evaluates community-driven and local responses in multiple regions of India in addition to a thorough analysis of susceptibility. This holistic approach has produced a comprehensive analysis of both institutional flaws and local adaptability throughout the global epidemic. The main goals are to: (1) identify and appraise the major economic and societal susceptibilities of COVID-19 found in Indian context; (2) investigate the nature and effectiveness of local remedies; and (3) generate concepts for bolstering medical services, social insurance, and resilience tactics to reduce future public health incidents.

8.2 Materials and Methods

The review may bring together important insights into the sociocultural and economic implications of the epidemic in India by selecting significant investigation issues, using methodical search techniques, and critically examining previous research. We have selected the academic databases e.g., Web of Science, Scopus, Google Scholar, and PubMed that provide comprehensive coverage of research on COVID-19 and societal impacts. We have limited the search to publications from 2000 onward to focus on research that addresses the societal implications of the COVID-19 pandemic. We have used several keywords like "COVID-19 societal impact," "Community response to COVID-19," "Economic vulnerability and COVID-19," "Healthcare disparities in COVID-19" to identify the leading journals publishing in this domain. Common keywords such as "vulnerability," "pandemic," "community health," "economic impact," "unemployment," "health disparities," and "government response" were used to identify overall trends in the published literature. Published between 2000 and the present, peer-reviewed articles and reports from reputable sources are considered as inclusion criteria in this study. The analysis focuses on societal and community-level responses, with retrieved articles screened according to their relevance. Use the abstract and introduction sections to filter articles that directly address the research questions. Figure 8.1 illustrates the step-by-step process for conducting a literature review of the study.

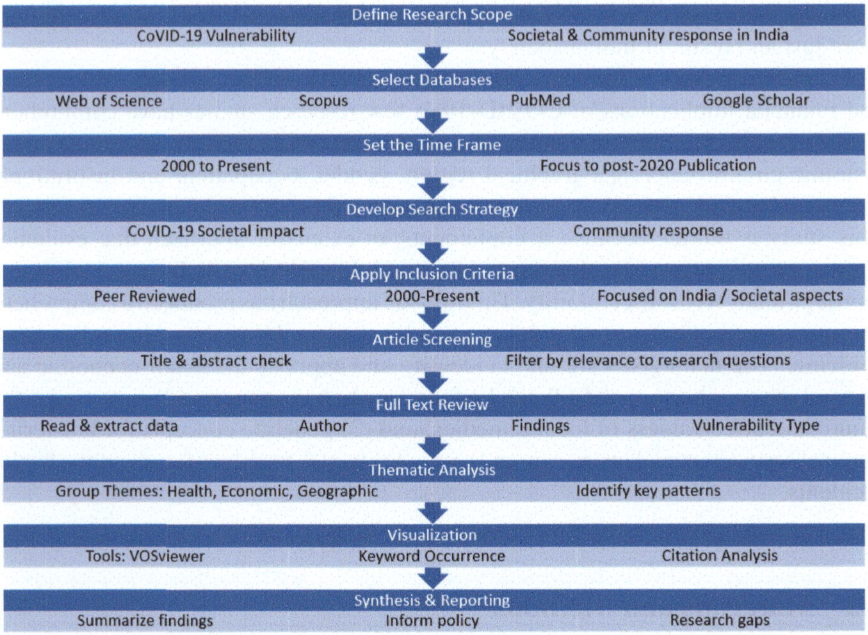

Fig. 8.1 Flow chart of details methodology of the study

8.3 Results

8.3.1 Societal Implications of COVID-19

The societal effects of COVID-19 are revealing venerable discriminations and exposures in India and worldwide. There is a necessity for comprehensive and equity-fixated policies that highlight the health and well-being of downgraded classes. This is also augmenting public health arrangements, economic resistance, and other significant issues. Reviewing the continuing impacts of the pandemic on society and the numerous approaches retained to promote recovery and resilience will complement this research.

Worldwide, the studies have pointed out that healthcare systems have failed, which led to the brief shortages in supplies and staff (Ghosh & Gupta, 2020; WHO, 2021). Besides that, Paltiel et al. (2020) and Bhatia et al. (2021a, 2021b) have demonstrated that in the case of disadvantaged populations, the lack of proper access (Table 8.1) caused these populations to have higher rates of death and diseases. In India, the condition in the villages was even worse, as Yadav et al. (2021) explains it is because of the absence of sufficient healthcare facilities.

Table 8.1 Summary of COVID-19 pandemic in association with societal implications in India

Year	Location	Objectives	Methods	Outcome	Source
2020	Nationwide	Assess the impact of lockdown on social behaviors	Surveys and observational studies	Significant reduction in social interactions and increased reliance on digital communication	Sarkar and Chowdhury (2020)
2020	Nationwide	Analyze mental health consequences of isolation	Cross-sectional surveys, psychological assessments	Increased anxiety and depression reported across various demographics	Banerjee and Rao (2020)
2020	Rural and urban areas	Evaluate the impact of online education on learning outcomes	Mixed methods: surveys and case studies	Widespread learning losses; significant disparities in access to digital education resources	Jena (2020)
2021	Delhi, Maharashtra	Explore political narratives and public compliance	Content analysis of media and public speeches	Political polarization influenced adherence to health guidelines	Chakraborty and Bhatia (2021)
2021	Nationwide	Evaluate learning losses and the role of digital learning in mitigating these	Mixed methods: field studies, surveys	Significant learning loss in rural and under-resourced areas; digital learning not accessible to all	Azim Premji Foundation (2021)
2021	Nationwide	Investigating educational policy changes due to COVID-19	Policy analysis and interviews	Emergence of NEP 2020 with a focus on digital education but challenges in implementation	Sharma (2021)
2021	Bihar, Uttar Pradesh	Examine migration issues and health impacts during lockdown	Qualitative interviews, surveys	Migrant crises highlighted inadequacies in social safety nets and public health responses	Kumar et al. (2022)

(continued)

Table 8.1 (continued)

Year	Location	Objectives	Methods	Outcome	Source
2021	Rajasthan, Tamil Nadu	Investigating the impact of COVID-19 on private sector education and governance challenges	Case studies and interviews with educators, parents	Private schools were quick to adapt but increased fees worsened inequalities	Bhandari and Yadav (2021)
2021	Kerala	Assess public health communication effectiveness	Surveys and media analysis	Mixed effectiveness in public health messaging; challenges in misinformation	Sathish (2022)
2022	Nationwide	Analyzing long-term impacts of COVID-19 on education	Longitudinal studies, educational assessments	Persistent learning deficits; need for targeted interventions to support vulnerable groups	Sundararaman (2020)
2022	Nationwide	Explore socioeconomic recovery post-COVID-19	Case studies and economic analyses	Uneven recovery patterns; increased inequalities and varying access to resources	Jain and Chatterjee (2024)
2023	Urban areas	Investigate urban mental health trends post-pandemic	Cross-sectional surveys, qualitative interviews	Continued mental health challenges; increased demand for mental health services	Basu et al. (2024)
2023	Nationwide	Assess the digital divide in education during recovery	Surveys and statistical analysis	Digital divide persists; calls for equitable access to technology and resources	Agarwal (2022)

(continued)

Table 8.1 (continued)

Year	Location	Objectives	Methods	Outcome	Source
2024	Nationwide	Evaluate governance effectiveness in managing health crises	Policy reviews and stakeholder interviews	Improved governance strategies but challenges in execution and public compliance remain	Mittal (2024)
2024	Various states	Study the resilience of social networks post-COVID-19	Mixed-methods approach: surveys and case studies	Emergence of new community support systems; resilience in social ties	Vandrevala et al. (2024)

8.3.1.1 Economic Impacts

This pandemic has severely impacted the economy across the globe and India is no exception. There is a lot of economic difficulty in terms of job losses and volatility in small business. In April 2020, the Centre for Monitoring Indian Economy (CMIE) conveyed that the unemployment rate reached 23.5%, up from 8.4% in March 2020. It was estimated by the ILO (2020) and NITI Aayog (2021) that the pandemic resulted in the loss of 140 million jobs because of lockdowns across the developing world and informal daily wage workers have been disproportionately affected as these jobs are often unstable and workers frequently lack protections associated with formal employment. While the economic impact of COVID-19 is evident, the reliance on informal work in India combined with the lockdown resulted in millions of individuals in the informal sector losing their employment (and thus their income), exacerbating poverty and inequality (Chakraborty & Maity, 2020; Mukherjee et al., 2021). Similar trends have been observed across the globe, with the IMF (2020) reporting severe contractions in GDP on a global scale. Countries increased unemployment rates and a greater reliance on social support measures (Kumar & Choudhury, 2021) due in large part to the pandemic.

In the fiscal year 2020–2021, India's GDP shrank by 7.3%, one of the largest declines of a major economy (Ministry of Statistics and Programme Implementation, 2021). Collaboration has significantly weakened, bringing about the most severe impacts within the manufacturing, hospitality, and travel sectors (Chakraborty & Maity, 2020). The All India Manufacturers Organization (AIMO) carried out a survey of small businesses consisting of 75% of the respondents reported lower revenues while a number of them indicated a need to shut their businesses permanently as a result of prolonged lockdowns (AIMO, 2020). Mukherjee et al. (2021) confirmed the existence of small businesses' greater susceptibility as a result of their modest degree of financial cushions. Small businesses that depended on internet and delivery systems sustained lesser impacts (Sinha & Prakash, 2021) but this also provided highlight on the digital resource gap (UNESCO, 2020). FICCI (2021) indicated that most entrepreneurs that were surveyed showed a lot of ambivalence regarding the future of their ventures, and a lot of them discussed the possibility of either downscaling their operations or even totally ceasing the operations of their businesses. The economic impact of COVID-19 triggered unconventional closures of small enterprises, high levels of unemployment, and economic uncertainties. Bihar and Uttar Pradesh were rated as the most economically fragile states because of their informal sectors' overwhelming dominance (approximately 85–90% of the workforce). Such conditions resulted in poor employment opportunities and income shocks during lockdowns. These states reported extremely high job loss rates, while household income plummeted by about 45–50% and there was virtually no access to personal savings, especially the low-income and marginalized communities. This economic hardship was provoked further by intense food insecurity, where daily wage workforces and the rural deprived reported the incapability to entrance basic food because of missing livelihoods and absence of social protection techniques (Table 8.2).

Vaccination coverage and the number of hospital beds per 10,000 population characterize the encouraging health infrastructure in Kerala and Tamil Nadu. There is less economic strain achieved in Kerala since it registered the highest health infrastructure index at 0.85 among the states compared due to sustained investments in decentralized healthcare and public health preparedness, leading to improved health indicators and a relatively lesser degree of economic strain during the pandemic.

They noted how crises unreasonably affect states of different socioeconomic infrastructures and pointed out the problems that states like Bihar and Uttar Pradesh face because they have more populations who are rural-based plus also have low literacy rates. The state gave an account of difficulties in healthcare messaging, management of community transmission, and access to online welfare schemes thereby increasing their vulnerability. Furthermore, literate and urbanized population groups in Kerala and Tamil Nadu states have more potential for mobilizing community response to health resources. This shows how important the function the socio-educational infrastructure has is for minimizing the impacts of such crises. These issues must be focused adequately through policy interventions that will not only help build recovery efforts over the short term but also stimulate flexibility as well as inclusion over longer terms. This emphasizes the socio-educational infrastructure and plays an important role in assuaging the consequences of these crises. It is critical that these issues are effectively addressed through policy intervention, which will not only build recovery efforts in the short term but also motivate resilience and inclusivity in the longer term. Key areas of investment include skills development, financial support, and digital infrastructure to empower small business owners to rebuild the economy after the pandemic.

Table 8.2 Economic vulnerability analysis of COVID-19 of some important states in India

State	Job loss Rate (%)	% Informal workers	Household income drop (%)	Access to savings (%)	Food insecurity (1–10)	Hospital beds/ 10 K	Vaccination (%)	Testing rate (/ 1000)	Maternal care (%)	Comorbidity (%)	Urban pop. (%)	Literacy rate (%)	SC/ ST/ OBC (%)	Pop. density (/km^2)	Health infra index (0–1)
Uttar Pradesh	35	85	45	20	8	5	65	350	60	18	22	67.7	70	829	0.45
Maharashtra	30	70	40	35	7	8	70	400	70	22	45	82.3	65	365	0.6
Bihar	40	90	50	15	9	4	60	300	55	15	11	61.8	75	1106	0.4
Kerala	20	60	30	60	3	15	90	800	95	30	48	94	40	860	0.85
Tamil Nadu	25	65	35	50	4	12	85	750	90	28	48	80.1	60	555	0.75

Source CMIE; CEDA; NFHS

8.3.1.2 Health Impacts

The COVID-19 epidemic has greatly affected mental health and access to healthcare services among vulnerable populations. Early in the pandemic, Wang et al. (2020) reported that 53.8% of its respondents in China experienced moderate to severe psychological effects, of which 28.8% were due to anxiety and 16.5% to depression. Pfefferbaum and North (2020) report that the prolonged pandemic, economic issues, and social isolation worsened mental health concerns, even after the peak, due to ongoing disruptions. Banerjee and Rai (2020) documented the devastating consequences of social isolation among elderly individuals and recognized increased loneliness-related disorders, depression, cognitive decline, and increased rates of suicide. Optimistically, Loades et al. (2020) indicated that the issues from prolonged social distance and school closures adversely affected children's and adolescent's mental health from increased anxiety, feeling of abandonment, and deteriorating emotional well-being. Lai et al. (2020) documented that healthcare workers experienced tremendous levels of burnout, depression, and anxiety due to overwhelming pressure, infection exposure, and high mortality rates, which created a mental health crisis in the field.

Moynihan et al. (2021) highlighted a significant disruption of routine and elective services as resources shifted toward COVID-19 management. This disruption meant that chronic illness patients were denied the care they needed, worsening their health status and worsening mortality from any cause other than COVID-19. Norris et al. (2021) identified fear of COVID-19 in medical facilities as a significant barrier that deterred individuals from obtaining care. That fear manifested itself into terrible health circumstances—particularly with regard to morbidities like diabetes, heart disease, and cancer which were presented late to providers, resulting in worse prognoses treatment pathways. Patel et al. (2021) highlight inequalities in healthcare access during the pandemic, showing that racial and ethnic minorities, economically disadvantaged individuals, and other vulnerable populations experienced elevated risks. This population was already at greater risk of negative health outcomes of the pandemic, including higher vaccine testing and vaccination rates, lack of insurance coverage, and existing disparities in health outcomes. Xue and McMunn (2021) observed that women had higher levels of anxiety and depression during the pandemic and reported on gender transformations in mental health. Among the reasons cited by some contributors were the greater burden on women of caring duties, higher job loss in female-dominated sectors, and rises in domestic violence under lockdown. Singh et al. (2020) conducted a review of the mental health effects on children and adolescents, and reported that school closure, isolation from society, and stress among parents were negatively associated with their relationships and development. They said many children were suffering from raised anxiety and depression as well as the behavioral problems which could have long-term effects on mental and emotional development.

Bihar and Uttar Pradesh have shown high levels of vulnerability in healthcare, as Table 8.3 indicates during the COVID-19. Both of these states were particularly exposed with severe shortages in terms of beds per 10,000 population and would have had limited surge capacity to handle additional load of cases due to COVID-19. In all these regions, access to maternal care was very low and in terms of the proportion of women who received full ANC, it reflected broader systemic weakness in primary HR systems. In addition, the burden of both communicable and non-communicable diseases (NCD) was high among these communities which further predisposed them to severe outcomes due to COVID-19. Compromised infrastructure, poor maternal care, and healthcare burden left these states quite fragile during a health emergency.

In contrast, the fall from the south, including Kerala and Tamil Nadu, was dramatically resisted by its much better developed health systems. Kerala was able to achieve vaccination coverage and maternal health care assessments that were substantially greater than the national average owing to a decentralized public health system that referred to high educational attainment and focused on basic medical services. Furthermore, Tamil Nadu had a strong immunization program, a large network of public and private hospitals, and improved COVID-19 testing capabilities. Higher readiness and generally better results were indicated by all variables. It is indicated that funding health systems and providing services fairly are important.

Assessments of health parameters in areas like West Bengal and Maharashtra show excellent capacity provision, as evidenced by their affordable demand for testing and beds, especially within urbanized and industrialized districts. Moreover, there is a substantial difference in accessibility to medical services between rural or peri-urban areas and urban areas. The COVID-19 pandemic was efficiently restricted to urban areas, but the rural hinterlands faced significant difficulties due to logistical obstacles and underfunded primary health systems. This indicates the essential for a more unbiased distribution of healthcare resources across these states.

Simultaneously, Assam and Jharkhand threw up the continued rural-urban discrepancy and universal infrastructure deficiency that clipped ability to fight back amid the pandemic. In these communities, rural populations were particularly disadvantaged with few laboratory resources, decreased visibility of vaccination activity, and little support for maternal health. These findings emphasize the importance of context in public health, highlighting that strategies must be tailored to specific socio-geographical conditions, particularly in peripheral or rural areas that often operate outside formal health systems.

8.3.1.3 Social Impacts

The COVID-19 pandemic has deep effects on the social structures, interactions, and education system of India. Desai and Vanneman (2021) and UNICEF (2020) emphasize that the most vulnerable groups, in particular low-income families, women, and marginalized communities, were put in a more challenging position, which was made worse by the problems of food insecurity and poor access to healthcare. According

Table 8.3 State-wise healthcare disparities: COVID-19 impact in India

State	Hospital beds/10,000	Vaccination coverage (≥1 dose %)	COVID-19 testing rate (tests/1,000)	Maternal care access (% ANC)	Comorbidity prevalence (% of pop.)
Bihar	~1.1 beds/ 1,000	~61.7% (NFHS-5)	Low (12 labs/ 1 M, low per capita)	~3.3% full ANC	High burden (TB/malaria)
Uttar Pradesh	~0.9 beds/ 1,000	~69.9%	Moderate (UP had >4 K hospitals, older data)	64% unable to access minimum ANC	Elevated NCD burden
West Bengal	~2.3 beds/ 1,000	~84.4%	78,566 public beds; high lab capacity	~70–80% estimated (NFHS-5 norm)	Standard national levels
Kerala	~1.05 beds/ 1,000	~82.1%	High testing (20+ labs/98 tests/lakh early)	~99% institutional births; strong ANC	Elevated comorbidity detection (30%)
Tamil Nadu	~1.1 beds/ 1,000	~89.4%	High (77,843 private beds; robust testing)	High ANC uptake (~>90)	Higher NCD prevalence (~28%)
Maharashtra	~2.3 beds/ 1,000	~73.6%	400+ tests/ 1,000; high ICU capacity	~80 + % ANC estimated	High prevalence NCDs
Assam	Not specified	~66.7%	Low (6 labs/ 1 M)	Lower ANC (>60?)	Noted rural comorbidity issues
Punjab	Modest beds (~0.6/10 k)	~89.1%	Good testing capacity	High ANC (~85%)	Moderate
Telangana	~1.3 beds/ 1,000	~79.1%	High (78 K private beds)	~80% ANC	~27% comorbidity
Jharkhand	Low beds (~0.4/10 k)	~74.1%	Low (labs underserved)	~60% ANC	Elevated
Mizoram	Small population, high frame	>90% immunization (pre-COVID)	Not specified	<5 per 1,000 IMR	~90% immunization indicates strong primary care

Source MoHFW COVID-19 Dashboard; NFHS-5 state/UT factsheets; National Health Profile data

to Rai and Singh (2021), their research shows that the caste and gender factors influenced the distribution of resources and support. The shutting down of the public places limited the communal activities to a great extent and hence, people got more isolated (Arora et al., 2020). Banerjee and Rao (2020) observed that due to social distancing people felt lonelier and that the community support system was breaking down, thus, these changes had an impact on the emotional state of individuals, especially in rural India. Besides that, they argued that community support is a source of an individual's happiness, particularly in rural India. Bhatia et al. (2021a, 2021b) documented instances of resilience when individuals shifted to online platforms for various social gatherings. Gupta and Khanna (2021) focused on gender-related issues. The evidence pointed to women being the most affected as a result of the increased caregiving and household chores during the lockdowns. Over 250 million students were said to have been impacted by school closures across the country, thus causing enormous disruption and learning loss (Jena, 2020). Urban and private schools were able to rapidly adopt online modes of instruction, whereas students from government schools and rural areas, constrained by limited access to devices, internet connectivity, and digital literacy, faced significant disadvantages (Chaturvedi et al., 2021). Dhar (2020) reported that the closing of schools caused the likelihood of child labor and early marriages, with girls being more susceptible to dropping out of school permanently because of economic pressures. The COVID-19 crisis revealed extreme differences between the Indian states in various aspects, especially healthcare and social vulnerability. Putting side by side the states, Bihar and Uttar Pradesh seemed the most precarious, being very vulnerable because of a mixture of low literacy rates (74.3% in Bihar and 69.7% in UP), poor health infrastructure, and large proportions of the marginalized population. More than 70% of the population in both states are from SC, ST, or OBC groups, and a significant number of them reside in rural and less developed areas. The shortage of hospital beds, limited access to maternal care, and weak public health delivery systems have slowed down pandemic mitigation efforts in these regions. These structural deficiencies have deepened the effects of COVID-19; thus, the hardest-hit have been daily wage earners, migrants, and socially disadvantaged communities (Table 8.4).

While the three states of Kerala, Tamil Nadu, and Maharashtra showed stronger resilience than other regions, Kerala, which has a literacy rate of 94% and the highest Health Infrastructure Index (HII) among large states (77.5), went far and beyond by adopting a proactive and decentralized approach in health management. Both Tamil Nadu and Maharashtra also kept their urbanization rates (above 45%) at a good level, had strong healthcare systems, as well as excellent testing and vaccination coverage. Altogether these factors allowed for a faster identification of the outbreak, provision of treatment access, and the better coordination of public communication. The sequence of high literacy, strong institutional competence, and infrastructure readiness made it promising for these states to react essentially and show persistent societal resilience. Furthermore, Delhi and Maharashtra as well, being highly urbanized areas, could tackle the fast-spreading interventions in the city center due to the availability of resources that were concentrated in urban metropolitan settings with institutional presence. In a way, these states had problems in rural and peri-urban areas, but the health systems in rural and deprived areas had been left untapped, besides there were digital exclusions that led to interruptions in care, delays in vaccines, and delivery of relief. The crisis of the concentration of health resources in metropolitan hubs pointed to a proper preparedness gap from one geography to another within one state.

Table 8.4 State-wise societal impact indicators of COVID-19 in India

State	Urban population (%)	Literacy rate (%)	SC/ST/OBC population (%)	Population density (per km^2)	Health infrastructure index (HII)
Bihar	11.3	74.3	~78	1,106	39.1 *(NITI)*
Uttar Pradesh	22.3	69.7	~74	828	34.4 *(lowest among large states)*
West Bengal	31.9	77.1	SC 28.6, ST 5.8, OBC ~45	1,029	39.7
Maharashtra	45.2	82.3	~65	365	61.8
Tamil Nadu	48.5	80.1	~68	555	64
Kerala	47.7	94	~40	860	77.5 *(highest among large states)*
Karnataka	38.6	75.4	~72	319	58.4
Rajasthan	24.9	66.1	~72	201	46
Madhya Pradesh	27.6	69.3	~78	236	43.5
Jharkhand	24.1	66.4	~75	414	48.3
Assam	14.1	72.2	SC/ST ~34, OBC ~30	398	45.8
Punjab	37.5	76.7	SC 31.9, OBC ~32	551	65.8
Chhattisgarh	23.2	71	SC/ST ~50, OBC ~30	189	42.6
Haryana	34.9	75.6	SC ~20, OBC ~38	573	58.6
Odisha	16.7	72.9	SC 17.1, ST 22.8, OBC ~30	270	50.5
Gujarat	42.6	78	SC 7.1, ST 14.8, OBC ~40	308	55
Delhi (UT)	97.5	88.7	SC 16.8	11,297	71.0 *(urban-focused infrastructure)*

Source Census 2011, NFHS-5, NITI Aayog; MoSPI

Moreover, the tribal areas that were mostly under the influence of tribal-dominated states, such as Chhattisgarh, Odisha, and Jharkhand, showed different kinds of problems during the coronavirus period. The tribal regions of these states are heavily comprised of the ST population, who are mostly dwelling in remote areas/forest. So, it was quite hard to carry out the instructions of the health department as there was the lack of transport and communication. Demographically, many of these tribal zones exhibited certain strengths such as communalism and traditional means of health, but the overall shortage of primary health facilities and low health literacy levels were barriers to effective national preparedness and coordinated responses. To ensure effective action in addressing vulnerabilities in these states, developing substantial interventions is crucial and these will need to evaluate both geographical and social-cultural contexts of tribal populations. Interventions should navigate the topography of tribal communities, mobile health services, multi-lingual communication, and empower community health workers.

8.3.1.4 Political and Governance Implications

The political and governance ramifications of COVID-19 were an enormous concern facing India and had a very significant impact on both the changing nature of social relations and, in particular, the education sector. Chakraborty and Bhatia (2021), for example, identified how public communications by the government, particularly during the early stages of the pandemic, projected the public health crisis in political terms. Political leaders, more specifically the Indian Prime Minister, would issue public messages about collective sacrifice, and these pronouncements, particularly at the height of the crisis, affected social behavior, with the wearing of masks, social distancing, and quarantine taken as acts of patriotic duty and service. Bose and Malhotra (2021) further explained that the politicization of other relief measures related to COVID-19 was recognized by them and, consequently, political partisanship influenced how states reacted to the pandemic, thus, in some cases, there was a local governance restructuring. Jena (2020) claimed that in the case of education, the government, among other things, recommended online platforms such as DIKSHA, ePathshala, and Swayam Prabha for providing digital resources to students. Sharma (2021) argued that it was particularly important that the National Educational Policy (NEP) had its main point in digital literacy and online education, which was the inevitable because schools all over the country had to switch to virtual platforms. Sundararaman (2020) conveyed that the loss of learning for a massive number of students, especially children in primary and secondary education, would not only hamper educational outcomes but also affect the economy for a very long time. Bhandari and Yadav (2021) noted that many private schools moved quickly to online platforms, often without the involvement of the government, creating high fee structures and different levels of access for students.

Although the measures were created in response to lessen the effects of the pandemic, inequalities were often deepened, rather than reduced. Restrictions put in place by the state worsened issues of isolation and mental health, especially within poor groups, for example, migrant workers and rural poor. The experience of education being shifted to an online environment emphasized the scale of the digital divide, with the most disadvantaged students suffering the largest impacts. Most of the governance strategies that were well-intentioned were poorly executed. This has resulted in social and educational issues taking much longer to remedy, as new policies will need to operate within social policy for sustained interventions.

8.3.2 Community Responses to COVID-19

However, in the face of adversity, the pandemic has also produced examples of community solidarity and resilience. These included grassroots initiatives and mutual aid networks that formed to assist vulnerable groups and communities, showing the power of collective action (Bhattacharya & Haldar, 2021; Patel & Haldar, 2021). According to some studies, it is especially important for communities to come together and provide assistance to meet the needs of the marginalized during pandemics (Mohan & Sharma, 2021). Table 8.5 provides several case studies that highlight how local communities and local governments from various regions in India were able to respond to COVID-19 by developing creative approaches that addressed challenges unique to each region.

8.3.2.1 Role of Community Organizations and NGOs

The COVID-19 pandemic has revealed very clearly that community organizations and non-governmental organizations (NGOs) are the most necessary pillars during crisis times. Their ability to quickly respond, convince communities to get involved, and influence policies for the better is what made most people realize how important their presence is in public health emergencies. NGOs, such as Feeding America in the U.S. and Goonj in India, took the lead to fight food insecurity by means of food banks and itinerant-packs distribution to provide food needs and curb economic insecurity in such a tough time of hardships (Sharma & Sharma, 2020). On the other hand, special attention was paid to the mental health of the population by providing access to counseling services and mental health support. In this respect, the Samaritans, a UK-based organization, have expanded their telephone lines in response to people's growing demand (Smith et al., 2022).

A study conducted by Kumar and Singh (2021) is a good example of how swiftly local groups were able to change their mode of operation to take care of new issues rising amid the pandemic. Anderson et al. (2021) stressed the point that trust is a pivotal thing in the relationship between community organizations and their members, which in turn, leads to more participation in health campaigns. However, it is important to note that the authors have also pointed out that some difficulties were faced as well. According to Ranjan (2021), the situation was such that organizations had to ramp up their activities while fundraising was purposely put on hold, which in turn limited the organization's capacity. The researchers Mohan et al. (2022) observed that the increase in collaboration between NGOs and government agencies was instrumental in the development of the responsive strategies and would further facilitate the smooth implementation of the strategies. Nevertheless, Patel et al. (2021) argued that, on the whole, the presence of strong NGO representation in communities was one of the most important factors that contributed to the pandemic response success and these communities were also more adaptable.

Table 8.5 Case studies on community responses to COVID-19 in India

Year	Location	Objectives	Methods	Outcome	References
2020	Kerala	Early detection and containment of COVID-19	– Early implementation of lockdowns	– Reduced spread of infection in the early stages	WHO report on Kerala's COVID-19 management, 2020
			– Effective contact tracing	– High recovery rate	
			– Local quarantine centers	– Globally recognized as a model response	
2020	Dharavi, Mumbai	To prevent the outbreak in the dense slum population	– Community-based testing	– Significantly reduced spread in one of the largest slums	The Lancet, 2020
			– Setting up local quarantine facilities	– Flattened the curve within weeks	
			– Mobilization of local leaders and healthcare workers		
2020	Odisha	Minimize disruption to livelihood and food security during the lockdown	– Community kitchens providing meals to the poor	– Reduced food insecurity	Government of Odisha Report, 2020
			– Distribution of dry rations through PDS	– Sustained support for marginalized communities	
			– Coordination with local NGOs		
2020	Rajasthan	Increase public awareness and healthcare access in rural areas	– Telemedicine initiatives for rural outreach	– Increased healthcare accessibility in rural areas	National Health Mission Rajasthan Report, 2020
			– Distribution of masks and hygiene kits through local self-help groups (SHGs)	– Enhanced awareness of COVID-19 hygiene practices	

(continued)

Table 8.5 (continued)

Year	Location	Objectives	Methods	Outcome	References
2020–21	Delhi	Protect vulnerable groups, especially migrant workers, from the economic impact	– Establishment of shelter homes for migrants	– Helped mitigate the mass migration crisis	Ministry of Home Affairs, Government of India, 2020
			– Free transport to native states	– Temporary shelters for thousands of workers	
			– Financial assistance and ration distribution		
2021	Pune	To provide mental health support to COVID-19 affected communities	– Digital mental health counseling through community-based apps	– Improved mental health outcomes for isolated individuals	Indian Journal of Psychological Medicine, 2021
			– Tele-counseling centers	– Wide-reaching access to counseling services	
2021	Tamil Nadu	Strengthen health infrastructure and local governance for COVID-19 management	– Localized health camps	– Increased vaccination coverage	Tamil Nadu Health and Family Welfare Department Report, 2021
			– Community-level vaccination drives	– Reduced hospitalization rates	
			– Decentralized COVID-19 care centers		

8.3.2.2 Grassroots Initiatives and Volunteerism

A insight-pieces put forth the pivotal role of local/grassroots movements and volunteers in the community-building process during the COVID-19 pandemic. Primarily, this report's research illustrates the fact that these movements and volunteers constituted the most excellent forms of support because of their ability to swiftly mobilize, be flexible, and respond to the needs of the most vulnerable people in the community. O'Donnell et al. (2021) gave the example of community groups stepping in to provide on-the-spot assistance when the government's reaction was inadequate, while Fisher and Lutz (2021) argued that besides helping the vulnerable, these groups also strongly contribute to social cohesion and collective resilience strengthening.

Moreover, volunteers exposed that their commitment brought about a reinforced awareness of purpose and community connectedness (Holland & Phelps, 2021). McCabe et al. (2022) portrayed the scenario where volunteers changed their skills to suit the community's evolving needs as a demonstration of their openness and inventiveness. Nash et al. (2021) documented the fast adaptability of these entities to supply the vacancies left by the disappearance of formal institutions. Baker et al. (2021) showed the triumph in the endeavors to undertake food insecurity, dearth of healthcare, and mental health challenges because of the disparities that had been exacerbated by COVID-19. Kumar et al. (2022) asserted that better results could be achieved through closer cooperation between grassroots organizations and government agencies. Finally, volunteering in activities served to enhance social capital, an important element of community resilience (Williams et al., 2022). Lastly, work of Taylor (2021) indicated that initiatives provided immediate support to the community as well as giving community members an opportunity to speak freely about their issues and needs.

8.3.2.3 Government and Public Health Responses

Different strategies led to appreciable progress. Ferguson et al. (2020) illustrated that early and strong lockdowns were associated with substantial decreases in transmission rates in multiple countries. Guan et al. (2020) found that travel bans could delay outbreaks in newly affected locations in a timely manner. Baker et al. (2020) stated that mitigation efforts were of primary importance to stabilize the economy and control employment. Paltiel and Zheng (2020) promoted the idea that rapid testing could identify cases and move toward isolation of those cases preventing more transmission. Fischer et al. (2021) emphasized the value of trust in the public and effective means of communicating about vaccines to drive vaccination uptake. Nikolov et al. (2021) indicated that misinformation clouded trust in the health authorities and complicated efforts to promote adherence to the health guidelines. A comparative analysis by Vranckx et al. (2021) accentuated that countries with established public health infrastructure and plans fared better in managing the pandemic. Blumenthal et al. (2021) expressed that investment in infrastructure and training of public health workforce must be prioritized for future. Bollyky et al. (2021) indicated that enhanced global cooperation and health information sharing must take priority for a more effective collective public health response to the next pandemic.

8.3.2.4 Innovation in Response Strategies

The COVID-19 pandemic has resulted in transformational innovation in many areas, including health, health technologies, and governance. Innovation has been at the forefront of responses across the globe and allowed the need for rapid change across responses for health delivery, policy implementation, community engagement, and education. Research has shown important innovation in a number of applications: Effective use of telemedicine and telehealth technologies greatly expanded healthcare access and minimized exposure to viruses (Smith & Jha, 2021). Contact tracing applications were helpful collectively with robust testing (Wymant et al., 2021), and other technologies and techniques were worthy of early detection of serious outcomes (Iliescu et al., 2021). The integrated approach to healthcare has also led to improved patient outcomes and satisfaction (Gonzalez & Maul, 2021). Supply-chain innovations made the system more responsive and resilient during times of uncertainty (Sharifi et al., 2020). From a political and governance point of view, one of the most significant impacts was the fast-tracking of traditional regulatory pathways that led to changes in delivery, enabling the manufacture and distribution of vaccines at the unheard-of pace (Wang et al., 2021). In addition to that, remote work policies allowed employees to maintain their productivity and well-being (Moreno López et al., 2021). Universal Basic Income (UBI) programs provided the much-needed economic relief to the people living in vulnerable situations and may serve as a model for sustainable economic viability in the future (Murray et al., 2021).

In the domains of community and education, grassroots campaigns played a crucial role in providing essential resources to those facing inequalities and in strengthening a shared sense of community and belonging (Norris et al., 2021). Social media platform allowed for community engagement and the dissemination of essential health information (Gollust et al., 2020). Flexible learning models met the need for learning solutions that were available to accommodate various needs (Del Río-Lozano et al., 2022). Encompassing these new attempts and innovative solutions will be critical for intensified resilience in future circumstances (Lal et al., 2021). This change will also require more funding and better collaboration between government, academia, and private enterprise (Rodrigues et al., 2025).

8.3.3 Case Studies

This comparative analysis clarifies the distinction in the societal pattern in different Indian states which were the societal pattern of the states' experience during the pandemic due to their healthcare systems, education, economic strain, and community-organized activities. Those states which had not only an already better health care system but also a better governing structure or state bureaucracy (Table 8.6), e.g., Kerala, had a comparatively lesser severe societal impact of COVID-19-arising situation in effect of which they overcame faster whereas the rest of the states particularly Uttar Pradesh, Bihar, and those with limited health care systems in rural areas that encountered significant barriers and challenges faced drastically worse situations.

8.4 Discussion

These will be central to fulfilling our first aim, which is to examine vulnerability. The socioeconomic consequences of COVID-19 in India demonstrate how systemic invincibility exhibit in the economic, health, social, and governance arenas. They also provide important conceptual frameworks for understanding vulnerability, the first objective of this research. The economic ramifications of the situation are

Table 8.6 Comparative analysis of societal implications and community response of COVID-19 vulnerability in different states of India

Components	State/region	Objective	Outcome	Tools and techniques	References
Healthcare infrastructure impact	Maharashtra	To assess the strain on healthcare services	Severe shortage of hospital beds, ventilators, and oxygen supplies. Significant pressure on both public and private healthcare systems	Health data tracking, hospital surveys, government records	Government of Maharashtra Health Dept. Reports
	Kerala	To evaluate healthcare preparedness and response	Strong healthcare response through decentralized management and better infrastructure, lower fatality rates compared to other states	Decentralized public health system, testing and tracing	Kerala State Government Health Reports
	Uttar Pradesh	To understand healthcare infrastructure challenges	Significant rural health infrastructure gaps. Over-reliance on urban centers. Struggles with oxygen supply and access to care	Healthcare availability mapping, resource tracking	Ministry of Health and Family Welfare Reports
Social impact (migration and employment)	Gujarat	To assess the impact on migrant workers	Large-scale migration of workers from industrial zones. Challenges in implementing state-level relief for unemployed migrant labor	Migration tracking systems, employment surveys	National Institute of Labour Economics Research (NILERD)

(continued)

Table 8.6 (continued)

Components	State/region	Objective	Outcome	Tools and techniques	References
	Bihar	To study reverse migration and employment impacts	Massive reverse migration. High unemployment and strain on local economies, reliance on state relief packages and job guarantee schemes (MGNREGA)	Labor data analytics, state relief distribution records	Bihar State Government Employment Data
Education disruptions	Tamil Nadu	To analyze the impact of school closures and online education	Significant impact on rural students, digital divide worsened learning outcomes. Government initiatives for online education helped urban but not rural students	Surveys, digital education analytics	Tamil Nadu State Education Dept. Reports
	Karnataka	To study digital learning and access during the pandemic	Widened digital divide between rural and urban students. Urban areas benefited from online learning, while rural students faced access issues and dropout concerns	Online education assessments, digital access surveys	Karnataka State Education Data

(continued)

Table 8.6 (continued)

Components	State/region	Objective	Outcome	Tools and techniques	References
	Rajasthan	To evaluate access to education and alternative learning methods	Large-scale impact on rural education. Implementation of mobile learning initiatives in remote areas was moderately successful	Educational surveys, mobile learning platforms	UNESCO Report on India Education, Rajasthan Govt. Data
Economic impact	Delhi NCR	To study the effect on informal workers and small businesses	Significant loss of income for informal sector workers and small businesses. Slow economic recovery, with government support schemes offering limited relief	Business impact surveys, employment tracking	NITI Aayog Reports, Delhi Economic Survey
	West Bengal	To analyze economic vulnerabilities due to lockdowns	Severe economic impact on agriculture and small scale industries. Government relief was slow but targeted sectors like agriculture and small traders	Economic impact studies, agriculture impact assessments	West Bengal Govt. Economic Review Reports
	Telangana	To study small business recovery post-lockdowns	Partial recovery in urban areas, severe challenges in rural regions. State-specific schemes helped recovery, but economic vulnerabilities persist	Economic recovery surveys, government relief tracking	Telangana State Government Post-Lockdown Recovery Data

(continued)

Table 8.6 (continued)

Components	State/region	Objective	Outcome	Tools and techniques	References
Mental health consequences	Punjab	To assess mental health impacts of COVID-19 on urban/rural divide	Increase in anxiety, depression, especially among youth and elderly. Rural areas lacked access to mental health services	Mental health surveys, telemedicine tracking	Punjab State Mental Health Surveys, WHO Reports
	Assam	To evaluate mental health support mechanisms	Limited mental health infrastructure. NGOs played a key role in providing emotional support through helplines and counseling services	NGO reports, mental health helplines	Assam State Health and Family Welfare Reports
Community response and NGO involvement	Tamil Nadu	To analyze the role of NGOs in COVID-19 relief	NGOs played a crucial role in food distribution, healthcare assistance, and awareness campaigns, especially in underserved communities	NGO field reports, government collaboration records	Tamil Nadu State NGO Collaboration Report
	West Bengal	To assess the grassroots response to pandemic challenges	Strong grassroots initiatives helped distribute essentials and healthcare information. Local governance bodies collaborated effectively with NGOs	Community initiative records, local governance data	West Bengal Government and NGO Collaboration Reports

(continued)

Table 8.6 (continued)

Components	State/region	Objective	Outcome	Tools and techniques	References
	Kerala	To evaluate community participation in relief efforts	High community participation. Localized relief efforts were efficient in handling healthcare and livelihood support, especially through Kudumbashree and similar groups	Local community participation surveys	Kerala Kudumbashree Reports, WHO Collaboration Data
Government response (relief and recovery)	Maharashtra	To study state-level COVID-19 relief measures	Comprehensive lockdowns, financial relief schemes for poor families and small businesses. However, large sections of the population remained vulnerable economically	State government relief data, financial aid distribution	Maharashtra Government COVID-19 Response Reports

(continued)

Table 8.6 (continued)

Components	State/region	Objective	Outcome	Tools and techniques	References
	Uttar Pradesh	To analyze the effectiveness of relief distribution	Relatively slow rollout of relief measures, especially in rural areas. Struggles with implementation of MGNREGA and food distribution schemes	Government records, rural economic surveys	Uttar Pradesh Govt. COVID-19 Response Data
	Kerala	To evaluate state-level coordinated response	Kerala's decentralized governance model led to better coordination in managing healthcare services, food distribution, and public awareness campaigns	State governance records, COVID-19 relief tracking	Kerala Govt. COVID-19 Response, WHO Analysis Reports

widespread job losses, disruption of small business, and fragilities in the informal sector workforce. Such events represent weak economic conditions. They also offer built-in familiarity of at-risk demographic and professional groups. The impact on informal sector workers represents over 80% of the labor force in India. According to ILO, lockdowns and mobility restraints interrupted livelihoods, forcing millions into poverty and food insecurity. Additionally, according to Dev and Sengupta, in 2020, small and micro-enterprise business structures faced a severe challenge often with little or no financial resources and access to relief packages. Further, health and social challenges of inequity accessing health care, mental health challenges, education access, and risks for marginalized communities provide a wider canvas for understanding vulnerability in terms of geography, gender, and class. Health inequalities included access to health care as there were urban/rural disparities in the availability of infrastructure and medical personnel for COVID-19 responses (Yadav & Subhashini, 2021). The burden was compounded for people with comorbidities and chronic health conditions and abandoning non-COVID service made general public health worse (Banik et al., 2023). In addition, the pandemic has taken a significant social toll—from increased gender-based violence and school closures for girls (UNICEF, 2021), to heightened incidence of mental disorders induced by isolation and uncertainty (Rajkumar, 2020). While lockdown measures would have value—and as would the decentralization of the public health response and public trust in relative institutions—lessons learned about institutional preparedness and accountability would have value as well. These considerations overlap with examination of structural vulnerabilities. Finally, common political and governance issues included issues with enforcement of lockdowns, lacking coordination or conversation among central and state governments, and ineffective delivery of protections to the social safety net (Mehrotra, 2021).

Community organizations and NGOs were instrumental in delivering critical goods like food, medicine, and hygiene kits to vulnerable populations, especially in disadvantaged communities where government assistance was either slow or nonexistent (Paliwal et al., 2023). As an example, organizations like Goonj and SEWA gave extensive support to migrants and informal workers. Local leaders and grassroots initiatives, including women's self-help groups (SHGs), were essential for health communications, community kitchens, and mask production, all of which were examples of localized resilience (Agarwal, 2022). Volunteers, mostly children and youth from the civil society, were very instrumental in providing information, vaccination drives, and telemedicine services, to the last mile to the weak sections of the society (Kushwaha et al., 2022). Aside from that, government-led responses, though somewhat inconsistent, have made a big plunge in upgrading the healthcare infrastructure, facilitating the use of digital health platforms like Aarogya Setu, and carrying out one of the biggest vaccination drives in the world.

The research explores the widespread miscellany of ingenious emergency-response procedures revealed during the emergency, such as mobile health units, oxygen concentrator banks, and digital data dashboards which not only lighten the disaster but also allow a sturdy engraving for the farther future's fortitude..Moreover, these actions help achieve the main objective of the study by identifying the most affected areas and the responses taken, while also suggesting practical ways to build a more equitable, flexible, and resilient public health and social system in India. The reactions of both the community and the government reveal the flexibility of Indian society, and, at the same time, they provide essential clues on how to manage a crisis properly. The outcomes go straight to the study's third goal, which is to get the real workable strategies from the present situation for the future public health system, community mobilization frameworks, and resilience at both the policy and grassroots levels. Some of the effective instances demonstrate the viability of collective governance and public participation to establish a response system that is not only more comprehensive, but also flexible.

8.5 Limitation

This study may encounter different limitations especially because the COVID-19 situation is changing and is complicated and affects many different areas of society.

- **Data Availability and accuracy**: Because of the unprecedented nature of the crisis, data related to health outcomes, economic impact, and social responses have been volatile. Even though our studies may illuminate the national scene of the COVID-19 pandemic in India, detailed data depicting what is happening at a community or district level may not always be available. This restrains our capacity to explore particular differences and disadvantages of various regions.
- **Generalizability of Findings**: The behavior of urban and rural populations in the face of the pandemic may vary substantially because of a different pattern of healthcare, economic opportunities, and government support. The gap may cause us to have insufficient knowledge of community response strategies.
- **Methodological Limitations**: The pool of data through field surveys, particularly if they were done online, may imply a sample bias. Several rural regions and low-income groups may not have approach to digital platforms; consequently, their ideas might be ineffectively represented in the research.
- **Economic and Political Context**: The political left or right might affect the governmental responses and the spirit of the community, thus influencing the data interpretation or the participants' provided responses in the study.
- **Community Response Documentation:** Community responses at the grassroots level and informal mutual aid initiatives may lack formal documentation and may not be readily accessible. Hence, it is difficult to ascertain the extent of community involvement.
- **Limited Access to Healthcare Data:** Details on healthcare access, especially in a situation of a crisis, might be only partially available. These kinds of data are frequently fragmented, and I assume that our investigation will encounter the problem of insufficient disclosure of the different degrees of healthcare access that are able to be found in various states, the countryside, and among the socioeconomic classes in India.

8.6 Recommendation

- **Enhance Data Collection and Transparency**: To counteract pandemics effectually, government bodies and health agencies should not only generate but also preserve vigorous data collection arrangements that are able to track health products, economic influences, and social responses at a granular level. Local community data should specifically be contemplated to increase the accuracy and breadth of the response to pandemics.
- **Real-Time Monitoring**: The local health and socioeconomic situation as well as the emergence of mental health problems could be monitored in real-time through digital platforms. This kind of a system would provide a fast reaction and decision-making, which is extremely important in the case of an ongoing or new crisis.

- **Address Digital Divide**: The focus of this study is on the need to invest in digital infrastructure to close the gap in terms of accessibility in technology, especially in the remote areas and among disadvantaged populations. More affordable internet and devices for students are prerequisites for uninterrupted education during crisis situations.
- **Empowering Community-Based Organizations**: NGOs, grassroots organizations, and local community groups are the pillars which support societies in times of crisis. The formalization of cooperation between governments and community organizations for crisis response can be a means to more localized interventions that are also effective.
- **Decentralized Decision-Making**: The strategies for the future should consider the element of decentralized decision-making whereby authority to adapt the response to specific regional needs and vulnerabilities would be given to local government and community leaders.
- **Extended Analysis of Vulnerabilities**: Future research should focus on the long-term socioeconomic effects of the pandemic that concentrate on the disadvantaged communities, the urban-rural divide, and small businesses. Following economic recovery, mental health outcomes, and social inequalities through longitudinal study will open more comprehensive views.
- **Public Health Preparedness**: The research of the future should be directed toward assessing the resilience of India's healthcare system against pandemics. Such studies may include resource allocation, healthcare worker burnout, and healthcare infrastructure. Further study of the effectiveness of telemedicine and the role of community health services during the pandemic is also worthwhile.
- **Cross-Sectoral Collaboration**: Future research may pivot on this point to study how cross-sectoral collaboration (between government, private sector, NGOs, and international agencies) was instrumental in lessening the COVID-19 impact. Exploring such cooperation can provide an array of comprehension for encouraging the next generation of multi-stakeholder stratagems that are more comprehensive and better prepared for catastrophes.

8.7 Conclusion

Present study provides an extensive insight into the varied effects of the pandemic in health, social, economic, and political spheres. Present healthcare infrastructure and socioeconomic systems that existed in India unraveled several fault lines of COVID-19 pandemic and revealed the differential way in which the virus hit the marginalized, tribal, and backward segments of the society. This study also highlighted multi-dimensional crisis of COVID-19 pandemic. Furthermore, it steered to the unmasking of systemic inequities not only in healthcare in India but also in the Indian society. The research reveals that the immediate response was helpful in averting the crisis to some extent, however, the challenges remain. It is imperative first to fortify the healthcare systems, tackle mental health issues, bring down educational disparities, and create more stable economic and social support systems. The result of the analysis also showed that community organizations and NGOs play a significant role in the elimination of COVID-19 pandemic. This success can be attributed to localized and grassroots interventions that have been given due importance in crisis handling. The future scenario will demand a binding interaction between the government, private sector, and civil society to create a society that is not only resistant but also inclusive enough to handle the incoming shocks of life. The upcoming policies need to focus on narrowing down the digital gap, providing easy healthcare access, and enhancing the social protection system especially for those who are at the bottom of the social pyramid. Implementing these efforts jointly, India can emerge out of the COVID-19 pandemic in a better shape with are more efficient to deal with health emergencies and their wider social impacts.

The lessons and experience attained from India for grassroots resilience, community-led responses, and the role of civil society are highly pertinent and can be adopted with minimal adaptation in other developing and middle-income countries that share similar structural settings. The study pivots on characteristics of local innovation, digital elasticity, and volunteering, which are indicative of solutions with a probable for scalability that can be pragmatic to the determining of the global preparedness strategies and the disaster response policy. The endorsement of participatory governance, public–private partnerships, and social protection reform serves as a source of the worldwide discussion on fair recovery and green development of COVID-19 mitigation strategy. These are substantial building blocks of the United Nations Sustainable Development Goals (SDGs), mainly SDG 3 (Good Health and Well-being), SDG 10 (Reduced disparities), and SDG 16 (Peace, Justice, and Strong Institutions).

Present study provides an integration between public health, development studies, political science, and disaster risk reduction to develop remedial measures of COVID-19. The broad range of policy propositions over the years such as making the digital world accessible to everyone, strengthening mental health services, and planting equity deeply in recovery planning creates a valuable source of practical capability for both academic communities and public health professionals. The outcome of the study aids in implementing a universally significant agenda: the rebuilding of societies after the pandemic in a way that is socially just, more participatory, and structurally resistant to the coming of any new crisis.

References

Agarwal, B. (2022). Imperatives of recognising the complexities: gendered impacts and responses to COVID-19 in India. *Economic and Political (Bologna), 39*(1), 31–53. https://doi.org/10.1007/s40888-021-00242-8. Epub 2021 Sep 25. PMID: 35422596; PMCID: PMC8475308.

AIMO. (2020). Impact of COVID-19 on Small and Medium Enterprises in India. All India Manufacturers Organization.

Anderson, A., Brown, C., & Lewis, J. (2021). Trust and community engagement in COVID-19 responses: Lessons from NGOs. *Journal of Community Health, 46*(2), 350–357.

Arora, A., Bansal, R., Kandpal, C., Aswani, N., & Aswani, C. (2020). Social impact of COVID-19 pandemic lockdown on elderly population and their mental health. *Indian Journal of Health Sciences and Biomedical Research, 13*(4), 345–350.
Azim Premji Foundation. (2021). Impact of COVID-19 on school education in India: Field studies on learning loss and inequality. Azim Premji University Report.
Baker, L., Smith, J., & Patel, R. (2021). Food security and grassroots initiatives during COVID-19: Addressing disparities. *Food Security, 13*(2), 263–274.
Baker, S. R., Farrokhnia, R. A., Meyer, S., & Pagel, M. (2020). The COVID-19 crisis and the economy: A review of the policy responses. *The Review of Economic Studies, 87*(3), 1242–1266.
Banerjee, D., & Rai, M. (2020). Social isolation in Covid-19: The impact of loneliness. *International Journal of Social Psychiatry, 66*(6), 525–527.
Banerjee, D., & Rao, T. (2020). Social isolation in COVID-19: The impact of loneliness. *International Journal of Social Psychiatry, 66*(6), 525–527.
Banik, R., Rahman, M., Sikder, M. T., Rahman, Q. M., & Pranta, M. U. R. (2023) Knowledge, attitudes, and practices related to the COVID-19 pandemic among Bangladeshi youth: A web-based cross-sectional analysis. *Z Gesundh Wiss. 31*(1), 9–19. https://doi.org/10.1007/s10389-020-01432-7. Epub 2021 Jan 16. Erratum in: Z Gesundh Wiss. 2023;31(1), 165. https://doi.org/10.1007/s10389-021-01488-z. PMID: 33489718; PMCID: PMC7811151.
Bhandari, R., & Yadav, S. (2021). COVID-19 and private sector education in India: Governance challenges and inequalities. *South Asian Journal of Education, 14*(2), 99–118.
Bhatia, R., Singh, R., & Pal, A. (2021a). Social transformations during COVID-19: Virtual gatherings and the digital divide in India. *South Asia Journal of Social Sciences, 16*(1), 76–88.
Bhatia, S., Bhatia, R., & Ghosh, A. (2021b). COVID-19 and health inequities in India: A review. *Global Health Action.*
Bhattacharya, S., & Haldar, D. (2021). Community responses to COVID-19: Evidence from India. *Journal of Community Health.*
Blumenthal, D., Fowler, E. J., Abrams, M. K., & Collins, S. R. (2021). Covid-19—Implications for the future of health care. *New England Journal of Medicine, 383*(2), 156–159.
Bollyky, T. J., Brown, C., & Kahn, D. (2021). The importance of global cooperation in the response to COVID-19. *Health Affairs, 40*(7), 1141–1147.
Bose, A., & Malhotra, R. (2021). The politicization of pandemic relief measures and the role of local governance in India's COVID-19 response. *Governance in Crisis, 9*(1), 77–91.
Chakraborty, I., & Maity, P. (2020). COVID 19 outbreak: Migration, effects on society, global environment and prevention. *Science of the Total Environment, 728*, 138882.
Chakraborty, I., & Bhatia, R. (2021). Public communication and political narratives during the COVID-19 pandemic: The case of India. *Journal of Public Affairs, 21*(3), Article e2587.
Chatterjee, P., & Purohit, R. (2021). COVID-19 vulnerability: A societal implication and community response in India. *Journal of Public Health Research, 10*(2), 1–8.
Chaturvedi, K., Vishwakarma, D. K., & Singh, N. (2021). COVID-19 and its impact on education, social life, and mental health of students: A survey. *Children and Youth Services Review, 121*, Article 105866.
Desai, S., & Vanneman, R. (2021). The Impact of COVID-19 on vulnerable populations in India: Evidence from the COVID-19 household survey. *Economic and Political Weekly.*
Del Río-Lozano, M., García-Calvente, M., Elizalde-Sagardia, B., & Maroto-Navarro, G. (2022). Caregiving and caregiver health 1 year into the COVID-19 pandemic (CUIDAR-SE Study): A gender analysis. *The International Journal of Environmental Research and Public Health, 19*(3), 1653. https://doi.org/10.3390/ijerph19031653. PMID: 35162675; PMCID: PMC8835117.
Dhar, N. S., Nupur, S., & Dutta M. (2022). COVID-19 induced income loss among migrant workers: Evidence from eight villages of Bihar. *Economic Papers: A Journal of Applied Economics and Policy, 41*(4), 325–46. https://doi.org/10.1111/1759-3441.12368. Epub 2022 Oct 31. PMCID: PMC9874837.
Ferguson, N. M., et al. (2020). Impact of non-pharmaceutical interventions (NPIs) to reduce COVID-19 mortality and healthcare demand. Imperial College London.

FICCI. (2021). The impact of COVID-19 on Indian Business: An assessment. Federation of Indian Chambers of Commerce & Industry.

Fisher, M., & Lutz, A. (2021). Mutual aid in the time of COVID-19: A grassroots response to community needs. *Journal of Community Practice, 29*(3), 218–235.

Fischer, M., et al. (2021). Public trust and the vaccination rollout: A global perspective. *The Lancet Public Health, 6*(5), e321–e322.

Ghosh, A., & Gupta, K. (2020). The challenges of public health in India during COVID-19: A review. *International Journal of Health Sciences.*

Gollust, S. E., Nagler, R. H., & Fowler, E. F. (2020). The COVID-19 pandemic and the Infodemic: The role of public health communication. *American Journal of Public Health, 110*(9), 1248–1250.

Gonzalez, A., & Maul, J. (2021). Mental health implications of the COVID-19 pandemic: A global perspective. *Journal of Mental Health.*

Guan, W. J., et al. (2020). Clinical characteristics of 2019 novel coronavirus infection in China. *New England Journal of Medicine, 382*(18), 1708–1720.

Gupta, S., & Khanna, P. (2021). Gendered impact of COVID-19 on social interactions and domestic violence in India. *Journal of Gender Studies, 28*(2), 144–157.

Holland, D., & Phelps, C. (2021). Volunteerism and community resilience during COVID-19: A qualitative study. *Nonprofit and Voluntary Sector Quarterly, 50*(6), 1283–1301.

Iliescu, F. S., Ionescu, A. M., Gogianu, L., Simion, M., Dediu, V., Chifiriuc, M. C., Pircalabioru, G. G., & Iliescu, C. (2021). Point-of-care testing-the key in the battle against SARS-CoV-2 pandemic. *Micromachines (Basel), 12(12)*, 1464. https://doi.org/10.3390/mi12121464. PMID: 34945314; PMCID: PMC8708595.

International Monetary Fund. (2020). *World economic outlook: A long and difficult ascent.* IMF.

International Labour Organization. (2020). *COVID-19 and the world of work: Impact and policy responses.* Geneva: ILO.

Jain, R., & Chatterjee, T. (2024). The politics of state-level COVID-19 responses in India: Evidence from the first wave in 2020. *The Indian Economic Journal, 72*(2), 243–258.

Jena, P. (2020). Impact of COVID-19 on education in India. *International Journal of Current Research (IJCR), 12*(7), 12582–12586.

Kumar, A., & Choudhury, S. (2021). Economic impact of COVID-19 in India: Challenges and opportunities. *Asian Economic Policy Review.*

Kumar, A., & Kumar, A. (2021). Social inequalities and COVID-19: Implications for vulnerable populations in India. *Global Health Action, 14*(1), 1857325.

Kumar, A., & Singh, R. (2021). Rapid mobilization of community organizations during the COVID-19 pandemic: A case study. *Community Development Journal, 56*(4), 618–634.

Kumar, S., Verma, A., & Desai, K. (2022). Grassroots organizations and COVID-19 response: Challenges and opportunities. *Community Development Journal, 57*(1), 72–88.

Kushwaha, A. S., Banerjee, S., & Bandyopadhyay, K. (2022). Community involvement in controlling COVID-19 pandemic in Maharashtra, India: Best practices and missed opportunities. *Medical Journal of Dr. D. Y. Patil Vidyapeeth, 15*(3), 309–312. https://doi.org/10.4103/mjdrdypu.mjdrdypu_125_21

Lal, A., Abdalla, S. M., Chattu, V. K., Erondu, N. A., Lee, T. L., Singh, S., Abou-Taleb, H., Vega Morales, J., & Phelan, A. (2022). Pandemic preparedness and response: Exploring the role of universal health coverage within the global health security architecture. *The Lancet Global Health, 10*(11), e1675–e1683. https://doi.org/10.1016/S2214-109X(22)00341-2. Epub 2022 Sep 27. PMID: 36179734; PMCID: PMC9514836.

Lai, J., Ma, S., Wang, Y., Cai, Z., Hu, J., Wei, N., et al. (2020). Factors associated with mental health outcomes among health care workers exposed to coronavirus disease 2019. *JAMA Network Open, 3*(3), Article e203976.

Loades, M. E., Chatburn, E., Higson-Sweeney, N., Reynolds, S., Shafran, R., Brigden, A., et al. (2020). Rapid systematic review: The impact of social isolation and loneliness on the mental health of children and adolescents in the context of COVID-19. *Journal of the American Academy of Child & Adolescent Psychiatry, 59*(11), 1218–1239.

McCabe, M., Kelly, J., & Reddy, P. (2022). Adapting volunteer roles in response to COVID-19: Lessons learned. *International Journal of Volunteer Administration, 38*(2), 1–14.

Mehrotra, S., & Mehrotra, S. K. (2021). Joblessness and informalisation: Challenges to inclusive growth in India. *Economic and Political Weekly, 49*(37), 7–11.

Minstry of Statistics and Programme Implementation. (2021). National accounts statistics. Government of India.

Mittal, N., Steinert, J. I. & Vollmer, S. (2023). COVID-19 pandemic, losses of livelihoods and uneven recovery in Pune, India. *Humanities and Social Sciences Communications, 10*, 852. https://doi.org/10.1057/s41599-023-02343-0

Mohan, A. (2021). Impact of COVID-19 on mental health of vulnerable groups in India. *Indian Journal of Psychiatry, 63*(1), 7–10.

Mohan, A., & Sharma, R. (2021). Community resilience in response to COVID-19: Insights from India. *Global Social Policy.*

Mohan, A., Gupta, S., & Verma, P. (2022). Coordination challenges among NGOs during the COVID-19 pandemic in India. *Nonprofit Management & Leadership, 32*(1), 91–107.

Moreno López, J. A., Arregui García, B., Bentkowski, P., Bioglio, L., Pinotti, F., Boëlle, P. Y., Barrat, A., Colizza, V., & Poletto, C. (2021) Anatomy of digital contact tracing: Role of age, transmission setting, adoption, and case detection. *Science Advances, 7*(15), eabd8750. https://doi.org/10.1126/sciadv.abd8750. PMID: 33712416; PMCID: PMC8034853.

Moynihan, R., Sanders, S., Michaleff, Z. A., Scott, A. M., Clark, J., To, E. J., et al. (2021). Impact of COVID-19 on healthcare delivery for non-COVID-19 conditions: A rapid review. *British Medical Journal Open, 11*(1), Article e043387.

Mukherjee, A., et al. (2021). Economic consequences of COVID-19: Evidence from India. *Indian Journal of Labour Economics.*

Murray, C. J. L., GBD 2021 Collaborators. (2024). Findings from the global burden of disease study 2021. *Lancet, 403*(10440), 2259–2262. https://doi.org/10.1016/S0140-6736(24)00769-4. PMID: 38762327.

NITI Aayog. (2021). *India's response to COVID-19: Policy and socio-economic implications*. New Delhi: Government of India.

Nanda, P., & Choudhury, S. (2021). COVID-19 and its impact on Rural Communities in India: A societal response. *Journal of Rural Studies, 87*, 181 190.

Nash, H., Walker, T., & Smith, D. (2021). The role of grassroots initiatives in pandemic response: Flexibility and impact. *Global Health Action, 14*(1), 1858456.

Nikolov, D., et al. (2021). The impact of misinformation on the public's health response to COVID-19. *International Journal of Health Services, 51*(2), 212–223.

Norris, M., Anderson, D., Lin, J. A., & Gallagher, J. C. (2021). The impact of COVID-19 on healthcare for older adults. *Journal of Gerontological Nursing, 47*(1), 8–13.

O'Donnell, M., Lee, S., & Jones, R. (2021). Community mobilization during COVID-19: A case study of grassroots initiatives. *Journal of Community Health, 46*(5), 971–981.

Oxfam. (2021). The inequality virus: Bringing together a world torn apart by coronavirus through a fair, just and sustainable economy. Oxfam International.

Paliwal, M., Raj, R., Kumar, V., Singh, S., Sharma, N. K., Suri, A., & Kumari, M. (2023). Informal workers in India as an economic shock absorber in the era of COVID-19: A study on policies and practices. *Human Systems Management, 43*(1), 17–36. https://doi.org/10.3233/HSM-220155

Reserve Bank of India. (2020). Consumer confidence survey: October 2020. RBI.

Paltiel, A. D., & Zheng, A. (2020). Assessment of SARS-CoV-2 testing in the United States. *Journal of the American Medical Association, 324*(2), 193–194.

Paltiel, A. D., Zheng, A., & Zheng, A. (2020). Assessment of SARS-CoV-2 screening and control measures at a University. *JAMA Network Open.*

Pfefferbaum, B., & North, C. S. (2020). Mental health and the Covid-19 pandemic. *The New England Journal of Medicine, 383*(6), 510–512.

Patel, P., & Haldar, S. (2021). Community resilience and collective action during COVID-19 in India: A case study. *Journal of Development Studies.*

Patel, S., Chatterjee, P., & Rao, K. (2021). Building resilience through community organizations: COVID-19 responses in urban India. *Global Social Policy, 21*(3), 328–347.

Rajkumar, R. P. (2020). COVID-19 and mental health: A review of the existing literature. *Asian Journal of Psychiatry, 52*, 102066.

Rai, A., & Singh, R. (2021). Social inequalities and COVID-19: The Indian experience. *Social Science & Medicine.*

Ranjan, R. (2021). The funding crisis of NGOs during COVID-19: Challenges and strategies. *Journal of Nonprofit Organizations, 12*(3), 233–248.

Rathi, R., & Prakash, A. (2020). Community engagement and response to COVID-19 in India: Lessons learned from the first wave. *Health Policy and Planning, 36*(9), 1206–1213.

Rodrigues, S. P., Gomes, L. C., Peres, F. A. P., Correa, R. G. F., & Baierle, I. C. (2025) A framework for leveraging digital technologies in reverse logistics actions: A systematic literature review. *Logistics, 9*(2), 54. https://doi.org/10.3390/logistics9020054

Sanghvi, S. (2020). Vulnerability of urban slum dwellers during the COVID-19 pandemic in India. *Environment and Urbanization, 32*(2), 459–472.

Sarkar, A., & Chowdhury, R. (2020). Governance responses to COVID-19 in India: Lockdowns, social restrictions, and the challenges of marginalized communities. *Asian Journal of Political Science, 18*(2), 100–114.

Sathish, C. (2022). Social impact of COVID-19 first wave in India: Issues and challenges. *Journal of the Anthropological Survey of India, 71*(2), 202–219.

Sahni, J., & Chaudhary, A. (2020). COVID-19 and vulnerable groups: A study on impact and response in India. *Social Science Research Network (SSRN).*

Sharma, R. (2021). The National Education Policy 2020 and the shift to digital education during COVID-19: Opportunities and challenges. *Educational Review, 73*(4), 469–482.

Sharma, R., & Sharma, N. (2020). Food security initiatives by NGOs during the COVID-19 pandemic in India. *Food Security, 12*(4), 895–908.

Sharifi, A., & Khavarian-Garmsir, A. R. (2020). The COVID-19 pandemic: Impacts on cities and major lessons for urban planning, design, and management. *Science of the Total Environment, 749*, 142391. https://doi.org/10.1016/j.scitotenv.2020.142391. Epub 2020 Sep 18. PMID: 33370924; PMCID: PMC7499053.

Sinha, A. (2020). COVID-19: The impacts of pandemic on vulnerable groups in India. *International Journal of Health Sciences, 14*(4), 115–121.

Sinha, A., & Prakash, A. (2021). The digital divide and online education in India during COVID-19. *Journal of Educational Technology & Society.*

Singh, S., Roy, D., Sinha, K., Parveen, S., Sharma, G., & Joshi, G. (2020). Impact of COVID-19 and lockdown on mental health of children and adolescents: A narrative review with recommendations. *Psychiatry Research, 293*, Article 113429.

Smith, R., & Jha, S. (2021). Funding challenges faced by grassroots organizations during COVID-19: A survey. *Nonprofit Management & Leadership, 32*(3), 367–384.

Smith, J., Brown, K., & Green, M. (2022). Mental health responses to the COVID-19 pandemic by NGOs: A review. *International Journal of Mental Health Systems, 16*(1), 10.

Sundararaman, V. (2020). COVID-19 and learning loss: Policy interventions for addressing the educational impact of the pandemic in India. *International Journal of Educational Development, 74*(6), Article 104108.

Taylor, K. (2021). Empowerment through grassroots initiatives: Advocacy during the pandemic. *Social Movement Studies, 20*(4), 505–521.

UNICEF. (2020). The impact of COVID-19 on Children in South Asia: Evidence from a rapid assessment. UNICEF.

UNICEF. (2021). *Gender-based violence and education disruption during COVID-19.* UNICEF Report.

UNESCO. (2020). Education: From disruption to recovery. UNESCO.

Vandrevala, T., Morrow, E., Coates, T., Boulton, R., Crawshaw, A. F., O'Dwyer, E., & Heitmeyer, C. (2024). Strengthening the relationship between community resilience and health emergency communication: A systematic review. *BMC Global and Public Health Journal, 2*(1), 79. https://doi.org/10.1186/s44263-024-00112-y. PMID: 39681938; PMCID: PMC11622909.

Vranckx, M., et al. (2021). Comparative effectiveness of COVID-19 public health measures: A systematic review. *BMJ Global Health, 6*(7), Article e006386.

Wang, C., Pan, R., Wan, X., Tan, Y., Xu, L., Ho, C. S., & Ho, R. C. (2020). Immediate psychological responses and associated factors during the initial stage of the 2019 coronavirus disease (COVID-19) epidemic among the general population in China. *International Journal of Environmental Research and Public Health, 17*(5), 1729.

Wang, Q., Su, M., Zhang, M., & Li, R. (2021). Integrating digital technologies and public health to fight covid-19 pandemic: Key technologies, applications, challenges and outlook of digital healthcare. *International Journal of Environmental Research and Public Health, 18*(11):6053. https://doi.org/10.3390/ijerph18116053

WHO. (2021). COVID-19: Impacts on global health and well-being. World Health Organization.

Williams, A., Green, M., & Parker, R. (2022). Building social capital through volunteerism: The role of grassroots organizations during COVID-19. *International Journal of Community Well-Being, 5*(1), 45–61.

Wymant, C., Ferretti, L., Tsallis, D., Charalambides, M., Abeler-Dörner, L., Bonsall, D., Hinch, R., Kendall, M., Milsom, L., Ayres, M., Holmes, C., Briers, M., & Fraser, C. (2021) The epidemiological impact of the NHS COVID-19 app. *Nature, 594*(7863), 408–412. https://doi.org/10.1038/s41586-021-03606-z. Epub 2021 May 12. PMID: 33979832.

Xue, B., & McMunn, A. (2021). Gender differences in unpaid care work and psychological distress during the COVID-19 lockdown. *Journal of Epidemiology & Community Health, 75*(7), 664–670.

Yadav, S., & Yadav, D. (2021). Health system response to COVID-19 in India: Lessons learned. *Health Policy and Planning*.

Chapter 9
Covid-19 Vulnerability: A Societal Implication and Community Response

Deepika Chandra Verma, Janardan Krishna Yadav, Srinivas Jangirala, Meenakshi Prasad, and Sandeep Singh

Abstract This chapter examines the community response of Indian urban dwellers to the uncertainty posed by Covid-19. The study uses the global footprint calculator developed by the Global Footprint Network as an instrument to collect personal ecological overshoot day for the respondents in normal times (based on recall) and during the Covid times. We find a significant difference between the two and thus see a significant reduction in the resulting ecological footprint during the new normal. Literature shows that stress on self-regulation, warm glow, education and awareness campaigns have had limited success in motivating people to reduce their ecological footprint. In comparison, our results show that the vulnerability due to the pandemic and expectancy of future shortage acts as an external shock that leads to the community response of regulating its ecological footprint significantly. This finding reinforces that consumption constraining environmental regulations by policymakers can be more effective than the ones raising environmental efficiency in production because there is no rebound accompanying it. The reduction in ecological footprint has been short lived like any change resulting due to a crisis. For a sustained reduction in ecological footprint, we recommend a policy mix focusing on environmental efficiency and sustainable consumption.

D. C. Verma (✉)
Indian Institute of Management Calcutta, Kolkata, West Bengal, India
e-mail: deepika11@iimcal.ac.in

J. K. Yadav · S. Jangirala · S. Singh
Jindal Global Business School, O. P. Jindal Global University, Sonipat, Haryana, India
e-mail: jkyadav@jgu.edu.in

S. Jangirala
e-mail: sjangirala@jgu.edu.in

S. Singh
e-mail: fpm14012@iiml.ac.in

M. Prasad
P.G. Department of Geography, Magadh University, Bodh Gaya, Bihar, India
e-mail: meenakshi.mprasad@gmail.com

G. K. Panda et al. (eds.), *Socio-Ecological Crisis of Development*, Sustainability Solutions, https://doi.org/10.1007/978-3-032-22171-1_9

Keywords Community response · Community resilience · Covid-19 · Ecological footprint · Environmental efficiency policy instruments · Sustainable consumption policy instruments

9.1 Introduction

A pandemic is likely to create immense negative repercussions at various fronts—economic, health, political, social and environmental (Abegunde & Stanciole, 2006; Smith et al., 2011). The same holds true for the recent pandemic, Covid-19, which has caused immense loss of life, social alienation (Majumdar et al., 2020; Marroquín et al., 2020), economic slowdown (Baldwin & Mauro, 2020; Blustein et al., 2020; McKibbin & Fernando, 2021) and uncertainty (Altig et al., 2020; Baker et al., 2020; Koffman et al., 2020; Szczygielski et al., 2021).

Countries followed a strategic preparedness and response plan to manage Covid-19, aiming to contain the infection and spread by curtailing movement, surveillance and risk assessment, research into possible cures; optimize care for patients within the given resources; improve mental health and resilience through communication, community engagement and maintenance of essential services (OECD, 2014). The realization of such plans can only occur through constant interaction between individuals, social groups, governments, businesses and natural/constructed environment by building community resilience (Haldon et al., 2020; Sakurai & Chughtai, 2020) by building on the capacity to mitigate and absorb shocks, the capacity to recover the ability to redevelop and compensate for damages, flexibility to adapt to the new normal and capacity to transform to improve the status quo (Bryce et al., 2022).

One way in which communities have chosen to respond to Covid-19 and be resilient is to address the supply shock and uncertainty by changing their overall lifestyle (De Haas et al., 2020; Di Renzo et al., 2020; Gössling et al., 2020; Jribi et al., 2020; Kaushik & Guleria, 2020; Sidor & Rzymski, 2020). In other words, this implies the need to regulate community consumption and changing its ways of production.

Studies have captured the social impact and community response on the production process during the Covid period (Campbell, 2021; Miao et al., 2021; Tittonell et al., 2021), as well as on consumption (Gordon-Wilson, 2022; Martin et al., 2020; Raman & Peng, 2021). This chapter takes the consumption route. Communities vary in their lifestyles and collective consumption behaviour (Veal, 1993). Therefore, it is imperative to check whether such a response is reflected in specific geographies or not, and to what extent. Further, an unparalleled reduction in the energy consumption was observed during the pandemic due to a shorter supply chain but was accompanied by the rebound due to structural changes like increased reliance on telework and e-commerce (Khanna, 2021). Thus, it is also urgent to investigate the net effect on consumption. Therefore, in the context of the pandemic and the slowdown, we examine whether the Indian urban community has mitigated and adapted its consumption in response to the expectancy of future shortage and was there a net reduction in consumption. If yes, by how much?

We operationalize this problem by capturing the change in the ecological footprint and the resultant personal overshoot day, during normal times and the Covid-19 period using the ecological footprint calculator developed by the Global Footprint Network. Ecological footprint is a more complete measure of consumption. Ecological footprint measures the ecological assets required to produce the market goods directly or indirectly. This includes fishery and livestock products; products obtained from forests and croplands; products from the built-up area; and nature's capacity to ingest the waste produced as negative externality, resulting due to the activities associated with the said products. Ecological footprint thus represents the demand for natural resources and is inclusive of carbon footprint.

We address our first research inquiry by comparing the results of the ecological footprint survey for usual times during the Covid-19 period. The second research question is addressed by measuring the effect size of the changed consumption of natural assets during the Covid period. Effect size quantifies the size of the difference between any two groups (Coe, 2002). This measurement is important because our concern is also to understand the practical significance of the results (Lakens, 2013) and effect size does that by measuring the magnitude of the change (Schafer & Schwarz, 2019).

We test our hypotheses using a curated dataset from survey responses of 380 urban Indian respondents. India is an ideal context for this study because of the way in which the lockdown was implemented prompting various sectors in the country to shift their practices online (IMF, 2021). Further, India suffers from ecological deficit that is much higher than several countries, a fact reflected in Fig. 9.1.

In the face of the Covid-19 crisis, the supply shock and curtailment of movement made communities vulnerable—obstructing the production structure and the consumption flow. One way to build community resilience is to resort to novel ways of production. However, consumption-related vulnerabilities also need to be addressed. This chapter demonstrates a quantitative case of how communities, to build resilience, re-examined and changed their consumption habits by addressing their extant over-draw on the natural resources which has been termed as a sustainability threat. Our findings show that during Covid-19, there was a significant reduction in consumption of resources (or ecological footprint) in India, reinforcing that consumption constraining environmental regulations by policymakers can be more effective than the ones raising environmental efficiency in production because there is no rebound accompanying it. Further, this shows that curtailing conspicuous consumption also contains the vicious cycle of stockpiling and artificial shortage that reduces the community's vulnerability. Thus, this chapter offers a different perspective on improving environmental wellbeing and showcases a novel way of building community resilience.

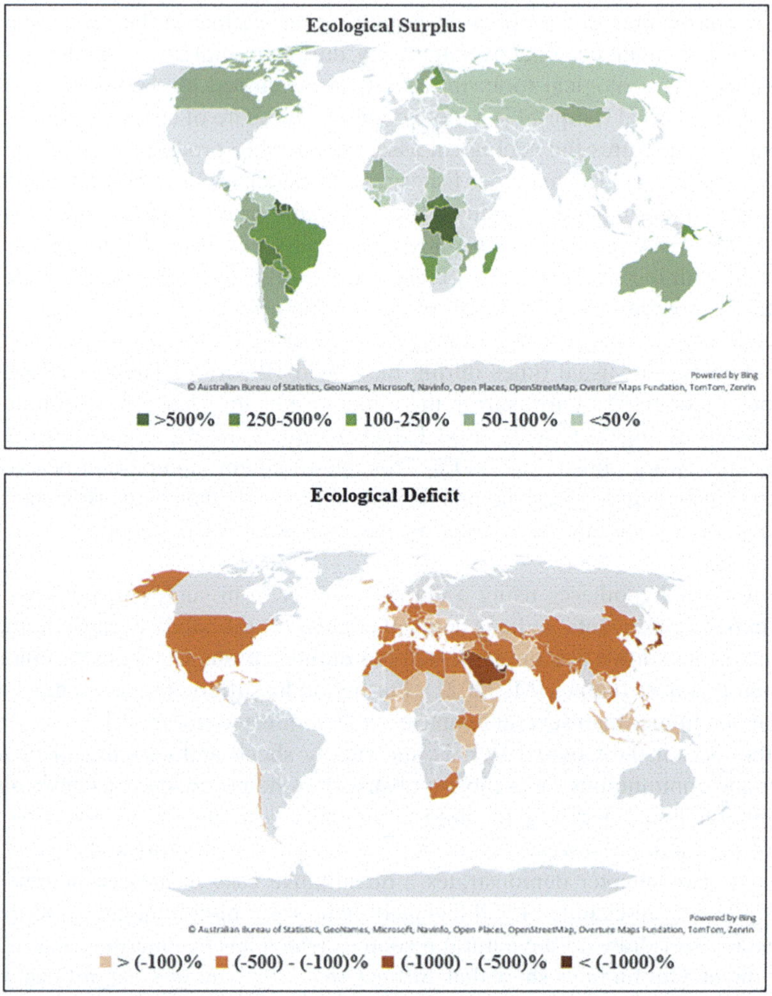

Fig. 9.1 World map showing ecological surplus and ecological deficit

9.2 Materials and Methods

9.2.1 Survey Instrument

Simply put, global biocapacity represents the supply of renewable resources and the systemic ability to absorb the population's spillover wastes. It is the earth's regenerative capacity to produce the natural assets and services. The ecological footprint represents the demand of the biocapacity by different communities. The ecological footprint analysis compares the economy's ecological footprint with respect to the

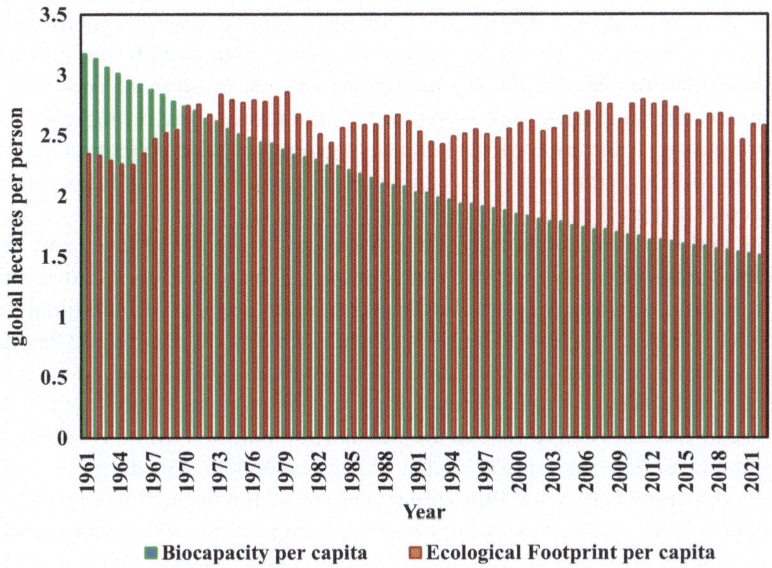

Fig. 9.2 World overshoot

biosphere's ability to regenerate resources and provide assimilative, life supporting and regenerating services at a given time. Global Footprint Network produces a comprehensive dataset to account for ecological footprint and biocapacity of countries the world over. Figures 9.1 and 9.2 are created using this database. An ecological overshoot occurs when the demand for natural assets (ecological footprint) exceeds the supply of natural capital (biocapacity). Figure 9.2 demonstrates this overshoot. The red bars (showing demand for natural assets) are becoming higher than the green bars (showing supply of natural assets). After 1972, the ecological footprint per capita of the world has overshot the per capita biocapacity every year. Further, the extent of the overshoot has also continuously increased.

Ecological footprint analysis can be used in various ways. First, at the macro-level, it can serve as a measure of sustainability for a nation (McIntosh & Pontius, 2016), enabling countries compare their ecological footprint with the rest of the world, and calculate their national biocapacity to balance their internal demand through corrective policy instruments. Second, it can act as a personal educator for individuals by helping them identify their highest segment of resource consumption, their ecological footprint vis-à-vis the average ecological footprint at the national level, or in comparison to the national biocapacity (Niccolucci et al., 2012). The analysis can also be used to make ecological footprint assessments for products and organizations (Global Footprint Network, 2009).

There are various calculators available for accounting the ecological footprint developed by the Global Footprint Network, Environmental Protection Agency and

Conservation International. We use the Global Footprint Network's footprint calculator as a survey tool as it is easily accessible and is aligned with the International Ecological Footprint Standards. Its alignment with the Ecological Footprint Standards ensures the credibility and consistency of the results. The ecological footprint calculator puts up a survey for individual respondents to indicate their consumption of food, housing and transport. This calculator is freely available at https://www.footprintcalculator.org/.

At the end of taking the survey, the respondent or the interviewer entering the respondent's response is led to the result summary. The results are divided into two parts, with other consecutive parts discussing the implications of individual results. The first part displays the respondent's personal earth overshoot days and the number of earths needed based on the respondent's consumption of natural resources. The number of earths needed implies how many earth-like planets (with the same biocapacity as earth) are needed to sustain the world's population if everyone chose to live at the same level of consumption as that of the respondent. Similarly, the personal overshoot day implies by what time of the year would the resources fall short of that consumption level, if everyone would live according to the consumption standards of the respondent. This calculation is done in reference to the per capita global biocapacity, which is a widely used indicator of environmental sustainability (McIntosh & Pontius, 2016). The second part of the results shows the respondent's ecological footprint values and the carbon footprint values. We have used this ecological footprint value for our analysis.

9.2.2 Research Context

The first Indian Covid-19 case was registered in January 2020 with rising cases from then on. To prevent the spread, a national lockdown was called in March 2020 with a phased and gradual reopening of the country through April to October 2020 as shown in Fig. 9.3. For the containment of the pandemic during the second wave of Covid-19, localized state-wide lockdowns were implemented in most Indian states in 2021 Q2. The economic impact of Covid-19 was substantial with the GDP contracting sharply to −24.4% year-on-year for 2020 Q2, −7.4% year-on-year for 2020 Q3, with overall, financial year 2020–21 GDP being −7.3% (IMF, 2021).

9.2.3 Research Design

The survey was carried out by research assistants who requested the participants to respond to on a telephonic survey twice. Their task was to explain the survey questions to the respondents who were then asked to keep in mind their consumption habits during normal times while taking the survey the first time and then their consumption habits during the new normal. The research assistants were asked to

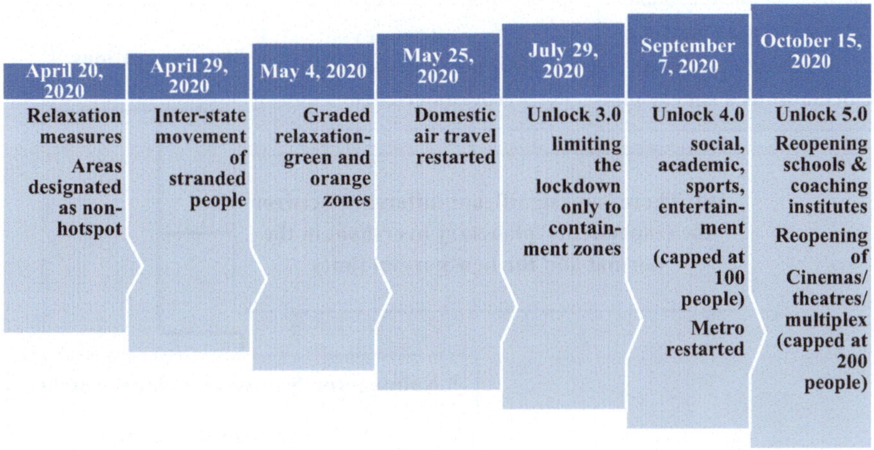

Fig. 9.3 Timeline of the phased unlock in India. *Source* Authors (compiled from IMF, 2021)

conduct these surveys over phone. Their services were used because of two reasons. First, we did not want the respondents to get fatigued due to the within pre-post research design by filling in the details of the survey themselves, especially those who were technically disabled. Second, we did not want the respondents from getting skilled to influence the survey. Further, it also avoided history and instrumental decay (Campbell & Stanley, 2015). The surveys were conducted in the month of November and captured the Covid-19 and lockdown experience from March 2020 to October 2020. A total of 380 respondents took the survey. We use this data to test whether there was a significant difference between the respondents' planetary overshoot (the number of earths needed to sustain their lifestyle if everyone lived like them) in the normal and the new normal times and to calculate the effect size—namely, Cohen's d and η^2. This process is diagrammatically reproduced in Fig. 9.4.

The within pre-post design is advantageous because it avoids heterogeneity between subjects and the issue of differential loss of respondents from the comparison groups (Campbell & Stanley, 2015). The concern in this type of research design is independence of the pre-post values, which invalidates ANOVA results and overestimates the effect size (Bakker et al., 2019). We conduct the Kolmogorov–Smirnov test for normality to check if the data for per capita ecological footprint in normal and Covid times is coming from a normal distribution. We find that both the variables in full dataset and the four subsets differ from normal distribution as expected given the nature of the data. The results are tabulated in Table 9.1. As the assumption of normality is not met, we perform Wilcoxon Signed-rank test and calculate the effect size by dividing the standardized z test statistic by the square root of the number of observations (Pallant, 2011).

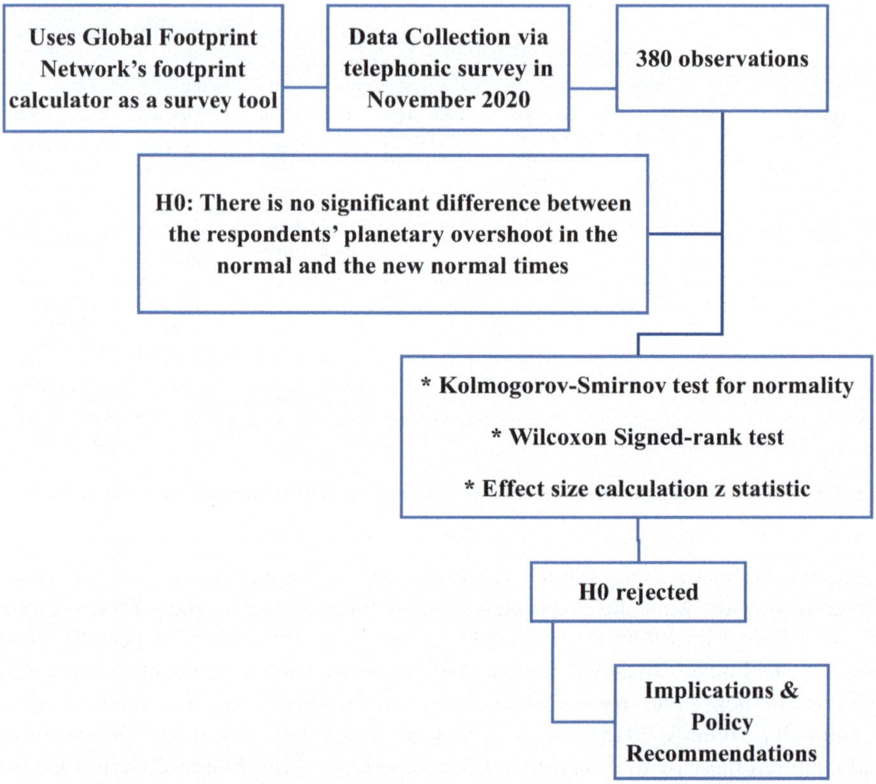

Fig. 9.4 Methodological flow chart

9.3 Results and Discussion

9.3.1 Descriptive Statistics

The descriptive statistics of the data is presented in Table 9.1. The total count of respondents was 380 with a mean of 4.333 earths consumed during normal times versus 4.235 earths consumed during the Covid-19 times. The number of female respondents was 160 with a mean consumption of 4.157 earths during normal times against the mean of 4.181 earths during the Covid-19 times. The total number of male respondents was 220 with a mean of 4.460 earths consumed during normal times versus 4.274 earths consumed during the Covid-19 times. Out of the 380 respondents, 263 were students with a mean of 4.282 earths consumed during normal times versus 4.391 earths consumed during the Covid-19 times. The other respondents had an average consumption of 4.447 earths in normal times and 3.884 earths during the Covid-19 times.

Table 9.1 Descriptive statistics

Dataset	Variable	Count	Average	Variance	One-sample Kolmogorov–Smirnov test significance
Full dataset	EF in terms of no. of earths (new normal times)	380	4.235	8.34	<0.001
	EF in terms of no. of earths (normal times)	380	4.333	5.33	<0.001
Female sample	EF in terms of no. of earths (new normal times)	160	4.181	7.93	<0.001
	EF in terms of no. of earths (normal times)	160	4.157	4.98	<0.001
Male sample	EF in terms of no. of earths (new normal times)	220	4.274	8.64	<0.001
	EF in terms of no. of earths (normal times)	220	4.460	5.58	<0.001
Student sample	EF in terms of no. of earths (new normal times)	263	4.391	8.34	<0.001
	EF in terms of no. of earths (normal times)	263	4.282	4.56	<0.001
Other groups' sample	EF in terms of no. of earths (new normal times)	117	3.884	8.34	<0.001
	EF in terms of no. of earths (normal times)	117	4.447	6.77	<0.001

Source Telephonic Survey

9.3.2 *Wilcoxon Signed Ranks Test*

As the paired differences are not normally distributed in this dataset, so we have conducted the Wilcoxon signed rank test. Further, as we are using a non-parametric test, we are also summarizing the medians. The median ecological footprint in terms

Table 9.2 Median ecological footprint

Dataset	Covid times (number of earths)	General times (number of earths)
Full dataset	2.751879699	3.420560748
Female sample	2.691176471	3.388888889
Male sample	2.859549533	3.452830189
Student sample	2.772727273	3.485714286
Other groups' sample	2.691176471	3.238938053

Source Telephonic survey

of no. of earths are reported for normal and new normal times in Table 9.2. The test results of Wilcoxon signed ranks test are reported in Table 9.3, and the effect size calculated values are reported in Table 9.4.

Note [a]denotes a p value of <0.001

The results of the Wilcoxon signed ranks test of the survey data are reported in Table 9.3. The Wilcoxon signed rank test revealed that the consumption of resources, namely number of earths, were significantly lower in the full in the new normal ($Md = 2.75$, $n = 380$) than in normal times ($Md = 3.42$, $n = 380$), $z = -6.602$ ($p < 0.001$), with a medium effect size, r = 0.34. Here, z represents the number of standard deviations a sample mean is away from a population mean (or the difference between two means), while r represents the strength and direction of a relationship between two variables, often in the form of a correlation. A z-score can be converted to an effect size r, and vice versa, allowing for a standardized measure of the magnitude of an effect. The results are similar for female, male and other groups' samples with significantly lower consumption in terms of number of earths in the new normal and a medium effect size. For student sample, the results are slightly different. For student sample, the consumption of resources, namely number of earths, were significantly lower in the full in the new normal ($Md = 2.77$, $n = 263$) than in normal times ($Md = 3.49$, $n = 263$), $z = -4.496$ ($p < 0.001$), with a low effect size, r = 0.28.

These results show that not only did the ecological footprint change in the new normal, but there was also a considerable fall in the ecological footprint for the urban Indians. These results are promising despite a medium effect size because they can have larger consequences on the planetary wellbeing if they are indeed replicated across countries and in larger samples.

9.3.3 *Discussion*

Our inquiry aimed to investigate whether the Indian urban community mitigated and adapted its consumption in response to the expectancy of future shortage and was there a net reduction in consumption. This research also aimed at measuring the extent of this net reduction, if any. We operationalized this inquiry by capturing the change in the ecological footprint and the resultant personal overshoot day, the overall

Table 9.3 Wilcoxon signed ranks

Dataset	EF in terms of no. of earths (new normal times)—EF in terms of no. of earths (normal times)	N	Mean rank	Sum of ranks
Full dataset	Negative ranks	271[a]	180.35	48,875.5
	Positive ranks	103[b]	206.31	21,249.5
	Ties	6[c]		
	Total	380		
Female sample	Negative ranks	111[a]	77.85	8641
	Positive ranks	48[b]	84.98	4079
	Ties	1[c]		
	Total	160		
Male sample	Negative ranks	160[a]	103.27	16,523.5
	Positive ranks	55[b]	121.75	6696.5
	Ties	5[c]		
	Total	220		
Student sample	Negative ranks	179[a]	122.57	21,940
	Positive ranks	78[b]	143.76	11,213
	Ties	6[c]		
	Total	263		
Other groups' sample	Negative ranks	92[a]	59.32	5457
	Positive ranks	25[b]	57.84	1446
	Ties	0[c]		
	Total	117		

Notes [a]EF in terms of no. of earths (new normal times) < EF in terms of no. of earths (normal times)
[b]EF in terms of no. of earths (new normal times) > EF in terms of no. of earths (normal times)
[C]EF in Terms of no. of earths (new normal times) = EF in terms of no. of earths (normal times)

Table 9.4 Effect size

Dataset	Standardized test statistic z	Number of pairs	Effect size r
Full dataset	−6.602[a]	380	0.338675
Female sample	−3.923[a]	160	0.31014
Male sample	−5.380[a]	220	0.36272
Student sample	−4.496[a]	263	0.277235
Other groups' sample	−5.455[a]	117	0.504315

consumption inclusive of carbon footprint, during normal times and the Covid-19 period using the ecological footprint calculator developed by the Global Footprint Network.

Our results as shown above prove that indeed there was a significant fall (demonstrated by a medium effect size) in the ecological footprint of the Indian urban dwellers in our curated dataset. These results are comparable to those researches undertaken during Covid times that demonstrate a fall in the carbon footprint, a measure of carbon emission equivalents (Muhammad et al., 2020; Rupani et al., 2020; Saadat et al., 2020). Our results add to these works by showing a reduction in the overall strain on the planet and by showing a net change in consumption.

In terms of the social impact of Covid-19, our results are useful in showing that vulnerability of the community during crisis can be reduced by curtailing conspicuous consumption because it can contain the vicious cycle of stockpiling and artificial shortage. Conspicuous consumption creates a vicious cycle of stockpiling and artificial shortage that can create panic and increase a community's vulnerability. Thus, this work offers a unique perspective of building community resilience through addressing consumption.

Our results also reflect community resilience showing how communities respond by adjusting their ecological footprint in the best possible way to maintain their standard of living demonstrating its adaptive and mitigation capacity. In fact, the medium to low effect size in different subsets of our dataset demonstrate community resilience in the face of uncertainty and vulnerability of Covid. This mediated impact can be explained by the state support during the Covid times. Documented evidence lays out the execution of state led response focused on benefits in kind (food and cooking gas), enabling the most vulnerable part of the population with cash transfers, health care and insurance coverage to maintain essential consumption levels and safety. Other crisis mitigating and adapting expenses included investment in healthcare infrastructure, interest-free loans, credit easing, etc., all culminating in a relief package of around 10% of GDP. The measures that were announced later (October–November 2020) promised additional public investment and support schemes targeting 13 priority sectors amounting to about 0.8% of GDP over 5 years to further production and reduce shortage. These earmarked funds were spent on a higher fertilizer subsidy allocation benefiting the agriculture sector (0.3% of GDP) and support for urban housing construction (0.1% of GDP). Further support where 1.9% of the GDP was handed out as credit support to businesses, 1.6% of GDP to the poor and migrant households, 0.4% of GDP to distressed electricity distribution companies and 0.7% of GDP was offered as targeted support for the agricultural sector. At the same time, tax relief was announced for a range of sectors. In February 2021, the budget for Financial Year 2021–22 expanded spending on health and well-being to the tune of Rs. 350 billion. With a surge in infection, free food grains worth Rs. 260 billion were announced to be provided to 800 million individuals along with the provision of interest-free loans for the Indian states earmarking Rs. 150 billion for capital expenditure.

9.4 Limitations and Future Research

These results need to be interpreted with caution because they do not reveal causal relationships. Possible strategy to prove a causal relationship between ecological footprint and Covid-19 would be to capture data for each response and not just the planetary overshoot. Another possible strategy could be to examine the fall in consumption during times of crises by separating the effect of unemployment led income fall and the effect of shortage or lack of access of goods and services due to mobility restrictions. We did not capture data in that much detail because we wanted to avoid respondents' fatigue even if there were research assistants to help. Otherwise, it would have taken longer to capture all the responses on a separate response sheet. Another improvement would have been to analyse time-series data at every point of the unlock period but that would again make the survey long and tedious and also lead to an issue of recall. Further, while, these results may show a medium effect size in the Indian data, results may need to be rechecked with a wider sample spread across the country.

Given our results, it is imperative that further research into investigating the social implications of the inequality and social unrest due to consumption reduction by a certain social group in crises should be undertaken possibly in the form of case studies accompanied with resilience strategies for building up mitigative and transformative capacity.

9.5 Policy Recommendations

Changes due to any crisis are often short lived, driven largely by fall in income or lack of choices, implying that society may revert to previous patterns of behaviour (here, specifically consumption) once the crisis is past (Rousseau & Deschacht, 2020). Given that the fall in the ecological footprint was a result of the external shock due to Covid, the changes were at best provisional and did peter out. Therefore, it is important to identify ways to sustain the positive changes in the long run. Further, given that consumption should not fall below the level of current consumption on the consumption path that maximizes the present discounted value of utility (Arrow et al., 2004), ways to create a balance between productivity inducing policies and sustainable consumption policies need to be identified.

9.5.1 Ecological Efficiency Inculcating Policy Instruments

Technology-Forcing: Technology-forcing is a regulatory strategy of specifying a standard that cannot be met either due to technical or economic infeasibility (Gerard & Lave, 2005). This cost may very well be related to the cost of learning new practices in favour of old ones (like a large move towards working online in the lockdown period where the technology was available but was not used so often or on the scale required during the lockdown). Technology-forcing is fulfilled either in the presence of current policy specifying technology and performance standards or in anticipation of such future regulations.

A set of environmental regulation scholars (Palmer et al., 1995; Jaffe & Palmer, 1997, Jaffe et al., 2002) claim that technology-forcing may not be successful because uniform standards from different agents could be counter-productive. The counter-productivity is because of various reasons, namely real or perceived unfairness; unattainability of the target; or low immediate productivity (due to high expense or need for relearning/upskilling by the employees). Further, technology-forcing policies may create a contested backdrop allowing agents to bargain according to their own vested goals or interests (Lee et al., 2010). Another possible reason to delay response from such a policy stimulus could be double externality of environmental innovation (Rennings, 2000). The inertia and cost of switching technology to reduce wastes from economic activities of production, consumption, distribution (inclusive of trade) may also make technology-forcing policies unpopular (Jaffe et al., 2005).

Yet, there are three reasons why we suggest technology-forcing as an effective policy instrument. First, historically, it has delivered well if we look at the environmental standard setting in the auto industry across the world—EURO Emission Standards in Europe, BHARAT Stage Norms in India, or Advanced Automotive Emissions Control in the USA. The possible reasons may have to do with competitive advantage (Gerard & Lave, 2005). Second, technology-forcing regulations have been found to stimulate innovation (Lee et al., 2010), which could mean increased environmental innovations given the ideational environment for closing the loop of value generation for not only profit but also the planet. Third, it is evident from our results that in response to an external shock—Covid-19, the ecological footprint has fallen substantially. Therefore, a restrictive policy instrument like technology-forcing in a transitioned ideational policy environment can be effective in leading an effective reduction even with a medium effect size.

Compulsory Use of Environmental Management System: Often, firms are unable to observe cost savings due increase in ecological efficiency. Making environmental management system mandatory makes firms use techniques like process mapping, root cause analysis and environmental accounting that enable them to identify and locate inefficiencies and to innovate ways to increase efficiency thus saving costs. These costs not only refer to the costs of pollution control (Khanna et al., 2009) but also cost savings from reducing energy and material use in production (Horbach et al., 2012).

We recommend compulsory environmental management because of its effectiveness in achieving cost savings (Morrow & Rondinelli, 2002). It would not only be effective in attaining for environmental improvements but also for adapting to the changed circumstances of new resource limitations imposed by the lockdown.

Incentive-Based Policy: Incentive-based policies like subsidies, state financing of R&D into eco-efficient technology, tax benefits, etc., may have some success in pushing the firms to innovate for environmental improvements as well as adopt ecologically efficient technology. Though empirical observation shows limited success of these measures (Kemp, 2000), they along with technology push measures may create a more balanced policy mix. Such a policy mix can reduce the perceived costs of pro-environmental behaviour in terms of time, convenience, effort, money, comfort, while increasing the perceived benefits (Steg et al., 2014).

9.5.2 Sustainable Consumption Inculcating Policy Instruments

The production of conspicuous goods reduces aggregate social wellbeing; however, taxing such goods is ineffective in reducing their demand (Corneo & Jeanne, 1997). Thus, policies that promote pro-environmental consumer behaviour are crucial (Jackson, 2004) and can nudge consumer behaviour in the right direction so that they choose green electricity, purchase organic goods, conserve water and energy, and opt for a low carbon lifestyle. Literature shows that stress on self-regulation and warm glow, education and awareness campaigns have had limited success in motivating people to reduce their consumption and therefore the relevant question becomes, how to persuade people to behave in an environmentally and socially responsible ways (Jackson, 2004). Thus, a well-balanced policy mix is required.

Command and Control Policies Items: Command and control policies that issue mandates and compulsory standards for consumers can play an important role in regulating conspicuous consumption and push the community towards sustainable consumption. Mandating recycling of household wastes and reuse as in Malaysia since 2012 (Moh & Manaf, 2014), policies pushing for the use of energy and water-efficient appliances as in USA, European Union (Hausman & Joskow, 1982; Khazzoom, 1980; Schleich et al., 2021) and Australia (Wilkenfeld, 2003), fuel-efficient norms for vehicles as in India (https://beeindia.gov.in/en/programmesenergy-efficiency-in-transport-sector/fuel-efficiency) and pushing the use of public transport like plying of odd–even private cars as in Delhi, India (https://indianexpress.com/article/what-is/what-is-delhi-odd-even-scheme-air-pollution-9016010/), can be quite effective in inculcating sustainable consumption behaviour. Creative policies mandating the division of the workweek into work-from-home and office-going days can also effectively reduce carbon footprint (Jackson, 2004).

Soft Policies: Using soft policies like persuading and nudging to change the consumption behaviour again could be quite effective. Using communication to persuade the consumer to reduce conspicuous consumption and consume responsibly has been in vogue since the 1950s. With the spread of digital social media, empirical evidence shows a positive impact on sustainable purchasing and towards environmental responsibility (Zafar et al., 2021).

A nudge refers to the 'purposeful changes in the choice architecture' aimed at influencing peoples' behaviour by altering the environment guiding individual choices

(Lehner et al., 2016). Nudges are being used to accompany other policy measures to increase the effectiveness of environmental policy (Lehner et al., 2016). For example, health discourse can effectively influence consumers to choose organic and locally sourced food reducing carbon footprint. Nudges can also be used to influence residential energy consumption and personal transport behaviour (see Table 9.5 for the effect size of these nudges).

Economic Incentives: Policymakers may use economic incentives to change consumer behaviour. If the consumer is charged for buying incandescent light-bulbs and subsidized for buying fluorescent ones, they will avoid incandescent and buy fluorescent bulbs (Jackson, 2004) if the economic benefit of using fluorescent bulbs is greater than the economic benefits of incandescent ones. The net gains of the sustainable choice over the non-sustainable ones in economic terms do matter and such policies should be accompanied by functionally superior sustainable choices to reduce the apparent costs of sacrificing for sustainability (Steg et al., 2014).

9.6 Conclusion

This chapter uses a quantitative case to demonstrate how communities can build resilience by re-examining and altering the way they consume and demand resources. Our results show that vulnerability of the community during crisis can be reduced by curtailing conspicuous consumption because it can contain the vicious cycle of stockpiling and artificial shortage. Conspicuous consumption creates a vicious cycle of stockpiling and artificial shortage that can create panic and increase a community's vulnerability. Thus, this work offers a unique perspective of building community resilience through addressing consumption.

Our findings also show that during Covid-19, there was a significant reduction in consumption of resources (or ecological footprint) in India, addressing the over-draw on the natural resources that is a threat to sustainability, reinforcing that consumption constraining environmental regulations are more effective than the environmental efficiency regulations because there is no rebound accompanying it.

Further, this chapter also contributes towards understanding the impact of external shocks in reducing negative environmental externalities of technological innovation. Just like Covid-19 acts as an external shock and can constraint conspicuous consumption, a consumption constraining policy can also act as an external shock to constraint conspicuous consumption. Thus, this chapter also recommends a policy mix of environmental efficiency regulations with sustainable consumption regulations.

Future research needs to further investigate and confirm the causal links and explore the nature of the inter-relationship between an external unanticipated shock and environmental sustainability. Extensions could be made to validate these results in a larger sample spread across the globe and through a time-series analysis.

Table 9.5 Effect size of nudges for influencing consumption behaviour

Nudge mechanisms used	Purpose	Applications	Effectiveness
Simplification and framing of information	Food consumption	– Simplifying information	– Controlled small-scale studies show large impact, impact varying for different social strata
	Residential energy consumption	– Feedback on energy consumption – Energy labelling of appliances and buildings	– Small-scale interventions show small to large impact, large-scale studies show small impact – No effect size to report
	Personal transport behaviour	– Decluttering streets – Providing clear information and maps – Changing framing to encourage cycling by offering cycling training and walking – Simplifying information on fuel consumption of cars	– 19–35% reduction in carbon emissions – 10% reduction of car use due to personal travel plans
Changes to the physical environment	Food consumption	– Change visibility and accessibility – Influence size	– High effect size in controlled environments – Large effect size with portion size and package size
	Residential energy consumption	– Sustainable design of homes and appliances, e.g. key card removal turns of lights – Prompts for sustainable behaviour	– no effect size to report
	Personal transport behaviour	– Road and lane planning	– Effective
Changes to the default option	Food consumption	– Positioning of product choice	– Wide use in retailing hints at medium to large effect size
	Residential energy consumption	– Opt-out green electricity offers – Opt-out from smart grid trial	– Large effect size

(continued)

Table 9.5 (continued)

Nudge mechanisms used	Purpose	Applications	Effectiveness
	Personal transport behaviour	– Auto-pilot decisions in cars, road planning, helmet wearing	– Effective
Use of social norms	Food consumption	– Comparing behaviour	– Effective for publicly visible behaviour
	Residential energy consumption	– Social comparison billing feedback	– Large effects in small-scale trials, smaller effects in large-scale trials
	Personal transport behaviour	– Walking feedback programme in smartphone apps and social media updates	– Mixed evidence

Source Adapted from Lehner et al. (2016)

Acknowledgements We are grateful to our research assistants who assisted us in the data collection. We also thank the respondents who took out their valuable time to complete the survey, helping us collect the data for proceeding with our analysis to test our hypothesis. We are also extremely thankful to Prof. Shashi Kant Srivastava in his reviews and inputs for developing this work.

Authors Contribution: All authors contributed to the study conception and design. Material preparation and data collection was done by Janardan Krishna Yadav, Srinivas Jangirala, Meenakshi Prasad, and Sandeep Singh. Data analysis was performed by Deepika Chandra Verma and Janardan Krishna Yadav. The first draft of the manuscript was written by Deepika Chandra Verma. The first and all successive drafts were commented on by all the authors. All authors read and approved the final manuscript.

Funding: The research was not supported by any funding.

Potential Conflicts of Interest: On behalf of all authors, the corresponding author states that there is no conflict of interest.

AI Discloser Statement: No AI tools have been used in this research.

Informed Consent: Informed consent was obtained from all individual participants included in the study.

References

Abegunde, D., & Stanciole, A. (2006). An estimation of the economic impact of chronic noncommunicable diseases in selected countries. *World Health Organization, Department of Chronic Diseases and Health Promotion, 2006.*

Altig, D., Baker, S., Barrero, J. M., Bloom, N., Bunn, P., Chen, S., & Thwaites, G. (2020). Economic uncertainty before and during the COVID-19 pandemic. *Journal of Public Economics, 191*, Article 104274.

Arrow, K., Dasgupta, P., Goulder, L., Daily, G., Ehrlich, P., Heal, G., & Walker, B. (2004). Are we consuming too much? *Journal of Economic Perspectives, 18*(3), 147–172.
Baker, S., Bloom, N., Davis, S., & Terry, S. (2020). COVID-induced economic uncertainty and its consequences. *VoxEU. org, 13.*
Bakker, A., Cai, J., English, L., Kaiser, G., Mesa, V., & Van Dooren, W. (2019). Beyond small, medium, or large: Points of consideration when interpreting effect sizes. *Educational Studies in Mathematics, 102*, 1–8.
Baldwin, R., & Di Mauro, B. W. (2020). Economics in the time of COVID-19: A new eBook. *Vox CEPR Policy Portal, 2*(3), 105–109.
Blustein, D. L., Duffy, R., Ferreira, J. A., Cohen-Scali, V., Cinamon, R. G., & Allan, B. A. (2020). Unemployment in the time of COVID-19: A research agenda. *Journal of Vocational Behavior, 119*, Article 103436.
Bryce, C., Ring, P., Ashby, S., & Wardman, J. K. (2022). Resilience in the face of uncertainty: early lessons from the COVID-19 pandemic. In *COVID-19* (pp. 48–55). Routledge.
Campbell, C. (2021). The impact of COVID-19 on local government stakeholders' perspectives on local food production. *Journal of Agriculture, Food Systems, and Community Development, 10*(2), 71–88.
Campbell, D. T., & Stanley, J. C. (2015). *Experimental and quasi-experimental designs for research* (pp. 1–71). Ravenio Books.
Coe, R. (2002, September). It's the effect size, stupid. In *British Educational Research Association Annual Conference* (vol. 12, p. 14).
Corneo, G., & Jeanne, O. (1997). Conspicuous consumption, snobbism and conformism. *Journal of Public Economics, 66*(1), 55–71.
De Haas, M., Faber, R., & Hamersma, M. (2020). How COVID-19 and the Dutch 'intelligent lockdown' change activities, work and travel behaviour: Evidence from longitudinal data in the Netherlands. *Transportation Research Interdisciplinary Perspectives, 6*, Article 100150.
Di Renzo, L., Gualtieri, P., Pivari, F., Soldati, L., Attinà, A., Cinelli, G., & De Lorenzo, A. (2020). Eating habits and lifestyle changes during COVID-19 lockdown: An Italian survey. *Journal of Translational Medicine, 18*, 1–15.
Gerard, D., & Lave, L. B. (2005). Implementing technology-forcing policies: The 1970 Clean Air Act Amendments and the introduction of advanced automotive emissions controls in the United States. *Technological Forecasting and Social Change, 72*(7), 761–778.
Global Footprint Network. (2009). *Ecological Footprint Standard 2009*. Global Footprint Network. https://www.footprintnetwork.org/resources/data/footprint-standards/
Gordon-Wilson, S. (2022). Consumption practices during the COVID-19 crisis. *International Journal of Consumer Studies, 46*(2), 575–588.
Gössling, S., Scott, D., & Hall, C. M. (2020). Pandemics, tourism and global change: A rapid assessment of COVID-19. *Journal of Sustainable Tourism, 29*(1), 1–20.
Haldon, J., Eisenberg, M., Mordechai, L., Izdebski, A., & White, S. (2020). Lessons from the past, policies for the future: Resilience and sustainability in past crises. *Environment Systems and Decisions, 40*, 287–297.
Hausman, J. A., & Joskow, P. L. (1982). Evaluating the costs and benefits of appliance efficiency standards. *The American Economic Review, 72*(2), 220–225.
Horbach, J., Rammer, C., & Rennings, K. (2012). Determinants of eco-innovations by type of environmental impact—The role of regulatory push/pull, technology push and market pull. *Ecological Economics, 78*, 112–122.
IMF. (July 2, 2021). Policy responses to COVID-19. https://www.imf.org/en/Topics/imf-and-covid19/Policy-Responses-to-COVID-19
Jackson, T. (2004). Negotiating sustainable consumption: A review of the consumption debate and its policy implications. *Energy & Environment, 15*(6), 1027–1051.
Jaffe, A. B., & Palmer, K. (1997). Environmental regulation and innovation: A panel data study. *Review of Economics and Statistics, 79*(4), 610–619.

Jaffe, A. B., Newell, R. G., & Stavins, R. N. (2002). Environmental policy and technological change. *Environmental and Resource Economics, 22*, 41–70.
Jaffe, A. B., Newell, R. G., & Stavins, R. N. (2005). A tale of two market failures: Technology and environmental policy. *Ecological Economics, 54*(2–3), 164–174.
Jribi, S., Ben Ismail, H., Doggui, D., & Debbabi, H. (2020). COVID-19 virus outbreak lockdown: What impacts on household food wastage? *Environment, Development and Sustainability, 22*(5), 3939–3955.
Kaushik, M., & Guleria, N. (2020). The impact of pandemic COVID-19 in workplace. *European Journal of Business and Management, 12*(15), 1–10.
Kemp, R. (2000). Technology and environmental policy—Innovation effects of past policies and suggestions for improvement. *Innovation and the Environment, 1*, 35–61.
Khanna, M. (2021). COVID-19: A cloud with a silver lining for renewable energy? *Applied Economic Perspectives and Policy, 43*(1), 73–85.
Khanna, M., Deltas, G., & Harrington, D. R. (2009). Adoption of pollution prevention techniques: The role of management systems and regulatory pressures. *Environmental and Resource Economics, 44*, 85–106.
Khazzoom, J.D. (1980). Economic implications of mandated efficiency in standards for household appliances.
Koffman, J., Gross, J., Etkind, S. N., & Selman, L. (2020). Uncertainty and COVID-19: How are we to respond? *Journal of the Royal Society of Medicine, 113*(6), 211–216.
Lakens, D. (2013). Calculating and reporting effect sizes to facilitate cumulative science: A practical primer for t-tests and ANOVAs. *Frontiers in Psychology, 4*, 863.
Lee, J., Veloso, F. M., Hounshell, D. A., & Rubin, E. S. (2010). Forcing technological change: A case of automobile emissions control technology development in the US. *Technovation, 30*(4), 249–264.
Lehner, M., Mont, O., & Heiskanen, E. (2016). Nudging–A promising tool for sustainable consumption behaviour? *Journal of Cleaner Production, 134*, 166–177.
Majumdar, P., Biswas, A., & Sahu, S. (2020). COVID-19 pandemic and lockdown: Cause of sleep disruption, depression, somatic pain, and increased screen exposure of office workers and students of India. *Chronobiology International, 37*(8), 1191–1200.
Marroquín, B., Vine, V., & Morgan, R. (2020). Mental health during the COVID-19 pandemic: Effects of stay-at-home policies, social distancing behavior, and social resources. *Psychiatry Research, 293*, Article 113419.
Martin, A., Markhvida, M., Hallegatte, S., & Walsh, B. (2020). Socio-economic impacts of COVID-19 on household consumption and poverty. *Economics of Disasters and Climate Change, 4*(3), 453–479.
McIntosh, A., & Pontius, J. (2016). *Science and the global environment: Case studies for integrating science and the global environment*. Elsevier.
McKibbin, W., & Fernando, R. (2021). The global macroeconomic impacts of COVID-19: Seven scenarios. *Asian Economic Papers, 20*(2), 1–30.
Miao, Q., Schwarz, S., & Schwarz, G. (2021). Responding to COVID-19: Community volunteerism and coproduction in China. *World Development, 137*, Article 105128.
Moh, Y. C., & Abd Manaf, L. (2014). Overview of household solid waste recycling policy status and challenges in Malaysia. *Resources, Conservation and Recycling, 82*, 50–61.
Morrow, D., & Rondinelli, D. (2002). Adopting corporate environmental management systems: Motivations and results of ISO 14001 and EMAS certification. *European Management Journal, 20*(2), 159–171.
Muhammad, S., Long, X., & Salman, M. (2020). COVID-19 pandemic and environmental pollution: A blessing in disguise? *Science of the Total Environment, 728*, Article 138820.
Niccolucci, V., Tiezzi, E., Pulselli, F. M., & Capineri, C. (2012). Biocapacity vs ecological footprint of world regions: A geopolitical interpretation. *Ecological Indicators, 16*, 23–30.
OECD. (2014). *Guidelines for resilience systems analysis*. OECD Publishing.

Pallant, J. (2011). *SPSS survival manual: A step by step guide to data analysis using IBM SPSS.* Routledge.

Palmer, K., Oates, W. E., & Portney, P. R. (1995). Tightening environmental standards: The benefit-cost or the no-cost paradigm? *Journal of Economic Perspectives, 9*(4), 119–132.

Raman, G., & Peng, J. C. H. (2021). Electricity consumption of Singaporean households reveals proactive community response to COVID-19 progression. *Proceedings of the National Academy of Sciences, 118*(34), Article e2026596118.

Rennings, K. (2000). Redefining innovation—Eco-innovation research and the contribution from ecological economics. *Ecological Economics, 32*(2), 319–332.

Rousseau, S., & Deschacht, N. (2020). Public awareness of nature and the environment during the COVID-19 crisis. *Environmental and Resource Economics, 76*, 1149–1159.

Rupani, P. F., Nilashi, M., Abumalloh, R. A., Asadi, S., Samad, S., & Wang, S. (2020). Coronavirus pandemic (COVID-19) and its natural environmental impacts. *International Journal of Environmental Science and Technology, 17*, 4655–4666.

Saadat, S., Rawtani, D., & Hussain, C. M. (2020). Environmental perspective of COVID-19. *Science of the Total Environment, 728*, Article 138870.

Sakurai, M., & Chughtai, H. (2020). Resilience against crises: COVID-19 and lessons from natural disasters. *European Journal of Information Systems, 29*(5), 585–594.

Schäfer, T., & Schwarz, M. A. (2019). The meaningfulness of effect sizes in psychological research: Differences between sub-disciplines and the impact of potential biases. *Frontiers in Psychology, 10*, 813.

Schleich, J., Durand, A., & Brugger, H. (2021). How effective are EU minimum energy performance standards and energy labels for cold appliances? *Energy Policy, 149*, Article 112069.

Sidor, A., & Rzymski, P. (2020). Dietary choices and habits during COVID-19 lockdown: Experience from Poland. *Nutrients, 12*(6), 1657.

Smith, R. D., Keogh-Brown, M. R., & Barnett, T. (2011). Estimating the economic impact of pandemic influenza: An application of the computable general equilibrium model to the UK. *Social Science & Medicine, 73*(2), 235–244.

Steg, L., Bolderdijk, J. W., Keizer, K., & Perlaviciute, G. (2014). An integrated framework for encouraging pro-environmental behaviour: The role of values, situational factors and goals. *Journal of Environmental Psychology, 38*, 104–115.

Szczygielski, J. J., Bwanya, P. R., Charteris, A., & Brzeszczyński, J. (2021). The only certainty is uncertainty: An analysis of the impact of COVID-19 uncertainty on regional stock markets. *Finance Research Letters, 43*, Article 101945.

Tittonell, P., Fernandez, M., El Mujtar, V. E., Preiss, P. V., Sarapura, S., Laborda, L., & Cardoso, I. M. (2021). Emerging responses to the COVID-19 crisis from family farming and the agroecology movement in Latin America–A rediscovery of food, farmers and collective action. *Agricultural Systems, 190*, Article 103098.

Veal, A. J. (1993). The concept of lifestyle: A review. *Leisure Studies, 12*(4), 233–252.

Wilkenfeld, G. (2003). *A mandatory water efficiency labelling scheme for Australia.* Final Report prepared for Environment Australia.

Zafar, A. U., Shen, J., Ashfaq, M., & Shahzad, M. (2021). Social media and sustainable purchasing attitude: Role of trust in social media and environmental effectiveness. *Journal of Retailing and Consumer Services, 63*, Article 102751.

Chapter 10
Indian Governance Toward Climate Change-Induced Redressal Mechanism in the Era of Green Energy

Gabriela Michael and Harshita Mishra

Abstract In the past few decades, there have been massive climatic shifts on the planet as a result of human activities, including overuse of resources and pollution, among other factors. Although these changes might not be a direct threat to life, they are gradually worsening the quality of life and exhausting natural resources. In reaction, the United Nations came up with 17 Sustainable Development Goals (SDGs) in order to encourage responsible development. In the chapter, SDG Goal 7, which is: "Ensure access to affordable, reliable, sustainable and modern energy to all," is discussed. With the world going green with energy, nations are experiencing issues such as the high cost of implementation and old laws that cannot keep abreast with the heavy technological changes. India, one of the world's largest energy consumers, has emerged as a global leader in renewable energy. It has been at the forefront in the establishment of the International Solar Alliance and has advanced domestic production of solar, wind, and green hydrogen power. Its legal and policy frameworks are increasingly aligned with sustainable development, emphasizing decarbonization and disaster-risk mitigation. This chapter examines India's transition to renewables in comparison with developments in other QUAD countries—Australia, Japan, and the USA—to assess collective progress toward net-zero emissions. Also, the chapter discusses the way in which the energy transition of India promotes climate governance and enhances community resilience. The study offers hints on a sustainable, inclusive energy future by connecting legal frameworks, environmental policy, and resilience strategies of the grassroots. It can provide a possible example to other developing countries that are also in need to stay economically viable and environmentally responsible, and sustainable over the long run.

Keywords Climate change · Renewable energy · India · Decarbonization · Green hydrogen · Governance · Sustainable development · Resilience

G. Michael (✉)
Indian Maritime University, Chennai, India
e-mail: gabrielamichael1998@gmail.com; scholar.gabrielam@ac.in

H. Mishra
Rashtriya Raksha University, Shivamogga, Karnataka, India
e-mail: harshita.mishra@rru.ac.in

G. K. Panda et al. (eds.), *Socio-Ecological Crisis of Development*, Sustainability Solutions, https://doi.org/10.1007/978-3-032-22171-1_10

10.1 Introduction

Recorded human activities can be dated back to around 2500 years, but the Earth has been a host to life forms for millions of years, and the Earth always has the ability to neutralize itself whenever there is a threat. The current threat that Earth is grappling with is the unprecedented changes due to global warming, a phenomenon strictly driven by human activities with heavy industrialization since the twentieth century. Studies show that the rate and magnitude of atmospheric warming had significantly risen during the industrial period due to the incessant burning of fossil fuels, excessive cutting down of trees leading to loss of forest ecosystem, etc. (Shaw et al., 2022). The impacts of such a swift industrial growth are manifested in the form of extreme weather patterns all world over like off season rains, cyclones, heat waves, and altering winds and ocean currents. The oceans have also not been left out as the dying marine life, and disappearing land-based ecosystems have caused the sea levels to rise hence altering world ocean temperatures. Such severe and drastic climatic change poses a severe threat as it becomes irreversible for thousands of years (Intergovernmental Panel on Climate Change, 2022). As the population of any country grows annually, these shifts pose to be a significant challenge to the national, regional, and global political climate that distorts the social framework because of the lack of resources causing the conditions of never-satisfied markets, the migration to new regions, instability, the infection of illnesses, and other problems.

The geography of India is unique and distinctive, and this probably is the reason as to why India is referred to as the peninsula subcontinent. Indo-Gangetic Plains, Central Highlands and Deccan Plateau, Thar Desert, Himalaya and the Northeast Mountain Ranges, the East Coast, the West Coast, the Bordering seas, and islands all add to the topographical nature of the country (Youdon, 2020). This extreme difference in the weather scenarios and climatic conditions just within a short-range distance makes it one of the biggest disaster zones in the world where "59% of the land is vulnerable to earthquakes, 8.5% of the land is vulnerable to cyclonic storms, and 5% of the land is susceptible to river basin floods." (Ministry of Environment, Forest and Climate Change, Government of India, 2015). India has recently overtaken China in becoming the world's most populous country and has a unique geo-climatic profile. Furthermore, the vast majority of people rely on livelihoods from climate-sensitive industries like forestry, agriculture, and coastal ecosystems. Extreme weather conditions, high precipitation, cyclonic storms, and sea level rise brought on by climate change make India more vulnerable to a wide range of hydrometeorological hazards that are worryingly increasing in intensity and frequency while becoming less predictable.

The Ministry of Earth Sciences (MoES) of the Government of India in its report titled "Assessment of Climate Change over the Indian Region" in 2020 had stated that summer heat waves were projected to increase 3–4 times in frequency; it is predicted that the duration was lengthened by the end of the century as compared to the 1976–2005 baseline period (Krishnan et al., 2020). Furthermore, one-third of India's population lives in coastal cities, where severe weather and sea level rise are expected to remain significant risk factors. The Indian coastline is home to a variety of human settlements, from megacities to fishing villages and hamlets. With the average global temperature rising, the rate of sea level rise will only quicken. By the middle of the twenty-first century, many low-lying areas along India's heavily populated coastline are predicted to be submerged by projections based on climate models. Significant increases in sea level will exacerbate the effects of extreme weather phenomena like cyclonic storms and coastal flooding. Sea level rise and extreme weather events will negatively affect coastal ecosystems like mangrove forests, seagrass beds, and coral reefs, which are vital socio-economic services to coastal communities, in addition to their effects on human settlements along the coast (Youdon, 2020). Coastal food and water security is also threatened by saltwater intrusion into agricultural lands and subterranean aquifers as a result of sea level rise (Youdon, 2020). Cyclonic storms, rising sea levels, excessive flooding, and heat waves are just a few of the climate change-related dangers that pose a serious threat to India's vital maritime infrastructure, which includes its energy, transportation, and national security systems. India has over 200 non-major seaports run by private companies or state governments, and 12 major seaports run by the central government. These seaports are important hubs for the country's maritime trade, which makes up nearly 70% of the country's total trade in terms of value and 95% of its volume.

With an eye toward the future, India has actively engaged in multilateral negotiations under the auspices of the United Nations Framework Convention on Climate Change (UNFCCC), voicing its long-standing concerns about the effects of climate change on both the national and global scales and recognizing the myriad challenges it poses (Johanson, 2009). Similar national initiatives, such as robust climate change policies, continuous efforts to increase the capacity of renewable energy sources, protection of natural ecosystems, and encouragement of disaster risk reduction and climate change adaptation strategies, have reinforced India's position at the UNFCCC negotiations. It's a timely reminder of the relevance of India's actions and its growing global influence, which is due to factors other than its size. Rising greenhouse gas emissions (GHGs) are a possibility because of the country's size and continued industrialization, which creates a huge demand for energy. In addition to trying to concurrently comply with international and national climate change targets, India is working hard to swiftly develop equitable economic growth in order to meet the changing demands and ambitions of its 1.4 billion people.

This chapter presents a novel interdisciplinary approach by connecting India's green energy policies with climate-induced disaster governance. Unlike existing works that focus only on policy or technology, this chapter integrates legal, institutional, and community resilience lenses to assess India's holistic approach to climate risk governance in the green energy transition era. It aims to: (1) examine India's national policies and international commitments; (2) compare with legal frameworks of QUAD nations; (3) evaluate the integration of community resilience strategies; and (4) propose scalable, sustainable solutions. It directly addresses the socio-ecological crisis by exploring how India's green transition integrates climate governance, legal reform, and sustainable development. It shows how local-level programs—such as the Green India Mission and State Action Plans—enhance community resilience, provide livelihood security, and reduce disaster risk. The chapter offers sustainable models replicable in other global south contexts.

10.2 Methodology

This chapter adopts a doctrinal and comparative approach involving a review of national and international legal instruments, policy documents, scientific reports (IPCC, MoES), and think tank analyses. It further examines implementation-level initiatives under India's National Missions and international partnerships. The research methodology followed in this chapter is illustrated in Fig. 10.1, which is developed by the authors.

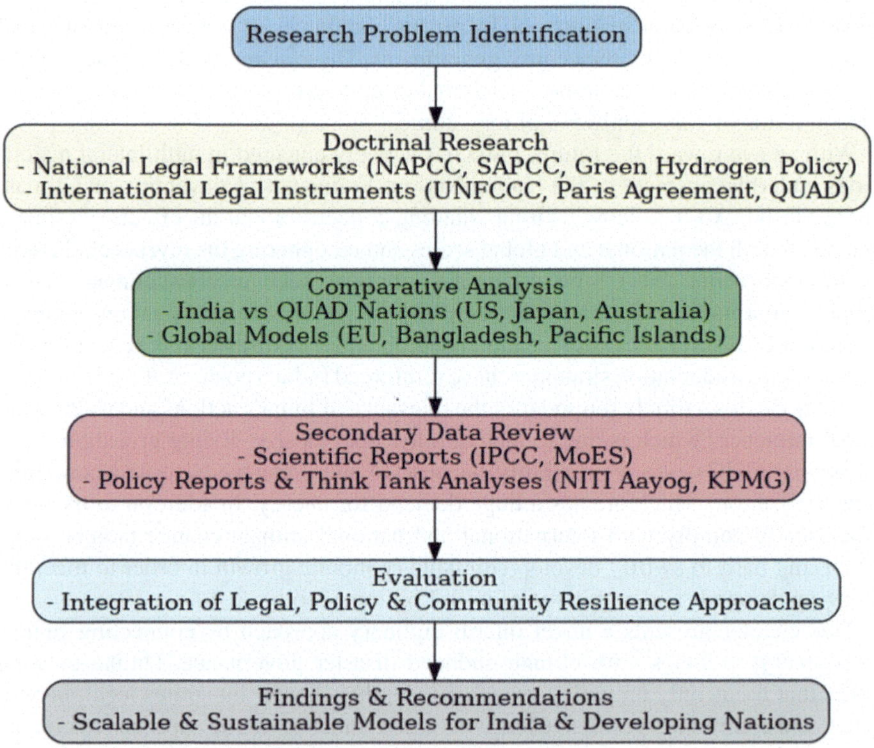

Fig. 10.1 Methodology flowchart (*Source* Authors' own creation)

10.3 Decarbonization: A Step to Counter Climate Change

Climate change stabilization aims to reduce net carbon dioxide (CO_2) emissions to zero. CO_2 concentrations in the atmosphere will continue to grow as long as people generate more than nature can absorb in its sinks (oceans, forests, and other plants), producing global warming. And the decisions we make now will have an impact on the planet's climate for generations to come (Fay et al., 2015). Governments have agreed on the United Framework Convention on Climate Change (UNFCCC) during the Rio Earth Summit for "stabilization of greenhouse gas concentrations in the atmosphere at a level that would prevent dangerous anthropogenic interference with the climate system." Developed countries agreed to return their emissions to 1990 levels (Blakemore & Yellen, 2021). According to the most recent scientific research, in order to stabilize climate change around the 2 °C threshold above pre-industrial temperatures—which has been agreed upon by governments as the maximum amount of warming that is acceptable—we must achieve zero net emissions by 2100.

Scientists' trust in the possibility of global decarbonization by 2100 is bolstered by the fact that they have studied potential solutions. These paths are based on a number of energy and economic models that look at what it would take to achieve decarbonization under various economic development and technological innovation scenarios. On the supply side, decarbonization would be accomplished through a rapid move to renewable energy (Fay et al., 2015).

10.3.1 What is Decarbonization?

A lot of words are associated with sustainability and climate change. Some of them are overused catchwords, while others are able to help the planet by meaningful description of latest innovative technologies and policies. One of the strongest words associated with addressing global warming is decarbonization (Cornell, n.d.). Decarbonization can be defined as either the process of eliminating carbon buildup and deposits from internal combustion engines or the transition away from the energy systems that emit CO_2 and other greenhouse gases. Although eliminating carbon is a part of both of these processes, they approach the task somewhat differently. Reducing the usage of fossil fuels and elevating the profile of low-carbon electricity generation are the two main components of decarbonization. This mostly consists of renewable energy sources, such as biomass, solar, and wind power. Along with "cleaner" technologies, the widespread use of clean ammonia and hydrogen (as determined by the recent QUAD summit in Japan) can help lower the usage of carbon power. Net-zero emission targets can be accomplished sooner and in accordance with international government requirements if the electricity and transportation sectors reduce their carbon intensity.

Decarbonization of the world's industrial processes, specifically energy production, is a hotspot nowadays due to declining fossil fuel resources, increasing oil prices, and high energy demands. Transportation is the second largest source of carbon emissions, behind the generation of heat and electricity. With time, there will be a greater need for transportation, necessitating greater measures to decarbonize the industry. Clean and green fuel is a dire need of the time in order to minimize the carbon emissions related to the transportation sector (Cornell, n.d.). Implementing decarbonization or bringing it for achieving net-zero emission have different methods like using clean hydrogen, clean ammonia, engine decarbonization, etc.

10.4 Green Energy Production for Reducing the Effect of Climate Change

Though there are some differences between green and renewable energy, green energy is frequently sourced from renewable energy sources like solar, hydro, etc. When it comes to low-carbon and independent economic pathways, green hydrogen—which is generated through renewable energy—has the potential to be quite important. Green hydrogen may feed numerous consumption streams, either as fuel or as an industrial feedstock, by enabling the application of domestically plentiful renewable energy resources across geographies, seasons, and industries. It can directly replace feedstocks obtained from fossil fuels in the fabrication of steel, fertilizer, and petroleum refinement, among other processes. Long-distance hydrogen-fueled cars and marine vessels can help the mobility industry become carbon-free. Green hydrogen is especially helpful as a flexible energy source for sustainably supplying isolated areas, such as islands, with the energy they need. High-energy hydrogen has the potential to be a solution to this issue because it can be produced with only electricity through the process of electrolysis and is readily available in the form of water (Akhtar & Liu, 2023). One of the major challenges in the implementation of the hydrogen economy is its transportation in bulk volumes due to its low energy density and massive investments (Blakemore & Yellen, 2021).

In order to meet climate targets, it is now acknowledged that a hybrid approach to energy policy is required; both the quick deployment of proven, clean technologies, and ongoing innovation in new technologies are required. Because of its adaptability, hydrogen has become a hot topic in energy policy discussions along with growing concerns about climate change and technological breakthroughs. Already, as an industrial feedstock, hydrogen is vital to our economy. In order to assist in balancing a net-zero energy system, it also has the potential to be a carbon-free fuel, feedstock, and energy storage technology. Hydrogen is versatile enough to be used in almost every area, including power, transportation, industry, and agriculture (KPMG, 2020). Energy is generated in the form of heat, when hydrogen is burnt, releasing water as a byproduct, making it a clean source of energy. However, it requires energy to make the hydrogen in the first place. Already, as an industrial feedstock, hydrogen is vital to our economy. In order to assist in balancing a net-zero energy system, it also has the potential to be a carbon-free fuel, feedstock, and energy storage technology. The Bush administration tackled energy in the early 2000s with an emphasis on innovation and R&D. Therefore, the administration's Energy Policy Act of 2005 included hydrogen breakthroughs as a major component to lessen the US's reliance on imported oil and its emissions of greenhouse gases. Green hydrogen is a new technology that separates hydrogen from water molecules by electrolysis. When powered by zero-emission electricity, green hydrogen produces no direct emissions, making it one of the cleanest methods of energy production. Moreover, the present cost is three to five times greater than the cost of producing blue or gray hydrogen. We require power in order to accomplish electrolysis. Green hydrogen refers to hydrogen produced via electrolysis powered by renewable energy source, including solar and wind. Due to the absence of CO_2 as a byproduct, green hydrogen, which comes from renewable energy sources, is the most environmentally friendly choice (as shown in Fig. 10.2).

In any decarbonized energy system, clean hydrogen serves three primary functions: (1) energy storage and balancing; (2) as a feedstock; and (3) as a fuel.

As part of larger initiatives pertaining to climate change and renewable energy, numerous major economies have announced their hydrogen agendas. The main goals of these national clean energy policies are to address the shared fundamental difficulties of increasing the production of green hydrogen, improving the usage of hydrogen in various industries, creating technologies, and creating supportive laws and regulations. The government's financing and support of R&D, initiatives to generate demand, and financial assistance for the development of infrastructure and manufacturing are clearly in the forefront. Green energy transition is an effective way to UN's Goal of Net-Zero Emission.

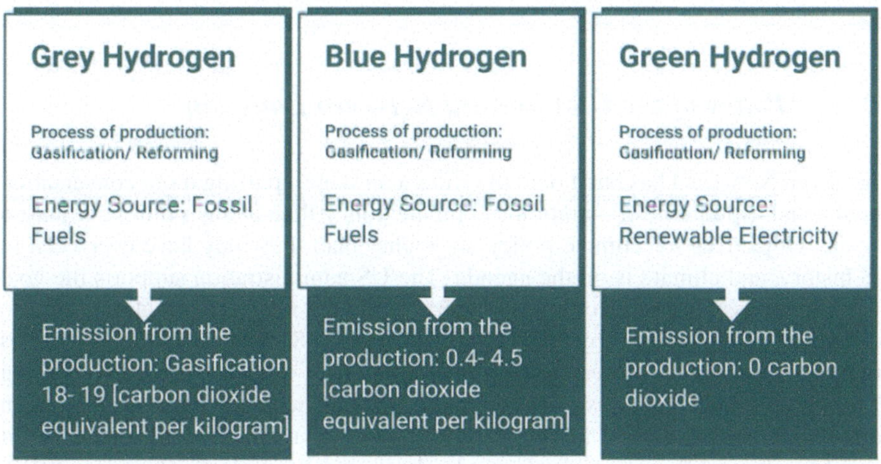

Fig. 10.2 Hydrogen production (*Source* Authors owns creation)

10.5 Policies of QUAD Nations for Sustainable Development and Climate Change

As the globe struggles to address climate change, a number of nations have taken the lead in their efforts to achieve net-zero emissions. These progressive countries have integrated sustainable development objectives into their plans in addition to setting aggressive targets to cut greenhouse gas emissions. The major nations driving the transition to a sustainable future will be closely examined in this article, along with how they are coordinating their efforts with the Sustainable Development Goals (SDGs) of the UN from ground-breaking renewable energy initiatives to revolutionary laws.

Climate change has always been one of the major concerns of the Quadrilateral Security Dialogue, the group though started with the aim to strengthen the maritime security in the region of Indo-Pacific, but later on, the purpose of the group shifted toward the overall development of the dedicated region giving special reference to the issue of climate change and its effect upon the region nations, and to achieve its aim, the member nations are coming in front with all the resources to bring such policies to meet the net-zero emission target by 2030 as decided by the world leaders in Paris Agreement COP 26 (The White House, 2021). The Quad Countries are committed to supporting our Indo-Pacific partners in light of their increasing demands for a practical transition to a net-zero economy and society as well as increased resilience to the effects of climate change. The three pillars of the Climate Working Group are "climate ambition, clean energy, and adaptation/resilience" which serve as the foundation for the Quad Countries' launch of the "Quad Climate Change Adaptation and Mitigation Package" (QCHAMP), which focuses on emission and adaptation/resilience.

10.5.1 Policy of the USA Toward Net-Zero Emission

The Green New Deal has burst onto the American stage, spurring more conversation about—and aspiration for—ambitious climate policy than at any point in at least a decade. Objectives for climate policy are higher than they may have ever been in US history, and climate is on the agenda. The US administration supports the goal of having emission-free electricity by 2035, while the majority of nations want to achieve net-zero emissions by 2050. The majority of U.S. decarbonization strategies are based on using CCUS technologies to capture CO_2 released during the burning of fossil fuels and prevent its emission into the atmosphere. Once the CO_2 has been captured, it can be utilized in an industrial process that either replaces the use of other CO_2 derived from fossil fuels or adsorbs it, such as in plastic (Lawson, 2018).

10.5.2 Policy of Japan Toward Net-Zero Emissions

In all, over 130 nations have committed to lowering their greenhouse gas (GHG) emissions to net-zero by the year 2050 or are in the process of doing so. When Prime Minister of Japan, Yoshihide Suga declared the government would embrace this goal in his first policy speech to the National Diet in October 2020, Japan joined those nations. In April 2021, Mr. Suga, the Japanese prime minister at the time, declared that Japan would cut emissions by 46% from 2013 levels by 2030. This was stated as an interim goal. This was a dramatic departure from the previous targets of reducing emissions by 26% by 2030 and 80% by 2050 under the Paris Agreement (McKinsey & Company, n.d.).

In 2020, the Japanese government committed to carbon neutrality 2050 and established hydrogen and ammonia as pillars of a new strategy to reach this environmental goal. Japan leads the world in patent applications for renewable energy; cutting-edge technology is essential to reaching the global objective of net-zero carbon emissions. Japan is actively encouraging technological innovation in renewable energy that will contribute to the global achievement of that objective (Bloomberg, n.d.). Japan has filed the highest number of renewable energy international patent applications than any other country in the world under the Patent Cooperation Treaty (PCT). Between 2010 and 2019, Japan filed 9,394 patent applications relating to solar, fuel cell, wind, and geothermal technologies (Mara et al., 2022).

10.5.3 Policy of Australia Toward Net-Zero Emission

The Australian energy market is advancing toward the Progressive Change Scenario (also known as Net-Zero 2050) where the country pursues an economy-wide net-zero carbon emission by 2050. Among the various countries, Australia indeed has an enormous potential to thrive in a low-carbon global economy. Leading Australian climate economist Ross Garnaut epitomized this best in his 2019 book Superpower: Australia's Low-Carbon Opportunity, in which he makes the case that the country has the potential to become a low-carbon "superpower" due to its abundance of wind, sunshine, and land (KPMG, 2020).

Australia's national hydrogen strategy seeks to rank the country among the top three exporters of hydrogen to Asia by 2030. Green ammonia offers the quickest path to clean energy transition in Australia in the near term, Achieving the net-zero objective and moving toward a low-carbon future depend on carbon capture and storage (CCS) and hydrogen. Over the next ten years, Australia has committed to investing over $300 million on CCUS projects. There are two CCS projects underway at the moment: the Moomba CCS project, whose final investment decision was made last year, and the largest in the world, the Gorgon Carbon Dioxide Injection project (GlobalData, 2022).

All four Quad member nations are putting their best foot forward to meet the net-zero emission by or before 2050. The actions and schemes launched by these nations are directly affecting the position of the net-zero emission targets by 2050 as decided by the world leaders at the Paris Agreement, not only in the Indo-Pacific region but also in the world.

10.6 Comparative Insights from Global Climate Governance Models

In comparison to India's climate governance efforts, several other regions and nations offer valuable insights into context-specific adaptation and resilience models. For instance, Bangladesh—despite being a low-income and disaster-prone country—has demonstrated remarkable success in implementing coastal adaptation strategies. These include the construction of cyclone shelters, community-based early warning systems, and afforestation programs involving local stakeholders. These measures have significantly reduced disaster mortality rates and enhanced community resilience, especially in low-lying coastal regions (Raihan et al., 2019). Unlike India's centralized missions, Bangladesh's adaptation approach is more localized and community-driven, providing a model of grassroots resilience that complements India's top-down policy frameworks.

Similarly, the European Union (EU) has developed an advanced legal framework for renewable energy integration through its Renewable Energy Directive (EU RED II), which mandates binding national targets for renewable energy consumption and emphasizes cross-border electricity trade and technology-sharing among member states (European Commission, 2021). In contrast, India's legal approach is more sectoral and mission-oriented, lacking a unified renewable energy statute. However, India's National Hydrogen Mission and Electricity Act reforms indicate a shift toward a more cohesive and incentivized legal environment akin to the EU's model. The EU also prioritizes citizen engagement and public investment in energy transitions, lessons that could be incorporated into India's national and state-level strategies.

The Pacific Island nations offer yet another compelling comparative perspective. Countries like Fiji, Vanuatu, and the Solomon Islands, which are acutely vulnerable to climate change, have adopted community-based resilience strategies integrated with traditional knowledge systems. These include participatory vulnerability assessments, integrated coastal management plans, and decentralized governance mechanisms for climate finance (Nunn et al., 2021). India's efforts under the State Action Plans on Climate Change (SAPCCs) and programs like the National Mission on Sustainable Agriculture similarly aim at local engagement, but the inclusion of indigenous knowledge and customary governance, as seen in Pacific Island models, remains limited. Adopting such practices could enhance the cultural and ecological legitimacy of India's climate interventions.

10.7 Various Government Initiatives in India for Eco-Friendly Development

India's profound dedication to the ambitious Climate Goals has been well recognized among the international community. Our goals have been met by our accomplishments. India is the country with the world's fastest-growing capacity for renewable energy. One of the most alluring locations for renewable energy investments is now India. India is aware of the crucial role that green hydrogen would play in helping it achieve its goals of attaining net-zero by 2070 and energy independence by 2047. India has the potential to create green hydrogen for the global market because of its abundant renewable energy resources. India has targeted to attain net-zero emissions by the year 2070. India's need for resources and energy is expected to increase as its growth story is revealed. The last 20 years have seen a doubling in energy demand, and by 2030, at least another 25% growth is predicted. Presently, India imports more than 40% of its basic energy needs, totaling more than USD 90 billion annually. Imported fossil fuels have a major role in industries like industrial production and transportation. This calls for a move away from fossil fuels and toward technology that allows for a larger proportion of renewable energy sources in the energy mix (Government of Odisha, 2023).

Global warming also presents India with economic opportunities since it necessitates the development, mass production, and deployment of new industries and technologies necessary for the decarbonization process. In his keynote address, NITI Aayog CEO Amitabh Kant said, "We need more such financial instruments to provide a further impetus to clean mobility in India. This is why we need to bring states, domestic and international financial institutions, manufacturers, and operators on the same page. We must come up with financing mechanisms that are widely applicable, acceptable, and, most importantly, sustainable. We must foster shared mobility by leveraging private sector investments and unlocking financing for e-buses, which form the core of public transportation in our cities."

The fundamental objective should be to strike a balance between the needs and aspirations of our people, increase liveability and productivity through improved connectivity, reduce the cost of logistics, and speed up clean mobility—all while adopting a strategy that is climate-centric and sustainable, both environmentally and financially. Low-interest financing for electric automobiles will be made possible by green financing. Dr. Stephen Koch said, "India requires a robust roadmap for the electrification of transport. Financing plays a crucial role here. Mobilization of capital is possible through multi-stakeholder collaborations. The NDC-TIA initiative brings together a wide spectrum of participants to facilitate peer-to-peer learning and information exchange on various themes, including 'Financing the Decarbonization of Transport'." For emerging nations like India to transition to a low-carbon economy, they must separate economic growth from greenhouse gas emissions. However, as these economies try to expand in a world with limited climate resources, they confront tremendous developmental obstacles. India faces a significant climate policy challenge as it strives to boost incomes, create millions of new jobs, and enhance public health over the next several decades. Relying on low-carbon power sources results in a decarbonized economy with reduced greenhouse gas (GHG) emissions released into the atmosphere.

India's climate governance directly addresses socio-ecological crises by empowering communities through decentralized renewable energy programs, disaster risk reduction policies, and sustainable agriculture missions under NAPCC.

10.7.1 India's Steps Toward Decarbonization

The Indian government unveiled the National Action Plan for Climate Change (NAPCC) on June 30, 2008. The Climate Change Action program has extended the initiative until 2025–2026. The "National Action Plan on Climate Change" (NAPCC) coordination, the "State Action Plan on Climate Change" (SAPCC), the "National Institute on Climate Change Studies & Actions," the "National Carbonaceous Aerosols Programme" (NCAP), the "Long Term Ecological Observations" (LTEO), international negotiations, and capacity building are its eight primary sub-segments.

The current climate governance approach in India as addressed in the National Action Plan on Climate Change (NAPCC) kind of climate change is a multidimensional approach that both looks at the technology to come up with innovations, and it also looks on the grass roots level to see how resilient they are. The National Solar Mission aims at encouraging decentralized solar power system and advocates rural electrification and rural employment in green jobs (Government of India, 2008). Enhanced Energy Efficiency Mission develops market-based interventions to minimize energy consumption and cut the price of energy consumption especially on small industries and urban poor (NAPCC, 2008). Through the National Mission on Sustainable Habitat, India has addressed its goals regarding urban resilience by identifying energy efficiency in buildings, municipal waste management, and green transport, thereby promoting the adaptive capacity of the vulnerable population of the cities.

The National water mission is being put in place to enhance the efficiency of water use by 20% including the practices such rain water harvesting and efficient irrigation which are crucial in the drought-prone agricultural communities (Ministry of Water Resources (MoWR), 2011). Among the Himalayan Ecosystem Mission aspects, local administration organizations like Panchayats will be enable to maintain vulnerable mountain ecosystem besides sustaining native livelihoods (Ministry of Environment, Forest and Climate Change (MoEFCC), 2010). The green India mission is a drive that encourages afforestation by use of Joint Forest Management Committees in which communities are directly engaged in the process of carbon sinks development and increasing biodiversity (DST, n.d.). In the same fashion, the Mission on Sustainable Agriculture is aiding climate-resilient agriculture by the use of drought tolerant crops and agri-insurance, which builds food security in a rural setting (Ministry of Agriculture (MoA), 2014). The Strategic Knowledge Mission promotes both research and popular awareness, and it is involved in motivating the communities to contribute to climate governance (DST, n.d.).

The National Green Hydrogen Mission that is being introduced in 2023 is complementary to these missions and seeks to decarbonize heavy industry and the transport sector and generate green energy employment opportunities and reduce regional energy disparities. The mission also aims to promote local manufacturing and foreign trade liaisons, therefore, connecting the net-zero transition in India with economic growth and the empowerment of local communities and, of course, 5 MMT annual production until 2030 (Government of Odisha, 2023).

India is one of the world's largest carbon dioxide emitters, behind China and the USA. The per capita emissions in the country are rising at an alarming rate calling an urgent need to bring more proactive steps by the authorities as well as the private players to adopt decarbonization using clean energy. Nevertheless, India's shift to a net-zero carbon energy economy will be difficult and may even be the biggest challenge to global decarbonization due to the country's rapidly increasing energy demand and the political and economic inertia slowing a shift away from coal reliance. Even though renewable sources are getting more affordable every day, India will need to make large amount of investments for creating infrastructure for electricity transmission and increasing clean power generation capacity in order to both meet future demand and phase out its current coal-fired power plants. India is already the third-largest emitter in the world, and from now until 2050, it will be responsible for the lion's share of the increase in global energy demand. This is because of the country's massive and growing population, its widening energy access, and the rising per-capita energy use that comes with rising temperatures (Blakemore & Yellen, 2021). The government of India currently owns and operates one of the biggest coal mining company in the world, and the sector generates a sizable amount of revenue for both the executive and the legislature. The coal industry is also vital to India's political and economic system.

Prior to COP26 in Glasgow, it was crucial to fortify relations between the two countries. The recently officially confirmed "US-India Climate and Clean Energy Agenda Partnership and Special Envoy for Climate John Kerry's trip to India to launch the Partnership's Climate Action and Finance Mobilization Dialogue" both signaled the US growing importance of climate commitment and assistance for developing economies' energy transitions. However, given the ongoing effects of COVID-19 and the strain that the dual priorities of decarbonization and development are placing on the region's resources, a number of issues at the heart of India's energy transition—including market design, grid development, and the empowerment of a nascent clean energy technology sector—will prove to be arduous challenges that call for a more rigorous approach (Blakemore & Yellen, 2021).

Studies carried out by top think tanks in the nation are providing more and more proof that there are quicker routes to a prosperous, low-carbon India. India actually has a great chance to select a developmental path that depends on lower-emission technologies at this crucial juncture in its history. India has focused heavily on electrification—a large percentage of renewable energy—and energy efficiency in its decarbonization efforts thus far. However, this wouldn't be sufficient to remove carbon emissions from industries that produce steel and iron. Green hydrogen-based interventions will be necessary for the deep decarbonization of these and other industries, such as methanol, ammonia, and refineries. In order to sustain momentum toward geostrategic and climate goals, the Quad needs to support India's shift to clean energy more heavily. Implementing clean energy in India will require funding, which is an essential part of this. With Quad-backed capital and access to financing opportunities, an empowered India would significantly aid in South Asia's decarbonization. It would also strengthen the Quad's standing as a crucial strategic partner in clean energy technology and market expansion, particularly as China continues to dominate the sector (KPMG, 2020). Even though India's transition will require a significant time and resource investment, the potential consequences would be much greater for the planet and for the global political interests of the USA and the Quad.

India has a extensive experience in framing policy, contract processes, and the implementation of renewable energy resources. Presently, the lowest long-term levelized costs for producing wind and solar energy have been achieved by India. We believe that the trend will continue in future. The National Green Hydrogen Mission, on the other hand, suggests expanding a number of regulatory frameworks for connectivity, finance, transmission, open access, and energy storage for Green Hydrogen production projects in order to assure low cost in the projects related to renewable energy. In furtherance of the Paris Agreement and COP 26 which was held in 2021, the government of India launched the first phase of the Green Hydrogen Policy in February 2022, as part of the National Hydrogen Mission which was launched by PM Narendra Modi on the occasion of Independence Day celebration in 2021. The basic aim of the Green Hydrogen Policy (under the National Hydrogen Mission) is to promote the production and usage of green hydrogen while aiming at India's climate targets (The White House, 2022).

10.7.2 Functioning of Green Hydrogen Policy

The announcement of the Policy on Green Hydrogen is noteworthy because India is currently experiencing a power crisis and a shortage of coal. By encouraging the production of green hydrogen and green ammonia, the "future fuels," the policy seeks to lessen India's reliance on fossil fuels while simultaneously achieving its climate goals. The policy tackles various obstacles that India is encountering in its pursuit of cleaner fuel. It has a number of characteristics.

1. It provides a number of incentives to draw in investors, such as granting distribution licensees the ability to purchase and supply renewable energy at a discounted rate. For 25 years, interstate transmission fees for hydrogen and ammonia plants that were put into service before June 2025 will not be charged, granting space in renewable energy parks for the production of green ammonia and green hydrogen.
2. It makes business easier (including, but not limited to) by suggesting a single portal for all clearances associated with all of the company's operations. By

providing the renewable energy plant and the producers of green hydrogen/ammonia with priority grid connectivity.
3. It promotes export by storing green hydrogen close to ports to facilitate transportation. Producers should erect bunkers close to the port for green hydrogen storage before exporting it.

10.8 Conclusion

India has been steadfast in its transition to clean energy, achieving the highest pace of renewable capacity addition among all major nations and outlining aggressive transition goals at COP26 through Prime Minister Modi's Panchamrit declaration. India is in the forefront of tackling the global challenge of climate change. It has agreed to ambitious Nationally Determined Contributions (NDCs) aimed at reducing emission intensity by 33–35% by 2030 compared to 2005 levels. Over the last nine years, India has witnessed the Indian struggle against climate change, as we have already surpassed our pledge of 40% renewable energy output, well above the 2030 Paris Agreement target. It is one of the few countries in the world to have created a Cooling Action Plan (CAP) with a 20-year vision (2017–2018 to 2037–2038) that addresses cooling demands across industries. CAP investigates potential measures to reduce cooling demand from residential and commercial buildings, cold chains, and other sources, with a focus on building design and technological developments that do not sacrifice energy efficiency.

With a population of 1.4 billion, India has a large energy need to support its rapidly expanding economy. India, which had a power shortage ten years prior, has been pursuing energy independence since 2014. India's power generation mix is rapidly shifting toward a higher proportion of renewable energy in light of the objectives of sustainable development. With 40% (157.32 GW) of installed power capacity, India is currently the third-largest producer of renewable energy globally. Being the world's fifth-largest economy, it has the potential to attain energy independence by 2047 with sustainable technology. The Atma Nirbhar Bharat project of Prime Minister Narendra Modi will reduce imports by billions of dollars through the use of electric vehicles and a major increase in renewable energy sources. Renewable energy, electric vehicle (EV) batteries, and hydrogen infrastructure are capital assets with rapidly dropping costs, therefore clean energy will reduce and shield India's energy expenditure from inflation.

India's trajectory toward net-zero by 2070 is ambitious yet necessary. The legal and policy innovations, coupled with community-focused interventions, illustrate a multifaceted approach to climate change. India's international partnerships under the Quad and its investments in green hydrogen, renewable energy, and clean mobility show promising potential for climate leadership. This study demonstrates the integration of environmental governance, energy policy, and resilience-building. Its implications go beyond national interest, providing a blueprint for countries striving to align development with sustainability. Future research could explore localized implementation and citizen participation in climate adaptation. This chapter highlights how India's transition policies are not only environmental solutions but socio-economic interventions aimed at empowering vulnerable populations. The research provides a replicable model for other developing nations facing similar ecological and governance challenges.

10.9 AI Disclosure

No AI tools were used in the writing or content generation of this chapter.

References

Akhtar, M. S., Khan, H., Liu, J. J., & Na, J. (2023). Green hydrogen and sustainable development – A social LCA perspective highlighting social hotspots and geopolitical implications of the future hydrogen economy. *Journal of Cleaner Production, 395*, 136438. https://doi.org/10.1016/j.jclepro.2023.136438

Blakemore, R., & Yellen, D. W. (2021, September 20). *India is key to US climate ambitions, and the Quad can help.* Atlantic Council. Retrieved April 3, 2022, from https://www.atlanticcouncil.org/blogs/energysource/india-is-key-to-us-climate-ambitions-and-the-quad-can-help/

Bloomberg. (n.d.). *5 ways Japan is accelerating global decarbonization efforts.* https://sponsored.bloomberg.com/article/jco/5-ways-japan-is-accelerating-global-decarbonization-efforts

Cornell, P. E. (n.d.). *Energy and three levels of national security: Differentiating energy concern within a national security context.* Retrieved April 24, 2022, from http://connections-qj.org/system/files/08.4.04_Cornell.pdf

Department of Science & Technology (DST), Government of India. (n.d.). *Climate change programme.* https://dst.gov.in/climate-change-programme

European Commission. (2021). *Renewable energy directive (EU) 2018/2001 (RED II).* https://energy.ec.europa.eu/topics/renewable-energy/renewable-energy-directive-targets-and-rules_en

Fay, M., Hallegatte, S., Vogt-Schilb, A., Rozenberg, J., Narloch, U., & Kerr, T. (2015). Decarbonizing development: Three steps to a zero-carbon future. In *Climate change and development.* World Bank. https://doi.org/10.1596/978-1-4648-0479-3

GlobalData. (2022, May 31). *Australia steps up investments and decarbonizing policies to meet 2050 net zero emissions, observes GlobalData.* https://www.globaldata.com/media/oil-gas/australia-steps-investments-decarbonizing-policies-meet-2050-net-zero-emissions-observes-globaldata/

Government of India. (2008). *National action plan on climate change (NAPCC).* Prime Minister's Council on Climate Change. https://moef.gov.in/uploads/2018/04/NAP_E.pdf

Government of Odisha. (2023). *Odisha economic survey 2022–2023.* https://cdnbbsr.s3waas.gov.in/s3716e1b8c6cd17b771da77391355749f3/uploads/2023/01/2023012338.pdf

Intergovernmental Panel on Climate Change. (2022). *Climate change 2021: The physical science basis* (The Sixth Assessment Report). https://www.ipcc.ch/assessment-report/ar6/

Johanson, H. (2009). Selected case studies illustrating the three streams of action. In M. van Staden & F. Musco (Eds.), *Local governments and climate change* (Advances in global change research, vol. 39). Springer. https://doi.org/10.1007/978-1-4020-9531-3_6

KPMG. (2020). *Decarbonizing growth: Managing the transition.* Retrieved June 7, 2022, from https://assets.kpmg/content/dam/kpmg/in/pdf/2020/12/decarbonising-growth.pdf

Krishnan, R., Sanjay, J., Gnanaseelan, C., et al. (Eds.). (2020). *Assessment of climate change over the Indian region: A report of the Ministry of Earth Sciences (MoES), Government of India.* Springer Open. https://doi.org/10.1007/978-981-15-4327-2_1

Lawson, A. (2018, June). *Decarbonizing U.S. power.* Center for Climate and Energy Solutions. https://www.c2es.org/wp-content/uploads/2018/06/innovation-power-background-brief-07-18.pdf

Mara, D., Nate, S., Stavytskyy, A., & Kharlamova, G. (2022). The place of energy security in the national security framework: An assessment approach. *Energies, 15*(2), 658. https://doi.org/10.3390/en15020658

McKinsey & Company. (n.d.). *How Japan could reach carbon neutrality by 2050.* https://www.mckinsey.com/capabilities/sustainability/our-insights/how-japan-could-reach-carbon-neutrality-by-2050

Ministry of Environment, Forest and Climate Change. (2008). *National Action Plan on Climate Change.* Government of India. Ministry of Environment, Forest and Climate Change. https://moef.gov.in/uploads/2018/04/NAP_E.pdf

Ministry of Agriculture (MoA). (2014). *National mission for sustainable agriculture: Strategies for implementation.*

Ministry of Environment, Forest and Climate Change (MoEFCC). (2010). *National mission for sustaining the himalayan ecosystem.*

Ministry of Environment, Forest and Climate Change, Government of India. (2015). *India: First biennial update report to the United Nations Framework Convention on Climate Change, India's third national communication to the UNFCCC.* https://unfccc.int/resource/docs/natc/indbur1.pdf

Ministry of Water Resources (MoWR). (2011). *Strategic plan for water resources development and management.*

Nunn, P. D., Kumar, R., Eliot, I., & McLean, R. (2021). Climate-induced migration and traditional knowledge systems in the Pacific Islands. *Environmental Science & Policy, 120*, 57–66. https://doi.org/10.1016/j.envsci.2021.02.009

Raihan, S., Hossain, S. M., & Islam, M. R. (2019). Coastal adaptation strategies and community resilience in Bangladesh. *Climate and Development, 11*(3), 189–199. https://doi.org/10.1080/17565529.2018.1442803

Shaw, R., Luo, Y., Cheong, T. S., et al. (2022). Asia. In *Climate change 2022: Impacts, adaptation and vulnerability* (IPCC WGII Sixth Assessment Report, Chapter 10). Intergovernmental Panel on Climate Change.

The White House. (2021, March). *Fact sheet: Quad summit.* Statements and Release.

The White House. (2022, May). *Fact sheet: Quad leader's Tokyo summit.* Statements and Release.

Youdon, C. (2020, November 29). *Impact of climate change on coastal cities: An integrated adaptation approach.* National Maritime Foundation. https://maritimeindia.org/impact-of-climate-change-on-coastal-cities-an-integrated-adaptation-approach-part-i/

Part III
Socio-ecological Sustainability and Community Resilience

Chapter 11
Integrating Tourism and Agriculture for Sustainable Livelihoods of Villages Surrounding the Ciletuh-Palabuhanratu UNESCO Geopark

Saraswati Soegiharto, Sri Najiyati, Rizal Akbar Aldyan, Danarti, Rukmini Nugroho Dewi, Siti Khoiriyah, and Dian Purwasantana

Abstract Rural communities in disaster-prone areas face intersecting socioecological challenges, including climate-induced hazards, economic instability and environmental degradation. This chapter examines the integration of tourism and agriculture as a sustainable livelihood strategy to enhance community resilience in the Ciletuh-Palabuhanratu UNESCO Global Geopark, Indonesia. Applying the Sustainable Livelihood Framework and supported by qualitative fieldwork in nine villages, the study identifies key livelihood assets, vulnerability contexts, institutional dynamics and community strategies. Findings reveal that livelihood diversification through agrotourism and the integration of small-scale enterprises reduces dependency on single-sector economies, enhances adaptive capacity and supports local economic regeneration. The study underscores the pivotal role of regional institutions, social capital and village-level governance in bridging policy frameworks

S. Soegiharto · S. Najiyati · R. Akbar Aldyan (✉) · Danarti · R. Nugroho Dewi · D. Purwasantana
National Research and Innovation Agency (BRIN), Jakarta, Indonesia
e-mail: riza014@brin.go.id

S. Soegiharto
e-mail: sara010@brin.go.id

S. Najiyati
e-mail: srin009@brin.go.id

Danarti
e-mail: dana021@brin.go.id

R. Nugroho Dewi
e-mail: rukm008@brin.go.id

D. Purwasantana
e-mail: dian066@brin.go.id

S. Khoiriyah
Doctoral Program of Environmental Science, Universitas Sebelas Maret, Surakarta, Indonesia
e-mail: siti.khoiriyah12@staff.uns.ac.id

G. K. Panda et al. (eds.), *Socio-Ecological Crisis of Development*, Sustainability Solutions, https://doi.org/10.1007/978-3-032-22171-1_11

with grassroots initiatives. The chapter highlights the necessity for inclusive development planning, disaster risk management and education-based adaptation to ensure long-term sustainability. By integrating the Geopark's core pillars of conservation, education and local economic development into community livelihood strategies, the research presents a replicable model for other vulnerable rural areas. This work contributes to addressing the gap in the literature on integrated rural development under socioecological stress. It demonstrates how geopark landscapes can function not only as conservation zones but also as platforms for fostering resilient, self-reliant communities. The chapter positions integrated livelihood planning as a crucial pathway for achieving the Sustainable Development Goals in ecologically sensitive regions.

Keywords Sustainable livelihoods · Community resilience · Geopark · Agrotourism · Climate vulnerability

11.1 Introduction

Environmental issues, development and social crises are intertwined and have a major impact on the well-being of people around the world, especially in developing countries. Environmental degradation exacerbates poverty, non-inclusive development creates injustice and social inequality worsens the response to environmental disasters. The impacts of the three global issues are particularly evident in areas that are vulnerable to natural disasters, such as geopark areas and the villages within them.

Geoparks, a UNESCO initiative, highlight the intricate and important relationship between natural and cultural heritage and sustainable development. A geopark is a single or combined geographic area that has a geological heritage site and a valuable landscape related to aspects of geological heritage, geological diversity, biodiversity and cultural diversity (UNESCO, 2023). Geopark development aims to improve the quality of geopark management that has an impact on the welfare of the local community, and is managed holistically through protection, education and sustainable development. UNESCO has designated 195 UNESCO Global Geoparks (UGG) in 48 countries, and as many as 10 geoparks are in Indonesia.

Geological heritage sites and landscapes in geoparks generally have tourist attractions, so they have the potential to be developed as a source of livelihood. Various studies have shown that tourism has the potential to increase the village economy (Luo, 2023; Wang, 2023). However, in Indonesia, most of the villages in geoparks are in disaster-prone areas. Natural disasters, such as floods and landslides due to climate change, impact village tourism activities (Handayani, 2024; Susanto et al., 2020). This propensity to experience natural disasters shows that the dependence of villages on one economic sector has the potential to create vulnerability to community livelihoods. Conversely, diverse livelihoods can strengthen the community's adaptability to the risks of climate change. Integrating various sources of livelihood can be a strategy for empowering communities to respond effectively to the challenges of climate change.

This study aims to analyse the potential and strategy for integrated livelihood development in the agriculture and tourism sectors to realize sustainable livelihoods for rural communities living around the Ciletuh-Palabuhanratu UNESCO Global Geopark (CP-UGGp), Indonesia. This study offers a novel contribution by analysing the integration of tourism and agriculture through the lens of the Sustainable Livelihood Framework (SLF) within a geopark context vulnerable to climate-related risks. In contrast to previous studies that primarily address the integration of tourism and agriculture from a land-use or marketing perspective, this research adopts a socioecological and vulnerability-based framework, specifically within the context of the CP-UGGp, Indonesia. The study's innovation lies in its application of both global discourses on climate resilience, sustainable rural development and UNESCO geopark missions, alongside the regional contextualization of livelihood dynamics in climate-vulnerable rural areas. By incorporating insights from both foundational sources (e.g., Chambers & Conway, 1992; DFID, 2000) and more recent literature (e.g., Luo, 2023; Togaymurodov et al., 2023; Wang, 2023), this research bridges global and local perspectives to inform international audiences about strategies that can enhance resilience and diversify rural livelihoods. The aim is to propose an integrated livelihood strategy that aligns with the three pillars of geopark sustainability: conservation, education and local economic development, through the intersection of tourism and agriculture in disaster-prone regions.

11.2 Rationale for the Study

11.2.1 Sustainable Livelihood Framework

The Sustainable Livelihood Framework (SLF), developed by DFID, is among the most extensively utilized tools in development for analysing livelihood systems (GLOPP, 2008). It conceptualizes how individuals navigate vulnerability contexts, such as seasonal shifts, economic shocks and long-term trends, by employing various combinations of livelihood assets to achieve desired outcomes (De Stagé et al., 2002). The SLF comprises five types of capital: human, social, natural, physical and financial (DFID, 2000), which individuals utilize to formulate livelihood strategies. The framework also considers vulnerability as shaped by exposure to risks, uncertainty and limited capacities to cope. Structures and processes, including cultural factors, determine access to assets and influence strategy choices. Transforming these systems towards equity and inclusiveness is essential for fostering sustainable livelihoods. The SLF promotes local participation, internal capacity building and contextualized responses, rendering it a robust and comprehensive tool for strategy design.

SLF's adaptability is further augmented by its compatibility with diverse development approaches and its applicability across various local contexts (Kollmair &

Gamper, 2002). It facilitates community resilience, which is defined as the capacity to recover and adapt to shocks, whether environmental, economic or political (Nguyen, 2020), through the integration of preparedness (Kim & Kim, 2022), adaptation (McNamara, 2013), social capital (Nyahunda & Tirivangasi, 2021) and governance (Adger, 2006). Despite its strengths, the framework is not without challenges. SLF must address issues such as poverty, inequality, infrastructure deficits, gender disparities and environmental stress (Fairer-Wessels, 2019; Tesema, 2021). Additionally, critiques have pointed out its capitalist bias and the necessity to incorporate relational and environmental dimensions, including power, gender and socioecological dynamics (Natarajan et al., 2022). Therefore, a critical and contextual application of SLF is essential to avoid reductionism and to ensure sensitivity to local realities.

11.2.2 Tourism and Agriculture Integration

Numerous rural villages depend on agriculture and possess significant tourism potential, rendering the integration of these sectors through agrotourism a strategic approach to bolster local economies and achieve sustainable livelihoods. Livelihood strategies encapsulate the diverse choices individuals make to fulfil their objectives (GLOPP, 2008), and agrotourism has emerged as a pertinent tool for diversifying rural income sources (Pratama et al., 2019). Beyond economic advantages, agrotourism addresses urban demand for rural experiences and organic lifestyles (Togaymurodov et al., 2023), while simultaneously promoting sustainable farming practices through investments in rural education and health (Wang, 2023). It contributes to cultural preservation and agricultural education (Flanigan et al., 2015), with its long-term success contingent upon farmers' managerial competence and strategic training (Arizo & Apritado, 2022). Collaboration among stakeholders ensures adaptability to tourism market dynamics while safeguarding natural resources (Paniccia & Baiocco, 2020; Wang, 2023).

The integration of tourism and agriculture has been demonstrated to enhance eco-efficiency and increase farmers' income, particularly in economically developed regions with robust tourism capabilities (Luo, 2023; Wang et al., 2022). Additionally, it provides smallholder farmers with a viable means to remain competitive in the global market (Khanal & Mishra, 2014). Nevertheless, challenges such as technological readiness and labour shifts must be addressed to ensure the sustainability of agrotourism (Luo, 2023). Furthermore, the application of technology affects tourist preferences and the appeal of agrotourism destinations (Ha, 2024). Consequently, the tourism-agriculture nexus functions not only as a mechanism for income generation but also as a comprehensive strategy to empower rural communities and secure sustainable livelihoods.

11.2.3 Relevance of Tourism-Agriculture Integration to UNESCO Geoparks

Integrating tourism and agriculture within the geopark framework is pivotal in achieving the missions of the UNESCO Global Geopark (UGG), namely conservation, education and local economic development. Geoparks, recognized for their significant geological heritage, advocate for sustainable practices, community engagement and educational outreach (Chen et al., 2021; Fauzi et al., 2018). This integration enhances biodiversity conservation by aligning sustainable agricultural practices with geological preservation (Yuliawati et al., 2021; Zuvara et al., 2022). Local farming methods, adapted to minimize environmental impacts, can contribute to the maintenance of natural landscapes and geological features (Farsani et al., 2011). Furthermore, promoting local agricultural products that reflect unique geological and cultural contexts supports both economic sustainability and conservation efforts (Farsani et al., 2011; Yuliawati et al., 2021). Thus, the integration of tourism and agriculture serves as a strategic tool not only for economic benefit but also for the protection of landscapes, conservation of biodiversity and preservation of cultural values.

In the educational sphere, geoparks present opportunities to integrate geological knowledge with sustainable agriculture through community and visitor education (Fernandes et al., 2021). Educational initiatives enhance awareness of soil health, agricultural techniques and ecological stewardship, while equipping local populations with knowledge that bolsters resilience and improves farming practices (Gulas et al., 2019; Herrera et al., 2018). Economically, agrotourism within geoparks augments local income by attracting tourists to experience both geological beauty and agricultural heritage (Ginting et al., 2021; González, 2022). This dual appeal increases visitor expenditure, generates employment and promotes local products to wider markets (Herrera et al., 2018; Triana & Ruhimat, 2018; Yuliawati et al., 2021). The synergy between tourism and agriculture not only supports livelihoods but also aligns with the core missions of geoparks, exemplifying environmentally conscious economic development (Bonachea, 2023). Consequently, the integration of tourism and agriculture within geoparks offers a comprehensive approach to sustainability, balancing heritage conservation with community welfare.

11.3 Material and Methods

This study utilized a qualitative case study methodology, employing the SLF as the primary analytical framework, and was further enhanced by the CNT model to elucidate relational and institutional dynamics (Collaborative for Neighborhood Transformation, 2023). The connections approach is used to strengthen the analysis with SLF, which requires relational elements for a more accurate livelihood strategy (Natarajan et al., 2022). The qualitative approach was chosen to explore in depth the lived experiences, perceptions and strategies of rural communities in the context of climate vulnerabilities within a geopark setting. SLF was applied systematically to map and assess five key livelihood capitals—human, natural, physical, financial and social—and how these interact within the vulnerability context of CP-UGGp. This framework guided both the design of the interview and focus group discussion (FGD) protocols as well as the thematic coding process during data analysis. Conversely, CNT was employed to analyse social connections, inter-village cooperation and the roles of institutions in either supporting or constraining livelihood strategies. The application of this dual framework facilitated the study in evaluating not only individual household strategies but also community-level collective actions and institutional alignment.

The research design was executed through a comprehensive multi-method data collection approach, encompassing the following components:

- Semi-structured interviews with village leaders, local farmers, tourism stakeholders and government officials to examine efforts towards sectoral integration.
- Focus group discussions with community groups, including women, youth, farmers and environmental activists, to gather collective insights on local assets and vulnerabilities.
- Field observations of agrotourism sites, fishing activities and disaster-prone areas to corroborate spatial and livelihood dynamics.
- Document analysis of regional and village planning reports, master plans and policy documents.
- Integration of secondary data from the Village Development Index 2023 (Desa et al., 2023), disaster risk maps and statistical data on agricultural and tourism production.

The iterative application of these methods facilitated the triangulation of findings, ensuring that insights obtained from one source could be corroborated or enhanced by others. The integration of conceptual frameworks with practical application across diverse field settings renders the methodological contribution of this study both robust and adaptable.

Data triangulation, achieved through interviews, focus group discussions, observations and secondary sources, ensured that the analysis accurately represented the complex nature of vulnerability and opportunity. The empirical richness facilitated the development of context-sensitive recommendations, thereby reinforcing the study's practical orientation and relevance for policy and planning. The research methodology described above is illustrated in the flow chart presented in Fig. 11.1.

The CP-UGGp area is in Sukabumi Regency, West Java, Indonesia. This area covers eight sub-districts, namely, Cisolok, Cikakak, Pelabuhanratu, Simpenan, Ciemas, Ciracap, Surade and Waluran, which include 74 villages. The geopark covers an area of 126,000 Ha, which is 30.3% of the area of Sukabumi Regency. The topography of the geopark consists of mountains, hills, lowlands, rivers and coastal bays that directly border the Indian Ocean. This topographic condition causes a high potential threat of natural disasters such as landslides and moving soil, extreme weather, floods and flash floods, droughts, earthquakes, extreme waves and the

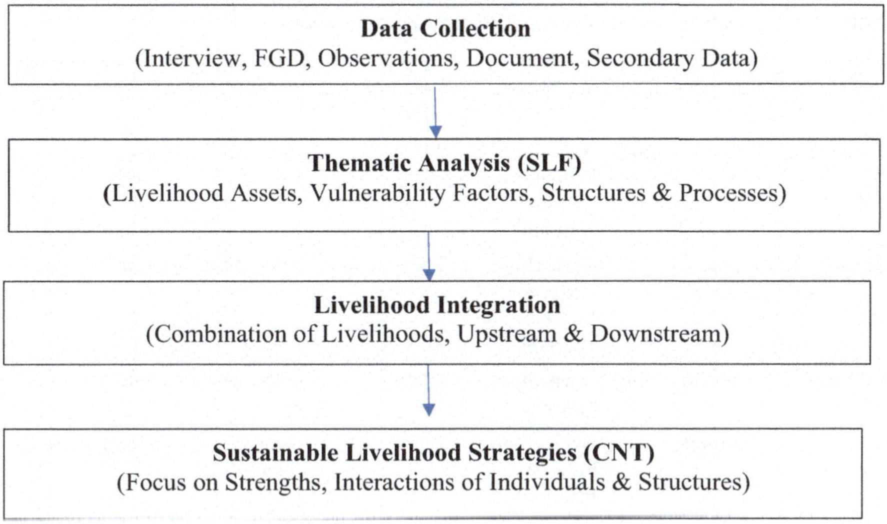

Fig. 11.1 Methodological flow chart

possibility of tsunamis. The research was conducted in nine villages in Ciemas District. The villages are Cibenda, Ciemas, Ciwaru, Girimukti, Mandrajaya, Mekarjaya, Mekarsakti, Sidamulya, Tamanjaya, which are located within the CP-UGGp area. The villages were selected because they have tourism and agricultural potential and are disaster-prone areas. The villages are also the centre of geopark activities, and there are several sites that are tourist attractions. The FGD was conducted in Girimukti Village.

11.4 Results and Discussion

11.4.1 Livelihood Assets and Its Potential Utilization

11.4.1.1 Agriculture and Fisheries

Ciemas district has abundance of agricultural land cultivated for rice fields, dry land agriculture and plantations. The land is generally flat or undulating, except in Mekarjaya and Girimukti, which have an undulating to hilly topography. The agricultural land is widely used for food crops, mainly rice, mango and chilli. Black and red rice grows in Girimukti. Mango, coconut, rubber, cloves and biopharmaceutical plants are cultivated in some villages, as presented in Table 11.1.

Rice is the main commodity for most villagers. The harvested area of rice fields in Ciemas District is 6,819 Ha, producing 38,268.23 tons. This harvested area is the

Table 11.1 Agricultural natural resources at the research location

Village	Regional topography	Livelihood potential	Agricultural land area (Ha)[a]	Main agricultural commodities
Tamanjaya	Flat to rolling	Agriculture, Plantation, Forestry, Tourism	297	Rice, Chilli, Coconut
Ciwaru	Flat	Agriculture, Fisheries/ Marine, Tourism	1.573	Rice, Mango, Chilli
Mekarsakti	Flat to rolling	Agriculture, Forestry	1.595	Rice, Mango, Chilli
Ciemas	Rolling	Agriculture, Forestry, Fisheries	2.494	Rice, Fruit, Biopharmaceutical Plants
Mekarjaya	Flat to rolling	Agriculture, Forestry	3.585	Rice, Cassava, Corn, Vegetables, Cloves, Coconut, Biopharmaceutical Plants
Mandrajaya	Flat	Agriculture, Forestry, Fisheries, Tourism	1.190	Rice, Mango, Chilli, Eggplant, Watermelon
Girimukti	Rolling	Agriculture, Forestry, Fisheries, Tourism	1.910	Lemongrass, Black And Red Rice
Sidamulya	Flat	Agriculture, Forestry, Fisheries	912	Rice, Secondary Crops, Rubber, Coconut
Cibenda	Flat	Agriculture, Forestry, Fisheries	920	Rice, Secondary Crops, Rubber, Coconut

[a] Rice fields, fields, plantations

Source Research data. Processed from Village Profile Year 2024 of each village (Pemerintah Desa Kecamatan Ciemas, 2024a, 2024b)

largest in Sukabumi Regency. In 2023, in Sukabumi Regency, the harvested area of rice fields was recorded at 113,191 Ha with a production of 647,302.74 tons (BPS Kabupaten Sukabumi, 2024). However, the existing agricultural land has not been optimally utilized due to various constraints such as limited irrigation water, capital and technology. In addition to rice, other agricultural products in Ciemas District are fruits and vegetables. Mango is the horticultural commodity with the highest production. From 2018 to 2021, mango production reached around 30,000 quintals/year. The community also uses a community forest to cultivate biopharmaceutical plants, such as ginger, turmeric, aromatic ginger and galangal. The various agricultural products show the diversity of production in the villages.

Other natural capital is water resources in the fisheries sector, as found in Ciwaru, Ciemas, Mandrajaya and Girimukti. These villages have beaches bordering the Indian Ocean and have river estuaries whose water comes from the hills. These water resources have not been widely used for capture fisheries and aquaculture (cage cultivation, floating nets). In addition, there is also inland fisheries cultivation in ponds (carp, tilapia, catfish, eel) and capture fisheries in rivers such as in Mekarsakti. However, these water resources have not been developed optimally for fisheries activities.

11.4.1.2 Tourism

The CP-UGGp has a natural landscape, such as a horseshoe-shaped plateau (amphitheatre) that opens to Ciletuh Bay. In the middle of the Amphitheatre, there is a distribution of the oldest rocks on Java Island. The geopark also has a natural diversity that offers beaches, views of hills and cliffs, waterfalls and peaks, forests and a diversity of flora and fauna. In this geopark area, there are 26 sites, including 49 geological objects, that have the potential to be tourist attractions. A total of 11 sites covering 26 geological objects are in Ciemas District (Table 11.2).

In addition to its natural attractions CP-UGGp also offers local traditions and culture that attract tourists. The local tradition of coconut tapping and processing coconut sap into sugar is an attraction for tourism activities in Ciemas District. This performance is free of charge, but the palm sugar products from the process are sold as souvenirs for tourists (Munawar, 2023). The supporting tourism facilities available are adequate, such as hotels, homestays, parking areas, restaurant, places of worship, public toilets and other recreational facilities to support tourism activities.

11.4.2 Vulnerability Context: Disaster Risk and Vulnerability

The vulnerability context discussed in this study is shock vulnerability. DFID (2000) defines shock vulnerability as a sudden change that is difficult to predict and can destroy assets directly. The factors that make up the vulnerability context are important because they directly impact people's capital status and the options open to them to pursue beneficial livelihood outcomes. In discussing the context of vulnerability, we see risk and vulnerability aspects. Villages in Ciemas District face multi-risk natural disaster threats, including floods, flash floods, extreme weather, extreme waves and seawater abrasion, earthquakes, forest and land fires, droughts, landslides and tsunamis, with low- to high risk levels (Chaidir et al., 2024).

The CP-UGGp area is prone to natural disasters. This condition can be seen from natural disasters caused by extreme weather that hit several sub-districts in the geopark. Extreme weather that hit 32 sub-districts in Sukabumi Regency on Tuesday,

Table 11.2 List of geosites in the CP-UGGp area in Ciemas District, Sukabumi Regency

Geosite	Geological object (tourism potential)	Regional administration (village)
Jampang Plateau Geosite	Loji-Sangrawayang Beach-Cisaar	Loji-Kertajaya-Girimukti (Simpenan and Ciemas Districts)
Ciletuh Bay Geosite	Karang Daeu	Girimukti
	Palangpang Beach	Ciwaru
	Mandra Island	Mandrajaya
	Manuk Island	
Badak Mountain Geosite	Mount Aseupan	
	Mount Badak	
	Kunti Island	
Legon Pandan Geosite	Legon Pandan	
Dragon Stone Geosite	Dragon Stone	
Mount Beas Geosite	Mount Beas	
	Tegal Pamakanan Ophiolite Complex	
Sodong Parat Geosite	Cikepuh Beach and River	
	Sodongparat	
Pasir Luhur Geosite	Pasir Luhur	
Citirem Beach Geosite	Citirem Beach	Sidamulya
Ciletuh Amphitheatre Morphology Geosite	Ciletuh Amphitheatre	Girimukti, Ciemas, Mekarsari, Tamanjaya, Cibenda
	Cimarinjung Waterfall	Ciemas
	Awang Waterfall and Tengah Waterfall	Tamanjaya
	Manik Peak Waterfall	Cibenda
	Sodong Waterfall	Ciwaru
	Cikanteh Waterfall	Ciemas
	Panenjoan Viewpoint	Tamanjaya
	Panyawangan Viewpoint	Tamanjaya
	Tugu Peak Viewpoint	Mekarjaya
Ciletuh Peak Waterfall Geosite	Jeruk Peak Waterfall	Mekarjaya and Mekarmurti (Ciemas and Waluran Districts)

Source Research Data. Processed from Pemda Provinsi Jawa Barat (2020)

December 3–5, 2024, resulted in floods, landslides, soil movement and strong winds in several areas, including in Ciemas District. On December 4, a regency decree was issued on the determination of emergency response status (Pemda Kabupaten Sukabumi, 2024). On December 6, a letter of temporary closure of tourist destinations in the regency was issued (Dinas Pariwisata Kabupaten Sukabumi, 2024). The extreme weather has an impact on the vulnerability of communities due to flooding and landslides. In Ciemas District, shops, houses and agricultural land were submerged in floodwaters, and public facilities were damaged (connecting roads to schools and places of worship were cut off). In Ciwaru, which is a tourist centre, 1,179 families were affected, 370 of whom had to evacuate because their homes were submerged in floodwater (Pendamping Desa Kecamatan Ciemas, 2024a, 2024b). As a result of the disaster, settlements became isolated, tourism activities stopped, public facilities were impaired, agricultural land was damaged and people's livelihoods were disrupted.

11.4.3 Structure and Processes

This section discusses the structural aspects that include the geopark development master plan, village medium-term plan and village cooperation; and the process aspects which include community assets that support sustainable livelihoods.

11.4.3.1 Structural Aspects

Geopark development is collaborative work by various agencies within the Sukabumi Regency. A master plan has been stipulated for geopark development (Pemda Kabupaten Sukabumi, 2021). Various programs have been established, including development of facilities and supporting infrastructure for the geopark, conservation of geological heritage, biodiversity conservation, socio-cultural conservation, education, research and development, sustainable community economic development, tourism destination development and geopark scientific value promotion. The implementation of the master plan is still far from ideal due to limited funds, not all villages in the geopark area are targeted for development, and the implementation by the relevant agencies is still low. Therefore, village involvement in supporting geopark development is highly expected. The village involvement is needed from the beginning of the geopark designation and the village development planning process, so that the geopark development program is inherent in each village development program.

The development of the Citamansakti Agrotourism Rural Area is one of the national rural development priorities, which was implemented to support the development of the CP-UGGp. This rural area was established through a decree of the Regent of Sukabumi, encompassing three villages (Ciwaru, Tamanjaya, and Mekarsakti) (Pemda Kabupaten Sukabumi, 2017a). The rural area is designed to be a centre for rice production in Sukabumi Regency. The leading activities developed are agrotourism, agriculture, fisheries and tourism, with leading commodities being mango, rice and marine products. A development coordination team was formed to carry out the development of Citamansakti area. The team consists of several elements, which are in charge of the regional secretary; regional development planning; community and village empowerment; food security; environment; public housing, residential areas and cleanliness; public works; communication and information; transportation; tourism; trade, cooperatives and SMEs; animal husbandry; maritime affairs and fisheries; agriculture; spatial planning and land; education; and health (Pemda Kabupaten Sukabumi, 2017b). However, the cross-agency programs had not been implemented at the time that the study was conducted. The coordination team had not had time to realize the Citamansakti area development plan because in year 2021, the development budget was largely allocated for handling COVID-19.

Most of the villages have programmed the *agriculture, fisheries and tourism development in their* village development plan. The villages have programmed the construction/rehabilitation of agricultural production facilities, including irrigation, village road irrigation foundations and rivers, milling production tools, utilization of appropriate technology, fishing ports, small rivers, village-owned inland fisheries ponds and fisheries assistance (seeds and feed). Village funds (APBDes and APBD II funds) are allocated to support the activities. The villages also have programmed village tourism, including rehabilitation of tourism facilities, appropriate technology for village economic development, training of SMEs management, environment, culture and arts, provision of trash bins, construction of studios and procurement of art equipment. Funding for activities comes from the government and the village. Three villages have planned the *inter-village cooperation*. The cooperation of Ciwaru, Tamanjaya and Mekarsakti villages is implemented through the establishment of the Citamansakti Joint Village-Owned Enterprise to support the development of the Citamansakti rural area. The joint village enterprise obtained the establishment certificate in December 2022. In this cooperation, each village plans to allocate funds for the enterprise management and activities during year 2020–2025. Other village cooperation is with third parties, generally in the form of physical works, such as construction, paving and asphalting of roads, as well as construction and rehabilitation of residential and agricultural roads. Funding for the projects comes from the Central, Provincial, District and Village Governments.

11.4.3.2 Processual Aspect

Processual aspect is shown from elements in society and villages that have the potential to drive interactions between structures and individuals to improve community resilience in carrying out their livelihoods, including aspects of education, social capital and village government policies related to disasters, technology and community economic activities, as presented in Fig. 11.2.

Figure 11.2 shows community assets that can boost or inhibit interactions between structures and individuals. Strong and potential community assets (shown in green

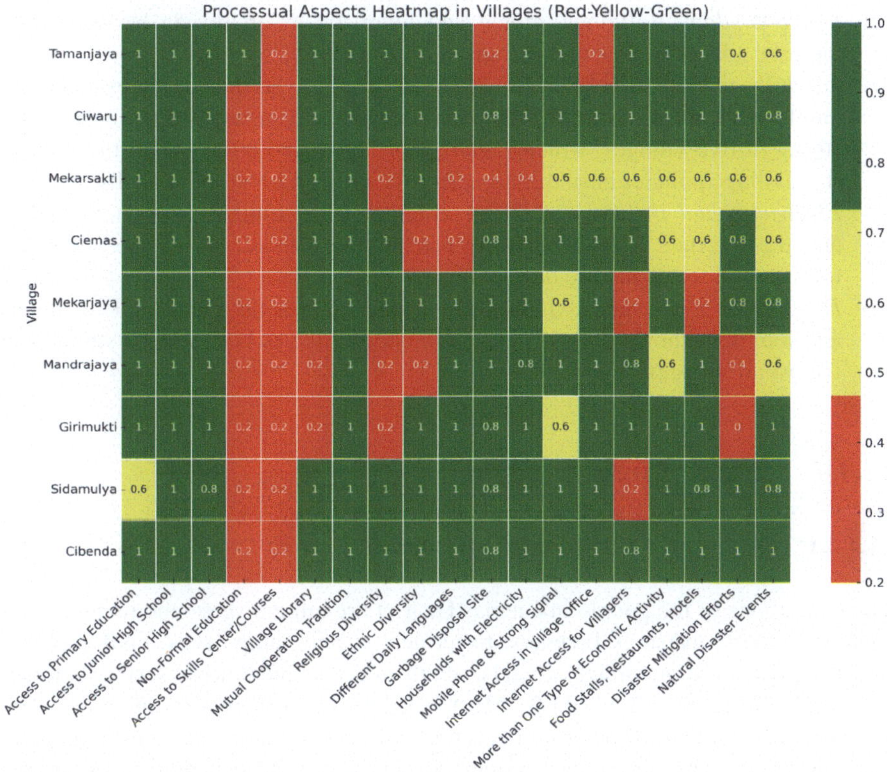

Fig. 11.2 Heatmap of process aspects in livelihood at research locations. *Description* Green (strong), Yellow (potential), Red (weak). *Source* Research data

and yellow) can be utilized to foster the interactions, while weak assets (shown in red) need to be strengthened so that they can function to realize structures—individuals' interactions. Efforts to utilize strong and potential assets in realizing interactions between structures and individuals to support sustainable livelihoods are as follows:

(1) *Education.* All villages, except Sidamulya, have strong capital for primary, junior and senior secondary education. Easy access to education can be utilized for long-term adaptation efforts by integrating adaptation efforts with the primary-secondary education sector. Access to community knowledge can be utilized through village libraries to create an adaptive society by utilizing local resources, developing entrepreneurship and strengthening sustainable livelihoods. To organize non-formal education programs, support from the Education Office is required.
(2) *Cooperation.* The tradition of mutual cooperation (*gotong royong*) in the community can increase collaboration between community members and collective efforts to overcome challenges.

(3) *Local Wisdom.* Religious, ethnic and linguistic diversity can be utilized to increase adaptive capacity relevant to local wisdom.
(4) *Landfills and Disaster Response Efforts.* Village government policies regarding landfills are strong in most villages. Meanwhile, disaster response efforts are strong in five villages, which require more intensive efforts from the village. Both village policies have the strength to be utilized for sustainable livelihoods.
(5) *Electricity and the Internet.* Most villages have strong access to electricity and the Internet. This technology can improve disaster response and other information for economic development.
(6) *Diversity of Economic Activities.* The existence of stalls, supermarkets, hotels and lodgings shows diversity in livelihood, which is becoming a strength in almost all villages.

11.4.4 Current Livelihood Practices

11.4.4.1 Diversity of Community Livelihoods

The livelihoods of the people in Ciemas District are quite diverse and include roles such as farmers, livestock breeders, fishermen, traders, employees (civil servants, military, private sector, village officials, health workers) and tourism services (homestays and tour guides) (Table 11.3). Most villagers are farmers, fishermen and farm labourers who cultivate food crops (rice, horticulture), forestry and fisheries (capture and cultivation). The agricultural system still uses a rain-fed system, forcing farmers to look for other side jobs during the dry season. On average, farmers become labourers and traders, and raise livestock (chickens, sheep) to fill the gap in the dry season.

Table 11.3 Percentage of livelihoods in ciemas district

Village	Agriculture	Fisheries	Tourism services	Others[a]
Cibenda	42.3	0	0	57,7
Ciemas	63.1	2.7	0.6	33,6
Ciwaru	39.6	10.8	4.1	45,5
Girimukti	60.1	2.1	1.5	36,3
Mandrajaya	48.6	0.7	0.1	50,6
Mekarjaya	37.0	0	0	63,0
Mekarsakti	41.2	0.8	0	58,0
Sidamulya	71.4	0	0	28,6
Tamanjaya	36.9	0	0.5	62,6

[a]Employees, traders, labourers, employees, Indonesian migrant workers, etc.
Source Research data. Processed from Village Profile Year 2024 of each village (Pemerintah Desa Kecamatan Ciemas, 2024a, 2024b)

Several agricultural/fishery-based economic activities are also developing, such as agricultural product processing businesses (cassava chips, palm sugar, salted fish) and fishing tourism on floating charts. The diversity of these businesses shows the diversity of skills in the community to support their lives.

In several villages, such as Ciwaru, Ciemas and Girimukti, livelihoods in the tourism sector are also developing. The impact of the development of the CP-UGGp, including the construction of roads to the geopark area and tourism promotion, has resulted in the emergence of various economic businesses that can support the development of geopark tourism. Among them are homestays, culinary businesses (restaurants), food processing home industries (such as cassava chips), craftsmen (such as bamboo crafts), village tour packages (experiencing village life and visiting tourist attractions) and tour guides.

In 2024, there were 20 villa units in Tamanjaya Village. However, according to residents, there is currently less economic activity in the village than when the Geopark was first developed. This is due to road construction in the Sukabumi Regency that has rerouted accessibility to the Geopark. As a result, the tourism centre has moved from Tamanjaya to Ciwaru Village. Economic activities in Tamanjaya declined, while in Ciwaru the activities grew. Livelihoods such as traders, tour guides, and lodging services have also flourished along with the increase in tourism in Ciwaru.

11.4.4.2 Livelihood Dynamics: Livelihood Integration Practices at the Household Level

Rural households adopt various strategies to maintain and improve their well-being in the context of social, economic and environmental changes. Livelihood integration is one of the main approaches used to face economic challenges and environmental uncertainties. This section discusses how households manage and combine various livelihoods, namely, livestock, SMEs and fisheries.

Chicken Farmers in Mekarjaya Village. A village resident works as a farmer and a chicken farmer on his substantial agricultural land. The farm he manages has 1,200 hens producing 50 kg of red-shelled eggs daily. The eggs are sold for Rp 29,000 32,000 per kg. He applies an integrated farming system (IFS), chicken waste and manure are used to fertilize plants on the farm. Egg production and farm produce are sold to meet the needs of restaurants and stalls in the geopark tourist destination, and in the sub-district market. Meanwhile, the hens that are no longer productive are sold to buyers around the village, for meat. The demand for the hens increases, especially during Eid al-Fitr, other religious holidays or local traditional events that attract tourists. In the practice of IFS, the farmer can reduce production costs, increase agricultural and livestock production, and increase income.

SMEs in Ciemas Village. The flash flood that hit the village on December 4, 2024, impacted one of the SMEs players in banana-based food. This SME produces various flavours of banana chips (cheese, green tea, chocolate, spicy, original flavours) and banana hump floss. Before the flood, the chip product could be marketed in at least

200 packages (size 15 × 24 cm) per month. The price is IDR 15,000/package. The flood, which began entering the house at around 03:00 a.m., significantly affected the sustainability of the business by sweeping away approximately 300 packages of chips ready for sale, damaging the kitchen and cooking equipment, and creating difficulties in obtaining raw materials as banana trees were washed away. Until January 15, 2025, the food production activities can no longer operate. Currently, it is only in the recovery stage. The existing business capital has been used to pay for repairs and cleaning the house due to the flood.

Fishery Business in Ciemas District. Some residents rely on the fisheries sector as their main livelihood. Fishery and tourism businesses in this district support each other, where fisheries provide products that support the tourism sector, while tourism opens new economic opportunities for the community. Tourists can enjoy the beauty of the beach and the sea by fishing on floating charts, enjoy fresh fish cuisine and buy fish products. On the other hand, tourists generate income to fishermen. If a tourist wants to fish in the sea, he can rent a boat and fish on a floating pontoon at a rate of around Rp. 150,000 per night per person. However, work at sea cannot be done all year round by most fishermen. During the west monsoon, between November and February, the waves are quite high, ranging from 5 to 7 m, and the wind is quite strong, so is the conditions are too risky for fishermen. Therefore, fishermen must find other jobs in the village, such as labourers, farmers, traders or migrate to find work in the city. The floods and strong winds also worsened the conditions of the fishermen. The water in the sea became murky, causing fish to move to the middle of the sea where the water was clearer, making it difficult for fishermen to find fish on the edges that were still within reach.

11.4.5 Strategies for Tourism-Agriculture Integration

Several studies demonstrate the weaknesses of both tourism and agriculture. The challenges faced by Indonesia's tourism sector are closely related to environmental vulnerability, uncertain economic impacts and limited community adaptive capacity. This circumstance considerably effects the community earnings, particularly in areas that depend on natural beauty and agrarian-based tourism (Idajati, 2024; Sugihardjo et al., 2023). Similarly, in coastal tourism, environmental vulnerability has impaired local coral reefs and fishery ecosystems, which are the primary tourist charms (Handayani, 2024; Susanto et al., 2020). The impairment decreases tourist interest and endangers the economic welfare of local peoples who hinged on the tourism sector for their incomes (Tanoamchard & Limjirakan, 2012). To ensure the sustainability of this sector, a thorough approach is required that incorporates boosting community resilience, investing in sustainable tourism attempts and developing integrated disaster risk management strategies.

The challenges facing Indonesia's agricultural sector as a source of income are closely linked to the effects of climate uncertainty and the occurrence of natural disasters. The vulnerability of smallholder farmers is worsened by inadequate adaptive capability, socio-economic aspects and increasing climate uncertainty. Studies reveal that the limited adaptive capacity of smallholder farmers, affected by socio-economic reasons such as access to resources, education and social safety nets, causes them highly vulnerable (Das et al., 2021). For example, irregular rainfall patterns and increasing temperatures caused by climate change have generated acute droughts in various regions, resulting in crop failures and lessening farmer earnings (Syakirin, 2024). These challenges are intensified by the shortage of sufficient infrastructure and support systems to help farmers in countering to the effects of climate uncertainty (Rozaki et al., 2021). Addressing these encounters entails an integrated attempt that relates local knowledge, modern agricultural technologies and effective disaster risk management strategies.

Livelihood strategies include the diversity and combination of activities and choices that people undertake to achieve their livelihood goals. In the research locations, various community activities were found in agriculture, fisheries and tourism, indicating livelihood diversity. Also found were the life practices of people who work on various livelihoods and those who do one type of work. Individuals who work on various livelihoods show resilience to the risk of natural disasters that have just occurred. At the same time, individuals who work on one type of work show vulnerability in their livelihoods.

Considering the challenges in the agricultural sector (including fisheries) and tourism, and the advantages of business diversity in dealing with disasters, tourism-agriculture integration can be one strategy for increasing the livelihood resilience of people who live and do business in villages in the geopark area. Tourism–agriculture integration is considered strategic for rural development (Togaymurodov et al., 2023), improving eco-efficiency (Wang et al., 2022), increasing farmer income (Luo, 2023), serving as an adaptation strategy for smallholder farmers (Khanal & Mishra, 2014), and preserving agricultural cultural heritage (Flanigan et al., 2015). Utilization of livelihood assets that consider geopark conservation and the local economy is the main framework for formulating tourism and agriculture integration in villages in the geopark area. The livelihood integration strategy for the sustainable livelihoods of people in these villages includes businesses in agriculture, fisheries, tourism, a combination of agriculture-tourism, fisheries-tourism and tourism support services (Fig. 11.3).

The development of the agricultural sector aims to increase the productivity of environmentally sound agriculture. The superior commodities include rice, horticulture and biopharmaceutical plants. The development of the fisheries sector aims to increase fishermen's income through marine fisheries and aquaculture. Tourism development aims to increase the number of tourists with the unique geopark attraction, village life and culture, and various tourism support service businesses. Furthermore, agrotourism and marine/fishery-based tourism aim to increase the marketing of agricultural/fishery products through integrated agricultural/fishery tourism. The diversity of livelihoods creates a synergy that evenly distributes economic benefits throughout the village. The development of livelihoods in the agricultural, tourism and fisheries sectors, a combination of the three, and supporting service

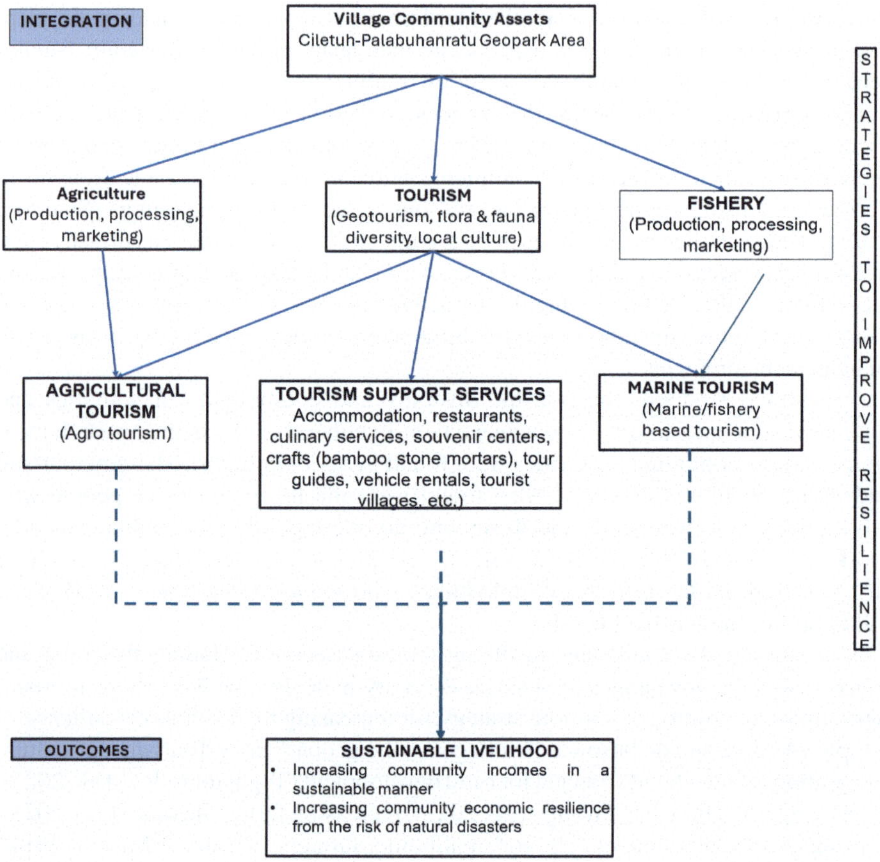

Fig. 11.3 Livelihood integration scheme for sustainable livelihoods. *Source* Research data

businesses is expected to increase the income of village communities in a sustainable and environmentally friendly manner. This integration will open wider market access, expand employment chances and eventually sustainably improve village communities' economic resilience.

An important principle in the livelihoods approach is to focus on strengths rather than needs and to link macro (policy and institution) and micro (individual and community livelihood options) aspects. These macro–micro links are relational aspects needed in implementing sustainable livelihoods. Community resilience is pivotal in sustainable livelihoods and is formed by preparedness, adaptation, social capital and governance. Efforts against disaster risks are still needed to be improved, especially when communities identify themselves as capable actor (Kim & Kim, 2022). The adoption of technology, especially in accessing ICT such as electricity, mobile phones and the internet accelerates immediate disaster response and enhances local governance structures (Macose, 2022). Combining indigenous and religious-cultural know-how supports culturally grounded adaptation approaches (Akuba, 2024; Ristroph, 2019). Inserting adaptation into school curricula advances continuing resilience and inhibits maladaptation (McNamara, 2013). Reinforcing communal trust and cooperation improves collective action and local adaptation (Nyahunda & Tirivangasi, 2021). Community gardens play significant part in food security and social cohesion (Clarke et al., 2019; Okvat & Zautra, 2011). Joint Village-Owned Enterprises and the incorporation of village planning with geopark development help align institutional frameworks and build economic resilience (MacDonald & Birchall, 2019) and effective governance (Adger, 2006). A combination of strategies for rural communities, such as entrepreneurship and non-formal education, helps address local vulnerabilities (Akinbami, 2021). This multidimensional approach underlines the significance of relating institutional systems with community-led resilience building.

11.4.6 Discussion

The integration of livelihoods in the fields of agriculture, fisheries and tourism in villages in the geopark is believed to stipulate benefits in the form of escalating community welfare, village independence and the accomplishment of the Geopark program. The integration of livelihood diversity expands employment opportunities in the three main sectors and the growth of other businesses as supporters of the main sector. The combination of these livelihoods will increase community resilience. Combining tourism activities with agricultural systems can diversify sources of income and provide alternative sources of income during crises and climate shocks, thereby increasing the resilience of rural communities (Ohe, 2023; Ngoma et al., 2023), and vulnerable coastal areas (Bernzen et al., 2023). At the same time, other efforts to increase resilience are also needed to implement sustainable livelihoods, including preparedness, adaptation, social capital and effective governance based on a focus on strengths and building relational aspects.

Improving community welfare will also accelerate the success of the UNESCO Geopark program because improving community welfare is an integral part of the success of the program, which emphasizes community development, ensuring that resident's benefit from geotourism and conservation efforts. By enhancing the people's welfare, they are more likely to support and actively contribute in geopark objectives, leading to sustainable practices that balance conservation with economic growth and improved quality of life (Komoo, 2010).

Each aspect examined throughout this study contributes immediately to achieving the principal objective: to formulate an integrated livelihood strategy that boosts the resilience and sustainability of rural communities in geopark regions encountering climate-related vulnerabilities.

(1) **Application of the Sustainable Livelihood Framework (SLF)** allowed a systematized appraisal of livelihood assets, vulnerabilities and current strategies. This framework facilitated a thorough insight of how rural communities organize human, social, natural, financial and physical capital. This immediately aligns with grounding the integration strategy in the lived realities and strengths of the communities examined.
(2) **The Incorporation of the Collaborative Neighbourhood Transformation (CNT) Perspective** deepened the analysis by describing the relational and institutional dynamics that are often unnoticed in SLF. It provided analytical understandings into how community cohesion, social networks and inter-village cooperation can be utilized to implement integrated and participatory development. This reinforces the study's objective to formulate a locally embedded and socially responsive strategy.
(3) **Data Triangulation Through Interviews, Focus Group Discussions, Observations and Secondary Sources** confirmed that the analysis revealed the multi-layered nature of vulnerability and opportunity. The rich empirical data allowed context-sensitive suggestions, strengthening the study's practical orientation and relevance for policy and planning.
(4) **Operational Mapping of Tourism and Agriculture Linkages** explained how upstream agricultural activities (production and cultivation) and downstream tourism services (attractions and accommodations) can be harmonized. This mapping showed the possibility for synergetic diversification, immediately answering to the objective of developing economically viable and environmentally sustainable rural livelihoods.
(5) **Integration of Disaster Risk Reduction and Adaptive Capacity Indicators** discussed the climate resilience component of the study. By relating preparedness, technological adaptation, local knowledge and institutional support with livelihood resilience, the study accomplishes its objective to offer a strategy that is not only integrative but also robust in the face of environmental shocks.
(6) **Positioning the Findings Within the UNESCO Geopark Framework** confirms that the proposed strategy is consistent with global sustainability directives. This heightens the transferability of understandings beyond the CP-UGGp, proposing a model that other geoparks and rural areas can adapt in comparable socioecological settings.

The elements discussed above are interrelated. They fulfil the study's overarching objective: to present a contextually grounded, scientifically rigorous and practically actionable model for sustainable livelihoods in geopark-based rural communities.

This study is unique in several key dimensions. First, it exercises a dual framework approach, combining the Sustainable Livelihood Framework (SLF) and Collaborative Neighbourhood Transformation (CNT), to analyse rural livelihood resilience in a geopark context, a methodology rarely employed in existing writings. Second, unlike conventional researches that focus on either tourism development or agricultural productivity in isolation, this study incorporates both sectors by mapping upstream and downstream processes and assessing their synergistic potential within a climate-vulnerable geopark area. Third, the study employs resilience not only through theoretical concepts but also through grounded, community-specific practices across nine villages in the CP-UGGp, connecting real-world dynamics with the global mission of UNESCO Geoparks. This localized, asset-based approach advances a practical model for livelihood diversification that maintains all three pillars of geopark sustainability: conservation, education and local economic development. Through this, the discussion consistently ties back to the main purpose of developing a replicable, community-centred strategy that addresses socioecological vulnerabilities and promotes sustainable, adaptive livelihoods under increasing environmental uncertainty.

The findings of this study have imperative implications for rural development, climate adaptation and geopark management strategies. The evidence reveals that livelihood integration especially through the synergistic arrangement of tourism and agriculture can substantially increase the adaptive capacity and economic resilience of communities in climate-sensitive landscapes. This multidimensional integration aligns with global best practices and supports the growing agreement that sectoral silos are unsatisfactory to deal with the complex realities of rural livelihoods under environmental stress.

Comparative insights from related initiatives in other regions offer both validation and contrast. For example, in the Imbabura Geopark in Ecuador, González (2022) discovered that integrating organic agriculture with geotourism increased local income while emphasizing conservation ethics, a pattern that closely mirrors the CP-UGGp experience. Similarly, a study from Portugal's Estrela Geopark (Fernandes et al., 2021) underlined how combining agricultural education with geotourism improves environmental consciousness and diversifies earnings sources. However, unlike those cases, this study uniquely combines the relational dimension through the CNT framework, offering a more dynamic understanding of how social networks and institutional interactions affect livelihood outcomes.

Likewise, studies in coastal Bangladesh (Bernzen et al., 2023) and the Philippine Cordilleras have revealed that livelihood diversification, though beneficial, often fails when institutional support and market access are weak. In contrast, this study emphasizes the importance of integrating efforts within a geopark governance structure and village-level planning systems, thereby enhancing their institutional legitimacy and sustainability. Therefore, the proposed model from CP-UGGp holds broader relevance, especially for other geopark areas and rural territories that face both ecological fragility and institutional fragmentation.

This chapter immediately aligns with the book's focal theme, addressing the socioecological crisis of development through an integrated analysis of community resilience and sustainable explanations. Focusing on villages within the geopark, the study shows how the strategic integration of tourism and agriculture can perform as an adaptive livelihood model in the face of escalating environmental risks and socioeconomic vulnerabilities. By employing the SLF and emphasizing local agency, social capital and institutional synergy, the chapter offers a holistic and empirically grounded approach to resilience building. It fills a critical gap in the literature by connecting environmental, developmental and social crisis viewpoints, showcasing how multi-sectoral integration can alter vulnerable rural settings into sustainable, self-reliant communities.

11.5 Limitations of the Study

This research has several limitations that should be recognized. First, the study is based on a qualitative case study carried out in nine villages within the CP-UGGp which may restrain the transferability of findings to other areas with different socioecological contexts. Second, the reliance on interviews, focus group discussions and field observations introduces potential subjectivity and response bias, despite efforts at triangulation. Third, the study was carried out in the aftermath of extreme weather occurrences, which may have temporarily changed livelihood conditions and community opinions, especially concerning vulnerability and adaptive strategies. Fourth, although institutional structures and village planning frameworks were studied, the analysis did not fully describe the dynamic evolution of governance and policy implementation over time. Lastly, while the research proposes a conceptual model for tourism-agriculture integration, it does not incorporate quantitative assessments of financial feasibility, market dynamics or long-term ecological consequences. These aspects entail further empirical exploration to strengthen the operational scalability of the proposed model.

11.6 Recommendations

This study illustrates that the villages have ample livelihood assets essential for the development in the agriculture, fisheries and tourism sectors. However, the livelihood assets still need reinforcement to foster the resilience of communities faced with crisis conditions and climate change through endeavours to enhance disaster preparedness, adaptation to climate change and community social capital. Execution of the strategy of integrating agriculture, fisheries and tourism to achieve sustainable livelihoods entails collaborative support from diverse stakeholders. The role of village governments and community institutions needs to be amplified. Villages require intensive assistance from related agencies in preparing integrated planning, reducing disaster risks, enabling community business diversification, strengthening village-owned enterprise cooperation and establishing collaboration between villages and third parties to realize sustainable livelihoods.

Sustainable development of geopark villages requires the implementation of the three main pillars of geopark development: conservation, education, and local economic development. Conservation actions aim to preserve geological, biological, and cultural diversity through ecotourism regulations, critical land rehabilitation, and environmentally sound agricultural practices. The education pillar focuses on increasing public and tourist awareness through geopark education centres, field school programs, and training for farmers and local business actors. The local economic development pillar emphasizes the development of community-based ecotourism, local products based on sustainable natural resources, and the strengthening of SMEs to enhance economic viability.

11.7 Conclusion

This study offers valuable knowledge to understand better how combining the interest of geopark management and village development can enhance community resilience through sustainable diversification of community livelihoods. The integration of community livelihoods in agriculture, fisheries and tourism and their supporting businesses has the potential to increase local community welfare, sustainable village development and the accomplishment of the geopark mission.

This study plays a role to the growing body of scientific literature on sustainable rural development by proposing a contextually grounded model for integrating agriculture and tourism in climate-sensitive geopark regions. By employing the SLF and the CNT approach, this study offers a multi-scalar understanding of how livelihood assets, institutional dynamics and socioecological vulnerabilities interrelate to shape resilience in rural communities. Scientifically, this research makes a unique contribution by explaining how relational and place-based analysis, supported by SLF and CNT, can inform bottom-up, community-led strategies for resilience. It moves beyond sectoral planning and proposes a livelihood-centred, ecosystem-aware methodology that links conservation, education and economic empowerment, as envisioned in the UNESCO Geopark mission. By offering both conceptual clarity and practical applicability, this study makes a substantial contribution to the global discourse on resilience building, sustainable tourism and agri-based livelihood transformation. It initiates new avenues for interdisciplinary research on socioecological systems. It proposes a valuable framework for policymakers, development practitioners and geopark managers looking for sustainable answers to livelihood vulnerability under climate uncertainty.

The proposed livelihood integration concept has implications for the significance of relationships between local communities, local village governments and geopark management so that achieving SDGs in villages around the geopark area is more effective. The implications of this study extend beyond the CP-UGGp. As climate change increasingly unsettles rural livelihoods worldwide, the integrated model proposed here serves as a replicable strategy for communities in other geoparks and disaster-prone landscapes. The findings highlight that diversification through agriculture and tourism synergy is not simply a local development approach, but a global adaptation imperative that aligns with international agendas such as the Sustainable Development Goals (SDGs), especially SDG 1 (No Poverty), SDG 13 (Climate Action) and SDG 15 (Life on Land).

Acknowledgements The authors thank the local communities and stakeholders in the CP-UGGp for their important insights in this study. Special thanks go to the village government officials and local business actors who contributed their time and knowledge to enrich our findings.

Potential Conflicts of Interest On behalf of all authors, the corresponding author states that there is no conflict of interest.

Informed Consent Informed consent was obtained from all individual participants included in the study. Ethical Committee on Social Studies and Humanities—National Research and Innovation Agency: Ethical Clearance Approval on Social Studies and Humanities Research. No: 240/ KE.01/ SK/03/2024.

Authors Contribution All authors have contributed equally to this work.

Data Availability Statement The datasets generated during and/or analysed during the current study are available from the corresponding author upon reasonable request.

AI Disclosure Statement The authors affirm that no generative artificial intelligence (AI) tools were used in the writing or analysis of this manuscript. Where AI tools (e.g., language models, grammar checkers) were used to support language refinement or formatting, the authors supervised all usage, and the intellectual content remains solely the product of human authorship. No AI-generated content was accepted without critical human oversight and revision.

References

Adger, W. N. (2006). Vulnerability. *Global Environmental Change, 16*(3), 268–281. https://doi.org/10.1016/j.gloenvcha.2006.02.006

Akinbami, C. (2021). Climatepreneurship: Adaptation strategy for climate change impacts on rural women entrepreneurship development in Nigeria. In W. Leal Filho (Eds.), African handbook of climate change adaptation (pp. 2143–2168). Springer, Cham. https://doi.org/10.1007/978-3-030-45106-6_191

Akuba, T. (2024). The sociological implications of climate change: Community adaptation and resilience. *Journal of Advanced Sociology, 5*(2), 30–42. https://doi.org/10.47941/jas.1858

Arizo, L., & Apritado, J. (2022). Level of satisfaction and impacts of agritourism sites applied in the Province of Cebu. *International Journal of Research Studies in Management, 10*(5). https://doi.org/10.5861/ijrsm.2022.59

Bernzen, A., Sohns, F., Jia, Y., & Braun, B. (2023). Crop diversification as a household livelihood strategy under environmental stress. Factors contributing to the adoption of crop diversification in shrimp cultivation and agricultural crop farming zones of coastal Bangladesh. *Land Use Policy, 132*, 106796. https://doi.org/10.1016/j.landusepol.2023.106796.

Bonachea, J. (2023). Suitability of valleys of cantabria area for a UGGp proposal. *Land, 12*(12), 2177. https://doi.org/10.3390/land12122177

Chaidir, N., Hindersah, H., & Rustidja, E. S. (2024). Pemetaan Multi Risiko Bencana Bagi Perencanaan Penyediaan Infrastruktur Mitigasi Di Kawasan Amfiteater Geopark Ciletuh-Palabuhanratu, *6*(2), 243–253. ISSN 2656-8705URSJ. https://doi.org/10.35965/ursj.v6i2.3982

Chambers, R., & Conway, G. (1992). *Sustainable Rural Livelihoods: Practical Concepts for the 21st Century*. IDS Discussion Paper No. 296.

Chen, G., Zhang, S., Yan, B., & Miao, S. (2021). Environmental safety evaluation of geopark based on CPTED concept and fuzzy comprehensive analysis. *PLoS ONE, 16*(11), Article e0260316. https://doi.org/10.1371/journal.pone.0260316

Clarke, M., Davidson, M., Egerer, M., Anderson, E., & Fouch, N. (2019). The underutilized role of community gardens in improving cities' adaptation to climate change: A review. *People Place and Policy Online, 12*(3), 241–251. https://doi.org/10.3351/ppp.2019.3396732665

Collaborative for Neighborhood Transformation. (2023). What is Asset Based Community Development (ABCD). https://resources.depaul.edu/abcd-institute/resources/Documents/WhatisAssetBasedCommunityDevelopment.pdf

Das, U., & Ansari, M. (2021). The nexus of climate change, sustainable agriculture and farm livelihood: Contextualizing climate smart agriculture. *Climate Research, 84*, 23–40. https://doi.org/10.3354/cr01648

De Stagé, R., Holloway, A., Mullins, D., Nchabaleng, L., & Ward, P. (2002). *Learning about livelihoods*. Insights from Southern Africa. Oxfam Publishing.

DFID. (2000). Sustainable Livelihoods Guidance Sheets. Department for International Development. http://www.livelihoods.org/info/info_guidancesheets.html (Accessed: 16.01.2025)

Dinas Pariwisata Kabupaten Sukabumi. (2024). Surat Edaran Nomor: 500.13/11157/DISPAR/2024 Tentang Penutupan Sementara Destinasi Wisata Kabupaten Sukabumi.

Fairer-Wessels, F. A. (2019). *Investigating sustainable development goals, livelihoods and tourism development in a rural protected area* (pp. 205–217). Routledge. https://doi.org/10.4324/9780429289422-15

Farsani, N., Coelho, C., & Costa, C. (2011). Geotourism and geoparks as novel strategies for socio-economic development in rural areas. *International Journal of Tourism Research, 13*(1), 68–81. https://doi.org/10.1002/jtr.800

Fauzi, N., Misni, A., Kamaruddin, S., & Ahmad, P. (2018). Geo-heritage conservation: The content analysis study of Kilim Geoforest Park, Langkawi. *Asian Journal of Quality of Life, 3*(12), 87–97. https://doi.org/10.21834/ajqol.v3i12.145

Fernandes, G., Castro, E., & Tracana, R. (2021). Tourism as a strategy for geo-education and cultural valorization of territories: Promoting ecocultural values and geoconservation in estrela geopark, Portugal. In *Handbook of research on the role of tourism in achieving sustainable development goals*. IGI Global Scientific Publishing. https://doi.org/10.4018/978-1-7998-5691-7.ch007

Flanigan, S., Blackstock, K., & Hunter, C. (2015). Generating public and private benefits through understanding what drives different types of agritourism. *Journal of Rural Studies, 41*, 129–141. https://doi.org/10.1016/j.jrurstud.2015.08.002

Ginting, N., Rahman, V., Nasution, A., & Dewi, N. (2021). Geotourism development through the public facilities in geotrail bakkara, toba caldera geopark. *Geojournal of Tourism and Geosites, 37*(3), 914–920. https://doi.org/10.30892/gtg.37324-726

GLOPP (2008). DFID's Sustainable Livelihoods Approach and its Framework.

González, Á. (2022). Marketing management and sustainable tourism in the Imbabura geopark, Ecuador. *Smart Tourism, 3*(2), 10. https://doi.org/10.54517/st.v3i2.2141

Gulas, O., Vorwagner, E., & Pašková, M. (2019). From the orchard to the full bottle: One of the geostories of the nature and geopark Styrian Eisenwurzen. *Czech Journal of Tourism, 8*(2), 143–155. https://doi.org/10.2478/cjot-2019-0009

Ha, N. (2024). The moderating role of technological knowledge in the relationship between perceived sustainable marketing and intention to agritourism. *International Journal of Analysis and Applications, 22*, 83. https://doi.org/10.28924/2291-8639-22-2024-83

Handayani, A. (2024). Fostering resilient communities and sustainable tourism against climate-driven disaster risk in Gili Islands. *IOP Conference Series Earth and Environmental Science, 1350*(1), Article 012022. https://doi.org/10.1088/1755-1315/1350/1/012022

Herrera, G., Carrión, P., & Briones, J. (2018). Geotourism potential in the context of the geopark project for the development of Santa Elena Province, Ecuador. *WIT Transactions on Ecology and the Environment, 217*, 557–568. https://doi.org/10.47696/adved.202228

Idajati, H. (2024). The role of social capital and individual competence on community resilience of the tourism industry against climate change. *IOP Conference Series Earth and Environmental Science, 1366*(1), Article 012023. https://doi.org/10.1088/1755-1315/1366/1/012023

Kementerian Desa, Pembangunan Daerah Tertinggal dan Transmigrasi. (2023). Indeks Desa Membangun. https://idm.kemendesa.go.id

Khanal, A., & Mishra, A. (2014). Agritourism and off-farm work: Survival strategies for small farms. *Agricultural Economics, 45*(S1), 65–76. https://doi.org/10.1111/agec.12130

Kim, Y., & Kim, M. (2022). Factors affecting household disaster preparedness in South Korea. *PLoS ONE, 17*(10), Article e0275540. https://doi.org/10.1371/journal.pone.0275540
Kollmair, M., & Gamper, St. (2002). The Sustainable Livelihood Approach. Input Paper for the Integrated Training Course of NCCR North-South. Development Study Group. University of Zurich.
Komoo, I. (2010). Geopark sebagai Peraga Pembangunan Lestari Wilayah (Geopark as a Model for Regional Sustainable Development). *Akademika, 80*(1). https://ejournals.ukm.my/akademika/article/download/518/4094
Luo, Y. (2023). Does the integrated development of agriculture and tourism promote farmers' income growth? Evidence from Southwestern China. *Agriculture, 13*(9), 1817. https://doi.org/10.3390/agriculture13091817
MacDonald, S., & Birchall, S. (2019). Climate change resilience in the Canadian Arctic: The need for collaboration in the face of a changing landscape. *Canadian Geographer/le Géographe Canadien, 64*(3), 530–534. https://doi.org/10.1111/cag.12591
Macose, V. (2022). Building a resilient community with an ICT-Powered Local Government Unit (LGU): Community-based disaster solution. *Jurnal Pendidikan Geografi, 27*(2), 219–227. https://doi.org/10.17977/um017v27i22022p219-227
McNamara, K. (2013). Taking stock of community-based climate-change adaptation projects in the Pacific. *Asia Pacific Viewpoint, 54*(3), 398–405. https://doi.org/10.1111/apv.12033
Munawar, R. (2023). Pengelolaan Geopark Untuk Mewujudkan Pariwisata Berkelanjutan (Studi Kasus di Geopark Ciletuh-Palabuhanratu). *Jurnal Bisnis Manajemen and Ekonomi (JBME), 21*(1), Juli 2023. ISSN: 1693-8305, e ISSN: 2797-5444, https://journal.widyatama.ac.id/index.php/jbme/article/view/1250/844
Natarajan, T., Parida, B. P., & Dhal, M. (2022). A sustainable livelihoods framework for the 21st century. *Frontiers in Environmental Science, 10*, 1–13. https://doi.org/10.1016/j.worlddev.2022.105898
Ngoma, H., Finn, A., & Kabisa, M. (2023). Climate shocks, vulnerability, resilience and livelihoods in rural Zambia. *Climate and Development, 16*(6), 490–501. https://doi.org/10.1080/17565529.2023.2246031
Nguyen, H. L., & Akerkar, R. (2020). Modelling, measuring, and visualising community resilience: A systematic review. *Sustainability, 12*(19), 7896. https://doi.org/10.3390/su12197896
Nyahunda, L., & Tirivangasi, H. (2021). Harnessing of social capital as a determinant for climate change adaptation in Mazungunye Communal Lands in Bikita, Zimbabwe. *Scientifica, 2021*, 1–9. https://doi.org/10.1155/2021/8416410
Ohe, Y. (2023). Exploring New Opportunities for Agritourism in the Post-COVID-19 Era. *Agriculture, 13*(6), 1181. https://doi.org/10.3390/agriculture13061181
Okvat, H., & Zautra, A. (2011). Community gardening: A parsimonious path to individual, community, and environmental resilience. *American Journal of Community Psychology, 47*(3–4), 374–387. https://doi.org/10.1007/s10464-010-9404-z
Paniccia, P., & Baiocco, S. (2020). Interpreting sustainable agritourism through co-evolution of social organizations. *Journal of Sustainable Tourism, 29*(1), 87–105. https://doi.org/10.1080/09669582.2020.1817046
Pendamping Desa Kecamatan Ciemas. (2024). Laporan Pendamping Desa Kecamatan Ciemas (11–1–2025) tentang Bencana Alam Tanggal 4–5 Desember 2024 di Kecamatan Ciemas Kabupaten Sukabumi. Unpublished.
Pemerintah Desa Kecamatan Ciemas. (2024). Profil Desa Cibenda, Ciemas, Ciwaru, Girimukti, Mandrajaya, Mekarjaya, Mekarsakti, Sidamulya, Tamanjaya Tahun 2024. Kecamatan Ciemas, Kabupaten Sukabumi.
Pemda Kabupaten Sukabumi. (2017a). Keputusan Bupati Sukabumi No. 141/Kep.48-DPMD/2017 tentang Kawasan Perdesaan Agrowisata Citamansakti.
Pemda Kabupaten Sukabumi. (2017b). Keputusan Bupati Sukabumi No. 141/Kep.46-DPMD/2017 tentang Tim Koordinasi Pembangunan Kawasan Perdesaan Kabupaten Sukabumi

Pemda Kabupaten Sukabumi. (2021). Peraturan Bupati Sukabumi No. 25 tahun 2021 tentang Rencana Induk UNESCO Global Geopark Ciletuh Palabuhanratu 2020–2029.

Pemda Kabupaten Sukabumi. (2024). Surat Keputusan Bupati Sukabumi Nomor: 300.2.1/Kep.930-BPPD/2024 Tentang Penetapan Status Tanggap Darurat.

Pemda Provinsi Jawa Barat. (2020). Keputusan Gubernur Jawa Barat Nomor 556/Kep.67-Rek/2020 tentang Kawasan Unesco Global Geopark Ciletuh-Palabuhanratu Kabupaten Sukabumi.

Pratama, Y. P., Samudro, B. R., Soesilo, A. M., Sarungu, J. J., & Irawan, B. B. (2019). Integrasi usahatani dengan pemanfaatan limbah ternak di Desa Sapen, Mojolaban, Sukoharjo. *Jurnal Berdaya Mandiri, 1*(2), 154–169.

Ristroph, E. (2019). Addressing climate change vulnerability in alaska native villages through indigenous community knowledge. *Sociology Study, 9*(1). https://doi.org/10.17265/2159-5526/2019.01.001

Rozaki, Z., Wijaya, O., Rahmawati, N., & Rahayu, L. (2021). Farmers' disaster mitigation strategies in Indonesia. *Reviews in Agricultural Science, 9*, 178–194. https://doi.org/10.7831/ras.9.0_178

Sugihardjo, S., Rusdiyana, E., Lestari, E., Widiyanto, W., Setyowati, R., & Samsita, N. K. B. (2023). Adaptation and mitigation in agrotourism development. *E3s Web of Conferences, 467*, 05008. https://doi.org/10.1051/e3sconf/202346705008

Susanto, J., Zheng, X., Liu, Y., & Wang, C. (2020). The impacts of climate variables and climate-related extreme events on island country's tourism: Evidence from Indonesia. *Journal of Cleaner Production, 276*, Article 124204. https://doi.org/10.1016/j.jclepro.2020.124204

Syakirin, S. (2024). Analysis of drought index with theory of run statistical method in dompu regency. *Journal Transnational Universal Studies, 3*(4), 231–243. https://doi.org/10.58631/jtus.v3i4.94

Tanoamchard, W., & Limjirakan, S. (2012). Impacts of climate change on water resources affecting tourism industry in a small Island: A case study of Koh Chang, Thailand. *International Journal of Environmental Science and Development*, 109–113. https://doi.org/10.7763/ijesd.2012.v3.198

Tesema, D. (2021). Critical linkages: Poverty, livelihoods, institutions, and community development in poverty reduction. *Journal of Economics and Sustainable Development, 12*(5). https://doi.org/10.7176/jesd/12-5-01

Togaymurodov, E., Roman, M., & Prus, P. (2023). Opportunities and directions of development of agritourism: Evidence from samarkand region. *Sustainability, 15*(2), 981. https://doi.org/10.3390/su15020981

Triana, I., & Ruhimat, M. (2018). Understanding of geography teacher towards geopark as learning resources. In *Proceedings of the Annual Civic Education Conference (ACEC 2018)*. Atlantis Press. https://doi.org/10.2991/acec-18.2018.113

UNESCO. (2023). UNESCO Global Geoparks. UNESCO. https://www.unesco.org/en/iggp/geoparks/About

Wang, Y. (2023). Agriculture-tourism integration's impact on agricultural green productivity in China. *Agriculture, 13*(10), 1941. https://doi.org/10.3390/agriculture13101941

Wang, J., Zhou, F., Xie, A., & Shi, J. (2022). Impacts of the integral development of agriculture and tourism on agricultural eco-efficiency: A case study of two river basins in China. *Environment Development and Sustainability, 26*(1), 1701–1730. https://doi.org/10.1007/s10668-022-02781-x

Yuliawati, A., Rofaida, R., & Pamungkas, B. (2021). Empowering Smes through digital marketing at unesco global geopark belitong, Indonesia. *International Journal of Community Service, 1*(3), 286–292. https://doi.org/10.51601/ijcs.v1i3.37

Zuvara, R., Herdiansyah, H., & Asteria, D. (2022). Environmental conservation in ciletuh geopark, west java. *IOP Conference Series Earth and Environmental Science, 1041*(1), 012043. https://doi.org/10.1088/1755-1315/1041/1/012043

Chapter 12
Balancing Socio-economic Livelihoods and Environmental Sustainability in Volcanic Regions: Insights from the Communities of Merapi and Kelud

Rachmini Saparita, Reza Amarta Prayoga, Eko Wahyono, Ulfa Sevia Azni, Lis Purbandini, Siti Fatimah, Nuzul Solekhah, Fatwa Nurul Hakim, and Djoko Puguh Wibowo

Abstract Volcanic regions present unique challenges and opportunities for local communities. This study examines the socio-economic livelihoods and environmental sustainability of communities in the Mount Merapi and Mount Kelud regions, focusing on the interplay between volcanic eruption patterns, community preparedness and livelihood systems. To examine the interactions between environmental pressures and community responses, this study integrates social-ecological

R. Saparita (✉) · R. Amarta Prayoga · E. Wahyono · U. Sevia Azni · L. Purbandini · S. Fatimah · N. Solekhah · F. Nurul Hakim · D. Puguh Wibowo
Research Center for Social Welfare, Village and Connectivity-National Research and Innovation Agency, Jakarta, Indonesia
e-mail: rach001@brin.go.id

R. Amarta Prayoga
e-mail: reza010@brin.go.id

E. Wahyono
e-mail: ekow005@brin.go.id

U. Sevia Azni
e-mail: ulfa007@brin.go.id

L. Purbandini
e-mail: lisp001@brin.go.id

S. Fatimah
e-mail: siti080@brin.go.id

N. Solekhah
e-mail: nuzu001@brin.go.id

F. Nurul Hakim
e-mail: fatw001@brin.go.id

D. Puguh Wibowo
e-mail: djok022@brin.go.id

G. K. Panda et al. (eds.), *Socio-Ecological Crisis of Development*, Sustainability Solutions, https://doi.org/10.1007/978-3-032-22171-1_12

systems theory and a Sustainable Livelihoods Framework through a mixed-methods approach. Key differences in eruption patterns, both frequent small eruptions at Mount Merapi and rare large eruptions at Mount Kelud, are examined, and their respective impacts on evacuation preparedness and community resilience. The socio-economic consequences of unregulated sand mining are also examined. Sand mining is driven by the abundance of volcanic material in the Mount Merapi region. High market demand for agricultural commodities in the Mount Kelud region has led to agricultural land degradation due to the excessive use of chemical fertilisers. These dynamics underscore the interconnectedness of ecological systems and livelihood strategies in volcanic regions. The research findings clearly demonstrate the need for an integrated social-ecological approach to address resource exploitation, promote sustainable development and build community resilience. Lessons learned from the communities of Mount Merapi and Kelud provide valuable insights into balancing socio-economic needs with environmental sustainability in high-risk volcanic regions.

Keywords Resource exploitation · Volcanic regions · Socio-ecological systems · Sustainable livelihood

12.1 Introduction

The volcanic activity and the fertility of agricultural land in surrounding regions are often subjects of debate. During eruptions, volcanic ash and lava flow damages agricultural commodities (Jenkins et al., 2017; Mihai et al., 2023; Troll et al., 2017; Wantim et al., 2018). However, over time, volcanic ash is considered to fertilise the soil (Arnalds, 2013; Mihai et al., 2023; Troll et al., 2017), potentially enhancing crop yields (Zakharikhina et al., 2022) and contributing to food security (Rozaki et al., 2022). Consequently, volcanic ash is regarded as a crucial element in sustaining agricultural activities in regions susceptible to volcanic hazards.

In addition to these debates, the challenges faced by communities living in volcanic regions are also unpredictable eruptions, which can damage infrastructure and settlements (Jenkins et al., 2017; Meredith et al., 2024; Wantim et al., 2018), and threats to public safety (Cuthbertson et al., 2020), socio-economic impacts, such as economic losses (Branca et al., 2015) and long-term recovery (Chester et al., 2015; Huhtamaa et al., 2022). On the other hand, the post-eruption impact that is the availability of sand and volcanic rock that can be exploited for development purposes (construction) thus encouraging excessive extraction (Kumar et al., 2022).

R. Amarta Prayoga
Faculty of Social and Political Sciences, University of Indonesia, Depok, Indonesia

N. Solekhah
Department of Anthropology, Gadjah Mada University, Yogyakarta, Indonesia

Managing disaster risks, preventing environmental degradation and maximising potential benefits require integrating the social-ecological systems (SES) approach with the Sustainable Livelihoods Framework (SLF). This integration enhances understanding of human-ecological interactions in sustaining livelihoods (2024b; Mozumder et al., 2024a; Shackleton, 2024), and emphasises resilience across social and ecological dimensions to support adaptive management strategies (Liu et al., 2024). The SES approach examines social-ecological interactions to identify vulnerabilities, resilience factors and strategies for balancing economic benefits with environmental conservation (Berrouet et al., 2018; Itzkin et al., 2021), focusing on dynamic interdependencies (Kumar et al., 2024). Meanwhile, the SLF assesses the role of assets (natural, financial, human, social and physical) in shaping livelihood outcomes amidst external shocks (Alexander, 2017; Lin & Chang, 2013), and emphasises livelihood diversification for resilience (Li et al., 2023).

Sustaining rural livelihoods requires governance structures that strengthen resilience through social, economic and environmental integration (Hossain et al., 2023; Liu et al., 2024). Holistic resource management (through stakeholder engagement, monitoring, scenario planning and adaptive strategies) is crucial due to the interdependence between ecological health and community well-being (Kolahi et al., 2024). While the integration of SES and SLF provides a comprehensive approach to sustainability, challenges persist in balancing ecological integrity with socio-economic development. This balance requires ongoing research and adaptive management strategies.

The present study is unique in its adoption of a comparative socio-ecological approach, which is utilised to examine two volcanic regions (Mount Merapi and Mount Kelud) that experienced major eruptions in the same period but exhibited divergent patterns of livelihood recovery and environmental management. Although prior studies have addressed disaster recovery and environmental management separately (Blaikie et al., 2014; Bryant & Wilson, 1998; Gupta, 1995; Quarantelli, 1999; Radcliffe et al., 2010), the present study is distinctive in its comprehensive approach. The novelty of this study lies in its integration of multiple theoretical frameworks, including social-ecological systems theory and Sustainable Livelihoods Frameworks. This integration facilitates a more holistic understanding of the interactions between disaster risk, resource use and community resilience.

This chapter examines the relationship between volcanic eruption patterns and community disaster preparedness, as well as the relationship between ecological-socio-economic systems and community livelihoods in volcanic areas. The aim of this research is to examine the post-disaster livelihood strategies of communities in the volcanic areas of Merapi and Kelud, which have different eruption patterns. The research also analyses the environmental impacts of local economic activities and their impact on long-term sustainability.

12.2 Rationale of the Study

Communities living near volcanic hazard zones often rely on complex livelihood systems. These systems can form the basis for designing disaster management strategies that not only respond to hazards but also strengthen local resilience and sustain daily livelihoods in the long term (Saparita et al., 2024). The concept of socio-ecological sustainability in volcanic hazard-prone areas is important because it reflects how geological, ecological, social and economic processes interact and are interdependent (Cosme et al., 2023).

Integrating a social-ecological systems (SES) perspective with the Sustainable Livelihoods Framework (SLF) offers a way to shape policies that link ecological priorities to community economic resilience. This is particularly relevant in situations where environmental change directly impacts household well-being (Itzkin et al., 2021; Takeuchi et al., 2016). An ecosystem-based approach to disaster risk reduction that links environmental restoration with livelihood enhancement can further support efforts to enhance community resilience.

This study focuses on Mount Merapi and Mount Kelud, two volcanoes that exhibit distinct eruption patterns. The differences between the two provide opportunities for comparative analysis and help deepen understanding of how communities adapt to recurring volcanic disturbances. Overall, this research adds to the discussion on managing volcanic landscapes through a balance of social, economic and environmental considerations.

12.3 Material and Methods

The research was conducted in the areas of Mount Merapi and Mount Kelud (Fig. 12.1). These locations were chosen because of the differences in eruption patterns. Mount Merapi is known for its frequent eruptions, while Mount Kelud has large eruptions, although they are rare. The study locations included villages located in high-risk hazard zones (within a 10 km radius of the volcano's peak) in both areas. From each region, three villages were selected as research samples from Mount Merapi region, Balerante Village, Kemalang District, Klaten Regency; Kemiren Village, Srumbung District, Magelang Regency and Klakah Village, Selo District, Boyolali Regency were selected. Three villages in Mt. Kelud region, namely: Kebonrejo Village, Kepung District, Kediri Regency; Pandansari Village, Ngantang District, Malang Regency and Pondok Agung Village, Kasembon District, Malang Regency. The units of analysis are households (as the main social actors) and ecological systems (agricultural land and surrounding environment). The stages of the research carried out are as shown in Fig. 12.2.

The research problem is defined by the tension between the economic needs of the community and environmental sustainability in areas prone to volcanic disasters, as evidenced by the examples of Merapi and Kelud. A comprehensive literature review

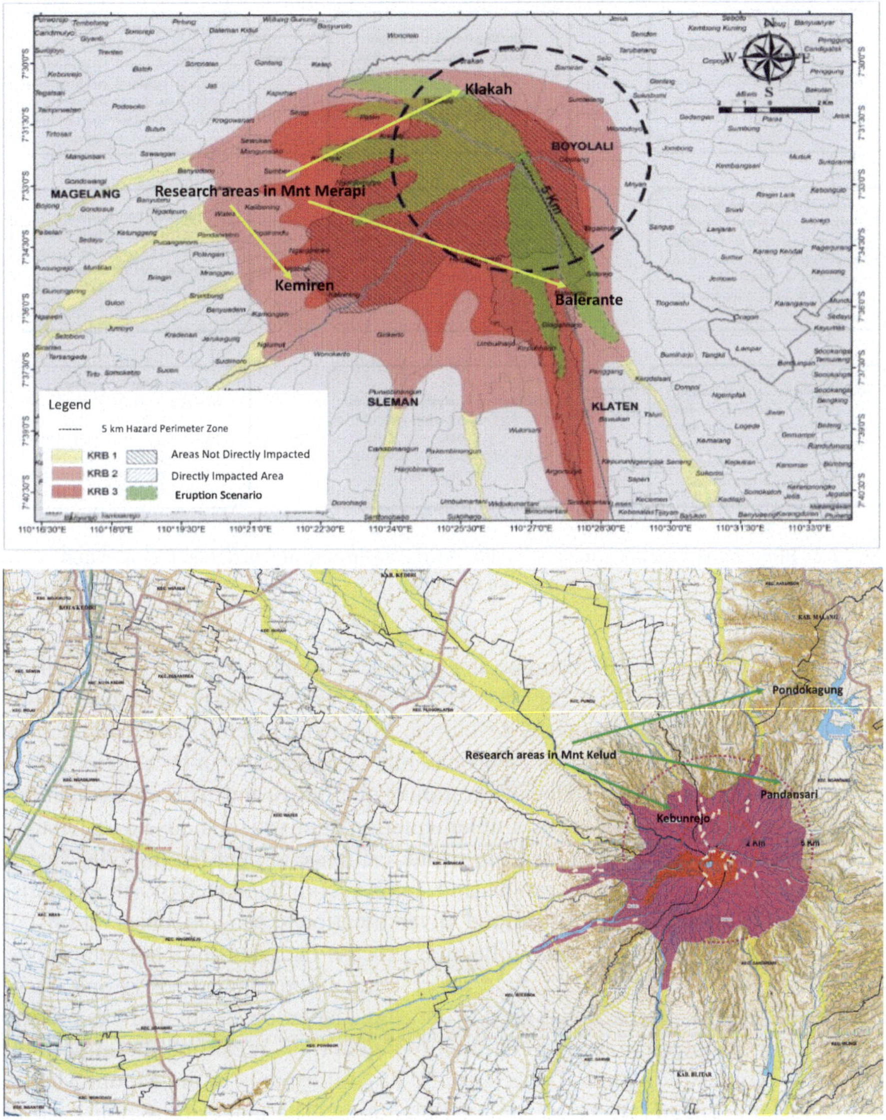

Fig. 12.1 Research areas. Caption: **a** Mount Merapi region; **b** Mount Kelud region. *Source* **a** The Governor of Central Java (2018); **b** National Disaster Management Agency (2014)

was conducted to examine the theory of social-ecological systems, the Sustainable Livelihoods Framework and previous studies on the impact of volcanic disasters on livelihoods and the environment. The subsequent phase of the study will entail the formulation of research questions. These inquiries will center on the post-disaster livelihood patterns in Merapi and Kelud, with a view to elucidate their evolution.

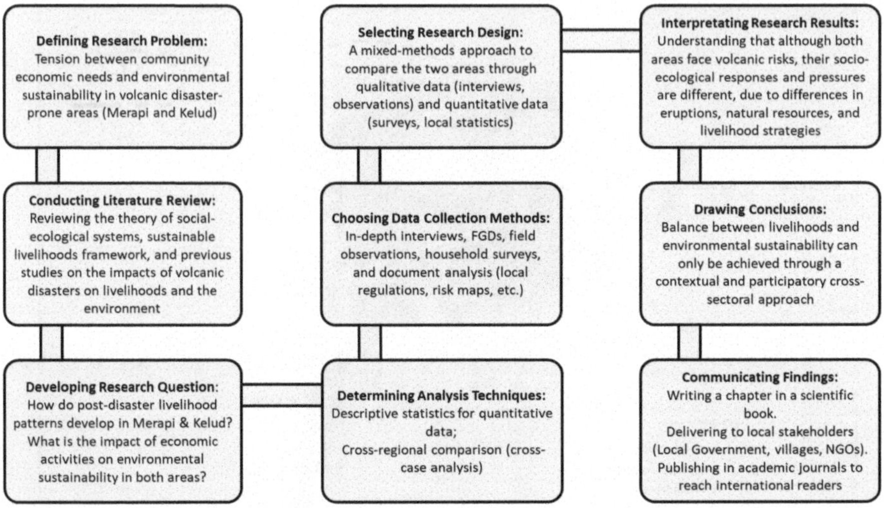

Fig. 12.2 Flowchart of socio-economic livelihood and environmental sustainability research in two volcanic areas. *Source* Research Team's own thoughts

Furthermore, the study will examine the impact of economic activity on environmental sustainability in these regions. The research questions proposed are: How do post-disaster livelihood patterns develop in Merapi and Kelud? and What are the impacts of economic activities on environmental sustainability in both areas?

The research design entailed the integration of qualitative and quantitative methodologies. Specifically, it involved the implementation of mixed methods, to understand the dynamics of community vulnerability and resilience in their livelihood systems, encompassing interviews, observations and surveys, complemented by village statistics. The data collected are livelihood diversification data, and observations of degraded environments. The data collection process included in-depth interviews, focus group discussions (FGDs), field observations, household surveys and document analysis. This approach was used to develop adaptation strategies by exploring socio-ecological interactions and community perceptions regarding the impacts of natural resource exploitation. There were 30 key informants in the Mount Merapi area and 21 informants from the Mount Kelud area. The latter group consisted of community leaders, farmers, miners and local village officials. Focus group discussions (FGDs) were conducted with local stakeholders (farmers' groups, women, and youth) to ascertain their perspectives on resource management and environmental risks. A total of six focused group discussions (FGDs) were conducted, with three FGDs conducted in each area. Field observations were also conducted to directly assess the condition of the ecosystem and the economic activities of the community. Structured surveys were conducted using stratified random sampling to ensure community representation. The survey data were collected from 90 respondents in the Mount Merapi area and 126 respondents in the Mount Kelud area. Document analysis encompasses a review of local regulations, risk maps and eruption maps of the two volcanoes (Merapi and Kelud). The research period extends from 2023 to 2024.

A comparison was made between the two research areas in terms of ecological vulnerability, defined as the level of environmental damage due to human activities (mining and agriculture), and social vulnerability, and defined as the economic and social impacts of livelihood systems. Additionally, risk perception was examined, defined as community understanding of disaster threats and economic sustainability. The analysis technique employed was a combination of descriptive statistics for quantitative data, while qualitative data processing was carried out using thematic analysis (Braun & Clarke, 2006) to identify narrative patterns and deep meanings from interviews and FGDs. Quantitative data validation was conducted by comparing secondary data sources, including village statistics and government reports, as well as community information on mining activities and environmental degradation. Qualitative data validation was conducted by cross-checking interviews with community members and key informants.

This study uses the Social-Ecological Systems (SES) approach and Sustainable Livelihoods Framework (SLF). SES is used to examine the interactions between communities and the environment in the face of external pressures, such as mining exploitation and changing livelihoods (Berrouet et al., 2018; Itzkin et al., 2021), while the Sustainable Livelihoods Framework (SLF) is a tool used to assess the capacity of assets to serve as a bulwark for community livelihoods and resilience (Alexander, 2017; Lin & Chang, 2013). The findings were disseminated to local stakeholders (i.e., local government entities, village representatives and non-governmental organizations [NGOs]) through public dialogue and published in scientific manuscripts in this book chapter to reach international readers.

12.4 Results and Discussion

12.4.1 Volcanic Eruption Patterns and Community Preparedness for Disasters

Living in the area of highly active volcanoes (Mount Merapi and Mount Kelud) faces the consequences of unexpected eruption disasters. Based on various references (Hidayati et al., 2019; Ministry of Energy & Mineral Resources, 2024; Nawiyanto & Sasmita, 2019; Nihayatul et al., 2017; Rinawati et al., 2018; Thouret et al., 1998; Voight et al., 2000), the two volcanoes (Merapi and Kelud) have different eruption patterns and magnitudes. The magnitude of the eruption is recorded through VEI (Volcanic Explosivity Index) data, which is a measure of how explosive a volcanic eruption is with a scale from 0 (non-explosive) to 8 (the most powerful in the world) (National Park Service, 2022). The magnitude of the VEI on the volcanoes (Merapi and Kelud) can be seen in Fig. 12.2 (Mount Merapi in Blue and Mount Kelud in

Orange). The eruptions of Mount Merapi occurred more frequently and consistently compared to the eruptions of Mount Kelud throughout the twentieth and twenty-first centuries, with a VEI ranging from 1 to 4 (Fig. 12.3). The largest eruption of Mount Merapi during the period 1900–2025 occurred in 2010 with a VEI of 4.

Mount Kelud has been known to erupt on various occasions throughout history, including in 1900, 1920, 1922, 1966, 1990, 2007 and 2014 (Hidayati et al., 2019; Nawiyanto & Sasmita, 2019; Nihayatul et al., 2017; Thouret et al., 1998). Mount Kelud eruptions are less frequent but often more explosive. In 2014, the largest eruption on record was characterised by high explosivity, with a VEI of 4 (Hidayati et al., 2019), leading to significant crop failure and substantial economic losses for the community (FGD, May 2023). The eruptions in the Mount Merapi region in 2010 and Mount Kelud in 2014 caused extensive damage to village facilities and infrastructure, houses, agricultural crops and livestock deaths (FGD, March and May 2023).

The difference in frequency and magnitude of eruptions between Mount Merapi and Mount Kelud leads to differences in community preparedness in the two areas. The basis for preparedness is the existence of evacuation facilities, including designated locations and evaluation routes (Patrisina et al., 2018). Additionally, the community's preparedness is enhanced through training programmes (Kako et al., 2020). The time spent at the evacuation site (Fig. 12.4), the knowledge of rescue routes (Fig. 12.5) and the selection of refuge locations vary. The community on Mount Kelud immediately selects the final evacuation site provided by the government (Fig. 12.6). As demonstrated by Figs. 12.4, 12.5 and 12.6, the Mount Kelud community is better prepared to face an eruption disaster. This increased level of preparedness can be attributed, at least in part, to the community's experience with the eruption of Mount Merapi in 2023 (FGD, May 2023).

The discrepancy in the frequency of eruptions has prompted the local government in the vicinity of Mount Merapi to undertake disaster risk reduction training with

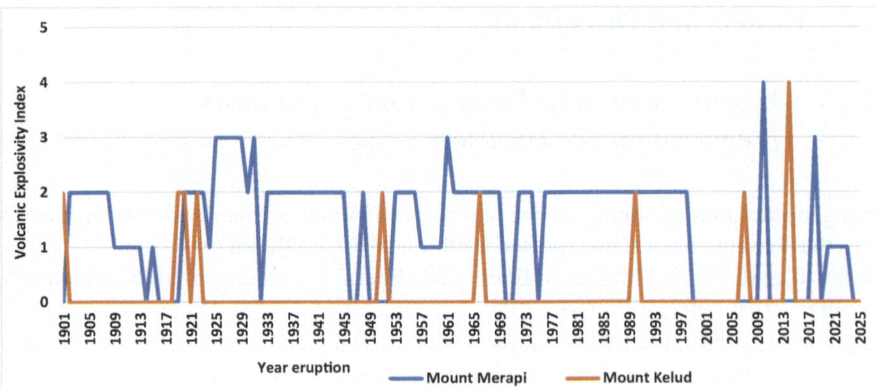

Fig. 12.3 Eruption level of Mount Merapi and Mount Kelud for the period 1900–2025. *Source* Voight et al. (2000), Nihayatul et al. (2017), Hidayati et al. (2019), Nawiyanto and Sasmita (2019), Thouret et al. (1998), Rinawati et al. (2018), Ministry of Energy and Mineral Resources (2024)

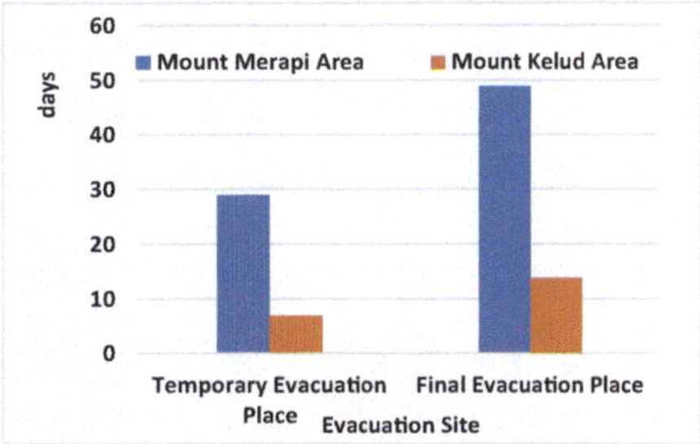

Fig. 12.4 Length of stay at evacuation site

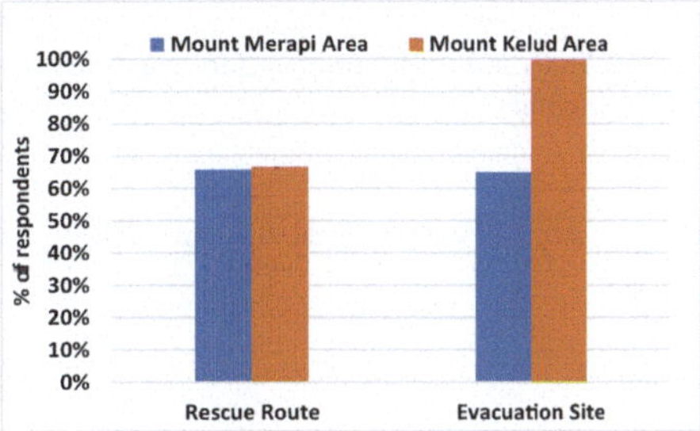

Fig. 12.5 Knowledge of the rescue path. *Data source* Original research data

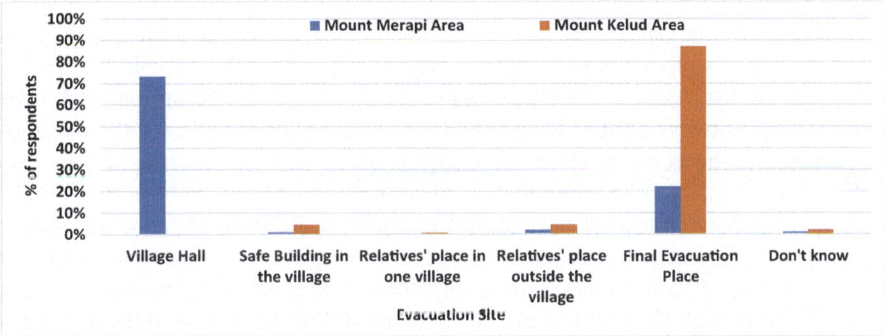

Fig. 12.6 Evacuation site. *Data source* Original research data

greater frequency in comparison to the region surrounding Mount Kelud (Fig. 12.7). Community training is implemented as a strategy to mitigate vulnerability within the community and enhance its capacity and preparedness in dealing with disasters. The regularity of training sessions has been found to encourage community members to volunteer for disaster risk reduction teams (Fig. 12.6). It is imperative to recognise that disaster risk reduction training constitutes a pivotal component of the central government's sustainable development programme, which is incumbent upon local governments to implement through the utilisation of knowledge and innovation to foster a culture of safety and resilience within communities (Noor et al., 2023; Pal

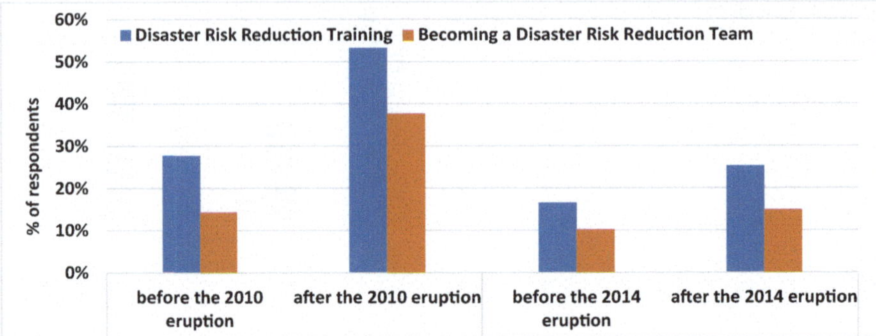

Fig. 12.7 Community preparedness as part of the risk reduction team. *Data source* Original research data

et al., 2021; Russo & Rindone, 2024; Sakurai & Sato, 2016; Tiepolo & Braccio, 2020).

12.4.2 Community Livelihood System

A comparative analysis of the two volcanic regions reveals notable parallels in the characteristics of livelihood systems, particularly within the agricultural sector, encompassing livestock, as illustrated in Fig. 12.8. The frequent eruptions of Mount Merapi, which always emit volcanic sand, are a source of income for the community, especially the younger generation, who work as sand miners (Fig. 12.8). This analysis encompasses both pre- and post-major eruption periods in both areas, highlighting the predominance of agricultural land use in the form of fields and yards. A salient transformation has been observed in the types of commodities produced by the agricultural sector, marked by a shift from food crops to horticulture as cash crop, driven by high market demand, as depicted in Fig. 12.9. A notable characteristic is the limited educational attainment of the population, which is predominantly elementary school graduates. This constraint impacts the adoption of technology, such as fertilisers and chemical pesticides, which is primarily driven by short-term productivity concerns (in-depth interviews with community leaders in Kebonrejo Village, November 2024).

The grass commodity in the Mount Merapi region, especially in Balerante Village (Fig. 12.9) is quite prominent after the major eruption (2010). Grass grows on empty land for feed because of the large number of livestock (cattle) as a community farming business. Pandansari Village still grows rice because it has an abundant water source from the Selo Reservoir, while other villages have stopped growing rice due to limited water resources and have switched to horticulture as a cash crop to meet market demand (Fig. 12.8).

Fig. 12.8 Livelihoods of communities in volcanic mount regions. *Data source* Original research data

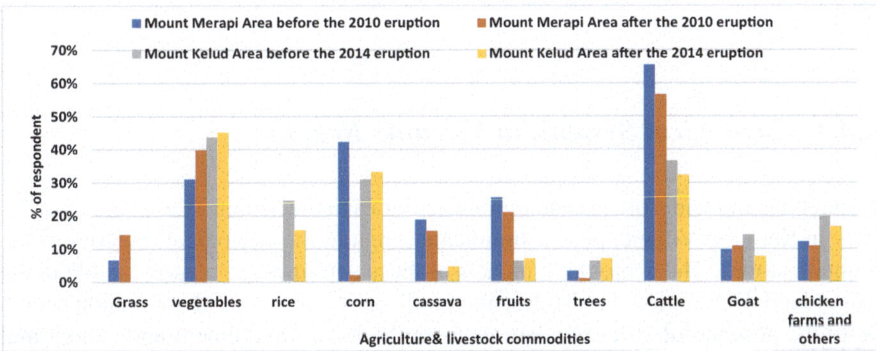

Fig. 12.9 Agricultural and livestock commodities in the volcanic mount regions. *Data source* Original research data

In the Mount Merapi region, frequent eruptions (both large and small) result in the consistent presence of volcanic sand, thereby ensuring an abundant supply of sand from cold lava floods in the rivers that flow into residential regions. This sand mining has become a source of transitional livelihood for the younger generation, who engage in the practice of sand mining. It has been reported that companies engaging in substantial sand mining operations are predominantly external to the region, with some even being based in the capital city itself (in-depth interviews with community leaders in Kemiren Village, September 2024). Concurrent with the occurrence of rare eruptions on Mount Kelud, the agricultural sector has experienced notable fluctuations in the region.

The 2010 Merapi and 2014 Kelud eruptions, both of which were categorised as VEI 4, had a significant impact on the populations of the affected areas. Eruptions represent a potential threat to the livelihoods of the population. The recurrent eruptions in the vicinity of Mount Merapi pose a dual threat and opportunity for the local populace. The volcanic material ejected during the 2010 Merapi and 2014 Kelud eruptions spread extensively, covering an area of approximately 10 km within the immediate vicinity of the villages. The volcanic sand and rocks have become a source of livelihood for some people, providing additional income, and in some cases, even serving as the primary source of income for young people in many villages in Mount Merapi Regions (FGD, March 2023; in-depth interviews with community leaders in Kemiren Village, September 2024).

The financial benefits of volcanic eruptions in terms of sand supply have been proven (Miller, 2022), and the potential for livelihood generation is significant (Farahani & Bayazidi, 2018), particularly given the high market demand for sand for infrastructure development (Shitima & Suykens, 2023; Tran et al., 2023; UNEP, 2023). However, the process of sand mining carries inherent risks, particularly in cases of uncontrolled and excessive extraction, which has been shown to result in environmental degradation on a global scale (Arjun et al., 2023; Rentier & Cammeraat, 2022).

12.4.3 Ecological Systems in Volcanic Regions

The ongoing and widespread sand mining activities in the Mount Merapi area, particularly in Kemiren Village, pose a significant threat to the ecological integrity of the mountain slopes. The ecological impacts of these activities are clearly visible in the form of land degradation, landslides, and biodiversity loss. This phenomenon occurs due to the absence of strict regulations from the local government and suboptimal regulations at the village level, resulting in rampant illegal mining activities (In-depth interviews with community leaders, September 2024). These activities are facilitated by foreign investors using large-scale equipment, while local residents act only as sand dredging operators and mechanics.

In the Mount Kelud region, sand mining can also be seen, but it is very limited (Fig. 12.8), yet the implementation of village regulations that restrict mining equipment to manual devices has resulted in a relatively lower rate of sand dredging and land degradation. Uncontrolled and excessive extraction poses significant environmental and socio-economic risks (Arjun et al., 2023). Unregulated sand mining often leads to illegal mining activities, which can result in conflicts, loss of government revenue and exploitation of workers. This illegal trade can destroy local economies and governance (Filho et al., 2021; Gavriletea, 2017).

The high market demand for agricultural commodities also triggers the use of chemical fertilisers and pesticides, which in the long term have the potential to damage the fertility of agricultural land (Mashamaite et al., 2024; Wang et al., 2019; Yan et al., 2019). In Kebonrejo Village, Gunung Kelud, the threat of long-term environmental problems may occur. The high demand for agricultural products, which is exacerbated by the widespread use of chemical fertilisers and pesticides, is driven by financial pressures faced by farmers, who are increasingly dependent on bank loans to finance their operations (In-depth interviews with community leaders, November 2024). Increased loans to financial institutions (banks) have the potential to weaken sustainable livelihood systems for farmers (Mozumder et al., 2024a, 2024b). However, it is important to note that consumer behaviour can have an impact on the characteristics of agrarian communities, due to the shift from local markets to industrial markets (Dolin, 2016), vulnerability to environmental and economic changes (Jackson, 2021), consumer preferences and market power (Posadinu et al., 2022).

The results of field investigations indicate that vulnerability in the community is attributable to its own actions. Capital loans from financial institutions intended for productive endeavours are frequently allocated for consumer purposes, such as acquiring vehicles and renovating residences. Effective management of production capital loans can foster a sustainable livelihood system (He & Li, 2024). Conversely, if these loans are utilised for consumption, they have the potential to become a chronic debt trap, thereby disrupting the sustainable livelihood system.

The phenomenon of ecological vulnerability can arise when a given ecological system is no longer capable of withstanding threats that have the potential to hinder human social life (Micheli et al., 2024). Communities encounter risks or challenges to their social welfare and social conflict due to social vulnerability. Ecological vulnerability has been compounded by the excessive utilisation of chemical fertilisers and pesticides, while social vulnerability has emerged as a consequence of social status and prestige.

Apart from the threat of eruptions, the most significant threat to livelihoods is the market system. The most salient of these is the inconsistency of commodity prices, which are notably low during harvests and significantly high during times of non-harvest. The high cost of fertilisers and medicines for plants, coupled with the difficulty of obtaining subsidised (chemical) fertilisers, further exacerbates the situation (FGD, March & May 2023). Furthermore, the community perceives that certain regulatory frameworks appear to be less favourable towards farmers, as they impede access to subsidised fertilisers.

Ecological systems have the capacity to withstand these threats without losing their capacity to support human life, provided that adaptation and strategies are carried out to balance ecological conditions (Berrouet et al., 2018). Formulation of policies that promote socio-ecological sustainability can play a pivotal role in ensuring environmental sustainability, particularly in the aftermath of disasters that result in community welfare shocks (Escolà-Gascón et al., 2023).

The restoration of community livelihoods in the two areas (Merapi and Kelud) was found to be significantly influenced by the presence of five assets. These five assets/capitals (human, natural, financial, physical and social) have been identified as crucial factors in the resilience of communities in the face of adversity (Department for International Development, 1999; Ellis, 2000; Serrat, 2008). These assets functioned as a safeguard against the repercussions of the volcanic eruption, thereby contributing to the community's resilience. The community has come to recognise the potential of these assets to serve as collateral for financial loans. This loan facilitates the

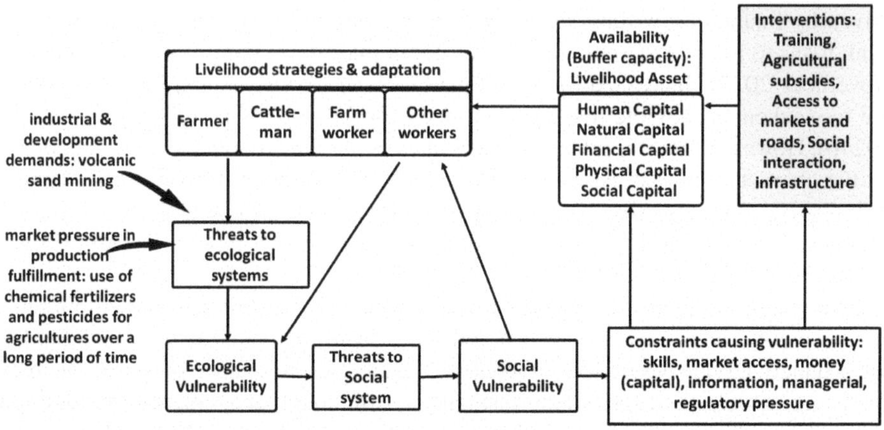

Fig. 12.10 Community livelihood system and its ecological system in the Volcanic region. *Source* Adopting explanatory thinking (Berrouet et al., 2018; Franco-Gaviria et al., 2022; Kadetz & Mock, 2018; Lecegui et al., 2022)

restoration of their source of income. The loan, designated as a financial asset, is utilised for the purpose of land rehabilitation and revitalisation following the volcanic eruption. The utilisation of loans has been demonstrated to enhance socio-economic resilience within the community (Mozumder, et al., Bridging the gap: enhancing socio-ecological resilience by breaking the debt cycle among small-scale hilsa fishers in Bangladesh, 2024). Referring to the explanations of Berrouet et al. (2018), Lecegui et al. (2022) and Kadetz and Mock (2018), examining the conditions in the two selected areas, the framework of the community's livelihood system and its ecological system in the volcano region is shown in Fig. 12.10.

It is clear that communities residing in areas susceptible to volcanic disasters invariably endeavour to sustain their livelihoods (Sarrazin et al., 2019). Within the context of the Mount Merapi and Mount Kelud regions, livelihoods are predominantly derived from the agricultural sector. This is a necessary strategy for socio-ecological sustainability, in order to exploit resources in their living environment (Ágústsdóttir, 2015; Escolà-Gascón et al., 2023; Saparita et al., 2024; Sarrazin et al., 2019).

The negative impacts of uncontrolled volcanic sand exploitation and soil degradation due to excessive use of chemical fertilisers indicate the need for an integrative approach. To formulate more effective strategies to support sustainable livelihoods in high-risk disaster areas, communities must adapt to volcanic eruption patterns (Andreastuti et al., 2023; Prayoga et al., 2024), and must be able to manage available resources through agricultural diversification as described (Purbandini et al., 2024; Saparita et al., 2024).

The agricultural livelihood system in the Mount Kelud area has shown signs of regeneration, with many young farmers continuing their parents' professions. Young people remain in the agricultural sector because they can earn a decent income. This is not the case in Kemiren, Mount Merapi. Because Mount Kelud is difficult to access from rivers carrying volcanic material, sand mining has not developed. Furthermore, village regulations restrict sand dredging in the Mount Kelud area.

Volcanic eruption patterns have been shown to significantly influence community responses and resilience development. Findings indicate that frequent eruptions of Mount Merapi have encouraged the normalisation of adaptive preparedness measures in affected communities. This has resulted in the development of community resilience, manifested through livelihood strategies and the implementation of organised evacuation protocols. In contrast, Mount Kelud exhibits a pattern of sporadic but substantial eruptions, resulting in cycles of agricultural expansion followed by sudden shocks. Communities respond by implementing intensive agricultural practices and market-based production strategies to mitigate the impacts of these natural disasters. This approach facilitates the assessment of how socio-economic livelihoods evolve under varying ecological pressures. Adaptation of livelihood systems to the use of chemical fertilisers and the cultivation of high-yielding crops has led to soil degradation, raising concerns about long-term sustainability (Itzkin et al., 2021).

Extensive and unregulated sand mining around Merapi has become a significant source of income. This activity has also resulted in substantial ecological problems, including erosion, deforestation and riverbed alteration. The absence of effective policy enforcement and governance mechanisms involving communities has exacerbated unsustainable practices in both areas. This phenomenon illustrates the tension within communities between immediate economic benefits and long-term environmental integrity (Miller, 2022). These findings are consistent with the proposition of pathways towards integrated socio-ecological resilience in volcanic areas, particularly in the context of local development planning.

12.4.4 Integration of Ecological–Socio-economic with Community Livelihood Systems

The integration of social, economic and physical elements in disaster risk management in areas prone to volcanic activity is a complex undertaking (Donovan, 2022). It is imperative for communities in such areas to be prepared with strategies to mitigate the adverse effects of volcanic activity, including the impact of volcanic ash (Saparita et al., 2024).

A re-examination of the dynamics that occur in the community (Fig. 12.10) is therefore essential, so that effective interventions can be proposed to improve socio-ecological sustainability amidst the uncertainty of volcanic hazards and their impacts on the community. The proposed interventions form part of effective ecosystem management to reduce vulnerability, increase resilience and support environmental recovery and stability in areas that still depend on natural resources (Ágústsdóttir, 2015). Interventions through improving rescue infrastructure, such as adequate evacuation routes and places and reforestation in forest areas, must be pursued as a buffer for the livelihood system while addressing the impact of the ecosystem on the livelihoods of local communities (Buot & Buot, 2022). The demonstration of community resilience through adaptive capacity, solidarity networks and resource diversification can serve as an effective measure of disaster preparedness and management in the face of persistent volcanic threats (Thouret et al., 2022).

The establishment of sustainable community livelihood systems necessitates the commitment of all involved parties. The influence of government policies on environmental sustainability on socio-ecological sustainability factors is significant, particularly in the aftermath of disasters that have an impact on community welfare shocks (Escolà-Gascón et al., 2023). However, in practice, the implementation of livelihood systems is often confronted by numerous challenges. These include a lack of knowledge regarding the utilisation of environmentally sustainable agricultural technologies (Guma et al., 2022; Maharjan & Issahaku, 2014), financial constraints (in the form of capital for agricultural activities) (Guma et al., 2022) and the aspiration to enhance social status through the acquisition of consumer goods (Maharjan & Issahaku, 2014), including the complexity of government regulations in obtaining capital assistance and fertiliser subsidies that are often misdirected. These challenges can hinder or disrupt sustainable livelihood systems that can meet decent needs.

Referring to the work of Datta et al. (2018), it is imperative that life needs and assets are interconnected, fostering a symbiotic relationship. Life needs must be met by the availability of assets. To address this imbalance, it is essential to implement interventions that span the entire lifecycle of the community, from the identification of activities to the execution of interventions (Datta et al., 2018). These interventions must be grounded in a comprehensive understanding of community needs and assets, serving as a foundation for the formulation of strategies and adaptations to the livelihood system. This approach, as illustrated in Fig. 12.11, is instrumental in ensuring the long-term sustainability of the community's livelihood system.

In order to enhance ecological and social resilience, it is imperative to cultivate continuous encouragement of adaptive capacity and resistance within the community through training initiatives (Kako et al., 2020; Zhang et al., 2023). The governance framework must be adapted to recognise the pivotal role of rural residents in enhancing resilience (Liu et al., 2024). Community resilience in socio-ecological systems is influenced by sustainable resource management; therefore, land management must align with sustainable development goals (Li et al., 2024). Furthermore, the integration of local knowledge into the livelihood resilience framework can enhance preparedness for ongoing environmental disturbances, such as volcanic eruptions and other threats (Caviedes et al., 2024). The integration of social and ecological resilience supporters is therefore pivotal in enhancing livelihood resilience through the implementation of informed management strategies that are underpinned by an adaptive framework (Franco-Gaviria et al., 2022).

These findings are consistent with the socio-ecological crisis of development (Liodakis, 2013), and the tensions between livelihoods and the environment in the Mount Merapi and Mount Kelud areas (Suša, 2019). In both contexts, the development trajectories shaped by post-disaster recovery have led to intensive resource

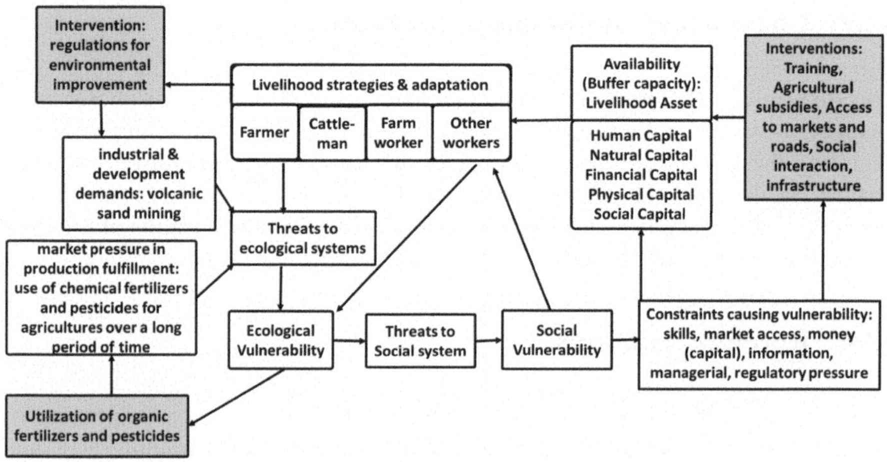

Fig. 12.11 Integration of social-ecological systems and sustainable livelihoods. *Source* Adopting explanatory thinking (Berrouet et al., 2018; Franco-Gaviria et al., 2022; Kadetz & Mock, 2018; Lecegui et al., 2022)

extraction (sand mining, chemical-dependent agriculture), which exacerbates environmental degradation and threatens long-term sustainability. These patterns reflect a classic socio-ecological crisis, where economic survival strategies collide with ecological integrity. When situated within the paradigm of a socio-ecological crisis, this research endeavours to elucidate the mechanisms through which local communities cultivate resilience and pursue sustainable solutions in the face of environmental disruptions and development pressures. The phenomenon of socio-ecological crises occurring in disaster-prone areas is a critical component of the socio-ecological crisis of development on community resilience and its sustainable solutions. The findings of this study illustrate how local resilience strategies, when bolstered by inclusive governance and sustainable resource management, can provide context-specific pathways to sustainable development.

The text elucidates elements of community resilience, including adaptive agricultural practices, collective disaster preparedness and locally rooted ecological knowledge that mitigate crises. A comparative analysis of the two sites provides practical and policy-relevant lessons for the development of sustainable solutions that integrate environmental management with livelihood security. Therefore, this study contributes to the exploration of how communities cope with the dual pressures of ecological vulnerability and development demands. This study is important for academics and policymakers seeking to design effective and context-sensitive strategies to enhance resilience in other disaster-prone areas.

This chapter significantly contributes to the international discourse on disaster risk reduction, climate adaptation and socio-ecological resilience. This is achieved by establishing a link between global issues of sustainable development (Wen et al., 2023) and local adaptation practices in volcanic regions.

12.5 Challenges/Limitations of the Study

This study did not explore specific ecological data, such as post-eruption microhabitat changes, as well as more specific socio-economic data. This can be an obstacle in analysing the relationship between ecological factors and community livelihoods further. The diversity of livelihood systems and the level of community preparedness can vary within a region, which can affect the accuracy of generalisation of findings.

12.6 Recommendations

Communities in the Mount Merapi and Mount Kelud areas need to improve their disaster preparedness capacity through local experience-based training and integration of easily accessible early warning system technology, as the unpredictable nature of volcanic activity requires continuous monitoring and preparedness. Effective disaster preparedness is essential to help mitigate the impact of large numbers of casualties and ensure that evacuation efforts are carried out smoothly given the large rural population in the volcanic areas. Some communities in these areas may have limited access to the information, training and resources needed. Building local capacity can reduce confusion during an eruption.

The devastating events of past eruptions are the reason for the need for a solid disaster preparedness strategy. Preparedness includes not only immediate response but also long-term recovery plans to restore affected ecosystems. Climate change also adds urgency to improving disaster preparedness, ensuring that communities are resilient not only to volcanic disasters but also to climate-related disasters such as floods and landslides.

Environmentally based economic diversification, such as agroforestry, community-based ecotourism and post-eruption land use for sustainable agriculture, needs to be developed more intensively by local governments and stakeholders in volcanic disaster-prone areas.

Volcanic eruption patterns and community adaptation, integrating social, economic and environmental aspects, can form the basis for formulating disaster management policies in volcanic eruption-prone areas.

This governance model can be implemented through regional policies based on community participation.

The government, academics, communities and the private sector need to work together to reduce disaster risk through sustainable natural resource management for post-eruption economic recovery.

12.7 Conclusions

Field data analysis reveals a clear interaction between volcanic eruption patterns and community preparedness. This interaction is further strengthened by livelihood systems, which play a crucial role in fostering socio-economic resilience and environmental sustainability.

The unique characteristics of Mount Merapi and Mount Kelud eruptions have implications for evacuation strategies and community resilience. In addition, resource exploitation, such as illegal sand mining in the Mount Merapi area and market-based agricultural pressures in the Mount Kelud area, exacerbate environmental degradation, particularly through the excessive use of chemical fertilizers.

This study identifies elements of community resilience, including adaptive agricultural practices, collective disaster preparedness and locally rooted ecological knowledge that mitigate crises. A comparative analysis of the two locations provides practical and policy-relevant insights for the development of sustainable solutions that integrate environmental management with livelihood security. Consequently, this study makes a direct contribution to the exploration of how communities navigate the dual pressures of ecological vulnerability and development demands.

The findings underscore the necessity of an integrated socio-ecological approach to managing resource exploitation, promoting sustainable development and fortifying community resilience. The insights garnered from communities in both areas offer valuable lessons regarding the balancing of socio-economic needs with environmental sustainability in high-risk volcanic areas.

This study offers transferable insights with global relevance, as many communities around the world face increasing environmental hazards due to climate change. Merapi and Kelud's experiences offer a valuable reference for navigating the complex interactions between coping strategies and ecological constraints. This research aligns with global priorities outlined in the Sendai Framework for disaster risk reduction, the Sustainable Development Goals (especially Goals 1, 11, and 13) and adaptation frameworks that call for community-based, integrated and ecologically sensitive approaches to building resilience. This study advances a social-ecological perspective grounded in real-world community experiences, contributing to the international debate on aligning disaster risk governance and development in the face of increasingly complex crises.

Acknowledgements This research activity is supported through RIIM Competitions funding from the Indonesia Endowment Fund for Education Agency, Ministry of Finance of the Republic of Indonesia and National Research and Innovation Agency of Indonesia according to the contract number: B-5768/III.12/HK.01/12/2022. We also thanks to the Research Center for Social Welfare, Village and Connectivity - Research Organization for Governance, Economy and Community Welfare National Research and Innovation Agency for supporting this research activities.

AI Discloser This study utilised artificial intelligence (AI) assistance in the preparation of certain sections of the manuscript. The AI tool used was OpenAI's ChatGPT (GPT-4o model), accessed on 2 July 2025, via the web interface at https://chat.openai.com. The tool was employed to support the refinement of the research questions, objectives and conclusion, particularly in clarifying the logical connections between discussion points and the main aims of the study.

The following prompt was used:

Strengthen the connection between the discussion points and the main objectives of the study.

The content generated by the AI was critically reviewed, edited and integrated by the author to ensure scholarly accuracy, relevance and originality. No AI tool was used to generate original data, analyse results or substitute human interpretation.

References

Ágústsdóttir, A. M. (2015). Ecosystem approach for natural hazard mitigation of volcanic tephra in Iceland: Building resilience and sustainability. *Natural Hazards, 78*(3), 1669–1691. https://doi.org/10.1007/S11069-015-1795-6

Alexander, B. (2017). Development and livelihoods for disaster risk reduction including climate change adaptation. In I. Kelman, J. Mercer, & J. Gaillard (Eds.), *The routledge handbook of disaster risk reduction including climate change adaptation* (1st ed., pp. 90–102). Routledge. https://doi.org/10.4324/9781315684260

Andreastuti, S. D., Paripurno, E. T., Subandriyo, S., Syahbana, D. K., & Prayoga, A. S. (2023). Volcano disaster risk management during crisis: implementation of risk communication in Indonesia. *Journal of Applied Volcanology, 12*(1). https://doi.org/10.1186/s13617-023-00129-2

Arjun, M. S., Panda, B. P., & Arun, P. R. (2023). Sand mining as a contemporary threat to sandbar nesting birds: A review. *Contemporary Problems of Ecology, 16*(2), 189–204. https://doi.org/10.1134/S1995425523020038

Arnalds, O. (2013). The influence of volcanic tephra (ash) on ecosystems. *Advances in Agronomy, 121*, 331–380. https://doi.org/10.1016/B978-0-12-407685-3.00006-2

Berrouet, L. M., Machado, J., & Villegas-Palacio, C. (2018). Vulnerability of socio—ecological systems: A conceptual Framework. *Ecological Indicators, 84*, 632–647. https://doi.org/10.1016/j.ecolind.2017.07.051

Blaikie, P., Cannon, T., Davis, I., & Wisner, B. (2014). *At risk: Natural hazards, people's vulnerability and disasters.* Routledge.

Branca, S., Azzaro, R., Beni, E. D., Chester, D., & Duncan, A. (2015). Impacts of the 1669 eruption and the 1693 earthquakes on the Etna Region (Eastern Sicily, Italy): An example of recovery and response of a small area to extreme events. *Journal of Volcanology and Geothermal Research, 303*, 25–40. https://doi.org/10.1016/j.jvolgeores.2015.07.020

Braun, V., & Clarke, V. (2006). Using thematic analysis in psychology. *Qualitative Research in Psychology, 3*(2), 77–101. https://doi.org/10.1191/1478088706qp063oa

Bryant, R. L., & Wilson, G. A. (1998). Rethinking environmental management. *Progress in Human Geography, 22*(3), 321–343. https://doi.org/10.1191/030913298672031592

Buot, M. M., & Buot, I. E. (2022). Volcanic seascapes: Disaster recovery for community wellbeing and habitat protection. In G. Pungetti (Ed.), *Routledge handbook of seascapes* (pp. 385–398). https://doi.org/10.4324/9780429273452-30

Caviedes, J., Ibarra, J. T., Calvet-Mir, L., Álvarez-Fernández, S., & Junqueira, A. B. (2024). Indigenous and local knowledge on social-ecological changes is positively associated with livelihood resilience in a globally important agricultural heritage system. *Agricultural Systems, 216*, Article 103885. https://doi.org/10.1016/j.agsy.2024.103885

Chester, D., Duncan, A., Kilburn, C., Sangster, H., & Solana, C. (2015). Human responses to the 1906 eruption of Vesuvius, southern Italy. *Journal of Volcanology and Geothermal Research, 296*, 1–18. https://doi.org/10.1016/j.jvolgeores.2015.03.004

Cosme, M., Bernardoff, O., Hély, C., Tiberi, C., Parat, F., Gautier, S., & Treydte, A. (2023). Risk assessment and recovery trajectories of a social-ecological system with a discrete-event model after a volcanic eruption. *International Journal of Disaster Risk Reduction, 92*, Article 103741. https://doi.org/10.1016/j.ijdrr.2023.103741

Cuthbertson, J., Stewart, C., Lyon, A., Burns, P., & Telepo, T. (2020). Health impacts of volcanic activity in Oceania. *Prehospital and Disaster Medicine, 35*(5), 574–578. https://doi.org/10.1017/S1049023X2000093X

Datta, N., Assy, A. E., Buba, J., Silva, S. J., & Watson, S. (2018). *Integrated youth employment programs: A stocktake of evidence on what works in youth employment programs.* Retrieved December 17, 2024, from https://documents1.worldbank.org/curated/ar/307301552636285526/pdf/Integrated-Youth-Employment-Programs-A-Stocktake-of-Evidence-on-What-Works-in-Youth-Employment-Programs.pdf

Department for International Development. (1999). *Sustainable livelihoods guidance sheet.* Retrieved July 01, 2022, from https://worldfish.org/GCI/gci_assets_moz/Livelihood%20Approach%20-%20DFID.pdf

Dolin, G. (2016). US food and drug administration. In T. R. Fulda, A. Lyles, & A. I. Wertheimer (Eds.), *Pharmaceutical public policy* (pp. 121–138). CRC Press. https://doi.org/10.1201/b19633

Donovan, A. (2022). Volcanoes and the human and physical geographies of risk. In *Oxford research encyclopedia of natural hazard science.* https://doi.org/10.1093/acrefore/9780199389407.013.367

Ellis, F. (2000). The determinants of rural livelihood diversification in developing countries. *Journal of Agricultural Economics, 51*(2), 289–302. https://doi.org/10.1111/j.1477-9552.2000.tb01229.x

Escolà-Gascón, Á., Dagnall, N., Denovan, A., Diez-Bosch, M., & Micó-Sanz, J. L. (2023). Social impact of environmental disasters: Evidence from Canary Islands volcanic eruption. *International Journal of Disaster Risk Reduction, 88*, Article 103613. https://doi.org/10.1016/j.ijdrr.2023.103613

Farahani, H., & Bayazidi, S. (2018). Modeling the assessment of socio-economical and environmental impacts of sand mining on local communities: A case study of villages tatao river bank in North-Western part of Iran. *Resources Policy*, 87–95. https://doi.org/10.1016/j.resourpol.2017.11.001

Filho, W. L., Hunt, J., Lingos, A., Platje, J., Vieira, L. W., Will, M., & Gavriletea, M. D. (2021). The unsustainable use of sand: Reporting on a global problem. *Sustainability, 13*(6), 3356. https://doi.org/10.3390/su13063356

Franco-Gaviria, F., Amador-Jiménez, M., Millner, N., Durden, C., & Urrego, D. H. (2022). Quantifying resilience of socio-ecological systems through dynamic Bayesian networks. *Frontiers in Forests and Global Change, 5*, Article 889274. https://doi.org/10.3389/ffgc.2022.889274

Gavriletea, M. D. (2017). Environmental impacts of sand exploitation. Analysis of sand market. *Sustainability, 9*(7), 1118. https://doi.org/10.3390/su9071118

Guma, I. P., Rwashana, A. S., & Oyo, B. (2022). Food security policy analysis using system dynamics: The case of Uganda. In *Research anthology on strategies for achieving agricultural sustainability* (pp. 301–320). IGI Global. https://doi.org/10.4018/978-1-6684-5352-0.ch017

Gupta, M. C. (1995). Environmental management and its impact on the operations function. *International Journal of Operations and Production Management, 15*(8), 34–51.

He, Y., & Li, W. (2024). The impact of livelihood capital on sustainable livelihood capacity of farmers: evidence from Yunnan Province, China. *Applied Economics Letters*, 1–6. https://doi.org/10.1080/13504851.2024.2332536

Hidayati, S., Triastuty, H., Mulyana, I., Adi, S., Ishihara, K., & Basuki, A., et al. (2019). Differences in the seismicity preceding the 2007 and 2014 eruptions of Kelud volcano, Indonesia. *Journal of Volcanology and Geothermal Research, 382*, 50–67. https://doi.org/10.1016/j.jvolgeores.2018.10.017

Hossain, M. S., Basak, S. M., Amin, M. N., Anderson, C. C., Cremin, E., & Renaud, F. G. (2023). Social-ecological systems approach for adaptation to climate change. *Sustainable Development, 32*(3), 2766–2778. https://doi.org/10.1002/sd.2801

Huhtamaa, H., Stoffel, M., & Corona, C. (2022). Recession or resilience? Long-range socioeconomic consequences of the 17th century volcanic eruptions in northern Fennoscandia. *Climate of the past, 18*(9), 2077–2092. https://doi.org/10.5194/cp-18-2077-2022

Itzkin, A., Scholes, M. C., Clifford-Holmes, J. K., Rowntree, K., Waal, B. V., & Coetzer, K. (2021). A social-ecological systems understanding of drivers of degradation in the tsitsa river catchment to inform sustainable land management. *Sustainability, 13*(2), 516. https://doi.org/10.3390/su13020516

Jackson, G. (2021). Critical junctures, agrarian change, and the (re) production of vulnerability in a marginalised Indigenous society. *World Development, 145*, Article 105538. https://doi.org/10.1016/j.worlddev.2021.105538

Jenkins, S. F., Day, S. J., Faria, B. V., & Fonseca, J. F. (2017). Damage from lava flows: Insights from the 2014–2015 eruption of Fogo, Cape Verde. *Journal of Applied Volcanology, 6*(6), 1–17. https://doi.org/10.1186/s13617-017-0057-6

Kadetz, P., & Mock, N. B. (2018). Problematizing vulnerability: Unpacking gender, intersectionality, and the normative disaster paradigm. In *Creating Katrina, rebuilding resilience: lessons from new orleans on vulnerability and resiliency* (pp. 215–230). Butterworth-Heinemann. https://doi.org/10.1016/B978-0-12-809557-7.00009-0

Kako, M., Steenkamp, M., Ryan, B., Arbon, P., & Takada, Y. (2020). Best practices for evacuation centers accommodating vulnerable populations: A literature review. *International Journal of Disaster Risk Reduction, 46*, Article 101497. https://doi.org/10.1016/j.ijdrr.2020.101497

Kolahi, M., Abdollahzadeh, A., & Noori, R. (2024). Holistic management approach for social-ecological systems in Iran. *Environmental Reviews, Ja.* https://doi.org/10.1139/er-2024-0028

Kumar, B. S., Zainuddin, M. F., Soon, D. C., & Bhatawdekar, R. M. (2022). Learning lessons from river sand mining practices in India and Malaysia for sustainability. In T. Roshni, P. Samui, D. T. Bui, D. Kim, & R. Khatibi (Eds.), *Risk, reliability and sustainable remediation in the field of civil and environmental engineering* (pp. 315–331). Elsevier. https://doi.org/10.1016/B978-0-323-85698-0.00002-2

Kumar, P., Fürst, C., & Joshi, P. K. (2024). Differentiated socio-ecological system approach for vulnerability and adaptation assessment in the Central Himalaya. *Mitigation and Adaptation Strategies for Global Change, 29*(7). https://doi.org/10.1007/s11027-023-10097-y

Lecegui, A., Olaizola, A. M., López-i-Gelats, F., & Varela, E. (2022). Implementing the livelihood resilience framework: An indicator-based model for assessing mountain pastoral farming systems. *Agricultural Systems, 199*, Article 103405. https://doi.org/10.1016/j.agsy.2022.103405

Li, T., Singh, R. K., Cui, L., Xu, Z., Pandey, R., & Liu, Y., et al. (2024b). Managing multiple stressors for sustainable livelihoods in dryland ecosystems: Insights and entry points for resource management. *Land Degradation and Development, 35*(3), 968–984. https://doi.org/10.1002/ldr.4964

Li, T., Singh, R. K., Pandey, R., Liu, H., Cui, L., & Xu, Z., et al. (2023). Enhancing sustainable livelihoods in the three rivers headwater region: A geospatial and obstacles context. *Ecological Indicators, 156*, 111134. https://doi.org/10.1016/j.ecolind.2023.111134

Lin, K.-H., & Chang, C.-Y. (2013). Everyday crises: Marginal society livelihood vulnerability and adaptability to hazards. *Progress in Development Studies, 13*(1), 1–18. https://doi.org/10.1177/146499341201300101

Liodakis, G. (2013). Considering (economic and ecological) crisis from a communist perspective. *Perspectives on Global Development and Technology, 12*(1–2), 194–218.

Liu, R., Zhang, L., Tang, Y., & Jiang, Y. (2024). Understanding and evaluating the resilience of rural human settlements with a social-ecological system framework: The case of Chongqing Municipality China. *Land Use Policy, 136*, Article 106966. https://doi.org/10.1016/j.landusepol.2023.106966

Maharjan, K. L., & Issahaku, Z. A. (2014). Communities and livelihood strategies: An overview. In K. L. Maharjan (Ed.), *Communities and livelihood strategies in developing countries* (pp. 1–11). https://doi.org/10.1007/978-4-431-54774-7_1

Mashamaite, C. V., Motsi, H., Manyevere, A., & Poswa, S. B. (2024). Assessing the potential of biochar as a viable alternative to synthetic fertilizers in Sub-Saharan Africa smallholder farming: A review. *Agronomy, 14*(6), 1215. https://doi.org/10.3390/agronomy14061215

Meredith, E. S., Jenkins, S. F., Hayes, J. L., Lallemant, D., Deligne, N. I., & Teng, N. R. (2024). Lava flow impacts on the built environment: Insights from a new global dataset. *Journal of Applied Volcanology, 13*(1), 1. https://doi.org/10.1186/s13617-023-00140-7

Micheli, F., Saenz-Arroyo, A., Aalto, E., Beas-Luna, R., Boch, C. A., Cardenas, J. C., & Giulio A. De Leo, E. (2024). Social-ecological vulnerability to environmental extremes and adaptation pathways in small-scale fisheries of the southern California current. *Frontiers in Marine Science, 11*, 1322108. https://doi.org/10.3389/fmars.2024.1322108

Mihai, R. A., Espinoza-Caiza, I. A., Melo-Heras, E. J., Cubi-Insuaste, N. S., Pinto-Valdiviezo, E. A., & Catana, R. D. (2023). Does the mineral composition of volcanic ashes have a beneficial or detrimental impact on the soils and cultivated crops of ecuador? *Toxics, 11*(10), 846. https://doi.org/10.3390/toxics11100846

Miller, M. A. (2022). A transboundary political ecology of volcanic sand mining. *Annals of the American Association of Geographers, 112*(1), 78–96. https://doi.org/10.1080/24694452.2021.1914539

Ministry of Energy and Mineral Resources. (2024). *Laporan Aktivitas Gunung Api (Volcanic Activity Report)* daily report. Retrieved December 2, 2024, from https://magma.esdm.go.id/v1/gunung-api/laporan

Mozumder, M. M., Schneider, P., Islam, M. M., Deb, D., Acharjee, M. R., & Washi, A. M. (2024a). Bridging the gap: Enhancing socio-ecological resilience by breaking the debt cycle among small-scale hilsa fishers in Bangladesh. *Maritime Studies, 23*(1), 10. https://doi.org/10.1007/s40152-024-00355-3

Mozumder, M. M., Schneider, P., Islam, M. M., Deb, D., Acharjee, M. R., & Washi, A. M. (2024). Bridging the gap: enhancing socio-ecological resilience by breaking the debt cycle among small-scale hilsa fishers in Bangladesh. *Maritime Studies, 23*(1). https://doi.org/10.1007/s40152-024-00355-3

National Disaster Management Agency. (2014). *Peta Kawasan Rawan Bencana Letusan Gunungapi Kelud Provinsi Jawa Timur (Map of Disaster-Prone Areas of Mount Kelud Eruption in East Java Province).* National Disaster Management Agency.

National Park Service. (2022, August 5). *Volcanic Explosivity Index (VEI).* Retrieved December 14, 2024, from Volcanic Explosivity Index (VEI): https://www.nps.gov/subjects/volcanoes/volcanic-explosivity-index.htm#:~:text=The%20Volcanic%20Explosivity%20Index%20(VEI,for%20the%20size%20of%20earthquakes.

Nawiyanto, & Sasmita, N. (2019). The Eruption of Mount Kelud in 1919: Its impact and mitigation efforts. In *1st international conference on social sciences and interdisciplinary studies (ICSSIS 2018)* (pp. 127–133). Retrieved December 14, 2024, from https://www.researchgate.net/profile/Nawiyanto-Nawiyanto-2/publication/333017935_The_Eruption_of_Mount_Kelud_in_1919_Its_Impact_and_Mitigation_Efforts/links/5cd6a265299bf14d9589d01e/The-Eruption-of-Mount-Kelud-in-1919-Its-Impact-and-Mitigation-Efforts.pdf

Nihayatul, U., Supriyono, A., & Rinardi, H. (2017). The Eruption of Mount Kelud and it's impacts in Blitar 1919–1922. *Indonesian Historical Studies, 1*(1), 67–77. https://doi.org/10.14710/ihis.v1i1.1164

Noor, M. T., Shahar, H. K., Ismail, S. N., Manaf, R. A., Said, S. M., Ahmad, J., & Muthiah, S. G. (2023). A systematic review on an optimal dose of disaster preparedness intervention utilizing health belief model theory. *Pertanika Journal of Tropical Agricultural Science, 31*(1). https://doi.org/10.47836/pjst.31.1.10

Pal, I., Raj, V., Pal, A., & Sukwanchai, K. (2021). Disaster risk reduction education (DRRE) and resilience in Asia-Pacific. In I. Pal, R. Shaw, & S. Shrestha (Eds.), *Disaster resilience and sustainability: Adaptation for sustainable development* (pp. 667–683). Elsevier. https://doi.org/10.1016/B978-0-323-85195-4.00004-4

Patrisina, R., Emetia, F., Sirivongpaisal, N., S. S., Alfadhlani, A., & Fatrias, D. (2018). Key performance indicators of disaster preparedness: a case study of a tsunami disaster. *MATEC Web of Conferences, 229*. https://doi.org/10.1051/matecconf/201822901010229

Posadinu, C. M., Rodriguez, M., Madau, F., & Attene, G. (2022). The value of agrobiodiversity: An analysis of consumers preference for tomatoes. *Renewable Agriculture and Food Systems, 37*(2), 237–247. https://doi.org/10.1017/S1742170521000491

Prayoga, R. A., Purbandini, L., Wahyono, E., Fatimah, S., Solekhah, N., & Hakim, F., et al. (2024). Resilience rising: Redefining livelihood systems in disaster-prone rural communities. *Progress in Disaster Science, 24*, 100391. https://doi.org/10.1016/j.pdisas.2024.100391

Purbandini, L., Wahyono, E., Prayoga, R., Fatimah, S., Solekhah, N., & Hakim, F., et al. (2024). Sustainable livelihood diversification in the Merapi volcano disaster-prone area. In *Climate change management* (pp. 621–644). Springer. https://doi.org/10.1007/978-3-031-58261-5_27

Quarantelli, E. L. (1999). The disaster recovery process: What we know and do not know from research. *Preliminary Paper #286*. Retrieved July 3, 2025, from chrome-extension: www.efaidnbmnnnibpcajpcglclefindmkaj, https://udspace.udel.edu/server/api/core/bitstreams/a6c54b9f-a684-4aca-b830-d5d01ca8b5b7/content

Radcliffe, S. A., Watson, E. E., Simmons, I., Fernández-Armesto, F., & Sluyter, A. (2010). Environmentalist thinking and/in geography. *Progress in Human Geography, 34*(1), 98–116. https://doi.org/10.1177/03091325093387

Rentier, E. S., & Cammeraat, L. H. (2022). The environmental impacts of river sand mining. *Science of the Total Environment, 838*(2022), Article 155877. https://doi.org/10.1016/j.scitotenv.2022.155877

Rinawati, D. I., Sari, D. P., Handayani, N. U., & Siwi, B. R. (2018). Predicting the probability of Mount Merapi eruption using Bayesian event Tree_Eruption forecasting. *MATEC Web of Conferences, 154*(154), 01050. https://doi.org/10.1051/matecconf/201815401050

Rozaki, Z., Rahmawati, N., Wijaya, O., Azizah, S. N., Pratama, A. C., & Pramudya, Y., et al. (2022). Farmers' food security in the volcanic area: A case in Mount Merapi, Indonesia. *Open Agriculture, 7*(1), 554–565. https://doi.org/10.1515/opag-2022-0122

Russo, F., & Rindone, C. (2024). Methods for risk reduction: Training and exercises to pursue the planned evacuation. *Sustainability, 16*(4), 1474. https://doi.org/10.3390/su16041474

Sakurai, A., & Sato, T. (2016). Promoting education for disaster resilience and the Sendai framework for disaster risk reduction. *Journal of Disaster Research, 11*(3), 402–412. https://doi.org/10.20965/jdr.2016.p0402

Saparita, R., Wahyono, E., Prayoga, R. A., Fatimah, S., Purbandini, L., & Solekhah, N., et al. (2024). Exploring the dynamics of community transformation in the Indonesian volcanic region: An analysis of socioeconomic metamorphosis. *International Journal of Safety and Security Engineering, 14*(4), 1169–1180. https://doi.org/10.18280/ijsse.140414

Sarrazin, C., Gautier, E., Hollé, A., Grancher, D., Belizal, E. D., & Hadmoko, D. S. (2019). Resilience of socio-ecological systems in volcano risk-prone areas, but how much longer? Assessment of adaptive water governance in Merapi volcano, Central Java, Indonesia. *Geo Journal, 84*(1), 183–213. https://doi.org/10.1007/S10708-018-9856-5

Serrat, O. (2008). *The sustainable livelihoods approach.* Asian Development Bank. Retrieved July 3, 2023, from https://www.adb.org/sites/default/files/publication/27638/sustainable-livelihoods-approach.pdf

Shackleton, R. T. (2024). Social-ecological systems (SES). In *Elgar encyclopedia of interdisciplinarity and transdisciplinarity* (pp. 483–486). Edward Elgar Publishing. https://doi.org/10.4337/9781035317967.ch106

Shitima, C., & Suykens, B. (2023). Formalization of sand mining in Dar es Salaam Tanzania. *Resources Policy, 82*, Article 103589. https://doi.org/10.1016/j.resourpol.2023.103589

Suša, O. (2019). Global risks and conflicts: The social, environmental, and political consequences. *Critical Sociology, 45*(6), 829–843. https://doi.org/10.1177/0896920518792299

Takeuchi, K., Nakayama, N., Teshima, H., Takemoto, K., & Turner, N. (2016). Ecosystem-based approaches toward a resilient society in harmony with nature. In F. Renaud, K. Sudmeier-Rieux, Estrella, M., & U. Nehren (Eds.), *Advances in natural and technological hazards research* (pp. 315–333). https://doi.org/10.1007/978-3-319-43633-3_14

The Governor of Central Java. (2018). Peraturan Gubernur Jawa Tengah Nomor 6 Tahun 2018 Tentang Rencana Kontingensi Erupsi Gunung Merapi (Regulation of the Governor of Central Java Number 6 of 2018 Concerning the Contingency Plan for the Eruption of Mount Merapi). Semarang: Pemerintah Provinsi Jawa Tengah.

Thouret, J., Abdurachman, K. B., & Bronto, S. (1998). Origin, characteristics, and behaviour of lahars following the 1990 eruption of Kelud volcano, Eastern Java (Indonesia). *Bulletin of Volcanol, 59*, 460–480. https://doi.org/10.1007/s004450050204

Thouret, J.-C., Wavelet, E., Taillandier, M., Tjahjono, B., Jenkins, S. F., Azzaoui, N., & Santoni, O. (2022). Defining population socio-economic characteristics, hazard knowledge and risk perception: The adaptive capacity to persistent volcanic threats from Semeru, Indonesia. *International Journal of Disaster Risk Reduction, 77*, Article 103064. https://doi.org/10.1016/j.ijdrr.2022.103064

Tiepolo, M., & Braccio, S. (2020). Mainstreaming disaster risk reduction into local development plans for rural tropical Africa: A systematic assessment. *Sustainability, 12*(6), 2196. https://doi.org/10.3390/su12062196

Tran, D. D., Thien, N. D., Yuen, K. W., Lau, R. Y., Wang, J., & Park, E. (2023). Uncovering the lack of awareness of sand mining impacts on riverbank erosion among Mekong Delta residents: Insights from a comprehensive survey. *Scientific Reports, 13*, 15937. https://doi.org/10.1038/s41598-023-43114-w

Troll, V. R., Carracedo, J. C., Jägerup, B., Streng, M., Barker, A. K., & Deegan, F. M., et al. (2017). Volcanic particles in agriculture and gardening. *Geology Today, 33*(4), 148–154. https://doi.org/10.1111/gto.12193

UNEP. (2023, February 06). *The problem with our dwindling sand reserves*. Retrieved August 19, 2024, from https://www.unep.org/news-and-stories/story/problem-our-dwindling-sand-reserves

Voight, B., Constantine, E., Siswowidjoyo, S., & Torley, R. (2000). Historical eruptions of Merapi Volcano, Central Java, Indonesia, 1768–1998. *Journal of Volcanology and Geothermal Research*. https://doi.org/10.1016/S0377-0273(00)00134-7

Wang, H., Xu, J., Liu, X., Zhang, D., Li, L., Li, W., & Sheng, L. (2019). Effects of long-term application of organic fertilizer on improving organic matter content and retarding acidity in red soil from China. *Soil and Tillage Research, 195*. https://doi.org/10.1016/j.still.2019.104382

Wantim, M. N., Bonadonna, C., Gregg, C. E., Menoni, S., Frischknecht, C., Kervyn, M., & Ayonghe, S. N. (2018). Forensic assessment of the 1999 Mount Cameroon eruption, West-Central Africa. *Journal of Volcanology and Geothermal Research, 358*, 13–30. https://doi.org/10.1016/j.jvolgeores.2018.06.007

Wen, J., Wan, C., Ye, Q., Yan, J., & Li, W. (2023). Disaster risk reduction, climate change adaptation and their linkages with sustainable development over the past 30 years: A review. *International Journal of Disaster Risk Science, 14*(1), 1–13. https://doi.org/10.1007/s13753-023-00472-3

Yan, B., Sun, Y., & Wei, Y. (2019). Long-term cultivation and fertilizer changes soil chemicals. In *IOP conference series: Earth and environmental science* (Vol. 384, p. 012201). IOP Publishing. https://doi.org/10.1088/1755-1315/384/1/012201

Zakharikhina, L. V., Litvinenko, Y. S., & Gainatulina, V. V. (2022). Volcanic ash application in agricultural practice. *Universal Journal of Agricultural Research, 10*(1), 77–87. https://doi.org/10.13189/ujar.2022.100107

Zhang, Y., Xie, X., Qiu, X., Jing, Z., Yu, Y., & Wang, Y. (2023). Study on livelihood resilience of rural residents under the rural revitalization strategy in ethnic areas of Western Sichuan, China. *Agriculture, 13*(10). https://doi.org/10.3390/agriculture13101957

Chapter 13
Organic Agriculture Toward Sustainable Agriculture in Supporting Food Security in Indonesia

Tri Bastuti Purwantini, Rosita Novi Andari, and Kurnia Suci Indraningsih

Abstract The implementation of technology after the Green Revolution increased the rate of food production by an unprecedented amount; however, this also risks worsening environmental damage. Therefore, it is important to maintain and protect the environment. The focus of this chapter is on the analysis of Indonesia's organic agriculture within the framework of sustainable agriculture and achieving food security by addressing the practices, potentials, and challenges. Indonesian organic agriculture practice emphasizes ecologically sustainable organic farming. Therefore, it can increase food security and achieve global goals in the framework of Sustainable Development (SDGs). The Organic Alliance of Indonesia (AOI) and the Research Institute of Organic Agriculture FiBL reported in 2023 land area of certified organic land was 71,947 ha (0.1% of total agricultural land) and had an average reduction in that land of 3.36% during the period, 2014–2023. Even with this projected decrease having been reported, organic farming continues to remain a vital component of climate-informed adaptation. The remaining vital component of climate-informed organic agriculture is the integration of local knowledge into the system of crop and livestock alteration. To optimize its contribution, government support is essential, especially facilities in technology deployment, access to capital, training, and incentives, so that organic agriculture can become a key pillar of sustainable food security in Indonesia.

Keyword Sustainable agriculture · Climate change · Food security

---

T. B. Purwantini (✉) · R. N. Andari · K. S. Indraningsih
Research Centre for Social Welfare, Village, and Connectivity, National Research and Innovation Agency, Jakarta, Indonesia
e-mail: trib005@brin.go.id

R. N. Andari
e-mail: rosi014@brin.go.id

K. S. Indraningsih
e-mail: kurn028@brin.go.id

G. K. Panda et al. (eds.), *Socio-Ecological Crisis of Development*, Sustainability Solutions, https://doi.org/10.1007/978-3-032-22171-1_13

13.1 Introduction

Practices of intensive conventional farming, such as those of the Green Revolution, have had their share of problems. Overall, these problems are a threat to the productivity and sustainability of the agricultural sector: of these are soil degradation and overall poor soil health, declining soil fertility and loss of biodiversity, the overuse of chemical fertilizers and pesticides, accompanied by pollution of water and the resulting imbalance of ecosystems (Ortega et al., 2024; Pereira et al., 2023). There is empirical evidence that, although dominated by conventional agricultural practices that rely on inorganic fertilizers, the agriculture of Indonesia is able to derive increased production from its agricultural systems. The question that arises is the sustainability of a system of such conditions. The concerns about the conventional agriculture system in the long run have been the driving force for the adoption of alternative agricultural systems, including organic, conservation, and integrated agricultural systems (Reganold & Wachter, 2016).

Despite the specific challenges that organic farming meets in practice, research about its benefits is encouraging organic farming practices amid the global shift toward sustainable and organic agricultural systems (Afaf, 2025). In recent decades, the interest in organic products on both domestic and international markets has been increasing, which has been largely driven by growing consciousness among consumers about the advantages of eating healthy foods that are also environmentally sustainable.

The organic sector generates in excess of USD 120 billion, with an annual increase of USD 9 billion from 2022 (Willer et al., 2023)-FiBL estimates. In Indonesia, the demand for organic products is increasing, and government and private sector support is essential. The demand has been buttressed by the increased awareness of health and the environment (Ministry of Agriculture, 2023). Interest from Consumers, along with technological innovations and export opportunities, positions Indonesia's organic Agriculture for enhanced development. To achieve this, joint action from the government, the private sector, and the involvement of farmers and rural communities is crucial.

Organic farming is a type of sustainable agriculture since it involves organic farming and the alternative agricultural practices of crop rotation, biological pest control, environmental enhancement, and the safeguarding of consumer and farmer health, and better health economic outcomes (Willer & Lernoud, 2019). It describes the organic farming as an even increased focus on the need for the environment to be protected, as well as the socioeconomic need for rural folk to be supported, as the even increased focus on sustainable farming practices (Pretty, 2008). There is no doubt about the need for the development of organic agriculture in Indonesia as it possesses the most remarkable opportunities for the production of quality and organic products from environmentally sustainable agricultural practices. To build a strong foundation of organic farming, it is critical to target the supportive government policies, to land stewardship, and prioritize biodiversity. Notable obstacles have been highlighted, such as the absence of suitable technology, a lack of consumers, and a lack of market access (Ministry of Agriculture, 2019). In Indonesia, the implementation of organic agriculture, although having great potential, is very complex. The amount of land and production has grown considerably, and so has the awareness of consumers to seek organic foods. The organic farming needs investation in adequate and coherent planning to encourage the sustainable and widespread use of organic farming in Indonesia. Global issues include public health, environmental protection, and food security.

In the last 20 years, due to the relevance related to human health, planetary well-being, and global food security and nutrition to the food health of the world's population, the phenomenon of organic farming has been attracting attention. The organic food sector is indicative of growing consumer interest, for health, product certification, and the global market (Brumă et al., 2023; Chopin et al., 2023; Savero Qomara & Noer, 2025). There is concern over potential health and environmental consequences as well as climate change, and consumer behaviors such as PAS (Pro-Environmental consumer behaviors), e.g., the purchased food, and PAS are core to the consumer choice discourse during the food crisis (Savero Qomara & Noer, 2025). Most of the research output during the period of 1945–2021 was from developed nations, largely America and Europe, whereas developed nations, including Indonesia, are relatively less involved (Chopin et al., 2023). The greatest hurdle facing organic farming in Indonesia is the need to change farmers' entrenched behaviors, especially their worries over marketing their products and the challenges of initial yield losses.

This chapter will examine the impact of organic farming practices in Indonesia through a literature review on organic farming and the development of sustainable agriculture and food security implementation in Indonesia while employing a broad set of secondary data from reputable sources nationally and globally. Addressing the research gap on organic agriculture in Indonesia through the study of the practices, development, potential, opportunities, and challenges, as well as the strategic role in the promotion of sustainable agriculture, will be the novelty of the chapter in the context of developing countries. Furthermore, this chapter is not only descriptive but also strategic, linking Indonesia's organic agriculture potential to broader goals of sustainable farming, environmental conservation, and rural socioeconomic improvement, offering a locally grounded yet globally relevant contribution. The chapter aims to present a model of sustainable agriculture and food security for developing nation by examining the potential, challenges, and strategies of organic agriculture in Indonesia.

13.2 Rationale of the Study

IFOAM explains that "organic farming is the system that sustains the health of people, the ecosystems, and the soils." IFOAM was first able to adopt this definition in the year 2005 in Adelaide, Australia, where it outlined organic agriculture to sustain the well-being of people and the ecosystems and soils. Organic agriculture as a method of sustainable farming reduces dependence on synthetic fertilizers and pesticides by substituting them with crop rotation, green manures, and biological pest management (Devi & Nengparmoi, 2023). The integrity of ecosystems and the health of soils, along with the production of safe, nutritionally adequate food, and the use of organic practices with a complement of modern techniques to enhance the welfare of people, are the goals of the aforementioned organic farming practices.

Instead of using the conventional, antagonistic approach to farming, there is an organic approach to farming which works with the nature surrounding it. The organic approach sees the soil as the heart of farming, which is why they keep it healthy and productive through crop rotation, composting, eco-friendly manuring, and biological pest control (Raj et al., 2024). Ecological and biological farming focuses on the traditional knowledge of the farmers and adds modern techniques to it (Reganold & Wachter, 2016). Organic farming can take various forms, including agroforestry, integrated farming, diverse mixed crop and animal systems, and conservation agriculture, are all forms the organic approach to farming can take. The approach nurt the natural balance and the soil using manure and compost, which is why other farming systems that use chemical fertilizers and pesticides are labeled organic farming. This approach also seems to have a lot of potential in Indonesia as it can greatly enhance sustainable farming in the areas of the country and also improve the long-term food security of the population.

Organic farming is an approach to food production that integrates natural resources with methodologies that are more ecological in nature to produce food (Reganold & Wachter, 2016). Conversely, sustainable farming emphasizes enhancing the system's productive capacity while using resources efficiently and preserving soil and water quality and providing the farmer with enhanced social and economic well-being (Altieri & Nicholls, 2017; FAO, 2020). The two concepts are interrelated, and while in agriculture management and production, the attention shifts more from one concept to the other, in sustainable agriculture, the economic, social, and environmental aspects are considered in their totality to achieve a certain level of output in farming while ensuring that the natural resources are not degraded (Akanmu et al., 2023; Wezel et al., 2020). It minimizes environmental impacts through efficient use of resources, maintains soil health, water, and biodiversity, and supports the well-being of rural communities (Pretty, 2008).

Creative approaches to managing land resources are critical to the practice of sustainable farming. Activities such as crop rotation, cover cropping, and applying organic fertilizers enhance the soil and act as an erosion barrier (Lal, 2020). Furthermore, the biological control of pests helps to preserve the farming environment (agroecosystem) and lowers the use of industrial pesticides (Peshin & Dhawan, 2019). Agricultural diversification and climate change adaptation are other facets of sustainable farming. In this context, the production of different crops and the farming of varied livestock enables the farmer to reinforce his/her farming system against a number of risks such as crop pests and disease outbreaks, climate change, and volatile prices (Altieri & Nicholls, 2017). Sustainable farming also contributes to poverty alleviation, though this is an indirect contribution. Nearly all the world's smallholder farmers depend on the livelihoods that sustainable farming has to offer. Evidence has shown that the adoption of sustainable farming practices increases the economic resilience of farmers (Altieri & Nicholls, 2017). Therefore, organic farming represents an integral component of sustainable agriculture. Practices of sustainable farming contribute to poverty alleviation, though this is in an indirect way, through the enhancement of the farming system. Sustainable farming mitigates threats to biodiversity, which is crucial to conserving the resiliency of the ecosystem, while guaranteeing a continuous supply of healthy food. Consequently, the non-agricultural sector is food securing. The productivity of organic farming has been demonstrated across various countries, and it is protective of healthy food (FAO, 2018).

13.3 Method

This chapter uses a literature review method drawn from various sources. To discuss the performance of organic farming, we utilize scientific papers such as journals and proceedings, policy documents, agency reports, secondary data, and primary statistical sources from AOI and FiBL. Descriptive data analysis using simple statistics. The review was conducted using content analysis without using systematic review tools. AI serves as a tool to assist in clarifying the framework and direction of this paper.

13.4 Results and Discussion

13.4.1 Indonesian Organic Agriculture: Current Situation and Progress

Growing awareness for the value of protecting the environment, consumer health, and smallholder farmers' financial stability has increased organic farming in Indonesia. In Indonesia, organic farming started to grow constructively in the 1980s (David & Ardiansyah, 2017; Tamtomo, 2021). As part of the pioneering social movement organizations, David and Ardiansyah attribute the socio-ecological impacts from the Green Revolution and the development of organic agriculture in the country through four distinct stages of its evolution organic agriculture in the country (David & Ardiansyah, 2017).

The early phase, Organic 1.0 (1980–1990); the Decade of Networks, Organic 2.0 (1990–2000); the Decade of Regulation (2000–2014); and Organic 3.0 (2014 to nowadays), which is Innovation and Research. During the Decade of Networks in Yogyakarta, fishers and farmer networks formed and catalyzed several local initiatives, especially in community-based rice projects. Also during this time, SKAL became the first to certify organic coffee in Gayo, Aceh.

In 1998, the Indonesian Organic Agriculture Network (JAKERPO) came into existence, as of Regulation Decree Organic 2.0 (2000–2014), as an umbrella organization for organic farming groups nationwide. In 2000, the Indonesian Organic Community (MAPORINA) came into existence. Due to the organic markets starting to become available to big business & farmers in the agribusiness sector starting in 2010, and in organic farming and the health benefits of Go Organic policies, the Indonesian government implemented policies to educate the citizens. The organic standards of the IoA (Indonesian Organic Alliance, 2015) are based on the country's standards on the principles of the Codex Alimentarius Commission and IFOAM (David & Ardiansyah, 2017). The National Standardisation Agency (BSN) established the Indonesian National Standard (SNI) 01-6729-2002, covering definitions, labelling, production procedures, processing, internal auditing, certification, and the organic food import and audit system. BSN revised this standard to SNI 01-6729-2010 in 2010.

The traits of Indonesia's organic agricultural system are reflected in Organic Decade 3.0 Innovation with Research (2014 onward). In Indonesia's early years, small-scale farmers embraced organic farming because of its welfare and justice tenets. The growth of organic farming has motivated many small-scale farmers to adopt it, as it facilitates collaborative and efficient work. As shown in Fig. 13.1, developments from 2016 to 2019 illustrate the government's attention to, and to a certain degree, the progress of organic agriculture via the Organic Farming Village Programme in the Ministry of Agriculture (Ministry of Agriculture, 2020).

Regarding the extent of organic agricultural land, over the decade 2014–2023, its development actually decreased. Statistical data (FiBL & IFOAM, 2025) presented in Table 13.1 shows that during this period, the area of organic land decreased (−3.36%/

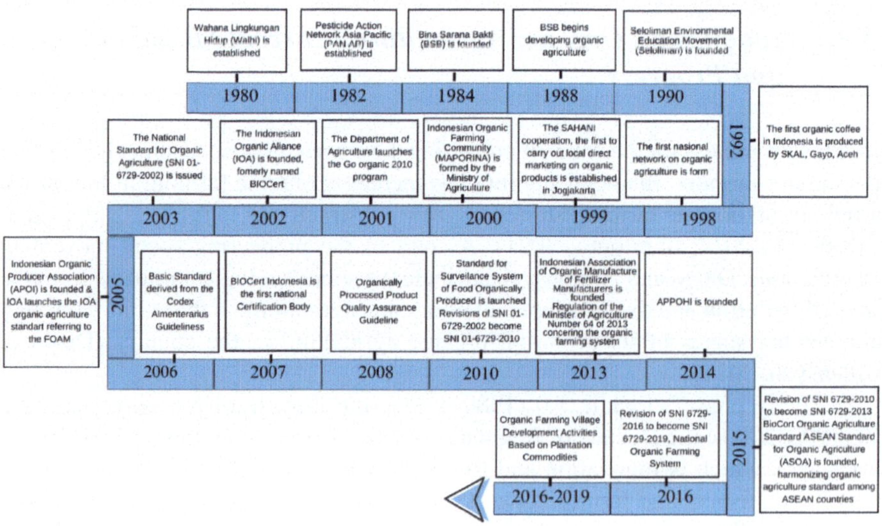

Fig. 13.1 The development of organic farming in Indonesia over time. *Source* David and Ardiansyah (2017), Ministry of Agriculture (2020)

year), and the share of total agriculture in Indonesia fluctuated and tended to decrease. The decline in organic land is because organic farming businesses are still at certain points and are inconsistent at times. Globally, the organic land area in the world in 2023 was recorded at 98.86 million hectares and tended to increase during the same period.

The dynamics of organic farming practices in Indonesia can be observed through data from the Indonesia Organic Agriculture Statistics (SPOI), published by civil society organizations such as AOI, which remains the main reference for information on the progress of organic farming nationwide. The data are likewise incorporated into the global organic agriculture statistics produced by the Research Institute of Organic Agriculture (FiBL, Forschungsinstitut für Biologischen Landbau). The statistics used here were obtained from surveys of all Organic Certification Institute (LSO) operators in Indonesia. Trends in Indonesia's organic agriculture between 2019 and 2022 are reflected in several key agricultural commodities (David & Alkausar, 2023).

a. Organic Rice

Organic rice has become one of the most sought-after commodities among consumers. The area of land certified for organic rice fluctuated from 3,350.415 ha in 2019 to 4,766.107 ha in 2022. The volume of organic rice farms from 2019 to 2022 shows an increase from 7,398 to 12,752, and organic rice production increased from 32,550,938 tons to 40,367.58 tons, while production remained inconsistent over the years.

b. Organic Coffee

The total area of organic coffee farms in Indonesia increased only slightly from 31,393.84 to 31,772.96 ha within the years 2019 to 2022, and the organic coffee

Table 13.1 Development of Farmland expansion and its share of Indonesia's total agricultural land, 2014–2023

Year	Organic area (farmland) (ha)	Share of total farmland (%)
2014	113.638	0,20
2015	130.384	0,23
2016	126.014	0,21
2017	121.535	0,20
2018	121.535	0,20
2019	121.535	0,20
2020	75.793	0,12
2021	83.362	0,13
2022	87.195	0,13
2023	71.947	0,10
Trend (%/year)	−3,63	

Sources FiBL (2025). Data processed

farmer population also increased from 12,676 to 13,666. But the total volume over the years demonstrated a drop according to the harvest output data, from 38,249.841 tons in 2019 to 33,302.805 tons in 2022. This condition was a result of newly planted coffee, and there was also a lack of data on the volume of coffee harvested within the data collection steps. However, in the context of the coffee consumption ratio, it was 1.15 kg/capita/year in 2019, and the growth of coffee consumption in Indonesia was also 1.7% per year.

c. Organic Oranges

The total volume of orange consumption reached 12.57 g/capita/day, that is about 3.6 kg per capita/year, in 2022. The area of orange plantations increased to 61 and recorded an annual growth rate of 2.3%. The area of organic orange plantations until 2022 was recorded at only 7.2 ha, down from 15 ha of organic land recorded in 2019. However, the 2023 SPOI cannot serve as a reference due to the absence of data from several LSOs, raising the possibility of unreported organic orange areas.

d. Organic Cocoa

The growth of organic cocoa production has remained relatively stable from 2019 to 2022. The land area, total production, and the number of producers show no significant increase in the past four years. Farmer productivity until 2022 has decreased from an average of 4.77 tons/farmer/year in 2019 to 5.75 tons/farmer/year in 2022.

e. Organic Vegetables

Organic vegetables are the most consumed organic products by the community. Based on surveys conducted in 2015, 2019, and 2022, it is evident that vegetables rank among the top three choices of the public in organic consumption. The production of organic vegetables has experienced fluctuations that tend to decline from 4,193.807 tons in 2019 to 3,043.58 tons in 2022.

13.4.2 Organic Agriculture's Role in Sustainable Agriculture

Previous research has stated that organic farming is crucial in a sustainable food chain, because it might be a key approach to the creation of sustainable food production and long-term agricultural sustainability. Organic farming can increase soil fertility, reduce dependence on synthetic inputs, and encourage ecosystem resilience (Afaf, 2025). In terms of agricultural management and production, organic farming and sustainable agriculture have different focuses, notwithstanding their close relationship. Since sustainable agriculture supports the livelihoods of the majority of smallholder farmers globally, it indirectly contributes significantly to the reduction of poverty. This methodological approach also takes into account long-term land use and the sustainable management of natural resources. Thus, smallholder farmers achieve consistent output and higher returns (Pretty et al., 2018). A recent study highlighted the value of customization to the locality in the practice of organic agriculture (Afaf,

2025). Several regions of Indonesia are implementing, within the practice of organic farming, different strategies designed to suit the local context. Table 13.2 provides case studies from different districts and regions in Indonesia, and the focus of these studies is organic agriculture.

Table 13.2 Case study of organic agriculture implementation in several regions in Indonesia

No	Case study area	Forms of organic agriculture implementation	References
1.	Yogyakarta City	· There are several new activities that have occurred within the scope of the Organic Movement of the World Food Day Work Ganjuran, Farmers' Collectives, Young Organic Farmer Leader from Kalongan Sleman, including a junior activist and member of the Sahani Organic Community · Developing a movement of organic farming by training farmer organizations on the marketing of organic agricultural products	Nindito and Tamtomo (2020), Tamtomo (2021)
2.	Yogyakarta City	· In 2006, the Rukun Farmers Group, the first-ever organic farming collective in Yogyakarta, began the journey toward organic certification and now specializes in the organic farming of native rice varieties, becoming fully organic certified in 2011 and 2015	Irham et al. (2020)
3.	Bandung City	· An organic farming business using community support (CSA) was created by a group of 5 young people in Community 1000, titled Seni Tani, who now cultivate organic vegetables in the SUTT (High Voltage Air Line) zone	Haniyah et al. (2022)
4.	Tasikmalaya Regency	· Growth and application of organic rice cultivation	Sujianto et. al. (2023)
5.	Gianyar Regency	· The Subak methodology has been included in organic rice cultivation methods. Farmers in Bali, particularly in Ubud, have transitioned to organic practices · The strategy has enhanced agricultural yields and farmer incomes, fostering sustainable tourism by appealing to ecologically concerned visitors	Sutarta et al. (2022)
6.	Temanggung Regency	· In light of the rising demand for organic goods in domestic and international markets, coffee growers in the region have adopted organic farming methods	Hidayat and Setiawan (2021)
7.	Sumba Island, NTT	· Location of the organic farm model · The region's indigenous agricultural practices, integrated with contemporary organic techniques, have produced a sustainable livestock production system	Kurniawan et al. (2023)

13.4.3 Challenges and Strategies in Organic Farming Development

Some of the issues contributing to food scarcity are weak domestic demand, low customer education, and lack of consumer demand, communication and publicity issues, and the need for supportive factors for derived landed systems. Support factors around the scarcity of organic food, publicity, and education focused on the consumer. Weak consumer education and poor publicity are issues within the supportive factors. Poor organic food publicity and communication related to food education are issues within the supportive factors (David & Ardiansyah, 2017). The following are some tactics that can be used to support organic farming and address some of those issues in Indonesia:

a. Farmer Awareness and Participation

The challenge of organic farming is obtaining organic seeds, so the attitude of farmers is important to encourage conversion to organic farming (David & Ardiansyah, 2017). An experimental study in Tasikmalaya Regency and Yogyakarta City demonstrated that providing farmers with extensive training is a viable strategy to boost the adoption of organic agricultural techniques. Farmers' understanding of organic farming improved, their perception of organic farming improved, and they experimented more with organic farming methods due to the training. Farmers become motivated not only by premium prices, which are usually higher than those of conventional agricultural products, but also by the opportunity to improve soil quality(Grimm & Luck, 2023). In addition, it is also important to encourage farmers' awareness of the use of organic fertilizers (Zainuddin et al., 2024).

b. Government Policy and Program Support

Since 2001, the national government has launched an organic agriculture campaign as known as *Organic Indonesia 2010*, which includes 23 certification bodies and 224 organic producers. Some local network communities include 16 organic training centers, 44 organic traders, five organic food processors, and 15 organic importers. Indonesia has also recommended the use of a single national Logo for Organic Food "Borganik Indonesia," hence a policy was made establishing an Organic Fertiliser Factory as a supplier of organic inputs for farming, called APOI (David & Ardiansyah, 2017). In the contemporary world, organic farming is a viable farming system as it has a certain degree of health, ecological, and economic value. The effectiveness of organic agriculture relies on a specific group of certain categories of farmers who are certified and trained, with the necessary incentive system and policies from the governing body for adoption and certification (Malhan & Ram, 2024).

c. Organic Product Market Growth, Organic Agriculture Growth Analysis: Influence of Organic Product Market

Partly due to the level of education and income, current consumers comprise educated middle-class households. Indonesia's organic food sector has potential, but it has been stunted due to the barriers of certification and exportation, even though there are still exports to be made (David & Ardiansyah, 2017). The positive effects of organic farming particularly lie in the diminished environmental effects due to the farming techniques used: natural cycles are maintained, and chemical inputs are avoided. While there are challenges, they can be integrated with new technologies and adaptable frameworks. This is the vision of sustainable agriculture (Gamage et al., 2023).

Consumers interested in organic goods are prevalent in the Czech Republic (Krause, 2017). Dairy products and milk, and especially vegetables and fruits, are the products consumers purchase. The willingness to purchase organic goods at a premium is there. Therefore, taking various factors into consideration, producers can target specific organic consumer markets. Additionally, organic food can contribute positively to sustainable food security (Hariyani & Sofwani, 2025). The changing of the consumption habits of people can largely be due to the positive and nutritious environmental impact that organic food has. Farming the organic way has many upsides, but there are some downsides to it as well. One of the downsides is that the cost is higher, and its productivity tends to fall short of that achieved through conventional farming methods. Yet, in the EU, subsidies, particularly for environmental services and disadvantaged regions, mitigate the income of organic farming from a loss to a gain (Krause, 2017). Organic farming is a sustainable approach to farming that resourcefully utilizes and allocates farming inputs, meets the required quality of production, and protects the ecosystem through environmentally friendly farming practices (Hariyani & Sofwani, 2025).

The differences in profit margins and productivity of agriculture in organic farming versus conventional farming, and the implications of food security that arise from this scenario, is still problems to solve. Against this backdrop, a major obstacle to achieving food security is ensuring that nutritious food reaches the appropriate populations available and accessible in correct location and at the necessary moment. Organic farmers, as well as the organic sector as a whole, must promote inclusive and systemic/holistic collaborations, a high degree of adaptive learning, and implementation of targeted interdisciplinary food system innovation regarding organic farming (Tsvetkov et al., 2018). Organic farming boosts food production sustainably, supports food security, and protects ecosystems from production to consumption (Maun & Kaleka, 2024). A far more important debate is how healthy food is sufficiently available and sustainable in the right place, time, and cost. Thus, farmers and the organic sector must continue inclusivity, fostering originality, a culture of knowledge, and in-depth publicity regarding new solutions and pioneering practices (Manida, 2021).

The influence of organic or agroecological practices on Africa's food security has drawn international controversy, as some research indicates organic farmers tend to manage their farms more passively (or ineffectively) than actively (or effectively) in their craft. Yet, agroecology's capability in resolving some of the social and environmental predicaments in food production has led to its acknowledgment in lower-income countries, both in policy and research. The impact of agroecology, however, leaves much controversy and sediment to be desired, particularly in its correlation to food security and overall nutrition. A review by Bezner Kerr et al. (2021) documents evidence that agroecological interventions can generate benefits for food security and household nutrition in low- and middle-income countries. This approach includes the practice of crop diversification, intercropping, agroforestry, integration of plants with livestock, and land management (Bezner Kerr et al., 2021).

Organic farming will develop in Indonesia if demand grows and consumer awareness rises, especially among educated youth and middle-class households, with social media as the main potential tool (David & Ardiansyah, 2017). The results of the Organic Consumer Survey conducted at SPOI 2023, which involved 1,002 respondents from various provinces in Indonesia and was conducted between January and December 2022, showed findings that can be used to increase consumer awareness of organic agricultural products, which tends to increase when viewed from how organic consumers shop, both online and offline (David & Alkausar, 2023). For those living in urban areas, offline shopping tends to involve supermarkets and specialty organic stores through communities that have begun to emerge in several places, and some also shop at organic restaurants. Furthermore, under organic regulations, some consumers have grown vegetables or fruit for their daily needs. The changes are insignificant because conditions and access can be adjusted to daily activities. The need for organic guarantees from 2019 to 2022 remains and continues to increase, along with consumers' increasing understanding of their importance. Products with guarantees are chosen based on several guarantees, including the Indonesian Organic label, the Foreign Organic label, and the Community Guarantee, as well as increasing trust in producers and product references. Increasing consumer awareness of organic agricultural products can be done by paying attention to product quality, price, marketing (online/offline), and organic guarantees.

13.4.4 Obstacles and Limitations to Developing Organic Agriculture

Organic agriculture provides multiple advantages, including enhanced soil quality and the production of environmentally friendly products. Enhanced attention to the following problems and challenges will facilitate the successful implementation and continued diffusion of this approach. Issues and challenges in the acceptance of organic farming include the following:

a. Limited Farmer Capacity

Another challenge to organic farming's development is the low levels of education and the insufficient knowledge and skills of many farmers in the developing world regarding organic farming arrangements. A farmer's capability includes knowledge of various methods of organic farming, optimal resource use, and problem-solving within the frame of organic systems. A lack of training and education in organic farming may restrict the acceptance and use of these practices (Pretty et al., 2018). The empowerment of farmers is possible through training programs and technical support (Willer & Lernoud, 2019). Consequently, for the development of organic farming, the financing of programs aimed at educating and training farmers is a prerequisite (Pretty et al., 2018).

b. Insufficient Government Support

Inadequate government assistance explains the lack of further encouragement for supporting the transition to organic farming. Independent of the support of farmers, this transition cannot take place. Farmers need supportive mechanisms as well as infrastructure, which are all government-dependent. When there is minimal assistance and support, and regulations are vague, farmers are reluctant to shift from conventional farming methods to organic approaches. Government support is essential. The government can lower the certification costs, offer access to the market, and increase subsidization of organic products. With this support, for farmers, organic farming is more viable and accessible (Kimhi & Tzur-Ilan, 2021; Willer & Lernoud, 2019).

c. Threat of Chemical Contamination from Neighboring Land and Land Maintenance Hazard of Chemical Pollution from Land Maintenance and Adjacent Land

The first challenge in the organic farming practice is the potential for chemical contamination from neighboring properties. The contamination of chemical and organic farms is a coexisting problem that would be ignored by organic farmers (Watson et al., 2002). It results in the organic systems of farming in order to lose the ability to maintain the status of certified organic. Substantial control of organic farming systems is required to protect and manage soil to optimize the desired soil conditions and to prevent other forms of contamination (Reganold & Wachter, 2016).

d. Substantial Uncertainties Pertaining to Starting Capital

The other significant barrier to entry is significant starting capital. The transition from traditional agriculture methods of organic practices entails great expenses to update the soil with fertilizers required to pass the organic certification process and obtain/facilitate the required organic inputs. For inexperienced farmers looking to start organic farming, the investments required to adapt to organic agriculture can be a serious barrier to entry (Crowder & Reganold, 2015; Home et al., 2017).

These expenditures, which are frequently more than those for traditional farming systems, include equipment, training, and upkeep. As a result of the insubstantial control of the law of the land, the psychosocial costs of the conversion, the economic costs of the changeover interval, the sustained difficulty of obtaining dependable marketing channels for organic outputs, and the ever-decreasing profit potentials, there remains a slew of challenges to organic farming. Therefore, it cannot be stressed enough how vital it is to give farmers the needed support in access to markets and subsidies, as well as a range of support. Nonetheless, it is impossible to ignore the capacity of organic farming to advance sustainable farming and enhance food security (Sharma, 2024).

13.4.5 How Organic Farming Supports Food Security

According to Indonesia's Law Number 18 of 2012 on Food, food security means households have sufficient, safe, nutritious, and affordable food. It is grounded in four core dimensions: availability, accessibility, utilization, and stability. Food security constitutes a critical issue within the Indonesian context. As an archipelagic country with a large population, it faces complex challenges in ensuring sustainable food security. Efforts to achieve food security face major constraints, including climate change, population growth, land conversion, and volatility in national and global food prices. Food security, in turn, is shaped by the availability, accessibility, and stability of food supplies.

Sustainable agricultural practices, including organic agriculture, aim to minimize agrochemical inputs and reduce environmental pollution (Abobatta & Fouad, 2024). This is one of the most contentious statements made about the SDGs, as the SDGs claim that the goals are interconnected, whereas many experts claim that goals such as sustainable agriculture have to be attained in isolation. Munné-Bosch & Bermejo focus on sustainable agriculture as the reference point in the SDGs and claim that sustainable agriculture is the primary approach to mitigate the pillars of food insecurity. We are trying to show this in Fig. 13.1 as depicting food security as the center point, and having sustainable agriculture as the answer to the food insecurity challenges (Munné-Bosch & Bermejo, 2024). Soni et al. have also reported the practices, benefits, and the transition that sustainable agriculture proposes from traditional farming to the organic farming system (Soni et al., 2022) (Fig. 13.2).

Nevertheless, there needs to be a modification in the activities of the farmers, and there is the need for a slow switch from traditional to organic farming, which is a

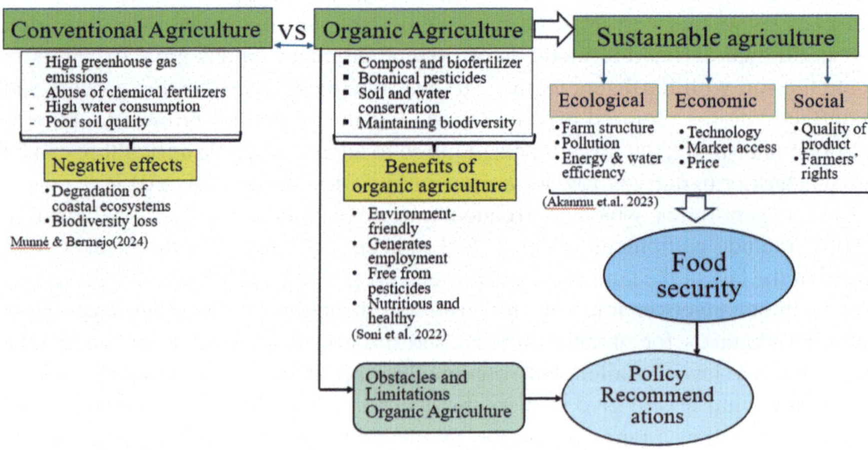

Fig. 13.2 The linkage of organic agriculture and sustainable agriculture supports national food security. *Source* Analysis from a combination of several sources (Akanmu et al., 2023; Meena et al., 2020; Owasa & Fall, 2024; Soni et al., 2022)

ladder to sustainable farming, and consequently boosts the food safety of a country and is in accordance with the world strategies like the Sustainable Development Goals and climate change adjustments (Akanmu et al., 2023; Meena et al., 2020; Owasa & Fall, 2024; Soni et al., 2022). Various organic farming practices need to be addressed, so they need government support, and to achieve food security, policies that favor farmers are also needed. Sustainable development principles, especially the development of good here means eco-friendly, have to be sustained, which translates into the long-term availability of food, plus how agricultural patterns endure and withstand shocks. There are positive feedback between sustainable food security and sustainable agriculture, and/or tackling the negative perishables of a growing population, environmental degradation/shocks, and climate change (Abobatta & Fouad, 2024). Sustainable agriculture aims to continue producing them to produce enough food without endangering the capacity of future generations to do the same.

Sustainable agriculture equips societies with the means and strategies required to fulfill the nutritional needs of the population without negatively impacting the resources and tools available to future populations. Food security means access to food for all and the ability to consume food. These concepts are necessary to build and design a food system that remains able to absorb and endure disturbances and shocks. Implementing both concepts provides the basis for a sustainable future. Global emergencies and inequalities and resource-poor economies, present challenges that are extremely complex and require a nuanced approach to these challenges (Sarma et al., 2024). Addressing challenges in agriculture in particular at a system level must focus on organic agriculture. Organic agriculture addresses socio-ecological challenges by integrating the safeguarding of the environment while promoting economic progress and social advancement. This chapter gives evidence on sustainable development to Indonesia while contributing to the discourse. It offers practical tools to policymakers, researchers, and actors at the international level and grounds these in the realities of the local to design sustainable, inclusive, and resilient food systems.

13.5 Recommendations

Policy support, favoring organic farmers, research investment, and sustainable agricultural techniques suitable for areas with limited resources are needed to implement organic farming (Afaf, 2025). Organic farming represents sustainable agriculture that helps protect the environment, but its high costs discourage farmers from adopting it. The growth of organic farming can be achieved by subsidizing farmers or retailers, which may simultaneously reduce the socially responsible requirements of the retailers as well as the social premium. Lastly, government subsidies for organic farming, whether given to retailers or farmers, shape production and purchasing dynamics, as explained by evolutionary game theory (Yang et al., 2024).

Organic farming supports green jobs, infrastructure, and skills, as emphasized in EU strategies like the Green Employment Initiative and Regional Policy 2014–2020. The focus is on low-carbon technology, energy efficiency, and renewable energy in rural areas, alongside programs to increase youth employment and skills for Not in Employment, Education, or Training (NEETs) aged 16–24 (Popovici et al., 2018). Therefore, the price risk arises from the underdeveloped local market for organic products and the significant price disparity between conventional and organic products. Therefore, the demand for urban land and land conversion, such as residential and commercial, is constantly increasing, which could limit the expansion of areas under organic agricultural management (Popovici et al., 2018).

The bottom line is that there is enough food to be had when climate change mitigation strategies are put in place; farming cooperatives and Non-Governmental Organizations (NGOs) help support organic farming through tech, finance, and market assistance. Government support in organic farming and these players in the market and consumers are essential to the collaborative organic farming approach (Lee, 2021). The incorporation of new age technology is the farming approach of choice to help diminish the constraints and difficulties of organic farming. This encourages the environmentally friendly transition to improved and more productive systems of agriculture, and in turn enhances the farming community's quality of life. In this context, organic farming is an example of agricultural sustainability at the global level (Gamage et al., 2023).

13.6 Conclusion

The trend of organic agriculture in the last decade in Indonesia shows a decrease in the area of organic farming. This condition indicates uneven adoption and tends to be sporadic in some regions. However, the potential area for organic farming has the opportunity to be developed and can increase regional and national food security. The agricultural system also supports a sustainable agricultural system by implementing organic farming practices. Of the various organic agricultural commodities, coffee and rice have the potential to develop, mainly supported by government policies.

Those who are involved in organic farming are farmers with small scale. Therefore, the agricultural business will continue depending on the farmer who consistently cultivates organic agriculture, so that a healthy and sustainable system and ecosystem is achieved.

On a small scale, the organic agriculture development is still scattered in certain parts of the region sporadically. For this reason, it is also necessary to develop an agricultural system on a larger scale.

Some of the obstacles to organic farming include: lack of training, weak farmer support, limited funding, and low market demand. Applying the Triple Bottom Line (TBL), a profit, people, and environment approach can help strengthen the economic and social benefits of organic farming, particularly for rural livelihoods and sustainable employment. Organic farming in the context of climate change also supports sustainable agriculture and can contribute to regional and national food security, especially as organic systems expand.

Potential Conflict of Interest On behalf of all authors, the corresponding author states that there is no conflict of interest.

Informed Consent Informed consent was received from all subjects participating in the study.

AI Disclosure The authors improved the document quality using Grammarly and also conducted AI-detection checks and plagiarism screening.

References

Abobatta, W. F., & Fouad, F. W. (2024). Sustainable agricultural development: Introduction and overview. In *Achieving food security through sustainable agriculture* (pp. 1–27). IGI Global.

Afaf, A. (2025). Sustainable food production through organic farming worldwide and locally in Kingdom of Saudi Arabia. *Edelweiss Applied Science and Technology, 9*(3), 1144–1158. https://doi.org/10.55214/25768484.v9i3.5439

Akanmu, A. O., Olowe, O. M., Phiri, A. T., Nirere, D., Odebode, A. J., Karemera Umuhoza, N. J., Asemoloye, M. D., & Babalola, O. O. (2023). Bioresources in organic farming: Implications for sustainable agricultural systems. *Horticulturae, 9*(6), 659. https://doi.org/10.3390/horticulturae9060659

Altieri, M. A., & Nicholls, C. I. (2017). Agroecology: A brief account of its origins and currents of thought in Latin America. *Agroecology and Sustainable Food Systems, 41*(3–4), 231–237. https://doi.org/10.1080/21683565.2017.1287147

Bezner Kerr, R., Madsen, S., Stüber, M., Liebert, J., Enloe, S., Borghino, N., Parros, P., Mutyambai, D. M., Prudhon, M., & Wezel, A. (2021). Can agroecology improve food security and nutrition? A Review. *Global Food Security, 29*, Article 100540. https://doi.org/10.1016/j.gfs.2021.100540

Brumă, I. S., Jelea, A. R., Rodino, S., Bertea, P. E., Butu, A., & Chițea, M. A. (2023). A bibliometric analysis of organic farming and voluntary certifications. In *Agriculture (Switzerland)* (Vol. 13, Issue 11). Multidisciplinary Digital Publishing Institute (MDPI). https://doi.org/10.3390/agriculture13112107

Chopin, P., Menegat, A., Bergkvist, G., Dahlke, S., Jäck, O., Karlsson, I., Lana, M., Ortman, T., Reumaux, R., Öborn, I., & Watson, C. A. (2023). The reflection of principles and values in worldwide organic agricultural research viewed through a crop diversification lens. A bibliometric review. In *Agronomy for sustainable development* (Vol. 43, No. 1). Springer-Verlag Italia s.r.l. https://doi.org/10.1007/s13593-023-00875-y

Crowder, D. W., & Reganold, J. P. (2015). Financial competitiveness of organic agriculture on a global scale. *Proceedings of the National Academy of Sciences, 112*(24), 7611–7616. https://doi.org/10.1073/pnas.1423674112

David, W., & Ardiansyah. (2017). Organic agriculture in Indonesia: challenges and opportunities. *Organic Agriculture*, 7(3), 329–338. https://doi.org/10.1007/s13165-016-0160-8

David, W., & Alkausar, S. (2023). Statistik Pertanian Organik Indonesia. In *Universitas Bakrie Press and Indonesia Organic Alliance* (Vol. 1, No. 1). B Press Penerbitan Bakrie Press.

Devi, O. R., & Nengparmoi, T. (2023). Organic farming-concepts. In *Recent trend in agriculture* (Vol. 5, pp. 351–366). Integrated Publication. https://www.researchgate.net/publication/370962767

FAO. (2018). *The future of food and agriculture—Alternative pathways to 2050.* FAO.

FAO. (2020). The state of food and agriculture 2020. Overcoming water challenges in agriculture. In *The state of food and agriculture 2020*. FAO. https://doi.org/10.4060/cb1447en

FiBL & IFOAM. (2025). Organics International the World of organic agriculture Statistics & Emerging Trends 2025. Helga Willer, Jan Trávníček and Bernhard Schlatter (Eds.). https://www.fibl.org/fileadmin/documents/shop/1797-organic-world-2025.pdf

Gamage, A., Gangahagedara, R., Gamage, J., Jayasinghe, N., Kodikara, N., Suraweera, P., & Merah, O. (2023). Role of organic farming for achieving sustainability in agriculture. *Farming System, 1*(1), Article 100005. https://doi.org/10.1016/j.farsys.2023.100005

Grimm, M., & Luck, N. (2023). Experimenting with a green 'green revolution'. Evidence from a randomised controlled trial in Indonesia. *Ecological Economics, 205*, 107727. https://doi.org/10.1016/j.ecolecon.2022.107727

Haniyah, D. N., Djuwendah, E., Judawinata, M. G., & Sadeli, A. H. (2022). Usaha pertanian organik berbasis CSA (Community Supported Agriculture) (Studi Kasus di "Seni Tani", Kota Bandung, Indonesia). *Mimbar Agribisnis: Jurnal Pemikiran Masyarakat Ilmiah Berwawasan Agribisnis, 8*(2), 962. https://doi.org/10.25157/ma.v8i2.7518

Hariyani, N., & Sofwani, A. S. (2025). Potential and challenges of organic vegetable development in supporting sustainable food security. *Jurnal Green House, 3*(2), 38–44.

Hidayat, A., & Setiawan, R. (2021). Sustainable coffee farming in Temanggung: Organic practices and market development. *Journal of Sustainable Agriculture, 9*(2), 125–137.

Home, R., Bouagnimbeck, H., Ugas, R., Arbenz, M., & Stolze, M. (2017). Participatory guarantee systems: Organic certification to empower farmers and strengthen communities. *Agroecology and Sustainable Food Systems, 41*(5), 526–545. https://doi.org/10.1080/21683565.2017.1279702

Irham, Wahyu Widada, A., Nurhayati, A., Anantasari, E., Yustika Devi, L., & Subejo. (2020). Indonesian organic farmers: The long journey of farmers' groups towards organic farming sustainability (A Case Study in Rukun Farmers Group, Yogyakarta). *IOP Conference Series: Earth and Environmental Science, 518*(1), 012030. https://doi.org/10.1088/1755-1315/518/1/012030

Kimhi, A., & Tzur-Ilan, N. (2021). Structural changes in Israeli family farms: Long-run trends in the farm size distribution and the role of part-time farming. *Agriculture, 11*(6), 518. https://doi.org/10.3390/agriculture11060518

Krause, J. (2017). Analysis of economic performance and opportunities for the development of organic farming: Case study of the Czech Republic. *Pakistan Journal of Agricultural Sciences, 54*(03), 717–724. https://doi.org/10.21162/PAKJAS/17.2527

Kurniawan, H., Supriyadi, A., & Wulandari, F. (2023). Organic livestock farming in Sumba: Integrating traditional knowledge and modern practices. *Agricultural Sciences, 12*(1), 45–59.

Lal, R. (2020). Regenerative agriculture for food and climate. *Journal of Soil and Water Conservation, 75*(5), pp. 123A–124A (Soil and Water Conservation Society). https://doi.org/10.2489/jswc.2020.0620A

Lee, S. (2021). In the era of climate change: Moving beyond conventional agriculture in Thailand. *Asian Journal of Agriculture and Development, 18*(1), 1–14. https://doi.org/10.37801/ajad2021.18.1.1

Malhan, P., & Ram, T. (2024). Farmers' perception towards the benefits of organic farming in Haryana. *Asia-Pacific Journal of Management Research and Innovation, 20*(1), 28–39. https://doi.org/10.1177/2319510X231179281

Manida, M. (2021). Organic farming—Current status and opportunities for future development. Agriculture & Food: E-Newsletter. www.agrifoodmagazine.co.in

Maun, M. Y. I., & Kaleka, M. U. (2024). Penerapan Pertanian Ramah Lingkungan Guna Menyediakan Pangan Sehat dan Aman di Indonesia. *Jurnal Ilmu Pertanian Tropis (JIPT), 1*(1), 14–17.

Meena, R. K., Meena, R., Naik, B. S. S. S., Meena, B. L., & Meena, S. C. (2020). Organic farming-concept, principles, goals & as a sustainable agriculture: A review. *International Journal of Chemical Studies, 8*(4), 24–32. https://doi.org/10.22271/chemi.2020.v8.i4a.9812

Munné-Bosch, S., & Bermejo, N. F. (2024). Fruit quality in organic and conventional farming: advantages and limitations. *Trends in Plant Science,* 29(8), 878–894. https://doi.org/10.1016/j.tplants.2024.01.011

Ministry of Agriculture. (2019). Strategic plan of the Ministry of Agriculture 2020–2024. https://pertanian.go.id

Ministry of Agriculture. (2020). *Pedoman Teknis Pengembangan Desa Pertanian Organik Berbasis Komoditas Perkebunan Tahun 2020 Refocusing [Technical Guidelines for the Development of Organic Agricultural Villages Based on Plantation Commodities in 2020 Refocusing].*

Ministry of Agriculture. (2023). *Laporan tahunan pertanian organik di Indonesia.*

Nindito, S., & Tamtomo, K. (2020). Revisiting Social Movement in Organic Agriculture Community in Yogyakarta, Indonesia. In *Annual International Conference on Social Sciences and Humanities (AICOSH 2020).* https://doi.org/10.2991/assehr.k.200728.025

Ortega, R., Miralles, I., Domene, M. A., Meca, D., & del Moral, F. (2024). Ecological practices increase soil fertility and microbial diversity under intensive farming. *Science of the Total Environment, 954*, Article 176777. https://doi.org/10.1016/j.scitotenv.2024.176777

Owasa, H. A., & Fall, A. F. (2024). Food security in developing countries: Factors and mitigation. *American Journal of Climate Change, 13*(03), 391–405. https://doi.org/10.4236/ajcc.2024.133018

Pereira, P., Bogunovic, I., Inacio, M., Zhao, W., & Barcelo, D. (2023). Agriculture intensification impacts on soil and water ecosystem services. https://doi.org/10.5194/egusphere-egu23-1423

Peshin, R., & Dhawan, A. K. (2019). In R. Peshin & A. K. Dhawan, (Eds.), *Integrated pest management: Innovation-development process.* Springer Netherlands. https://doi.org/10.1007/978-1-4020-8992-3

Popovici, E.-A., Grigorescu, I., Mitrică, B., Mocanu, I., & Dumitraşcu, M. (2018). Farming practices and policies in shaping the organic agriculture in Romania. A showcase of Southern Romania. *Romanian Agricultural Research, 35*, 163–175. https://doi.org/10.59665/rar3521

Pretty, J. (2008). Agricultural sustainability: Concepts, principles and evidence. In *Philosophical transactions of the royal society B: Biological sciences* (Vol. 363, No. 1491, pp. 447–465). Royal Society. https://doi.org/10.1098/rstb.2007.2163

Pretty, J., Benton, T. G., Bharucha, Z. P., Dicks, L. V., Flora, C. B., Godfray, H. C. J., Goulson, D., Hartley, S., Lampkin, N., Morris, C., Pierzynski, G., Prasad, P. V. V., Reganold, J., Rockström, J., Smith, P., Thorne, P., & Wratten, S. (2018). Global assessment of agricultural system redesign for sustainable intensification. *Nature Sustainability, 1*(8), 441–446. https://doi.org/10.1038/s41893-018-0114-0

Raj, J., Jat, S., Kumar, M., Reema, & Yadav, A. (2024). The role of organic farming in sustainable agriculture. *Advances in Research, 25*(3), 128–136. https://doi.org/10.9734/air/2024/v25i31058

Reganold, J. P., & Wachter, J. M. (2016). Organic agriculture in the twenty-first century. *Nature Plants, 2*(2), 15221. https://doi.org/10.1038/nplants.2015.221

Rekha, Naik, S. N., & Prasad, R. (2006). Pesticide residue in organic and conventional food-risk analysis. *Journal of Chemical Health & Safety, 13*(6), 12–19. https://doi.org/10.1016/j.chs.2005.01.012

Sarma, J., Rajkhowa, S., Bisharad, A., Roy, S., Kapil, N., & Kalita, A. (2024). Impacts of environmental factors upon food security. In *Food security in a developing world: Status, challenges, and opportunities* (pp. 29–52). Springer Nature Switzerland. https://doi.org/10.1007/978-3-031-57283-8_3

Savero Qomara, A., & Noer, L. R. (2025). Global trends in organic food research: A bibliometric analysis. *6*(3). http://jist.publikasiindonesia.id/1158

Sharma, S. (2024). Organic agriculture for sustainable food systems: A comprehensive review of benefits and constraints. *Turkish Journal of Agriculture—Food Science and Technology, 12*(8), 1476–1481. https://doi.org/10.24925/turjaf.v12i8.1476-1481.6482

Soni, R., Gupta, R., Agarwal, P., & Mishra, R. (2022). Organic farming: A sustainable agricultural practice. *Vantage: Journal of Thematic Analysis, 3*(1), 21–44.

Sujianto, Ermiati, & Sudjarmoko, B. (2023). Agricultural practice assessment to accelerate organic rice farming adoption: A case study in West Java, Indonesia. In *BIO Web of Conferences* (Vol. 69, p. 04002). https://doi.org/10.1051/bioconf/20236904002

Sutarta, W., Utami, A., & Raharjo, S. (2022). Integrating traditional subak system with organic rice farming in Bali: A pathway to sustainable agriculture. *Bali Agricultural Journal, 14*(3), 201–213.

Tamtomo, K. (2021). Unarticulated tensions in the marketization of organic agriculture: The case of pioneer organizations in Yogyakarta, Indonesia. *South East Asia Research, 29*(2), 195–213. https://doi.org/10.1080/0967828X.2021.1931422

Tsvetkov, I., Atanassov, A., Vlahova, M., Carlier, L., Christov, N., Lefort, F., Rusanov, K., Badjakov, I., Dincheva, I., Tchamitchian, M., Rakleova, G., Georgieva, L., Tamm, L., Iantcheva, A., Herforth-Rahmé, J., Paplomatas, E., & Atanassov, I. (2018). Plant organic farming research—Current status and opportunities for future development. *Biotechnology & Biotechnological Equipment, 32*(2), 241–260. https://doi.org/10.1080/13102818.2018.1427509

Watson, C. A., Atkinson, D., Gosling, P., Jackson, L. R., & Rayns, F. W. (2002). Managing soil fertility in organic farming systems. *Soil Use and Management, 18*(s1), 239–247. https://doi.org/10.1111/j.1475-2743.2002.tb00265.x

Wezel, A., Herren, B. G., Kerr, R. B., Barrios, E., Gonçalves, A. L. R., & Sinclair, F. (2020). Agroecological principles and elements and their implications for transitioning to sustainable food systems. A review. In *Agronomy for sustainable development* (Vol. 40, No. 6). Springer-Verlag Italia s.r.l. https://doi.org/10.1007/s13593-020-00646-z

Willer, H., & Lernoud, J. (2019). *The world of organic agriculture: Statistics and emerging trends 2019.*

Willer, H., Schlatter, B., & Trávníček, J. (2023). *The world of organic agriculture statistics and emerging trends 2023, Edited by*. https://doi.org/10.5281/zenodo.7572890

Yang, X., Dai, X., & Zhang, Y. (2024). The government subsidy policies for organic agriculture based on evolutionary game theory. *Sustainability, 16*(6), 2246. https://doi.org/10.3390/su16062246

Zainuddin, I., Subagjo, I., Ayunisa, W., Asfa Fuad, N., Tandirerung, M., Parmiyatni, S., Mumpuni Handayani, S., & Ngudiwaluyo, S. (2024). Strategy for developing the organic fertilizer industry to support sustainable agricultural practices in Indonesia. In *Proceedings of the 2nd International Conference on Nature-Based Solutions in Climate Change, RESILIENCE 2023, 24 November 2023*. https://doi.org/10.4108/eai.24-11-2023.2346483

Chapter 14
Climate Village Program in Indonesia: Concept and Implementation of Climate Change Adaptation and Mitigation Actions

Iwan Setiajie Anugrah, Badrun Susantyo, Carolina, Muslim Sabarisman, Hari Harjanto Setiawan, Rosita Novi Andari, and Rachmini Saparita

Abstract Climate change presents a growing challenge to Indonesia, reflected in rising temperatures, irregular rainfall patterns, and the increasing frequency of hydrometeorological hazards. In response, the Indonesian government introduced the Climate Village Program (CVP), nationally led by the Ministry of Environment and Forestry as a framework for community-based climate action. This chapter examines the conceptual foundations, field-level practices, and policy context of CVP through a combination of literature review, case analysis, and the use of national datasets. The program has a comprehensive scope covering adaptation, mitigation, and institutional capacity development in addition to local involvement, indigenous knowledge, and cross-cutting collaboration within sectors. Yet there are several challenges to implementing the program: regional disparities in capacity, low public

I. S. Anugrah (✉) · B. Susantyo · Carolina · M. Sabarisman · H. H. Setiawan · R. N. Andari · R. Saparita
Research Center for Social Welfare, Villages and Connectivity, National Research and Innovation Agency (BRIN), Jakarta, Indonesia
e-mail: iwan035@brin.go.id

B. Susantyo
e-mail: badr004@brin.go.id

Carolina
e-mail: caro001@brin.go.id

M. Sabarisman
e-mail: musl008@brin.go.id

H. H. Setiawan
e-mail: hari047@brin.go.id

R. N. Andari
e-mail: rosi014@brin.go.id

R. Saparita
e-mail: rach001@brin.go.id

G. K. Panda et al. (eds.), *Socio-Ecological Crisis of Development*, Sustainability Solutions, https://doi.org/10.1007/978-3-032-22171-1_14

awareness of climate change problems, insufficient cooperation between agencies and financial constraints. Within a global framework, the paper compares the CVP with international models such as CSV approach, suggesting ways to improve based on local innovation processes and better financing mechanism and adopting digital monitoring tool. Through experiences of Surabaya, Wonosobo (local government), Banjarmasin, and Wajo, they have shown both the opportunities as well as challenges in local climate resilience building. The CVP has a strong potential to serve as a model for participatory climate governance, particularly in the Global South. Its success, however, hinges not just on community engagement, but also on necessary reforms in governance, financing, and technical systems. The findings emphasize the need to align national climate goals with local capacities, and to foster adaptive institutions that can learn from diverse implementation experiences.

Keywords Climate village program (CVP) · Community-based governance · Environmental sustainability

14.1 Introduction

Climate change is one of the most significant challenges confronting humanity in the twenty-first century and has wide-ranging implications for environmental systems, economic security, and social well-being (IPCC, 2021). The indication of climate change is visible in phenomena such as higher global temperature, changes in patterns of precipitation, and an increase in the frequency and intensity of hydrometeorological disasters, alongside increases in sea levels (NASA, 2020). The impact of these changes in the environment will be felt not only all over the world but especially by developing countries like Indonesia, with its geographical and economic area already lying within its natural environments (Bappenas, 2019). As an archipelagic country with a long coastline, Indonesia is at a high risk of climate hazards and has been highly regarded for its rich biodiversity, which heavily relies on the sector sensitive to climate change that includes agriculture, fisheries, and forestry (KLHK, 2021).

It is getting harder to ignore the effects of climate change in Indonesia. Frequent floods, droughts, landslides, and coastal erosion are affecting people, their livelihoods, and essential infrastructure (Maharani et al., 2022). These urgencies not only harm the environment, but they are also compromising Indonesia's progress in critical development issues, including poverty reduction, food security, public health, and water management (Aggarwal et al., 2018). The impact on the rural and coastal communities is significant, and many of them are already marginalized and feel it hardest. They are generally not able to afford the necessary adaptation, and their ability to adapt is conditioned by how much access they have to knowledge, institutional support, and money (Measey, 2010; Mulyasari et al., 2020; Pawitan, 2018; Yusuf & Francisco, 2009).

In response to these escalating climate concerns, the Indonesian government—represented by the Ministry of Environment and Forestry (MoEF)—initiated

ProKlim/CVP in 2012 (KLHK, 2016). Ankaase is a national initiative that aims to build resilience at the local level and promote inclusive, participatory climate action in local communities. At village level, it encourages mix of mitigation and adaptation actions including water management systems, tree planting programs, practice of organic agriculture, renewable energy use, up to strengthening disaster preparedness (Ghina & Zunariyah, 2023; Santoso & Rahaju, 2020). CVP is built on the theoretical basis of four dimensions of legitimacy—economic, sociocultural, ecological, and institutional—and assesses implementation through three principal indicators: adaptation activities, mitigation actions, and community supports systems (Faedlulloh et al., 2019a, 2019b).

While CVP has a strong vision and is squarely in line with Indonesia's climate policy and Nationally Determined Contributions (NDCs), there remain many scale-up issues as well as benefits that are not consistently delivered across different regions. The government had an ambitious target of creating 20,000 climate villages by 2024, yet achieved only just over a quarter of that goal by the year 2020 (Budget Study Center, 2021). There are also a number of long-term barriers that continue to stall momentum, including poor public recognition of climate risks (Rinaldi et al., 2017), lacking financial and technical capacity in remote areas (Faedlulloh et al., 2019a, 2019b), and lackluster cooperation among institutions (Masripatin, 2017; MoEF: Ministry of Environment and Forestry Republic Indonesia, 2021). Compounding these difficulties is a steep decline in national funding for climate-related programs—from IDR 132.4 trillion in 2018 to only IDR 77.81 trillion in 2020—prompting worries over the initiative's future financial stability.

One of the shortfalls in the current literature is that there have been few large comparative studies assessing how CVP operates in different ecosystems and institutional settings throughout Indonesia. The majority of published literature has centered on the shaping of the program or reported success stories in isolation without systematically examining what facilitates or hampers successful roll-out across settings (Maharani et al., 2022; Rinaldi et al., 2017). With the geographic disparity across Indonesia, labeling all implementation as uniform and top-down is clearly inappropriate.

To fill this gap, in the present study, an innovative dual-layered analytical methodology to explore CVP's effectiveness across different settings is proposed. First, the article conducts a thematic policy analysis and reviews national regulations, government reports, and academic literature to map dispositive normative and institutional structure of CVP. Second, this paper presents case studies from four representative regions Surabaya, Banjarmasin, Wonosobo, and Wajo to illustrate Indonesia's ecological and socio-political diversity.

This double-perspective approach allows the study to link macro- with micro-disjunctures of climate governance. Through the investigation of local innovations, policy voids, stakeholder interactions, and implementation obstacles, the study provides theoretical insights while also being operationally relevant. In addition, by locating CVP within global debates around the Climate-Smart Villages (Aggarwal et al., 2018) and Community-Based Adaptation frameworks (Selje et al., 2024), the research contributes to international conversation on inclusive, decentralized, and sustainable climate governance in developing contexts.

As a whole, the proposed work will bring new knowledge about how national climate efforts like CVP can be transposed to succeed and scale in diverse local settings. It offers key policy and strategic options as well as empirical evidence in support of improving the design, implementation, and evaluation of such community-based climate programs to increase Indonesia's climate resilience.

14.2 Materials and Methods

14.2.1 Research Methods

The methodologies utilized in this study are the qualitative descriptive approach and based on a detailed literature review. By using this approach, we can delve into the complexities of the concept and practice of the CVP within a larger scenario that is climate adaptation and mitigation landscape in Indonesia. Consistent with Creswell (2018), qualitative research is an investigation into a phenomenon that looks at the depth, detail, and context of the situation, also known as thick description, to best understand phenomena. Document analysis is the central analytic approach that provides a history of the development and current expression of CVP in different geographic and institutional landscapes (Bowen, 2009).

14.2.2 Data Collection

The literature review is a data-driven review of peer-reviewed journals, policy reports, and governmental literature, as well as case studies that explain the emerging issues and best practices on climate change adaptation and resilience. Secondary data used in this research falls into four primary sources: (1) government official documents, such as regulations, annual reports, and program evaluations from Ministry of Environment and Forestry (KLHK), particularly on CVP; (2) scientific journals that cover peer-reviewed articles of community-based climate adaptation (CBA) approaches, climate policy, and empirical research on CVP; (3) empirical data based on experiences from different implementation sites in Indonesia, to include best practices and challenges faced; and (4) online resources, containing relevant web-based information including project websites, news articles, and digital publications. Together, these sources are a rich resource for studying the policy regime, organizational politics, and conditions of life on CVP.

14.2.3 Scope of Analysis

This chapter offers a descriptive analysis of a Indonesia's Climate Village Program (CVP), a community-based climate adaptation initiative which has been implemented in many regions across the country. As shown in Fig. 14.1, the analysis is arranged into three sections, namely:

(1) a summary of CVP's guiding principles and participatory governance model;
(2) an exploratory analysis of adaptation and local variation in a cross-site empirical study with diverse ecological and social settings;
(3) an examination of the major constraints and opportunities that have influenced the program's effectiveness.

The trends and implementation strategies of the initiative and the accumulated knowledge in various global contexts are described, which allows locating CVP in a wider theoretical and policy context. The insights are supplemented by statistical data, qualitative appraisals, and empirical case studies to provide strong analytic arguments. The chapter concludes with policy recommendations aimed at enhancing the sustainability and long-term impact of CVP. The analysis draws on examples from selected CVP villages, representing a range of geographic and socio-ecological contexts.

Conceptual Framework of CVP is based on the assumption that enhancing local resilience to climate-related hazards is a significant avenue to national climate resilience. The program puts the idea of climate villages into the foreground, i.e., localized governance systems that are supposed to organize collective action in response to climate-related threats. The classification of villages as climate villages, by which CVP members are late comers to a participatory decision-making process and are assigned new tasks linked to preparedness, response, and recovery in the event of climate-related shocks. In this participatory paradigm, there is a focus on intersectoral cooperation, the strategic use of information and communication technologies (ICT), and the co-generation of knowledge among members of the community and external technical experts.

Implementation Patterns and Localized Practices. CVP has been implemented in 206 villages across 28 districts in 10 provinces since its introduction in 2015. These regions form a geographically dispersed and heterogeneous population including low-lying coastal regions, upland parts in volcanic regions, and the outlying islands. Although the main logistics of the program are centrally planned, local implementation plans have been constantly modified to the socio-ecological settings. The localized projects with great promise are (a) the formation of Women Disaster Management Committees to provide women with more control over decision-making processes; (b) implementation of cloud-based farming management systems in high-altitude farming areas; and (c) the creation of early warning systems of coastal communities by adopting mobile applications and community radio broadcasting. Limitations and enablers: The program performance has been affected by constraints and enabling factors at various scales of governance. The barriers at the village level

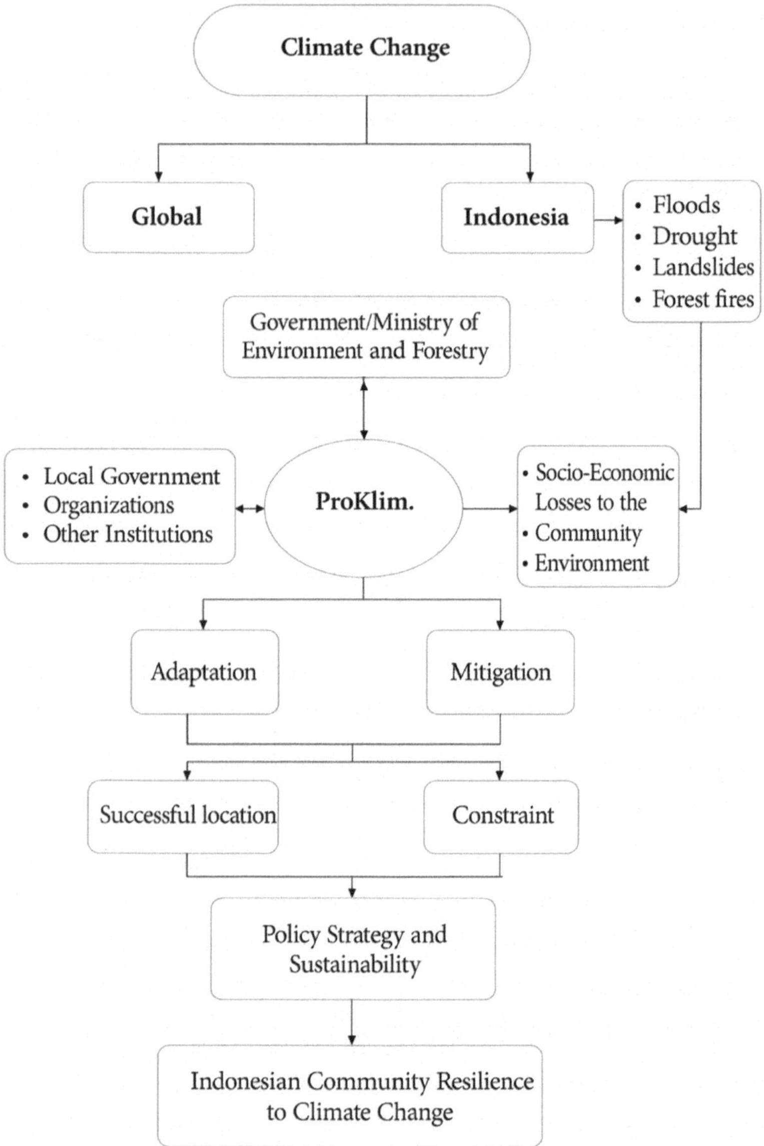

Fig. 14.1 Research methodology process of the climate village program (CVP) study

are low levels of institutional capacity to coordinate, inadequate levels of human and financial resources, and poor intersectoral connections. On the other hand, community-based organizations and the overlap of environmental and social issues have been enabling processes.

The National Level Limitations are lack of technical skills, regulatory, and financial gaps in the sectoral ministries and lack of policy coherence in the various departments within the government. However, the existence of strong program governance mechanisms, good inter-ministerial coordination, and provision of training facilities to build capacities constitute a set of enabling factors. Policy Recommendations: The presented evidence implies three areas where policy intervention should take place as priority. One, to enhance village-level governance networks, institutional capacity building, mobilization of human and financial resources, and comprehensive incorporation of intersectoral planning will have to be given priority. Second, to boost policy coherence on the national level, it is necessary to establish institutional units, which will coordinate cross-sector climate work, a single set of indicators to track program results, and inter-ministerial training courses. Lastly, the implementation of a hybrid financing model, that is, grant-based financing and results-based incentive schemes, would assist in maintaining and expanding the scope of the program.

14.2.4 Research Process

The chapter was written in a systemic progression of steps in order to have a cohesive and evidence based study. These phases will involve: (1) literature gathering, (2) data collection, (3) categorization of supporting materials and information, (4) descriptive qualitative analysis of content in each sub-section of the discussion, and (5) synthesis of the analysis into important findings (Fig. 14.2).

The steps taken are as follows:

1. Literature Collection: The sources of information which were relevant to the study were collected through numerous materials such as scientific journals, academic books, policy reports, and official government documents. The most important of them is the Regulation of the Minister of Environment and Forestry No. 84 of 2016 that lays the legal base of the Climate Village Program (CVP) (KLHK, 2016).
2. Data Classification: The information gathered was systematized by using major themes that were pertinent to the research. These areas are as follows: the basics of CVP, the strategies of climate change adaptation and mitigation, and the enablers and barriers that affect the implementation of the programs (Sugiyono, 2019). This division assisted in organizing the discussion and enabled narrow analysis in every part of the chapter.
3. Thematic categorization is used to sort and group data in an organized manner, allowing the conclusion-drawing process to proceed more clearly, systematically, and comprehensively.
4. Qualitative Analysis: Each thematic area was examined using a qualitative descriptive approach. The process involved reading through the materials carefully and piecing them together into a clear narrative. As the data was reviewed, patterns, connections, and key insights began to emerge, helping to clarify what

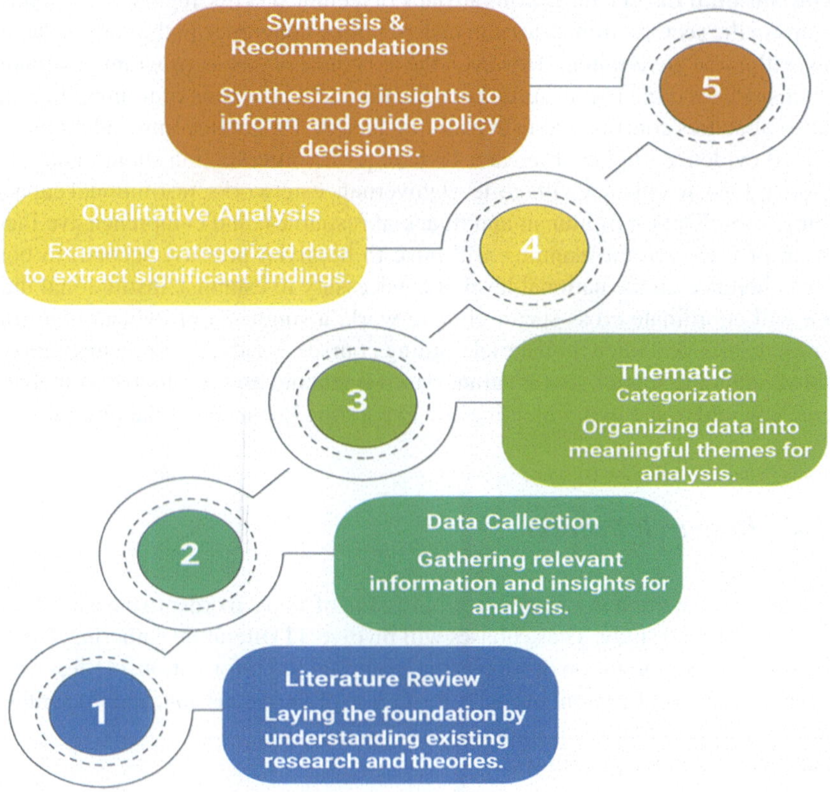

Fig. 14.2 Methodological framework of the study

the program needs and how it should be carried out. The discussion also took into account the local setting—looking at how institutional structures work on the ground, how communities participate, and whether existing policies align with the goals of the program.

5. Synthesis of Findings: In the final stage, the results from the analysis were brought together and distilled into a cohesive set of findings. This process combined insights from each thematic discussion to form an overall picture of the issues, opportunities, and the role that the CVP can play in Indonesia's climate change agenda. From this synthesis, conclusions were drawn and recommendations were developed to help strengthen the program's effectiveness and ensure its long-term sustainability (Fig. 14.3).

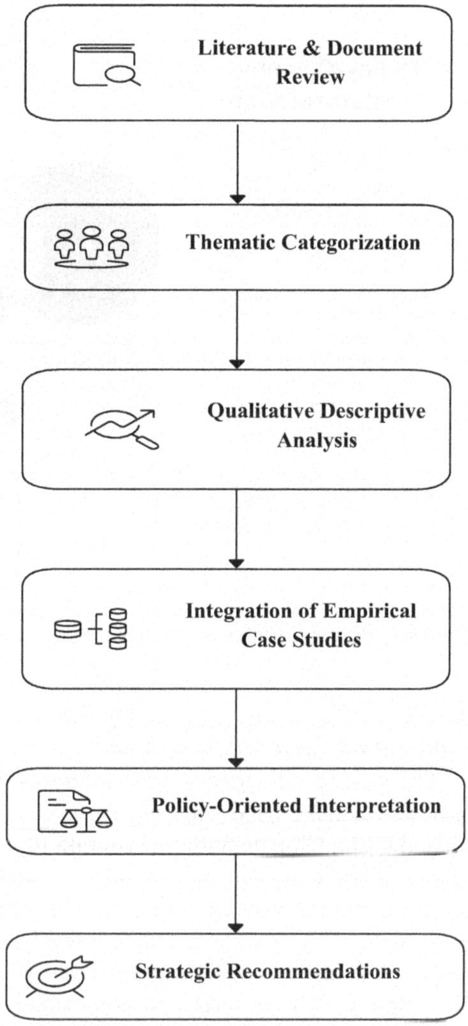

Fig. 14.3 Research process flowchart

14.2.5 Analytical Framework and Process

The collected qualitative data were analyzed and organized thematically, then synthesized through a comparison of findings to produce comprehensive conclusions. Key themes that emerged included the CVP policy framework, adaptation and mitigation strategies, community participation, and program outcomes. Findings from multiple sources were compared to highlight similarities and differences in program implementation across different regions and institutional contexts. Based on this process, evidence-based conclusions and recommendations were formulated to strengthen

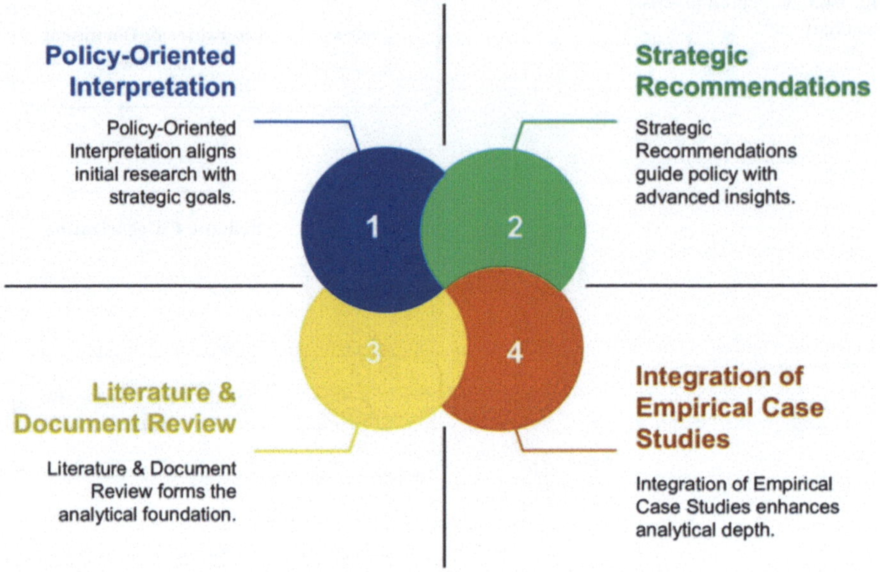

Fig. 14.4 Research methodology framework

the program's effectiveness and ensure the long-term sustainability of the Climate Village Program (CVP) (Fig. 14.4).

The process of writing the manuscript began with a comprehensive literature review, which included searching for scientific articles, policy documents, and reports related to the Climate Village Program (CVP). This was followed by the collection of secondary data from various sources, such as government publications, academic journals, and relevant case studies. The gathered data were then thematically categorized into key focus areas, including adaptation, mitigation, and institutional challenges. A qualitative descriptive analysis was subsequently carried out using a narrative approach to interpret and contextualize the information. Finally, all findings were synthesized to draw meaningful conclusions and develop evidence-based policy recommendations aimed at strengthening the strategic implementation of the CVP.

14.3 Results and Discussion

14.3.1 Concept of the Climate Village Program (CVP) in Indonesia

The Climate Village Program (CVP), locally known as the "Program Kampung Iklim (ProKlim)," is more than a policy initiative; it represents a dynamic socio-ecological framework that exemplifies community-based climate governance in the Global South. Introduced by Indonesia's Ministry of Environment and Forestry (KLHK) in 2012, the program aims to integrate local-scale adaptation and mitigation efforts with participatory mechanisms that actively involve communities (Ghina & Zunariyah, 2023; KLHK, 2016).

To enrich the analytical depth of this study, this section synthesizes empirical evidence derived from documented field practices, government performance reports, and independent reviews. Case studies conducted in four regions—Surabaya, Wonosobo, Banjarmasin, and Wajo—reveal varying outcomes in CVP implementation, highlighting region-specific success factors as well as persistent challenges. These differences underscore that CVP operates within diverse institutional and ecological contexts, necessitating locally tailored strategies to ensure effective program implementation (Maharani et al., 2022; Rinaldi et al., 2017).

This discussion applies a thematic framework focused on three main dimensions: adaptation, mitigation, and sustainability. The analysis is conducted qualitatively, taking into account regional differences in institutional capacity, community participation, and resource availability. In addition, comparative insights are drawn from international models, particularly the Climate-Smart Village (CSV) approach, which emphasizes the integration of climate science, sustainable livelihoods, and community resilience (Aggarwal et al., 2018).

The urgency of the Climate Village Program (CVP) is underscored by Indonesia's high vulnerability to climate-related hazards, including rising temperatures, shifting rainfall patterns, and increasing frequency of hydrometeorological disasters such as floods, droughts, and landslides (Measey, 2010; Mulyasari et al., 2020; Yusuf & Francisco, 2009). These challenges not only threaten environmental sustainability but also have direct impacts on public health, economic productivity, and the country's broader development priorities (Bappenas, 2019; IPCC, 2021; KLHK, 2021).

Conceptually, the Climate Village Program (CVP) focuses on two main strategies. The first, climate adaptation, includes efforts to reduce disaster risks, promote sustainable agricultural practices, conserve water resources, and restore ecological balance. All these measures aim to strengthen community resilience to climate change (Faedlulloh et al., 2019a, 2019b; KLHK, 2016).

The second, climate mitigation, emphasizes the implementation of low-carbon practices, such as 3R-based waste management (Reduce, Reuse, Recycle), energy efficiency, the use of renewable energy, and urban greening initiatives (Santoso & Rahaju, 2020).

While CVP shares conceptual similarities with the Climate-Smart Village (CSV) approach, it is distinguished by its stronger reliance on local knowledge and its alignment with national policy frameworks, compared to CSV, which is more technocratic and science-driven (Aggarwal et al., 2018). This distinction has direct implications for how the program is implemented, particularly in adapting strategies to the diverse social, cultural, and environmental contexts across Indonesia.

The government has set an ambitious target of establishing 20,000 Climate Villages by 2024 (DPR RI, 2021). However, several structural challenges remain significant, including fragmented coordination, limited technical and financial support in rural and remote areas, and underdeveloped monitoring systems (Faedlulloh et al., 2019a, 2019b; KLHK, 2021; Masripatin, 2017; MoEF, 2021).

Rather than making a definitive judgment on the program's overall success, this section focuses on providing a critical, evidence-based understanding of CVP's current status. The findings indicate that, while CVP holds considerable potential as a replicable model for participatory climate action, its effectiveness relies heavily on flexible governance, institutional learning, and the alignment between national ambitions and local realities.

14.3.2 CVP Implementation Patterns and Local Innovation

The implementation of the Climate Village Program (CVP) across different regions of Indonesia demonstrates a strong capacity for adaptation to local conditions, while also highlighting both innovation and systemic challenges. In major cities such as Surabaya, the CVP has successfully combined community-driven waste management with the development of green infrastructure. These initiatives align closely with the principles of urban ecological resilience and the circular economy, reflecting practices observed in similar programs internationally (Gunawati & Rejekiningsih, 2020; Santoso & Rahaju, 2020).

In contrast, rural and highland areas such as Wonosobo showcase local agroforestry systems alongside climate-sensitive irrigation infrastructure. These practices illustrate ecosystem-based adaptation, which harnesses natural resources to strengthen resilience, similar to the models applied within the Climate-Smart Village (CSV) framework (Aggarwal et al., 2018). The ability of such regions to internalize CVP goals is shaped by local leadership, environmental awareness, and external facilitation support.

However, in remote regions such as parts of Wajo, institutional fragility limits program uptake. This echoes findings from Maharani et al. (2022), who argue that policy diffusion without capacity-building mechanisms leads to "islands of success" rather than systemic transformation.

14.3.3 CVP Implementation and Challenges

CVP as a local program of climate resilience is an important role to track its execution. Considering the fact that CVP has been incorporated in the transition policy of Indonesia to a low-emission and climate-resilient future after Indonesia Nationally Determined Contribution (NDC) document. The implementation of CVP has elements and strategies that constitute the central framework referring to the Directorate of Climate Change Control, Ministry of Environment and Forestry. The structure is pegged on three points, which include adaptation, mitigation, and support of sustainability.

Specifically, these aspects are discussed as follows: First, this adaptation component addresses the processes of controlling, handling, and predicting the occurrence of different natural phenomena, i.e., drought, flood, sea level rise, abrasion, erosion, rob, and landslide. Besides, the adaptation component is also connected to the activities to augment food security and climate-related diseases control. Second, the mitigation component has numerous activities that include waste management; solid and liquid waste; use of new renewable energy and energy conservation; low GHG emission agricultural cultivation; increased vegetation cover; and forest and land fire prevention and control activities. Third, CVP implementers should critically look at some of the factors that will determine the implementation of the climate resilience program at the local level. ProKlim has joined the Indonesian transition policy to a low-emission and climate-resilient future as described in the Indonesia Nationally Determined Contribution (NDC) document. Referring to the Directorate of Climate Change Control, Ministry of Environment and Forestry, CVP has elements and strategies that constitute the main framework. This framework is anchored on three factors, that is, adaptation, mitigation, and support of sustainability.

In details, the aspects mentioned above are as follows: First, this adaptation component subsumes activities of controlling, handling, and anticipating different natural phenomenon like drought, flood, sea level rise, abrasion, erosion, rob, and landslide. Also, the adaptation element is also connected to the activities that are meant to improve the food security and reduce the impact of climate-related diseases. Second, there are a number of activities that are carried out in the mitigation component, such as waste management of solids and liquids, the utilization of new renewable energy and energy conservations, low-GHG-emission agricultural cultivation, enlarging vegetation cover, and forest and land fire prevention and control measures. Third, the CVP implementers should pay particular attention to a number of elements of the sustainability support component. Sustainability support consists of benefits analysis, development, other activities, capacity, policies, self-reliance, gender, community groups, and external support.

According to the findings of a research conducted by the Budget Study Center, Expertise Agency of the Secretariat General of the Indonesian House of Representatives (2021), it was discovered that the Indonesian government aims to achieve 20,000 climate villages in the year 2024. This is a big goal bearing in mind that by 2018, only 1,548 locations were realized in CVP, and by 2020, 2,775 climate villages were engaged in managing regional climatic change. The low level of the realization of this realization is occasioned by a number of difficulties in the realization of this target. Moreover, it was clarified that the climate change budget in the State Budget (APBN) in the last three years has also been reduced, and the effects of climate change are also increasing and are increasingly being experienced so that in the work plans of ministries/state institutions in the future; commitment has to be enhanced. Therefore, despite being rather high, the climate change budget in the APBN has been moving in the negative direction since 2018–2020. In 2018, climate change ministry/institution budget was IDR 132.4 trillion. This budget was reduced to the amount of IDR 97.66 trillion in 2019 and IDR 77.81 trillion in 2020. In the meantime, the mean mitigation expenditure was IDR 62.7 trillion annually and IDR 40.4 trillion annually to conduct adaptation.

The Fiscal Policy Agency also indicated that 88% of climate change budget is spent on green infrastructure. In the meantime, the rest 12% is applied to finance the supporting activities, including regulation and policy, research and development, capacity building, and community empowerment. Another sustainable financing instrument that is being prepared by the Ministry of Finance is the SDG Bond that is expected to be issued in 2021. It may be utilized in a number of projects or programs in the sphere of environment, renewable energy, energy efficiency, or social activity. One of the initiatives to adjust to climate change, CVP, is facing barriers to the implementation process at the grassroot level, which includes both positive and negative aspects in the community. The decision to take action in order to adapt to climate change or other environmental problems and reduce their effects is not taken easily by many people since the positive outcomes are not immediately experienced.

The Climate Village Program (CVP) is a community-based climate action program that combines the climate change adaptation and mitigation actions at the grassroots level. In implementing this program, local potential must be given first priority so that the communities are dealing with particular vulnerability to climate change. CVP implementation can be understood as the Climate Action Module developed by the Ministry of Environment and Forestry (KLHK) which is carried out in phases using coaching, mentoring, implementation, reporting, and evaluation systems. The activities involve two key pillars, namely adaptation action and mitigation action, and have an incentive and recognition system that will support the activities in enhancing program replication and sustainability.

Adaptation action refers to a sequence of activities aimed at increasing the ability of the people to reduce the impacts of climate change such as extreme weather, climatic variability, uncertain seasons, and hydrometeorological disasters. These are the following forms of implementation action adaptation in CVP: (1) Clean Water and Water Resources Management: Adaptation in most of the regions is done by development of infrastructure such as rainwater reservoir, well infiltration, drip irrigation, and conservation springs. This is especially applicable when there is drought and low water discharge as in the case of Wonosobo and Banjarmasin. (2) Watershed conservation should be done at the community level to rehabilitate the watershed through planting vegetation to hold the erosion, surface water flow arrangement, and natural embankment making. The practice makes the community flood and landslide-resistant. (3) Local Food Security Implementation and Diversification: Agriculture is dynamic and uses a range of stand climates, and CVP has applied agroforestry on-site in response to uncertain weather and reduced food productivity. (4) Manage diseases according to the Community Environment that is trained to handle waste, house stairs, guard sanitation, and control the vector diseases (e.g., malaria and dengue fever) that are likely to grow the consequences because of climate change. (5) Disaster and Social Adaptation Preparedness: Formation groups represent disaster locals, map risk communities, and early system warnings depending on the community, which are part of the adaptation measures in the vulnerable disaster regions.

CVP mitigation measures are geared toward cutting down on greenhouse gas (GHG) emissions by adopting efficient and environmental friendly energy activities. The major activities involve. (1) Management Waste: CVP recycling and composting promotes 3R (Reduce, Reuse, Recycle) practices at the house ladder level by establishing waste banks, composting organic wastes, and sorting wastes. The example of the cities, Depok and Surabaya, which actively participate in the integrated waste bank movement, can be considered as real-life examples. (2) Vegetation and Tree Planting: There is a joint effort in conservation, reforestation, and critical land conservation activities to boost the vegetation cover and absorb carbon. This is a significant test during evaluation of the Tropi CVP Lestari. (3) Renewable energy and energy-saving technology like biogas of livestock wastes, household solar panels, and energy-saving stoves have been adopted in several climate villages. In the areas of Wajo and Lombok, the communities apply biogas instead of LPG gas. (4) Green Transport and Home Energy Conservation: A number of regions have started to implement a sustainable mode of transport and lower the household use of electricity as a locally based form of mitigation contribution.

The execution of the CVP program is stage and methodically, made up of the Determination and Identification of CVP Locations. Depending on the indicators of climate vulnerability, local potential, and social readiness of the community, the places to apply CVP selectively are chosen. The government area conducted data collection with the help of the KLHK technical personnel. (1) Technical Facilitation and Community Assistance: CVP is based on the help of technical teams (environmental services, NGOs, universities) in preparing local climate action plans and creating environmental cadres-real action in the field. The community is performing activities on the basis of adaptation and mitigation modules adapted to the local conditions. Needs are evaluated on each activity. The form CVP is filled in by filling in the information about the activities and documentation on which the assessment and verification will be carried out by the KLHK team. This is the data which is used to award and report to the national system. Evaluation, Verification, and Awards; KLHK assesses every site by a field assessment and verification system. The places that comply with the requirements are presented with CVP certificates, CVP trophies, and recognition categories: Pratama, Madya, Utama, and Lestari.

The use of the Climate Village Program (CVP) shows that effective adaptation and mitigation of climate change are based on local resources, community, and regional capabilities. A number of regions in Indonesia have been used as a role model nationally in the implementation of ProKlim, and success has been measured by achievement action, involvement of many parties and an award by the Ministry of Environment and Forestry (KLHK). Four examples of the areas that stand out in the implementation of CVP will be as follows:

(1) One of the most consistent and progressive cities which have applied the principles of CVP is Surabaya City. Surabaya in 2023 set a record as the city with the most awards of acquisition CVP in Indonesia, more than 23 awards, including the category Main and Sustainable. Successful Keys: Waste Management Innovation is also famous with its integrated waste bank system that covers thousands of households and schools. This waste bank is a mitigation measure (it reduces the methane emission by landfills) and an economic generator to the community. Urban Farming and Green Open Space: As the means of adapting to urban heat and enhancing air quality, the city government promotes the creation of city parks, RT/RW parks, and vertical gardens. Adaptation measures include water conservation and recycling, household wastewater treatment projects, and mass infiltration well construction to minimize seasonal flooding and drought, and community and school engagement. Surabaya employs the power of schools and communities as the agents of climate education with the help of Adiwiyata School, Eco School, and Green Village programs.
(2) Banjarmasin, a city characterized by a dominant aquatic ecosystem, has implemented a community-based management approach. Awards Significant: Award CVP Lestari and 19 Certificates CVP Main Award Categories: The award is awarded to a number of RW and sub-districts actively operating conservation programs, spreading environmental awareness, and adapting to climate change. Practice Spotlighted: Community-based and watershed rehabilitation, rehabilitation of canals, and planting of vegetation by the government on the banks of the rivers and organization of the relocation of the affected settlements during the floods. The Management Waste and Energy RT/RW Community implements the system of rubbish processing and large-scale use of biogas, including small settlements. This strategy will encourage climate resilience, innovations, and sustainable architecture to establish a congenial atmosphere and use a drainage system, which will become a prototype of adaptation on local wisdom.
(3) Areas of mountains such as Wonosobo are exposed to the problem of land erosion, spring degradation, and susceptibility to food security impacts of climate change. Nevertheless, the Regency of Wonosobo managed to establish itself as one of the leaders in food resilience on the basis of community and conservation sources of water energy. Welahan Hamlet in Wonoroto Village is a successful example, having won the CVP Sustainability Trophy for its commitment to environmental management, innovation, and food security. Simple System of Water Saving Irrigation and Utilization Reservoir Usage Technology: Such a solution as the conservation of water can be adjusted to the season, especially during drought and long periods. Organic and Diversification in Agriculture: Farmers create more sustainable systems that minimize the use of pesticides and become resistant to harsh weather. Participation

of Young Farmers and Traditional Institutions: Training and mentoring of the young generation of the village to be involved in the role of environmental cadres and CVP facilitators.

(4) The actual case of the implementation of the CVP concept, which is based on the combination of proper technology, adaptation to agriculture, and environmental protection, is Liu Village in the Wajo Regency, Sabbangparu District. Award Trophy Liu Village Main CVP has 56 locations in Indonesia that were awarded Trophy based on the implementation of category action, mitigation, and adaptation integrated. Practice Featured: Waste Community Livestock uses farm waste to generate clean energy, which is used in cooking, thus less GHG emissions are produced when fossil fuels are burnt. Climate-adaptive agriculture applies agricultural models designed to address the impacts of climate change. Small Forest Rehabilitation and Conservation of Land: The aspect of planting and nurturing beneficial trees becomes part of the endeavor to safeguard the sustainable ecosystem in the village. As an illustration, the effective use of CVP in different areas proves the idea that the success of the program is defined by the policy center and local capacity to recognize risks, mobilize sources of power, and contextual innovation. Technical and sustainable government and partner development support becomes a vital driver of the replication practice expansion. Good. This is in the other parts of Indonesia.

The Climate Village Program (CVP) is a variation of the Climate-Smart Village (CSV) idea on a local scale, based on the idea of community involvement and environmental responsibility.

Experiences with Community-Based Adaptation (CBA) initiatives in Fiji and El Salvador likewise show a strong reliance on participatory approaches. However, the CVP distinguishes itself through its more comprehensive integration into national climate policy frameworks.

14.3.4 Governance and Institutional Challenges

One of the main challenges to strengthening the effectiveness of the Climate Village Program (CVP) is the persistent fragmentation across sectors. Although national policy requires CVP to be integrated into regional planning instruments such as the RPJMD and RAD-GRK, coordination between sectors on the ground remains limited. As noted by Luthfia and Alkhajar (2021), climate-related initiatives in Indonesia often operate independently within their respective sectors, which weaken overall policy coherence.

In several regions, the CVP has not yet been fully embedded in formal development planning documents. This lack of institutionalization affects budget allocation, staffing, and the consistency of monitoring efforts. Moreover, implementation frequently depends on NGOs or short-term donor funding rather than continuous, structural support from the government (Budget Study Center, 2021). These conditions raise concerns about the program's long-term sustainability and local ownership.

From a theoretical perspective, this situation reflects what Tompkins and Adger (2004) describe as an "institutional misfit," where policies formulated at the national level do not fully align with the governance dynamics operating at subnational levels.

14.3.5 National Strategy: Initiation of Climate Village Program (CVP)

Due to the prevalence of the effects of climate change, the Indonesian government has introduced numerous national strategies, such as the introduction of the Climate Village Program (CVP). The Ministry of Environment and Forestry (KLHK) started this program in 2012 and then reinforced it with Permen LHK No. P.84/MENLHK-SETJEN/KUM.1/11/2016, which governs a framework, criteria, and incentive mechanisms of its implementation. CVP is modeled as a national movement, community-based, which seeks to: (1) make climate action more grassroots by allowing the community to define risks and local responses; (2) build local socio-ecological resilience through strengthening technical, institutional, and social capital capacity of village communities; (3) integrate adaptation and mitigation actions in a system of implementation that is measurable. The program also belongs to the list of the Indonesian contribution to the implementation of the Paris Agreement and the achievement of the Nationally Determined Contribution (NDC) target, that is, a 29% reduction in GHG emissions independently and 41% with international support by 2030. The focus on the national policy regarding climate change, CVP plays a central role as a transformative program, which integrates technical, social, and institutional solutions on the local level. The success of CVP shows that climate change adaptation and mitigation cannot be successful without considering the communities as a central player. The capacity of the community, the support of the regional institutions, the inclusion of the development policies, and the technology innovation will play a key role in the future to make CVP the pillar of a resilient, inclusive, and sustainable climate policy.

The strengths of CVP are not only related to the coverage of the policy but also to the collaborative and participatory approach, as well as contextual one. CVP empowers communities to be central players in climate action by adopting a locally oriented agenda of potential. Social capital fueling success of this program are (1) cooperation and community organizations, e.g., RT/RW, PKK, youth groups, and farmer groups; (2) local knowledge and traditional wisdom, e.g., traditional planting patterns, customary water management, and forbidden forest systems; (3) informal community leadership, e.g., religious leaders, customary leaders, and environmental cadres as agents of change. Other sectoral programs incorporated in CVP are watershed (DAS) conservation, household waste management, energy efficiency, disaster risk management, and environmental education, which are all integrated across ministries and institutions. Such synergy allows CVP to become a cross-sector policy platform to develop climate-resilient villages/sub-districts.

14.3.6 Monitoring, Evaluation, and Technical Systems

The current monitoring and evaluation (M&E) system for CVP remains manual, fragmented, and lacking spatial precision. There is no integrated Measurement, Reporting, and Verification (MRV) platform to track adaptation and mitigation outcomes across villages. Consequently, it is difficult to assess the real performance and cost-effectiveness of interventions.

This weakness is particularly problematic for scaling the program or aligning it with international reporting standards (e.g., SDG or Paris Agreement targets). Several studies, including those by Aggarwal et al. (2018) and the Ministry of Environment and Forestry (MoEF, 2021), highlight that the effectiveness of climate policies depends greatly on the use of clear, spatially referenced indicators that focus on measurable outcomes.

Building on these insights, this study proposes the development of a cloud-based dashboard equipped with GIS integration and connected directly to local reporting systems. This approach draws on the success of digital monitoring tools used in South Asia's CSV programs, which have proven effective in enabling faster, more accurate, and evidence-driven evaluations.

As a policy recommendation, spatial MRV systems should be incorporated into the e-CPV platform, accompanied by training for local actors to strengthen their capacity in data reporting. Such steps are expected to enhance accountability and reinforce the program's legitimacy among communities and decision-makers.

14.3.7 Strategic Lessons from Comparative Cases

The analysis of cases in Surabaya, Banjarmasin, Wonosobo, and Wajo highlights not only the diversity of CVP implementation but also the factors that either support or constrain the program's success. In Surabaya, a techno-social model emerges, combining the use of digital tools with strong social capital within the community. Banjarmasin illustrates how traditional water management systems can be revitalized through active civic participation and the enforcement of local regulations.

In Wonosobo, the importance of traditional institutions, such as farmer groups, is evident in maintaining agroecological resilience. Wajo reveals the constraints of weak local government support and the need for external technical assistance. Each case illustrates a different dimension of what Aggarwal et al. (2018) describe as "scalable adaptation pathways," where solutions must be locally grounded but supported by national architecture.

14.3.8 Dynamics, Constraints, and Problems of Program Achievement

The local program of climate resilience CVP is a significant role to monitor its implementation. Because the CVP has been integrated into the transition policy, it plays a crucial role in the transition policy of Indonesia to a low-emission and climate-resilient future following the Indonesia Nationally Determined Contribution (NDC) document. The deployment of CVP has features and policies that form the main framework that refers to the Directorate of Climate Change Control, Ministry of Environment and Forestry. The framework is anchored on three pillars and they are adaptation, mitigation, and sustainability support.

Particularly, the specified aspects can be discussed in the following way: ***First***, the discussed adaptation component deals with the processes of controlling, handling and predicting the occurrence of various natural phenomena, i.e., drought, flood, sea level rise, abrasion, erosion, rob, and landslide. Other than this, the adaptation aspect is also related to the activities in order to increase the food security and the control of the climate-related diseases. ***Second***, the mitigation component contains a lot of activities such as Waste Management; solid and liquid waste; use of new renewable energy and energy conservation; low GHG emission agricultural cultivation; increased vegetation cover; and forest and land fire prevention and control activities. ***Third***, CVP implementers ought to critically examine some of the factors that will define implementation of the climate resilience program at the local level. CVP has become a part of the Indonesian transition policy to the low-emission and climate-resilient future as outlined in the Indonesia Nationally Determined Contribution (NDC) document. In line with the Directorate of Climate Change Control, Ministry of Environment and Forestry, CVP has components and plans that form the major framework. This framework is pegged on three aspects, namely, adaptation, mitigation, and support of sustainability.

In particular, the above aspects are as follows: The ***first*** is that this Adaptation component encompasses the activities of controlling, handling and predicting various natural phenomena such as drought, flood, sea level rise, abrasion, erosion, rob and landslide. Moreover, the element of adaptation is also linked to the activities that are supposed to enhance the food security and minimize the effects of climate-related diseases. ***Second***, several activities are undertaken in the Mitigation component, which include waste management of solid and liquid wastes, new renewable energy and energy conservation, cultivation of low GHG-emission agricultural production, expansion of vegetation cover, forest and land fire prevention and control. ***Third***, several aspects of the sustainability support component deserve close attention of the ProKlim implementers. Sustainability Support is composed of benefits analysis, development and other activities, capacity, policies, self-reliance, gender, community groups and external support.

Based on the results of a study carried out by the Budget Study Center, Expertise Agency of the Secretariat General of the Indonesian House of Representatives (2021), it was identified that the government of Indonesia intends to have 20,000 climate villages by the year 2024. It is a large ambition considering that as of 2018 only 1,548 sites had been implemented in CVP and as of 2020 2,775 climate villages were involved in the management of regional climatic change. The realization of this realization is not at a high level because there are several challenges in the realization of this target. In addition, it was explained that even the climate change budget in the State Budget (APBN) in the previous 3 years has been cut and the impacts of climate change are also rising and are being increasingly felt such that in the future work plans of ministries/state institutions, commitment will have to be strengthened. Thus, although the budget of climate change in the APBN is quite high, it is shifting toward negative values since 2018–2020. In 2018, budget on climate change ministry/institution was IDR 132.4 trillion. This budget declined by the value of IDR 97.66 trillion in 2019 and IDR 77.81 trillion in 2020. Meanwhile, the average mitigation spending was IDR 62.7 trillion every year and IDR 40.4 trillion every year to carry out adaptation.

The Fiscal Policy Agency also reported that 88% of budget on climate change is used up on green infrastructure. Meanwhile, the remaining 12% is used to fund the supporting activities such as regulation and policy, research and development, capacity building and community empowerment. The other sustainable financing tool is currently being developed by the Ministry of Finance is the SDG Bond which is projected to be issued in 2021. It can be applied to several projects or programs within the field of environment, renewable energy, energy efficiency, or social activity. CVP is one of the programs aimed to adapt to climate change, and the process of its implementation at the grassroot level has both positive and negative sides in the community. Many people do not take the decision to act in order to adapt to climate change or other environmental issues and decrease its impact easily because the results are not felt immediately.

The Climate Village Program (CVP) is a local-based climate action program which integrates the climate change adaptation and mitigation measures at the grass root levels. When carrying out this program, local potential should be given the priority of focus so that the communities will be addressing specific vulnerability to climate change. CVP implementation may be interpreted as Climate Action Module that is prepared by the Ministry of Environment and Forestry (KLHK) and implemented in stages with coaching, mentoring, implementation, reporting, and evaluation systems. Its activities have two major pillars, i.e., adaptation action and mitigation action, and an incentive and recognition system that will aid the activities in improving program replication and sustainability.

Adaptation action is a series of activities that are designed to enhance the capacity of the people to mitigate the effects of climate change like extreme weather, climatic variability, uncertain seasons as well as hydrometeorological disasters. Implementation action adaption in CVP has the following forms: (1) Clean Water and Water Resources Management: The adaptation in majority of the areas is through development of infrastructure like rainwater reservoir, well infiltration, drip irrigation, and conservation springs. This is particularly so when the situation is characterized by drought and reduced discharge of water like the one in Wonosobo and Banjarmasin.

(2) Watershed conservation is to be carried out at the community level to rehabilitate the watershed by planting vegetation to retain the erosion, the arrangement of the surface water flow and making the natural embankment. The practice renders the community flood and landslide resistant. (3) Implementation and diversification of local food security strategies aims to increase resilience to climate change. (4) Control illness in line with the Community Environment which is trained to handle waste, house stairs, guard sanitation, and control the vector illness (e.g., malaria and dengue fever) that are likely to increase the consequences due to climate change. (5) Preparedness to Disaster and Social Adaptation: The adaptation measures in the vulnerable disaster regions include the formation groups, who are the disaster locals, mapping of risk communities, and early system warnings based on the community.

CVP mitigation actions aim at reducing the emission of greenhouse gases (GHGs) through embracing energy activities that are efficient and friendlier to the environment. The key activities are: (1) Management waste CVP recycling and composting encourages the 3R (Reduce, Reuse, Recycle) activities at house ladder level through waste banks, composting the organic wastes, and separating wastes. The case of the cities, Depok and Surabaya, which are actively involved in the integrated waste bank movement, can be taken as real life experience. (2) Planting of Vegetation and Trees: Conservations, reforestation, and other land-saving efforts are being done collectively to increase the vegetation cover and soak carbon. This is a major test when carrying out the evaluation of the Tropi CVP Lestari. (3) Energy-saving technology, such as biogas of livestock wastes, household solar panels, and energy-saving stoves, has been introduced in a number of climate villages, as well as renewable energy. The communities in Wajo and Lombok areas use biogas as an alternative to LPG gas. (4) Home Energy Conservation and Green Transport: Some of the regions have begun to adopt a sustainable form of transportation and reduce the amount of electricity used in the household as a locally based form of mitigation contribution.

The implementation of the CVP program occurs in stages and methodically and consists of the Determination and Identification of CVP locations. The locations to use CVP selectively are determined depending on the indicators of climate vulnerability, local potential, and social preparedness of the community. The government area did data collection through the assistance of the KLHK technical personnel. (1) Community assistance and Technical facilitation. CVP is anchored in the assistance of technical teams (environmental services, NGOs, universities) in the preparation of local plans of action against climate and the formation of environmental cadres-real action in the field. The community is carrying out activities based on adaptation and mitigation modules which are modified to suit the locality. On every activity, needs are assessed. The Form CVP (LIP) is completed by completing the details of the activities and documentations where the assessment and verification will be conducted by the KLHK team. This is the information that is utilized to grant and report at the national system. Assessment, Accreditation, and Prizes: The field assessment and verification system evaluates all the sites by KLHK. The facilities that meet the standards are awarded CVP Certificates, CVP Trophies and recognition classes: Pratama, Madya, Utama, and Lestari.

The example of utilizing the Climate Village Program (CVP) demonstrates that the adaptation and mitigation of climate change are grounded on the local resources, community, and regional capabilities. Several areas in Indonesia have been utilized as a role model nationally toward the implementation of CVP and success has been determined through the achievement action, participation of various parties and an award given by the ministry of environment and forestry (KLHK). The four examples of the areas which stand out in the implementation of CVP will be as follows:

(1) Surabaya City is one of the most stable and developing cities that have implemented the principles of CVP. The city of Surabaya in 2023 became the record holder of the acquisition CVP awarding CVP in Indonesia, with more than 23 awards, including the category Main and Sustainable. Effective Keys: Waste Management Innovation also is renowned with its integrated waste bank system which has thousands of households and schools in its coverage. This waste bank is both a mitigation action (it lowers the landfill methane emission) and a source of income to the society. Green Open Space and Urban Farming. The city government encourages the establishment of city parks, RT/RW parks, and vertical gardens, as the methods of accommodating urban heat and improving the quality of air. The adaptation measures involve Water Conservation and Recycling, Household wastewater treatment projects, and mass infiltration well construction in order to reduce seasonal flooding and drought, and community and School Engagement. With the aid of Adiwiyata School, Eco School, and Green Village, Surabaya mobilizes the potential of schools and communities as the agents of climate education.
(2) The city which had the most common aquatic ecosystem, Banjarmasin, succeeded in carrying out a strategy which would resort to the community-based management of river basins (DAS) and settlement on the Coast River. Awards Significant: Award CVP Lestari and 19 Certificates CVP Main Categories of Awards: The award is given to several RW and sub-districts that are operationally running conservation programs, environmental awareness creation, and climate change adaptation. Highlighted practices include community-based watershed rehabilitation, canals rehabilitation, and government planting vegetation along the banks of the rivers, and the relocation of the affected settlements during the floods was organized. Management Waste and Energy RT/RW Community carries out the system of processing the rubbish and extensive utilization of the biogas including small settlements. The plan will promote climate resilience, innovations, and sustainable architecture to create a friendly environment and apply a drainage system, which will be a prototype of adaptation on the local wisdom.
(3) Mountainous regions like Wonosobo face the issue of land erosion, degradation of springs, and vulnerability to the effect of climate change on food security. However, the Regency of Wonosobo was able to position itself as one of the leaders in the food resilience based on the community and conservation sources of water energy. Welahan Hamlet in Wonoroto Village represents a successful case study. Wins Trophy Sustainable CVP as an award of its efforts in the

environment management, innovation, and food resilience at the local level. Water Saving Irrigation and Utilization Simple System of Reservoir Usage Technology: A solution such as water conservation can be adapted to the season particularly during dry and prolonged seasons. Organic and Diversification Agriculture: The farmers develop more sustainable systems, use less pesticides, and become resistant to bad weather. The participation of young farmers and traditional institutions is strengthened through training and mentoring programs, enabling them to act as environmental cadres and CVP facilitators.

(4) The real story of the application of CVP concept that lies in the combination of the appropriate technology, adjustment to the farming and environmental safeguard is the Liu Village in the Wajo Regency, Sabbangparu District. Award Trophy Liu Village Main CVP has 56 locations in Indonesia that received Trophy depending on the execution of the category action, mitigation, and adaptation integrated. Practice Featured: Waste Community livestock is the use of farm wastes to produce clean energy that is utilized in cooking thereby reducing the amount of GHG emissions that would be produced in the burning of fossil fuels. Adaptation Community agriculture employs an agriculture system that is contrary to climate change and founded on various adaptive measures in order to prevent drought as well as it is founded on plant rotation. Land Conservation and Rehabilitation of Small Forest: The element of planting and caring of useful trees will be included in the process of protecting the sustainable ecosystem within the village. To give an example, the successful application of CVP in other regions demonstrates the concept that the effectiveness of the program is determined by the policy center and local ability to identify risks, aggregate sources of power, and contextual innovation. The replication practice expansion is supported by the technical and sustainable development of the government and partners. Good. This is in the other Indonesian parts.

The Climate Village Program (CVP) is a local adaptation of the Climate-Smart Village (CSV) concept, which is a community-based action that operates on the principle of community participation and environmental stewardship. The similar reliance on the participatory approaches is also typical of experience with CBA initiatives in Fiji and El Salvador, yet CVP is unique in the manner in which it was integrated into national climate policy frameworks (Table 14.1).

14.3.9 Challenges in the Implementation and Sustainability of the Climate Village Program

Although CVP has proven quite a number of successes, its application is not without challenges. Some of the key issues include the shortage of resources particularly in remote places. Funding, technology, and skilled human resource remain a problem to most communities. Lack of consciousness in the population of some areas is also a significant obstacle because people are inclined to see climate change as something

Table 14.1 Key problems in CVP implementation and strategic solutions

No.	Problem	Description	Strategic solutions	Source
1	Limitations understanding public	Low climate literacy and understanding of the concepts of adaptation and mitigation	Multi-level training, community-based educational campaigns, and integration of local climate curriculum	Siwi (2021), UGM Press (2024)
2	Inequality capacity area	Advanced areas reach certification more quickly in the CVP 28 scenario 28g 3T region	Mentoring technical specials for 3T areas, zoning affirmative, and improvement access financing	KLHK (2017), UGM Climate Program (2024)
3	Limited and unsustainable funding	Relying on self-reliance, NGO assistance, and non-systemic CSR	The financing scheme is based on results (result-based), incentives 28 clear regions, and the establishment of the Green Fund 28 cena	KLHK (2021)
4	Cross-sector coordination is not yet optimal	In many regions, CVP has not been included in official planning (RPJMD, Renstra, RAD-GRK)	Integration of CVP in regional planning documents and the formation of regional cross-sector climate teams	Masripatin (2017), Ministry of Environment and Forestry (2020)
5	There isn't any precise monitoring yet based on integrated outcome-based MRV	The reporting process is still manual, non-spatial, and unrelated to the adaptation/mitigation indicator system	Development of GIS-based e-CVP and achievement indicator dashboard, local operator training, and use of cloud platforms	MoEF (2021), Aggarwal et al. (2018)
6	The ambitious target (20,000 CVP locations by 2024) is challenging to achieve	Disparities in regional readiness, weak reporting systems, and the absence of a national-scale financing strategy	Revision of the national roadmap CVP based on evaluation of annual and 28 scenario realistic regional achievements	National CVP Reflection

Source Various Publication

distant and not directly related to their lives in any way. The risk of policy change on the national and regional level is also possible to influence the sustainability of the program. Part of the CVP initiatives can change in the priorities of the policy since they are reliant on governmental budgets. Therefore, it needs some actions to improve the economic sustainability of the program, such as the integration of the private sector or the establishment of community-based funding systems. Monitoring and evaluation also require more attention to be given. CVP requires a more efficient system of evaluation so that whatever is being done can be practically applicable, and this will reduce the amount of greenhouse emissions and make the communities be climate change resilient.

Climate Village Program (CVP) in Indonesia is currently facing several issues that hamper its success in the achievement of its goals of emission reduction and enhanced climate resilience. All these difficulties may be classified into a few aspects:

(1) *Program Sustainability*. The sustainability of the CVP is a key issue that needs to be addressed, particularly through its integration into regional medium-term development plans. Such integration is essential to ensure program continuity and alignment with broader regional development objectives (Wiati et al., 2022). In addition, strengthening community capacity is crucial to enable local communities to effectively engage with the program, increase their awareness of climate change, and sustain climate resilience initiatives (Luthfia & Alkhajar, 2021).
(2) *Community Participation*. Mindset and Knowledge: Attitude toward climate change adaptation and mitigation measures is an uphill task all the time to change the mindset of the community and enhance their knowledge on the same. However, despite this enormous level of effectiveness of 80%, there is still need to educate and engage the community in climate action (Fitrina, 2024). Ownership and Participation: It is imperative to ensure the degree of participation and ownership among the members of the community. It is necessary in the planning and in the implementation and monitoring of the climatic measures (Fitrina, 2024).
(3) *Policy and Governance*. Process of Policy and Integration: Integration of CVP into the established governance structures is a significant challenge, and policy process is also a significant challenge. Effective policy integration is necessary in order to ensure the support of the program by the corresponding legislative and regulatory frameworks (Wiati et al., 2022). Financial Constraints: This is a major constraint of financial constraints. The majority of local communities remain heavily dependent on external funding, including projects co-financed by the European Union. These funding sources are irregular and vulnerable to political shifts, making such reliance a potential obstacle to the sustainability and scalability of the program (Hungary et al., 2019).
(4) *Institutional and Technological Barriers*. Technological Solutions: The approach of technological solutionism can pose both a threat and an opportunity, emerging alongside new technological innovations. Therefore, climate action strategies need to strike a balance, combining technology-based solutions with non-technological approaches (Haëntjens & Futuribles, 2015). Institutional Support: The adoption of climate-smart technologies and practices requires support from appropriate institutions, including local and national governments as well as international organizations. Such institutional backing is crucial to ensure that these innovations and practices are implemented effectively and remain sustainable over time (Aggarwal et al., 2018) (Figs. 14.5 and 14.6).

Addressing these challenges requires a multi-focused approach. This involves strengthening community capacity, establishing effective partnerships, actively

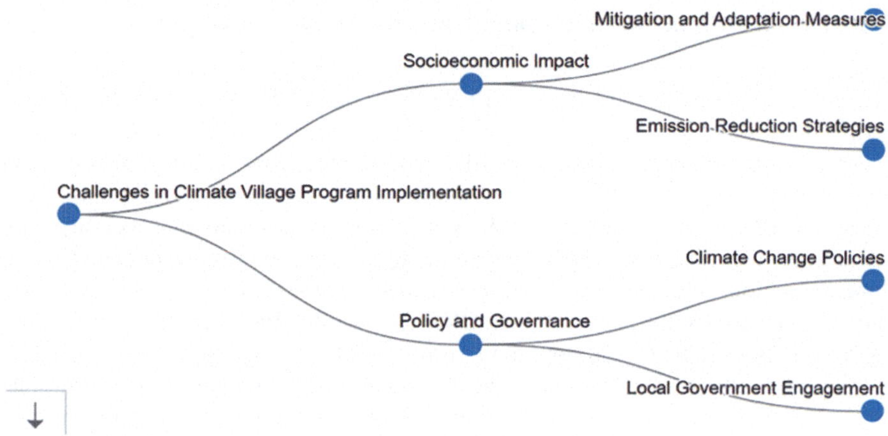

Fig. 14.5 Challenges in CVP implementation in Indonesia

Fig. 14.6 Policy flow climate change

engaging stakeholders, integrating the program into regional development strategies, and ensuring financial and institutional sustainability. By taking these steps, existing barriers can be minimized, allowing the Climate Village Program to more effectively achieve its goals of emission reduction and climate resilience.

Overall, managing these challenges is a complex process that demands coordination across several dimensions: enhancing local capacity, fostering solid collaboration with various actors, promoting proactive stakeholder involvement, embedding the program into development planning, and providing stable institutional and financial support. By overcoming these obstacles, the Climate Village Program will be able to achieve its core objectives more efficiently and sustainably.

14.3.10 Implications Comparative and Deeper Analysis

The implementation of the Climate Village Program (CVP) shows that the sustainability of community-based climate initiatives cannot rely solely on local efforts. Long-term success also requires coordinated support from national policies and international cooperation. Compared to similar initiatives, such as the Climate-Smart Village (CSV) programs in South Asia, CVP demonstrates a higher level of policy integration, with the government system acting as the primary driving force in its implementation. Nevertheless, CSV focuses on more robust technological innovation and market-based solutions, and thus, CVP can use the opportunities of extending its financial mechanisms and digital monitoring tools.

The payment of ecosystem services (PESs) and decentralization of institutions have been very supportive of local adaptation in Latin America, as, in the case of El Salvador and Mexico, ecosystem-based approaches have been promoted. CVP, however, continues to struggle with funding projects outside the pilot level. These results indicate that the incorporation of the community-based financing mechanism and decentralized governance into CVP might help to make it more resilient and scalable. Further, the implication of this research highlights the importance of Indonesia to strengthen the institutionalization of local knowledge, enhance multi-sectoral coordination, and make climate action part of regional development agendas. The case studies show that with the help of cross-sectoral synergy, the effect of CVP is more impressive and long-lasting, which can be associated with the same results of community resilience programs in Fiji and integrated village-based planning in Cambodia.

This discussion emphasizes the important role of each element within the CVP policy framework its implementation structure, the adaptation and mitigation pillars, and the support for sustainability in strengthening understanding of local climate resilience. By taking an integrative approach, this chapter goes beyond a simple descriptive overview, offering a structured analysis that aligns with research on policy and community development.

The findings also reinforce the book's main message by showing how national climate policies, such as CVP, can be adapted and applied at the local level to empower communities. The framework developed is both flexible and scalable, making it relevant not only for Indonesia but also for other countries in the Global South that face challenges in bridging the gap between policy and practice. Overall, this discussion reflects the book's central theme of promoting inclusive and transformative climate action.

14.4 Conclusion

The Climate Village Program (CVP) represents a strategic policy innovation in Indonesia's climate governance. As a nationally driven, community-based initiative, CVP aims to implement adaptation and mitigation efforts directly at the local level. This study shows that, while the conceptual framework of CVP aligns with international models such as Climate-Smart Villages and ecosystem-based adaptation, its implementation still faces significant challenges, particularly in institutional coordination, technical capacity, and monitoring systems.

Using a combined approach of thematic policy analysis and comparative case studies across four distinct regions—Surabaya, Banjarmasin, Wonosobo, and Wajo—this study identifies both enabling factors and persistent challenges in the implementation of the Climate Village Program (CVP). These factors include the presence or absence of local leadership, social capital, integrated planning, and the use of digital tools for monitoring and evaluation. The analysis also highlights a misalignment between national policy frameworks and local realities, reflecting broader challenges of decentralization in climate policy implementation.

This study offers a novel contribution through its proposal of a dual-scale diagnostic framework that links macro-level policy with micro-level innovations and constraints. This framework serves not only as a tool to assess program effectiveness but also as a guide for more targeted reforms. Specifically, the study recommends implementing spatially integrated MRV systems, institutionalizing CVP within local planning processes, and providing tailored support mechanisms for underserved regions.

In conclusion, the Climate Village Program (CVP) holds significant potential as a replicable model for participatory climate governance across the Global South. However, its success depends not only on community engagement but also on structural reforms in governance, financing, and technical capacity. The findings highlight the importance of aligning national ambitions with local capabilities and building adaptive institutions that can learn from diverse implementation contexts.

Acknowledgements The writing of this Book Chapter is the result of a collaborative effort by all authors listed in the chapter. We also express our appreciation and gratitude to the National Research and Innovation Agency (BRIN) for their support.

AI Disclosure The authors emphasize that generative artificial intelligence (AI) and AI-based technologies were used in the preparation of this manuscript solely to clarify language, improve grammar and writing style, and facilitate the management of references. AI did not play any role in the formulation of ideas, the analysis or interpretation of data, or the development of the discussion, which was carried out entirely through human judgment.

References

Aggarwal, P. K., Jarvis, A., Campbell, B. M. S., Thornton, P. K., Wollenberg, E., Hansen, J., Alvarez Toro, P., Aguilar-Ariza, A., Arango-Londoño, D., Patiño-Bravo, V., Rivera, O., Ouédraogo, M., & Bui Tan Yen. (2018). The climate-smart village approach: Framework of an integrative strategy for scaling up adaptation options in agriculture. *Ecology and Society, 23*(1), 14. https://doi.org/10.5751/ES-09844-230114

Bappenas. (2019). *National medium-term development plan 2020–2024*. Ministry of National Development Planning/Bappenas.

Bowen, G. A. (2009). Document analysis as a qualitative research method. *Qualitative Research Journal, 9*(2), 27–40.

Budget Study Center, Expertise Agency of the Secretariat General of the Indonesian House of Representatives. (2021). *Challenges of the 20,000 climate village program budget issue brief: Industry and development* (Vol. 01). ISSN: 2775-796X

Creswell, J. W. (2018). *Qualitative inquiry and research design: Choosing among five approaches* (4th ed.). SAGE Publications.

DPR RI. (2021). Pusat Kajian Anggaran, Sekretariat Jenderal DPR RI.

Faedlulloh, D., Irawan, B., & Prasetyanti, R. (2019a). Climate village flagship program (ProKlim) based on community empowerment. *PUBLISIA: Jurnal Knowledge Public Administration, 4*(1), 28–44. ISSN: 2541-2515. http://jurnal.unmer.ac.id/index.php/jkpp

Faedlulloh, D., Prasetyanti, R., & Irawan, B. (2019b). Village versus climate change: The dynamics of community empowerment through the climate village program (ProKlim). *Journal of Physics: Conference Series, 1424*(1). https://doi.org/10.1088/1742-6596/1424/1/012055

Fitrina, E. (2024). Implementation of the climate village program (ProKlim) related to gender-based. In *E3S web of conferences*. https://www.scopus.com/record/display.uri?eid=2-s2.0-85188345771&origin=scopusAI

Ghina, N. Y., & Zunariyah, S. (2023). *Climate village: Environmental management based on community empowerment*. DILEMMA, UNS Journal. https://jurnal.uns.ac.id/dilema/article/view/12394

Gunawati, D., & Rejekiningsih, T. (2020). Building ecological citizens through implementing climate village programs as climate change mitigation effort. In *ICLIQE 2019* (Vol. 397, pp. 1124–1131). https://doi.org/10.2991/assehr.k.200129.139

Haëntjens, J., & Futuribles. (2015). *Analyze et prospective, cities and climate—What visions*? https://www.scopus.com/record/display.uri?eid=2-s2.0-84942626608&origin=scopusAI

Hungary, P. C., Radics, Z., Tóth, JB, Kovács, E., Csorba, P., Fazekas, I., Szabó, G., & Tóth, T. (2019). Climate and energy governance perspectives from a municipal point of view in Hungary. *Climate*. https://www.scopus.com/record/display.uri?eid=2-s2.0-85071312844&origin=scopusAI

IPCC. (2021). *Climate change 2021: The physical science basis*. Cambridge University Press.

[KLHK] Ministry of Environment and Forestry. (2016). *Regulation of the minister of environment and forestry no. 84 of 2016 concerning the climate village program*.

[KLHK] Ministry of Environment and Forestry. (2020). *The important role of the young generation in the climate change agenda*. PPID KLHK.

[KLHK] Ministry of Environment and Forestry. (2021). *National report on climate change adaptation and mitigation*.

[KLHK] Ministry of Environment and Forestry Directorate General of Climate Change Control. (2017). *Road map of climate village program (ProKlim): Road map of national movement for community-based climate change control, December 2017, Jakarta*.

Luthfia, A. R., & Alkhajar, E. N. S. (2021). Reflections on climate change policy implementation. *IOP Conference Series: Earth and Environmental Science*. https://www.scopus.com/record/display.uri?eid=2-s2.0-85112445134&origin=scopusAI

Maharani, R., Hadiyan, Y., Naibaho, Y., Satria, W. I., Ngatiman, A., & Angi, E. M. (2022). Challenges to and Strategies for the climate village program plus: A lesson learned from Indonesia. *Sustainability, 14*(9), 5530. https://doi.org/10.3390/su14095530

Masripatin, N. (2017). Introduction to the climate village program road map (ProKlim). In *Road map ProKlim*. KLHK.

Measey, M. (2010). Indonesia: A vulnerable country in the face of climate change. *Global Majority E-Journal, 1*(1), 31–45.

MoEF. (2021). *ProKlim progress report*. Ministry of Environment and Forestry Republic of Indonesia.

Mulyasari, G., Irham, I., Waluyati, L. R., & Suryantini, A. (2020). Livelihood vulnerability to climate change of fishermen in the coastal area of Bengkulu Province, Indonesia. *Aquaculture, Aquarium, Conservation and Legislation, 13*(3), 1242–1254. http://www.bioflux.com.ro/aacl

NASA. (2020). *Global climate change: Vital signs of the planet*.

Pawitan, H. (2018). Climate change impacts on availability and vulnerability of Indonesia water resources. *IOP Conference Series: Earth and Environmental Science, 200*, 012003 https://doi.org/10.1088/1755-1315/200/1/012003

Rinaldi, R., Nulhaqim, S. A., & Gutama, A. S. (2017). The community development process in the climate village program in the village Siamese fighting fish Subdistrict Gempol Cirebon Regency. *Proceedings of Research and Community Service, 4*(2).

Santoso, A. P., & Rahaju, T. (2020). *The impact of the climate village program (ProKlim) in the community association (RW) 03, Jambangan village, Surabaya city* (Vol. 8, No. 1).

Selje, T., Schmid, L. A., & Heinz, B. (2024). Community-based adaptation to climate change: Core issues and implications for practical implementations. *Climate, 12*(10), 155. https://doi.org/10.3390/cli12100155

Siwi, N. (2021). Community-based adaptation and mitigation to climate change. *International Journal of Built Environment and Scientific Research, 5*(2), 135–150.

Sugiyono. (2019). *Research methods quantitative, qualitative, and R&D*. Alphabet.

Tompkins, E. L., & Adger, W. N. (2004). Does adaptive management of natural resources enhance resilience to climate change? *Ecology and Society, 9*(2), 10.

[UGM] Gadjah Mada University. (2024). *Indonesian ProKlim action movement 2020–2030. ProKlim Indonesia*. UGM Press.

Wiati, C. B., Dharmawan, I. W. S., Sakuntaladewi, N., Ekawati, S., Wahyuni, T., Maharani, R., Hadiyan, Y., Naibaho, Y., Satria, W.I., Ngatiman, N. and Abdurachman, A., & Khadka, C. (2022). Challenges to and strategies for the climate village program plus: A lesson learned from Indonesia. *Sustainability (Switzerland)*. https://www.scopus.com/record/display.uri?eid=2-s2.0-85130006024&origin=scopusAI

Yusuf, A. A., & Francisco, H. (2009). *Climate change vulnerability mapping for Southeast Asia. A report*. Economy and Environment Program for Southeast Asia (EEPSEA) (26 pages).

Chapter 15
Customary-Based Environmental Management as an Inclusive Solution for Sustainable Local Communities: A Case Study of Tenganan, Bali, Indonesia

Yumantoko, I. Wayan Rupa, I. Gusti Ngurah Jayanti, Carolina, Eko Wahyono, Tri Bastuti Purwantini, Herlina, Abdul Manan, and Rio Heykhal Belvage

Abstract In addressing socio-ecological crises linked to development-driven growth, this chapter investigates inclusive and sustainable human-environmental governance as practiced in Tenganan, Bali. Based on the interviews, observation, and documents' review conducted within a qualitative approach in a structured case study as guided by the Social-Ecological Systems (SES) framework in this chapter,

---

Yumantoko (✉) · Carolina · E. Wahyono · T. B. Purwantini · Herlina · R. H. Belvage
Reserch Center for Social Welfare, Village, and Connectivity, National Research and Innovation Agency, Jakarta, Indonesia
e-mail: yumantoko@brin.go.id

Carolina
e-mail: caro001@brin.go.id

E. Wahyono
e-mail: ekow005@brin.go.id

T. B. Purwantini
e-mail: trib005@brin.go.id

Herlina
e-mail: herl009@brin.go.id

R. H. Belvage
e-mail: rioh001@brin.go.id

I. Wayan Rupa
Research Center for Religious Heritage and Civilization, National Research and Innovation Agency, Jakarta, Indonesia
e-mail: iway024@brin.go.id

I. Gusti Ngurah Jayanti
Research Center for Environmental Archaeology, Maritime Archeology, and Cultural Sustainability, National Research and Innovation Agency, Jakarta, Indonesia
e-mail: igus019@brin.go.id

G. K. Panda et al. (eds.), *Socio-Ecological Crisis of Development*, Sustainability Solutions, https://doi.org/10.1007/978-3-032-22171-1_15

this chapter investigates how institutions of custom mediate shared values to concrete rules and ecological outcomes. The analysis argues that inalienable customary rights, awig-awig (customary law), habituation-based sanctions, and the subak irrigation system, integrated into a consolidated governance arrangement, are an institution supporting forest–agricultural landscape conservation through nurturing social capital. Embedded within the Balinese philosophy of *Tri Hita* Karana, participatory discussion and leadership oversight influence daily practices (for instance, tree cutting; water use; extraction of non-timber forest products) such that they remain congruent with collectively sanctioned norms. The contributions of this chapter include: (1) an SES-based map that explains how governance can shape interactions between users and resources and their outcomes; and (2) policy guidelines to formalize co-management with traditional institutions, strengthen intergenerational knowledge, and align development planing with existing structures at the local level. In conclusion, the case demonstrates that culturally embedded governance can endure and support livelihood security in response to spreading external threats while preserving ecological integrity and offers general sustainability design principles for other customary landscapes.

Keywords Local knowledge · SES · *Awig-awig* · Natural resources · Sustainability

15.1 Introduction

Preservation of the environment helps to safeguard local cultural heritage and livelihoods (Fa et al., 2020) and enables climate regulation and biodiversity conservation (Oldekop et al., 2020). By contrast, the degradation of environment and tendency to natural disasters tend to affect disproportionately poor and vulnerable groups (Armah et al., 2014). These challenges have been heightened by climate change mainly caused by human activities that tend to promote economic gains rather than ecological balance (Bashir et al., 2024; Olkiewicz, 2023; Sivakumar, 2021). In Indonesia, they depend on natural resources for food and livelihood which are rural communities (BPS, 2022; Kaskoyo et al., 2017). Local environmental sustainability is thereby an underpinning for social well-being as well as climate resilience, with a potential of orientation along community-level institutional structures impacting outcomes (Ofoegbu et al., 2017; Zeigermann & Böcher, 2020). However, not so many formal policies and legislation suitably support and recognize the effectiveness of traditional community-based governance systems (Nurlinda, 2016). Forestry- and coastal-related data illustrate this policy–practice gap: By 2020, Indonesia's forest cover was around 94.1 million hectares where more than 2 million of its peatland (out

A. Manan
Faculty of Economic, University of Mataram, Mataram, Indonesia
e-mail: abdmananfeb@gmail.com

of 14.9 million ha) and over 600 thousand of mangroves (approximately at 3.3 million ha) have been lost (Kemen LHK, 2021). Economic-based agendas often clash with conservation goals (Austin et al., 2019). It can also degrade ecosystems and increase disaster risk. Modern-day environmental pressures are a social-ecological development crisis in which growth-focused policies undermine ecosystem functioning with impacts on community vulnerability.

Economic-based agendas often clash with conservation goals (Rimba et al., 2019). It can also degrade ecosystems and increase disaster risk. Modern-day environmental pressures are a social-ecological development crisis in which growth-focused policies undermine ecosystem functioning with impacts on community vulnerability. The devastating floods in early September 2025 (Channelnewsasia.com, 2025) are best understood as a cumulative result of decades of environmental pressure. Overtourism has transformed arable land, particularly rice fields, into restaurants, resorts, and villas, and the pressures of activities upriver have led to deforestation and land degradation (Rimba et al., 2019). Over-dependency on government's role in waste management, along with low involvement of communities, accentuates environmental deterioration amidst mass tourism (Suhardono et al., 2024). This diagnosis sets the stage for the primary focus of this chapter as we consider the tension between centralizing, growth-first governance, and community practice that might work to support long-term ecological legacy.

However, Bali still has a strong socio-cultural basis for sustainability, symbolized by the philosophy of *Tri Hita Karana* and customary institution such as *awig-awig*, a set of traditional customary Balinese law, and *subak*, a Balinese farmers association that governs irrigation system (Saputro et al., 2024; Wiryawan, 2024). One village that practices traditional regulations in everyday life is Tenganan. This chapter explores how norms, organizations, and sanctions that are built into ritual–ecological practice affect resource use and resilience. Customary practices may serve as a check on purely rational models of development which promote economic return while ensuring places for communal life (Liana et al., 2024). In Bali, forests, water, and farmland remain important commons for which continued natural management is deeply entwined with continuity of Balinese society and environment.

As a chapter in this volume's theme—Socio-Ecological Crisis of Development: Community Resilience and Sustainable Development, it discusses how indigenous institutions addressed crisis using locally legitimate rules and practices. External economic and regulatory pressures, in turn, have potential to undermine traditional practices and social norms (Özgen Çiğdemli, 2022), highlighting the necessity of investigating how communities develop capacities to adapt or sustain local resource governance against changes (Oldekop et al., 2020). There is scarce evidence on reciprocal human–ecosystem dynamics particularly about the institutional mechanisms of conservation (Jiménez & Basurto, 2022), but such form of evidence is critical for durable policy designs (Jaung & Carrasco, 2022).

The chapter uses Ostrom's (2009) Social-Ecological Systems (SES) framework to integrate ecological and social variables, including Resource System (RS), Social, Economic, and Political Settings (S), Resource Units (RU), Users (U), Related Ecosystems (ECO), Actors (A), Governance System (GS), Interactions (I), and

Outcomes (O), into a single analytical model (Berkes, 2010). This perspective is particularly relevant for community-based resource governance, where local knowledge, norms, and institutions intersect with environmental change (Asteria et al., 2021; Theresia & Budiastuti, 2020). Ostrom's design principles of boundary clarity, rule enforcement, and collective action (Ostrom, 1990) help to explain why some community governance regimes persist despite strong external pressures (Collen et al., 2016). However, much of the SES literature remains at a macro-diagnostic level, using broad statements such as "overtourism reduces carrying capacity" or calls to "strengthen local wisdom." It often fails to specify how rules-in-use at smaller social scales shape user–resource interactions and ecological outcomes within an SES setting (Cole, 2012; Rimba et al., 2019). Recent SES work also highlights the lack of methods for mapping mechanisms and translating them into cross-scale operational strategies, particularly in customary contexts (Cox et al., 2010; Partelow, 2018). The literature on *subak* and *Tri Hita Karana* tends to highlight values and principles but does not often make explicit how rights, compliance architectures (monitoring–ritual–graduated sanctions), and institutional coordination result in instantiations of absorption, adaptation, or transformation resilience capacities in practice (Brooks et al., 2018). Our contribution is twofold: (i) a SES-based analysis that links customary rules, organizational roles, and sanctions to user–resource interactions and ecological outcomes; and (ii) operational design guidance that policymakers can apply to co-management in Bali. Based on these objectives, the study seeks to answer two central research questions: (1) applying the SES framework, how do specific governance components (GS), notably *awig-awig*, organizational roles and sanctions, interact with Actors (A), Resource Systems (RS), and Resource Units (RU) to produce Interactions (I) and Outcomes (O) that reflect sustainable use? (2) How do locally organized collective actions contribute to sustainable resource use, and what are the implications for Bali's sustainable development agenda?

15.2 Materials and Methods

This chapter is based on a research in Tenganan Pegrinsingan as one of the traditional villages in Karangasem, Bali, Indonesia. Tenganan village cover an area of 917,200 ha, consisting of forest and dryland farming (583,035 ha; 63.61%), rice fields (255,845 ha; 27.9%), and settlements areas (78,304 ha; 8.5%) (Pertiwi & Mardiana, 2020). Communities are dependent on the land for their survival; forests, for example, yield food resources and medicinal plants, construction materials, and textiles. Many are engaged in agriculture, growing crops such as rice, vegetables, and fruits, and such livestock as pigs, buffalo, and poultry. Agricultural products are both consumed locally and sold outside the village. The topography of Tenganan consists of a combination of lowlands and hills, ranging in elevation from approximately 70–400 m above sea level (Maria & Rupa, 2007). Three primary hills shape the village landscape: Bukit Kauh in the west, Bukit Kangin in the east, and Bukit

Fig. 15.1 Map of Tenganan (Google-Earth, 2025)

Kaja in the north (Maria & Rupa, 2007). The hilly areas of Tenganan provide various ecosystem services that support local livelihoods and its surroundings (Fig. 15.1).

This chapter utilized qualitative research (case study) as a method to enable the researcher to get close and gain an in-depth understanding of what informants think, how they feel, their attitudes toward and perspectives on research questions (Creswell, 2018). We aim to contribute to a better understanding of interactions on daily life, traditional governance, and community practices with the SES framework. Interviews, observations, and review of literature were performed to collect the data. Informants were interviewed based on their acquaintance or participation in the research problem. Field investigations recorded information with regard to forest status, livelihood resources in the village, and the existing community–environment relationship. A literature search was conducted to collect data regarding management of resources in Tenganan, using existing online materials. Purposive sampling was used to recruit informants that were relevant to the research topic. These key informants comprised the Village Chief, traditional leaders, farmers, local business people, relevant government departments, and those involved in the use and management of natural resources at village level. Standardized methods were used to collect data. In order to guarantee that the veracity, accuracy, and reliability of statements were confirmed properly all statements which could be controversial were cross-referenced.

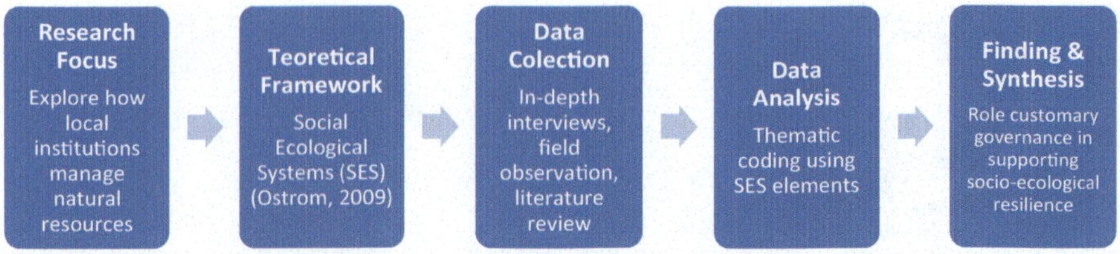

Fig. 15.2 Methodological flowchart

The data were analyzed and then categorized according to the research questions and the Social-Ecological Systems (SES) framework. These themes encompass Resource Systems (RS) (forest, irrigation systems, rice fields, and plantations); Resource Units (RU) such as timber, agriculture produce, and forest products; Governance Systems (GS), awig-awig, subak, and other customary institutions; Actors (A′) village residents "traditional leaders and government officials; Interactions and Outcomes (I and O) collaboration and coordination"; External Influences such as social economic and political background. The mechanisms of mass action supported by community institutions facilitate long-term resource management. Summary of the results is depicted in a flowchart adapted from Ostrom (2009) (Fig. 15.2).

15.3 Results and Discussion

This chapter examines environmental governance in Tenganan by investigating the functions of indigenous institutions as guardians of ecological well-being and unifiers of society. It aims to access two related dimensions: the relationship of a community with its environment through resource use practices which are rooted in cultural tradition and the collective action mechanisms provided by local institutions that support long-term management of resources.

Local knowledge and traditional institutions (awig-awig) and subak (irrigated agricultural systems involving the irrigation organizations themselves) have a role to play in maintaining an ecological integrity as well as a sense of mutuality and common spiritual commitment toward responsibility. It contains tradition rules that regulate social life, both in the form of do's and don'ts such as who is accountable for what action, which sanctions are imposed, and which village institutions may enforce it (Silviana et al., 2021). Subak is the traditional scheme for managing water for agriculture. Collectively, these dimensions represent a culturally situated model of sustainability that meets with global understandings and frames broader conversations on sustainable development. The analysis fulfills the first research objective in that it provides an account of locally embedded subsistence strategies on natural resources and contributes to the second by demonstrating how mutually reinforcing institutional processes shape long-term environmental outcomes. The Tenganan case therefore provides examples that can be adapted regarding how culture-based governance might be used to support inclusive and adaptive frameworks of environmental management.

15.3.1 Natural Resources as Gift and Responsibility

In Tenganan, land is held under customary land, which is collectively managed by the indigenous community for purposes according to customary regulations and by private right of ownership. Customary land is non-negotiable, but could only be used by customary law, while private land can be trade to someone else as long as they are among the village community following the rule from customary regulation (Sumunar et al., 2018). Village forests provide such economic, social, and cultural benefits as well as serve important ecological purposes including preventing soil erosion, generating oxygen, and providing other environmental services. They are collectively regulated through a system of so-called awig-awig (customary law), which is how forest stewardship and overall environmentally sustainable practices are organized to benefit the long-term well-being of the village.

The local natural resources are seen as a blessing from divine and should remain sacred. This spiritual perspective is consistent with the Balinese's *Tri Hita Karana* framework for living in harmony with nature and God. Nature is in this sense, a condition for life; not a thing that can merely be taken from but something that must be cared-for and kept (Sujarwo et al., 2020). As a fundamental principle of moral and ethical life in Bali society, *Tri Hita Karana* is used to address the balance and harmony with nature. Following such belief, the community of Tenganan sees a degradation in the environment as an imbalance of human–god relationship which might bring unexpected misfortune. This belief is the one that nourishes community's dedication to a responsible use of natural resources, in keeping with an harmonious relation with the divine.

The villagers are ever wary of rampant over utilization, the reason being that environment obviously adversely affects them. It is in agriculture, forestry, water, and the garden that this principle clearly reveals itself. The villagers work together to maintain the ecological balance while satisfying their basic needs such as food, medicines, construction items, household articles, and spiritual products that are collected from the local natural resource base. Short-term use may lead to unavailability in the long term, leaving future generations without the resources that they need. Therefore, environmental conservation is considered as a common obligation to ensure the prosperity of future generations.

Such close relationship between the community and nature is also evident through sustainable use of natural resources in Tenganan. Rivers and irrigation are necessary not only for agriculture, but also for domestic uses like drinking, cooking, washing, etc. These natural sources of water are consecrated with rituals of propitiation as acts of thanksgiving, while also serving to channel the rain-harnessing flow and acting as flood guards during the monsoon. Rice is grown following the traditional Balinese methods of "*subak*" cultivation, which meets strict social regulations, from handing out water to farmers (nobody in the community has more rights to irrigated fields than anybody else), as well as preserving and using all natural resources wisely.

Subak offers such a traditional irrigation management system that encompasses and upholds socio-cultural values to the routine practice of managing water. Food crops, etc., are grown in rice fields that can meet ceremonial needs and local food security for villagers and surrounding communities. To the east and west of the village is forest used for timber and NTFPs (bamboo; ketak grass *(Lygodium circinnatum* (Burm. f.) Sw.)), fungi, sugar palm (*Arenga pinnata*), and medicinal plants. A similar large comparative study of 8,000 households in 24 countries found that environmental income is crucial for prosperity among rural communities and can motivate biodiversity conservation under strong local leadership (Angelsen et al., 2014). In Tenganan, forests fulfill the same ritualistic and ceremonial needs as in other desa pat on Bali, maintain a cool microclimate with ecological and aesthetic value.

Bamboo, palm leaves, and sources of water in Tenganan support the villagers' material needs while preserving cultural, ritual, and symbolic value. Stewardship here is clearly intergenerational, to the benefit of those who would inherit. Local ecological knowledge develops as the environment changes (Ratnani et al., 2021), which allows for varying resources use yet without crossing ecosystem safety thresholds. For example, traditional use of medicinal plants by indigenous people spans a wide spectrum of species and utilizes the plants for different ailments. Several plants reported by Sujarwo et al. (2015) are frequently used, including sembung (*Blumea balsamifera*), jarak or castor (*Jatropha curcas*), sambiroto (*Andrographis paniculata*), and kecemcem (*Spondias pinnata*) for the treatment of diarrhea, fever, hypertension, and diabetes. This knowledge is transmitted orally among members of society as part of their day-to-day social interaction, representing cultural continuity; however, oral tradition also has the potential to be eroded by disinterest or lack of opportunity for younger members to learn.

Traditional construction methods are also supported by local resources. Bamboo (Bambusoideae) and jack wood (Artocarpus heterophyllus) are common materials, available locally, both for the walls and roof. The house is constructed in accordance with the Tri Mandala architectural concept which divides the house into three zones and intended to bring about inner peace, or harmony, for the inhabitants (Barliana et al., 2025). More modern construction materials such as zinc, cement, and steel are becoming increasingly popular; however, wood is the material of choice for community-customary buildings to maintain traditional/ancestral values. Those who cut timber for building purposes, especially for married couples, take precedence according to rigid customary regulations. These guidelines also serve the goal of equity and fairness in that they are facilitating the purchase of affordable housing for young families plagued by economic burdens.

Indigenous industries, including gringsing textile production and ketak grass (Lygodium circinnatum (Burm. f.) Sw.) weaving, show sustainable use of natural resources. Rituals are performed to have a successful shooting of the film. The gringsing have religious symbolisms and according to tradition it brings peace to the person who wears them, even thoroughly woven cloth, some sources also say that this really has a benefic effect on sick persons (Pebryani & Vogel, 2024). They're not based in America, but most of the materials for these textiles are! The ketak is also worked up into tissue box covers, bags, and colorful hats or used for ceremonial implements favored by tourists. The sustainable philosophy embedded in these handicraft activities is reflected in the management of Tenganan's farming environments.

In Tenganan, there are widely two types of agricultural land: irrigated wet-farming rice fields and dryland fields which also include stands of fruit trees (salak), cacao, coffee, cloves, coconut s, vanilla orchid, and sugar palm. Wet farming method is using traditional Balinese system irrigation of subak. In dryland areas, this type of farming involves crops that have economic and ecological values such as sugar palm (Arenga pinnata). The sap of sugar palms is transformed into palm wine or palm sugar, while the leaves are commonly used for thatching, and therefore lands a unique identity to local houses. The tree's roots penetrate deep into the soil and help manage erosion and protect the surrounding environment.

The harvesting of dryland fruits ensures a year round secure food supply and is conducive to dietary diversity. In this sense, the production of tropical fruits contributes to satisfy local demand of fiber and is an example on long-term use of biodiversity (Powell et al., 2015). There is evidence for similar relationships in other studies, which are reflected in Tenganan. Kamal and Lim (2019) also observed sustainable use of natural resources by the local indigenous communities among Orang Asli in Malaysia. Reliance on the use of surrounding natural resources, mainly NTFPs, has met the communities' daily household requirements. This utilization also contributes to environmental conservation efforts. Das et al. (2023) also identified a similar pattern in their study. The utilization of natural resources based on local knowledge not only succeeded in conserving key resources—such as agricultural land, forest areas, and irrigation systems—but also contributed to maintaining regional biodiversity at a level sufficient to sustain the livelihoods of local communities. People's ways of making a living based on NTFPs (e.g. bamboo, rattan, medicinal plants) resemble evidence from other parts of the world that there is considerable "environmental income" in rural livelihoods, and with strong local institutions, it does not necessarily lead to degradation (Angelsen et al., 2014). A study by Rankoana (2016) examined the Mantheding community in South Africa, highlighting their ability to utilize natural resources sustainably. The community meets their needs for food, medicine, housing, and energy directly from nature. They make use of various parts of plants, including leaves, fruits, wood, tubers, stems, roots, and bark. Several studies have shown that the management of natural resources by local communities can effectively fulfill their needs without causing environmental degradation.

The findings on resource management in Tenganan provide evidence that the bundle of inalienable customary land rights, *awig-awig* as collective regulations, and religious ethics from *Tri Hita Karana* together form a commons institution that fulfills some of the sustainability conditions (Ostrom, 2009) proposed under the SES framework (e.g., boundary demarcation, monitoring, graduated sanctions). The *subak* mechanism, a method that correlates plant and harvest activities through the water network, not only guarantees impartiality of its allocation but it has also been empirically demonstrated to have an efficiency gain through reflexive self-management with respect to ecological functions such as erosion control and regulation of water (Lansing et al., 2017; Okura et al., 2022). These embedded spiritual foundations, sometimes referred to as "cultural ecosystem services," enhance people's cooperation and rule-following behaviors for reasons other than material benefits, which helps self-sustaining conservation become an integral part of daily life (Chan et al., 2016). In terms of livelihoods, how people use NTFPs such as bamboo, rattan, and medicinal plants corresponds with broader evidence that "environmental income" is a significant component of rural household income. Such resources can also prevent environmental degradation when managed by efficient local institutions (Angelsen et al., 2014).

15.3.2 Collective Action of Local Institutions in Promoting Resource Sustainability

A leading aspect of environmental accountability at Tenganan is the absolute integration of belief and practice in daily living. Rivers, rice terraces, and springs, as well as certain sacred trees and forests, are spiritually significant to the community. This belief encourages residents to protect their environmental, at least partly because the conservation of nature will make people more religious and improve their spiritual relationship with the deity (Sujarwo et al., 2020). Stewardship also is rooted in the Balinese philosophy of *Tri Hita Karana*, which balances people, nature, and divinity and implies negative consequences when this harmony is disturbed.

Religious offerings made in the woods, for instance, are held to be sacred and important elements of people's spiritual lives. These don't contain something specific that's inherently special to the materials—typically flowers, leaves, and fruits picked from gardens or forests in the area—except for their sentimental meaning as an important and lasting part of these peoples' lives. A research study by Ratnani et al. (2021) reported that at least 130 plants species were used in rituals among the Tenganan, and 66% of them came from local areas in Tenganan. Conservation efforts that focus on heritage related rituals prove to be more efficient since the conservation of spiritual values increases community dedication toward maintaining a physically healthy and functional environment. The empirical study conducted by Geng et al. (2017) also demonstrated that ritual practices and the use of ritual plants have implications for plant conservation. Even loss of forests, pollution of water, and a reduction in food production can threaten these ceremonial requirements and disturb religious harmony and the village's relationship with that which is sacred. Religious practices at Tenganan are therefore not just religious devotion, but also an effective (top-priority) conservation practice that is naturally growing and exists within a local cultural context.

The traditional institution utilizes *awig-awig* as an instrument in controlling the communities which was heavily supported by indigenous community members, *jus sanguin*. In Tenganan, *awig-awig* or customary law has an important role to protect their environment (Maria & Rupa, 2007). The rules are the result of a consensus, affecting all activity such as cutting trees, selling or buying land, harvesting fruit, and conserving water and vegetation; extension; and forestry control. For example, tree cutting is permissible only for tree owners under clearly defined circumstances, such as constructing a home for newly married couples. These activities are tightly controlled by traditional authorities, with an emphasis on the removal of unhealthy trees, so that live forest stands are impacted minimally. These control measures prevent over harvesting and allow trees to continue regenerating. Local people are also prohibited from collecting unripe fruits to ensure seeds reach maturity for future regeneration (Aryawan, 2021). And land sales to non-villagers are limited, in a bid to keep important assets under community control. In the absence of those safeguards, there would be a gradual transfer of land into non local hands with loss of valuable resources to the people, erosion or dilution of indigenous cultural power, and also a corrosive effect on social cohesion as new cultures are introduced.

At Tenganan, accepted custom or village regulations (awig-awig) provide an hierarchical structured custom sanction from mild to severe like sikang, penging, sapa sumaba, and the ultimatum kesah or expulsion (Maria & Rupa, 2007). Such sanctions include not just a tangible or physical punishment, but social estrangement, which is considered especially shameful within the community. The nature of the sanction and its severity vary according to the situation where violators are spotted: "in the forests," "in rice fields," or "in gardens/in town/downtown" (for details, see Table 15.1). This system of sanctions contributes to the resilience of a common set of norms and hence cooperation in natural resource management (Kristiono, 2017; Sumunar et al., 2018). Especially customary sanctions can work because people very much do not want to be the recipient of social exclusion (Wiessner, 2005). One type of hard social control in Tenganan is temporally social isolation when people intentionally do not speak to those who violate certain rules.

Local myths also add to local norms that keep people scared and on their guard. Aryawan (2021) narrates the fable of *Lelipi Selan Bukit*, a large serpent thought to appear whenever the forest is destroyed. Such beliefs promote the avoidance of damaging behavior. Since everyone is afraid of something really bad happening, social control is strong and compliments formal rules established in the *awig-awig*.

Local people have come to believe that they need to be more sensitive to the environment not just for practical reasons, but also because not doing so could incur bad spiritual luck. This idea is part of daily customs and reinforces social stability. According to Jensen (2016), myths in traditional societies can shape collective behavior in deep and lasting ways, going beyond storytelling to affect how members of the society actually behave.

After political reform in Indonesia, the Provincial Government of Bali enacted Regional Regulation No. 4 of 2019 on Customary Villages to recognize and empower local indigenous communities (Nurlinda, 2016; Pemerintah Provinsi Bali, 2019). This regulation grants customary villages a legal basis to regulate their internal

Table 15.1 Prohibitions and sanctions in Tenganan customary community

Prohibited action	Sanction
Selling prohibited timber	Fine of 200 *kepeng* (*Kepeng* refers to traditional Balinese coins used as currency) plus twice the timber's price, or once its price if not yet claimed by buyer
Timber theft	Fine equal to timber's price plus 2,000 *kepeng*. If timber is transported out, the fine doubles. Community members involved face potential expulsion
Cutting branches from protected trees	Fine of 10 *catu* rice (1 *catu* = 2.5 kg); branches returned to the tree's owner
Collecting fallen fruits	Non-residents fined 10 *catu* rice
Picking fruits directly from trees	Expulsion from the village; customary officials involved are removed from office following village deliberation
Felling sugar palm trees (*enau*)	Fine of 400 *kepeng*; the felled trees become village property

Source Maria and Rupa (2007)

matters, in particular customary forests and water sources, as well as communal lands. Each village has a right to frame its own laws. In Tenganan, it's helped reinforce local autonomy—especially in the arena of internal government, or selection of traditional leaders. The prerequisites for leaders are formulated in local terms: maturity, local-born, monogamous, and good knowledge of adat (Kristiono, 2017). Consensus appointed agents are supposed to guide and make strategic decisions that align with the community's best interest. The quality of "traditional leadership" is a determinant for collective action toward environmental sustainability, due to the fact that community's actions are partly decided by leader's embodiment and promotion of environment protection.

Beyond the subak system, this landscape practices adaptive leadership and collective governance. Consisting of the area's farmers, it coordinates agricultural irrigation and incarnates a democratic, dynamic, and inclusive system of governance. Farmers' concerns are taken seriously at the organization. If a member's plot is water short, they can take up their concerns with subak leaders and discuss the issue openly and respectfully. This setup brings consensus among the community members which in return help to minimize water disputes. The complaint gets worked through on a community discussion and agree a solution. Through that practice, the decision-making process in subak presents a distinct local democratic culture based on discussion within group.

Effective coordination is critical. A loss of coordination could lead to the collapse of the agro-system, with serious implications for Tenganan and surrounding villages. Having another indication of subak's flexibility is that it seems prepared to listen, amend, and also take support the affected to farmer. Its way of deliberating creates social capital and collective action (Meinzen-Dick et al., 2002). Subak has been called an ism in Bali that embodies the spirit and the values of Tri Hita Karana concept (the harmony between people, nature and spiritual life) (Wijana et al., 2022). In this scenario, we can formulate a hypothesis that Tenganan farmers will face constraints when they try to displace short-run harvest to other food except for export products as well as long-run resources. This cycle of exchange in care reaffirms to members and the land that they are for each other and is essential to keep agricultural productivity strong and consistent.

People in Tenganan have detail communality in various aspects of their life, including when it comes to the use of this age old decision-making process. Regulations for every citizen, whether male or female and rich or poor stand equal right to express their opinion at a generally gathering. Community is not a monolithic, unified actor, but rather the mode of decision-making, participation, and sharing of power among its members are relevant for collective action and conservation to be successful (Agrawal & Gibson, 1999). These decisions are intended to be based on mutual interests and to be mutually supportive. A strong focus on equality contributes to that the results of deliberation reflect wide participation and real collective consent (Sumunar et al., 2018).

There are already very good examples of participatory governance in the traditional ceremony of community buildings. Everything from traditions and ceremonies is integrated into environmental management. Together and individually they are locations where all sorts of voices converge to make those local decisions. And then folks can bring burdens and hopes around the use of natural resources, whether it's logging or agricultural, and talk about that. In addition, consensus functions as a means to head off conflicts that may grow into higher levels of conflict. Both men and women regard the village community as an extended family, intermediaries in cases of discord and disharmony between men, leading them to good neighborly relations and societal harmony.

Some researchers have found that the involvement of Indigenous community decisions is important to environmental sustainability. For example, the research on Native Knowing and Education by Black and McBean (2016) reveals that in indigenous communities it was not simply a matter of consulting communities but actually investing power with them to fashion their own policies. This autonomy allows native denizens to spare their neighbors in the ecosystems from many forms of devastation. Similarly, Maclean et al. (2015) emphasize that strong local institutions are key to managing natural resources. Cumulatively, these studies highlight the critical nature of local participation in indigenous community-based conservation area management in the preservation of sensitive environmental knowledge and the development of site-specific bylaws.

In Tenganan, ecological concerns are the result of ancestral belief, community-based laws, and leaders who keep up with the times. All of this is part of what manages to keep humans and the landscape in balance. Villages have built and reinforced local groups' shared interest in nature predominantly through customary obligations, village assemblies, and cultural ceremonies. It is not only an ecological space but also a cultural and spiritual sanctuary. There are important lessons to be drawn from this experience with respect to sustainable community-based management.

This chapter elaborates the common strand presented in this book—inclusive environment governance, by describing how an indigenous community applies sustainability including participatory modes and systems located within culture. In the specific case of Tenganan where traditional leadership satisfies community consent and customary rules (awig-awig), we observe that local institutions are engaged in the creation of equitable and sustainable governance. The conclusion that sustainable ways of natural resource management require well-managed local knowledge and power governance, particularly in cultural complex settings, is supported by the findings of this chapter. The chapter further offers an extended interpretation of the wider, socio-ecological crisis, showing how the Tenganan village is subjected to threats of eco-destruction generated by economic pressures stemming from tourism and changes in cultural proprioceptions. The study bridges the environment-development dichotomy by showing that ecological conservation in Tenganan is deeply tied to cultural rituals and inclusive governance. In this way, it helps fill an important gap in the literature, where most sustainable development models still overlook Indigenous worldviews and spiritual-ecological values as core foundations (Fig. 15.3).

The village of Tenganan in Bali is an interesting example for human-ecological resilience, where institutionalized locale practices and institutional triggers mediate pathway scaling between state-driven policy-oriented vis-a-vis community led environmental ethos. Faced with a major alien threat, Tenganan actually showed that traditional local government could work to permit non-violent use even under dogma as venerable and when trained institutions were ready to greet Manifestation of Conservation (who will come introducing himself as a friend then) at two places along the path. Such overlapping also causes the increase of value added managing and many more contributions toward a sustainable diversity of nature. Although the importance of local knowledge is acknowledged for natural resource management (Berkes, 2010; Pilgrim et al., 2009), in Tenganan it demonstrates its formal role in decision-making. It is proof that harmonization of state and indigenous institutions can be achieved and is imperative for sustainable long-term success (Folke et al., 2005; Ostrom, 2009). Tenganan offers a culturally situated and contextually dependent model for reconciling modern environmental governance with traditional Indigenous value of homelands in an era dominated by accelerating ecological change.

15.4 Recommendation

Strategic lessons on sustainable natural resource management from the Tenganan case are as follows:

- Formalizing Governance with Customary Law. The use of customary law is not just rhetoric in pro-community development. The law needs to be meaningfully enforced in official policies. This will diminish the complacency of voices that have hitherto lent themselves to policies which are little more than slogans. The

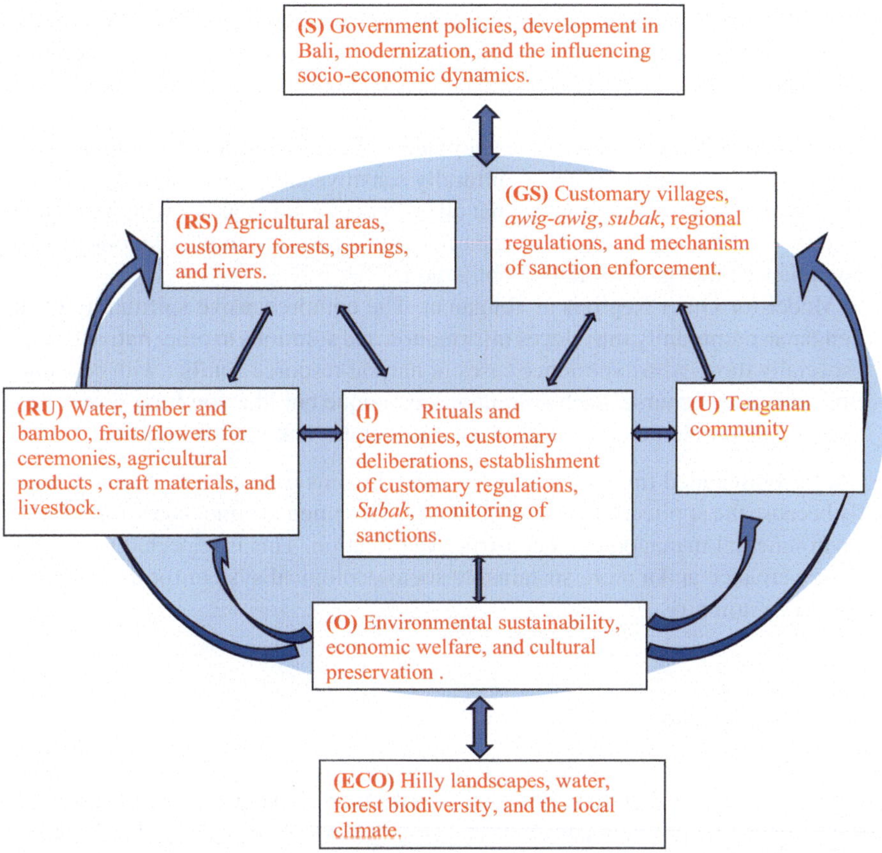

Fig. 15.3 Social-ecological system of the Tenganan community (adapted from Ostrom (2009)), illustrates how the Tenganan community governs natural resources—agricultural land, customary forests, and springs—through local institutions (customary village, *awig-awig*, *subak*, and customary village regulation) and social activities such as forest rituals, tree-felling deliberations, and *subak* water allocation. These resources, including irrigation water, wood and bamboo, fruit, ceremonial materials, and agricultural produce, are managed and utilized by various stakeholders (residents, farmers, and artisans) under customary rules informed by local wisdom. Their interactions are subject to external influences, such as government policies, development, and modernization, leading to diverse outcomes (forest conservation, sufficient water supply, improved welfare, and cultural preservation). The whole system is intimately connected with the wider ecosystem, its hilly landscape, and water sources, and it illustrates how environmental processes along with social and cultural factors are intertwined within Tenganan's mutually supporting socio-ecological system

hope is that traditional institutions will be strengthened and redeployed to the benefit of society and the environment.

- Nurturing the Next Generation in Learning Local Knowledge. In order to not let local knowledge get lost, we need to pass it on to the next generation. In addition, this approach also includes recording all indigenous wealth in a recording media so that coming generations can be learned from them.
- Emphasizing Participation in Development. Encouraging local communities to collaborate on the promotion of culturally sensitive sites is essential for community involvement. It is important that all sectors of society along with indigenous leaders, women, and young people participate to ensure smooth relationships and compliance among other agreed obligations.
- A Model for Other Regions at Tenganan. The comprehensive spirituality of the Tenganan community introduces information and solutions to other native groups, especially those who experience cases of natural resource conflict with economic imperatives. Of course, each community is unique but like you I am sure we can expect a favorable outcome with similar principles and modifications.

The recommended implementation of these lessons can serve as learning, especially because the application on tradition and environmental ethics can afford players in environmental management a constructive direction. This is expected to result in better governance and a more sustainable socio-ecological system for the benefit of the world community.

15.5 Conclusion

This chapter is a contribution to broader discussions within the environmental governance literature, by providing an empirical case of a customary system that integrates spiritual values, indigenous knowledge and institutional arrangements. As evidenced by Tenganan, Bali, such laws continue to be upheld in many communities around the world, even as we live through perpetual socio-ecological crises. Tenganan demonstrates that social institutions can manage natural resources well and maintain social order, even in a context pressured by tourism where there is much to be gained from "a quick buck." While the relationships between humans and nature are maintained in equilibrium, fundamental needs could be met in a manner that gives rise to inner peace and well-being. The chapter makes a plea for greater acknowledgement, in international sustainability debates, of the legitimacy and efficacy of traditional institutions for maintaining biodiversity and fostering social equity.

At the same time, this chapter illustrates that locally based resource management can address ecological, social, cultural, and economic considerations simultaneously. Influenced by *Tri Hita Karana*, a Hindu belief in the three harmonious relationship among humans and gods, humans themselves, and human's with nature, Tenganan also believes that nature is to be revered and not over used. It expresses itself through such material practices as selective tree-cutting and forest access, water control via subak, or careful preservation of gringsing weaving. Through awig-awig and religious rite in this way, these uses act as a conservation tool and they also serve to strengthen tradition.

The spirit of community solidarity Tenganan can be realized through participatory and justice that is holded by respecting and applying eco-friendly activities. The governance system in Tenganan has generated a customary natural resources management that constituted belief, regulation, and collective action needed for environmental sustainability. The Tenganan case offers principles that other regions can adapt, even if their local contexts differ. It also requires synergy in the formation of multi-stakeholder coalition between government, local networks, businesses, and others to manage environment based on the concept of local governance.

Acknowledgements We would like to express our sincere gratitude to everyone who supported the development of this chapter.

Authors Contribution All authors have contributed equally to this chapter.

Funding The research received no external funding.

Potential Conflicts of Interest On behalf of all authors, the corresponding author states that there is no conflict of interest.

Informed Consent Informed consent was obtained from all individual participants included in the study.

Data Availability Statement The datasets during and/or analyzed during the current study available from the corresponding author on reasonable request.

AI Disclosure Statement This study utilized AI on July 15, 2025, specifically ChatGPT (chat.openai.com), to assist in improving the writing structure and translating content into academic English. The prompts used included: "improve the writing using proper and formal Indonesian language" and "translate into English according to academic standards."

References

Agrawal, A., & Gibson, C. C. (1999). Enchantment and disenchantment: The role of community in natural resource conservation. *World Development, 27*(4), 629–649. https://doi.org/10.1016/S0305-750X(98)00161-2

Angelsen, A., Jagger, P., Babigumira, R., Belcher, B., Hogarth, N. J., Bauch, S., Börner, J., Smith-Hall, C., & Wunder, S. (2014). Environmental income and rural livelihoods: A global-comparative analysis. *World Development, 64*, S12–S28. https://doi.org/10.1016/j.worlddev.2014.03.006

Armah, F. A., Luginaah, I., Yengoh, G. T., Taabazuing, J., & Yawson, D. O. (2014). Management of natural resources in a conflicting environment in Ghana: Unmasking a messy policy problem. *Journal of Environmental Planning and Management, 57*(11), 1724–1745. https://doi.org/10.1080/09640568.2013.834247

Aryawan, I. W. (2021). Cerita Rakyat Lelipi Selan Bukit Sebagai Media Pendidikan Aspek Palemahan Tri Hita Karana. *Widya Accarya, 12*(2), 186–194. https://doi.org/10.46650/wa.12.2.1102.186-194

Asteria, D., Brotosusilo, A., Negoro, H. A., & Sudrajad, M. R. (2021). *Contribution of customary law in sustainable forest management for supporting climate action* (Vol. 940, No. 1). https://doi.org/10.1088/1755-1315/940/1/012080

Austin, K. G., Schwantes, A., Gu, Y., & Kasibhatla, P. S. (2019). What causes deforestation in Indonesia? *Environmental Research Letters, 14*(2), 024007. https://doi.org/10.1088/1748-9326/aaf6db

Barliana, M. S., Ilhamdaniah, Adhi, N., Mardiana, R., & Dwidayati, K. H. (2025). Contemporary urban design in Denpasar based on Balinese local culture and architecture. *City and Built Environment, 3*(1), 2. https://doi.org/10.1007/s44213-025-00043-w

Bashir, M. F., Bashir, M., & Shahzad, L. (2024). Natural resource abundance, digitalization and institutional governance: The critical challenges towards net-zero transformation. *Resources Policy, 99*. https://doi.org/10.1016/j.resourpol.2024.105412

Berkes, F. (2010). Restoring unity: The concept of marine social-ecological systems. In *World fisheries: A social-ecological analysis* (pp. 9–28). https://doi.org/10.1002/9781444392241.ch2

Black, K., & McBean, E. (2016). Increased indigenous participation in environmental decision-making: A policy analysis for the improvement of indigenous health. *International Indigenous Policy Journal, 7*(4). https://doi.org/10.18584/iipj.2016.7.4.5

Brooks, J., Reyes-García, V. & Burnside, W. (2018). Re-examining balinese subaks through the lens of cultural multilevel selection. *Sustainability science, 13,* 35–47.https://doi.org/10.1007/s11625-017-0453-1

BPS. (2022). *Jumlah Penduduk menurut Wilayah, Daerah Perkotaan/Perdesaan, dan Jenis Kelamin, Indonesia, 2022.* https://sensus.bps.go.id/topik/tabular/sp2022/187/1/0

Chan, K. M. A., Balvanera, P., Benessaiah, K., Chapman, M., Díaz, S., Gómez-Baggethun, E., Gould, R. K., Hannahs, N., Jax, K., Klain, S., Luck, G. W., Martín-López, B., Muraca, B., Norton, B., Ott, K., Pascual, U., Satterfield, T., Tadaki, M., Taggart, J., & Turner, N. (2016). Why protect nature? Rethinking values and the environment. *Proceedings of the National Academy of Sciences, 113*(6), 1462–1465. https://doi.org/10.1073/pnas.1525002113

channelnewsasia.com. (2025). *Poor waste management, land conversion worsened deadly Bali floods, says Indonesia's environment minister.* https://www.channelnewsasia.com/asia/indonesia-bali-flash-floods-waste-land-conversion-death-toll-5345896

Cole, S. (2012). A political ecology of water equity and tourism. *Annals of Tourism Research, 39*(2), 1221–1241. https://doi.org/10.1016/j.annals.2012.01.003

Collen, W., Krause, T., Mundaca, L., & Nicholas, K. A. (2016). Building local institutions for national conservation programs: Lessons for developing reducing emissions from deforestation and forest degradation (REDD+) programs. *Ecology and Society, 21*(2). https://doi.org/10.5751/ES-08156-210204

Cox, M., Arnold, G., & Villamayor Tomás, S. (2010). A review of design principles for community-based natural resource management. *Ecology and Society, 15*(4), art38. https://doi.org/10.5751/ES-03704-150438

Creswell, J. W. (2018). *Research design: Qualitative, quantitative, and mixed methods approaches.* SAGE Publication.

Das, A., Gujre, N., Devi, R. J., Rangan, L., & Mitra, S. (Eds.). (2023). Traditional ecological knowledge towards natural resource management. In *Sustainable agriculture and the environment* (pp. 275–294). Elsevier. https://doi.org/10.1016/b978-0-323-90500-8.00019-1

Fa, J. E., Watson, J. E., Leiper, I., Potapov, P., Evans, T. D., Burgess, N. D., Molnár, Z., Fernández-Llamazares, Á., Duncan, T., Wang, S., Austin, B. J., Jonas, H., Robinson, C. J., Malmer, P., Zander, K. K., Jackson, M. V., Ellis, E., Brondizio, E. S., & Garnett, S. T. (2020). Importance of indigenous peoples' lands for the conservation of intact forest landscapes. *Frontiers in Ecology and the Environment, 18*(3), 135–140. https://doi.org/10.1002/fee.2148

Folke, C., Hahn, T., Olsson, P., & Norberg, J. (2005). Adaptive governance of social-ecological systems. *Annual Review of Environment and Resources, 30*(1), 441–473. https://doi.org/10.1146/annurev.energy.30.050504.144511

Geng, Y., Hu, G., Ranjitkar, S., Shi, Y., Zhang, Y., & Wang, Y. (2017). The implications of ritual practices and ritual plant uses on nature conservation: A case study among the Naxi in Yunnan Province, Southwest China. *Journal of Ethnobiology and Ethnomedicine, 13*(1), 58. https://doi.org/10.1186/s13002-017-0186-3

Google-Earth. (2025). *Tenganan.* https://earth.google.com/web/search/Tenganan,+Karangasem+Regency,+Bali/@-8.47409273,115.5802523,146.17387343a,13807.88311893d,35y,0h,0t,0r/data=CpEBGmMSXQolMHgyZGQyMDhhZTA3YWI3ODBkOjB4ZDVlMDhhYTU0NjY2NTc3NhnegyG8h_EgwCEBmq1hK-RcQCoiVGVuZ2FuYW4sIEthcmFuZ2FzZW0gUmVnZW5jeSwgQmFsaRgCIAEiJgokCWD_3cxG5yDAEYKmscclCyHAGYcW-WZo6VxAIQwbHPTk3lxAQgIIATIpCicKJQohMXNPQWtaRmFBRlR0VG9kUFFJN2dOMk96Z1FlNGw1blZZIAE6AwoBMEICCABKBwiS1MEbEAE

Jaung, W., & Carrasco, L. R. (2022). A big-data analysis of human-nature relations in newspaper coverage. *Geoforum, 128*, 11–20. https://doi.org/10.1016/j.geoforum.2021.11.017

Jensen, J. S. (2016). *Myths and mythologies: A reader*. Taylor and Francis Ltd.

Jiménez, I., & Basurto, X. (2022). An organizational framework for effective conservation organizations. *Biological Conservation, 267*, 109471. https://doi.org/10.1016/j.biocon.2022.109471

Kamal, S., & Lim, V. (2019). Forest reserve as an inclusive or exclusive space? Engaging Orang Asli as stakeholder in protected area management. *Journal of Tropical Forest Science, 31*(3), 278–285. https://doi.org/10.26525/jtfs2019.31.3.278

Kaskoyo, H., Mohammed, A., & Inoue, M. (2017). Impact of community forest program in protection forest on livelihood outcomes: A case study of Lampung Province, Indonesia. *Journal of Sustainable Forestry, 36*(3), 250–263. https://doi.org/10.1080/10549811.2017.1296774

Kemen LHK. (2021). *Indonesia long-term strategy for low carbon and climate resilience 2050.* Kementerian Lingkungan Hidup dan Kehutanan Republik Indonesia.

Kristiono, N. (2017). Pola Kehidupan Masyarakat Adat Desa Tenganan Pegringsingan Bali. *Integralistik, 28*(2), 158–175. https://doi.org/10.15294/integralistik.v28i2.13734

Lansing, J. S., Thurner, S., Chung, N. N., Coudurier-Curveur, A., Karakaş, Ç., Fesenmyer, K. A., & Chew, L. Y. (2017). Adaptive self-organization of Bali's ancient rice terraces. *Proceedings of the National Academy of Sciences, 114*(25), 6504–6509. https://doi.org/10.1073/pnas.1605369114

Liana, I. K. A., Jenyani, N. P. D. P., Putra, I. M. D. C. P., Pradnyani, P. S. W., & Parwati, N. P. E. (2024). The urgency of awig-awig customary villages in Bali regarding subak regulations in saving land conversion and local wisdom of Balinese. In *Proceedings of the 6th international conference on law, social sciences and education, ICLSSE 2024, 17 October 2024, Singaraja, Bali, Indonesia.* https://doi.org/10.4108/eai.17-10-2024.2353696

Maclean, K., Robinson, C. J., & Natcher, D. C. (2015). Consensus building or constructive conflict? Aboriginal discursive strategies to enhance participation in natural resource management in Australia and Canada. *Society & Natural Resources, 28*(2), 197–211. https://doi.org/10.1080/08941920.2014.928396

Maria, S., & Rupa, I. W. (2007). *Desa adat Tenganan Pegringsingan kabupaten Karangasem Propinsi Bali.* Direktorat Kepercayaan Terhadap Tuhan Yang Maha Esa, Departemen Kebudayaan dan Pariwisata.

Meinzen-Dick, R., Raju, K. V., & Gulati, A. (2002). What affects organization and collective action for managing resources? Evidence from canal irrigation systems in India. *World Development, 30*(4), 649–666. https://doi.org/10.1016/S0305-750X(01)00130-9

Nurlinda, I. (2016). Kebijakan Pengelolaan Sumber Daya Alam Dan Dampaknya Terhadap Penegakan Hukum Lingkungan Indonesia. *Bina Hukum Lingkungan, 1*(1), 1–9. https://doi.org/10.24970/jbhl.v1n1.1

Ofoegbu, C., Chirwa, P., Francis, J., & Babalola, F. (2017). Assessing vulnerability of rural communities to climate change: A review of implications for forest-based livelihoods in South Africa. *International Journal of Climate Change Strategies and Management, 9*(03), 374–386. https://doi.org/10.1108/IJCCSM-04-2016-0044

Okura, F., Budiasa, I. W., & Kato, T. (2022). Exploring a Balinese irrigation water management system using agent-based modeling and game theory. *Agricultural Water Management, 274*, 107951. https://doi.org/10.1016/j.agwat.2022.107951

Oldekop, J. A., Rasmussen, L. V., Agrawal, A., Bebbington, A. J., Meyfroidt, P., Bengston, D. N., Blackman, A., Brooks, S., Davidson-Hunt, I., Davies, P., Dinsi, S. C., Fontana, L. B., Gumucio, T., Kumar, C., Kumar, K., Moran, D., Mwampamba, T. H., Nasi, R., Nilsson, M., … Wilson, S. J. (2020). Forest-linked livelihoods in a globalized world. *Nature Plants, 6*(12), 1400–1407. https://doi.org/10.1038/s41477-020-00814-9

Olkiewicz, M. (2023). Changes in the natural surroundings are a determinant of the implementation of the environmental management system. *Rocznik Ochrona Srodowiska, 25*, 357–366. https://doi.org/10.54740/ros.2023.036

Ostrom, E. (1990). *Governing the commons: The evolution of institutions for collective action*. Cambridge University Press. https://doi.org/10.1017/CBO9780511807763

Ostrom, E. (2009). A general framework for analyzing sustainability of social-ecological systems. *Science, 325*(5939), 419–422. https://doi.org/10.1126/science.1172133

Özgen Çiğdemli, A. Ö. (2022). Overtourism: Issues, risk and responsibility. In *Overtourism, technology solutions and decimated destinations* (pp. 135–156). https://doi.org/10.1007/978-981-16-2474-2_9

Partelow, S. (2018). A review of the social-ecological systems framework: Applications, methods, modifications, and challenges. *Ecology and Society, 23*(4), art36. https://doi.org/10.5751/ES-10594-230436

Pebryani, N. D., & Vogel, M. (2024). Weaving webs of protection: Gringsing textiles. *Textile, 22*(1), 128–141. https://doi.org/10.1080/14759756.2023.2165049

Pemerintah Provinsi Bali. (2019). *Peraturan Daerah Provinsi Bali Nomor 4 Tahun 2019 Tentang Desa Adat di Bali*.

Pertiwi, P. R., & Mardiana, R. (2020). Dinamika Awig-Awig dan Pengaruhnya terhadap Keberlanjutan Tanah Adat. *Jurnal Sains Komunikasi Dan Pengembangan Masyarakat [JSKPM], 4*(1), 125–136. https://doi.org/10.29244/jskpm.4.1.125-136

Pilgrim, S., Pretty, J., Adams, B., Berkes, F., De Athayde, S. F., Dudley, N., Hunn, E., Maffi, L., Milton, K., Rapport, D., Robbins, P., Sterling, E., Stolton, S., Tsing, A., & Vintinnerk, E. (2009). The intersections of biological diversity and cultural diversity: Towards integration. *Conservation and Society, 7*(2), 100. https://doi.org/10.4103/0972-4923.58642

Powell, B., Thilsted, S. H., Ickowitz, A., Termote, C., Sunderland, T., & Herforth, A. (2015). Improving diets with wild and cultivated biodiversity from across the landscape. *Food Security, 7*(3), 535–554. https://doi.org/10.1007/s12571-015-0466-5

Rankoana, S. (2016). Sustainable use and management of indigenous plant resources: A case of Mantheding community in Limpopo Province, South Africa. *Sustainability, 8*(3), 221. https://doi.org/10.3390/su8030221

Ratnani, D. A. S., Junitha, I. K., Kriswiyanti, E., & Dhana, I. N. (2021). The ethnobotany of Ngusaba ceremonial plant utilization by Tenganan Pegringsingan community in Karangasem, Bali, Indonesia. *Biodiversitas Journal of Biological Diversity, 22*(4). https://doi.org/10.13057/biodiv/d220457

Rimba, A. B., Chapagain, S. K., Masago, Y., Fukushi, K., & Mohan, G. (2019). Investigating water sustainability and land use/land cover change (LULC) as the impact of tourism activity in Bali, Indonesia. In *IGARSS 2019—2019 IEEE international geoscience and remote sensing symposium* (pp. 6531–6534). https://doi.org/10.1109/IGARSS.2019.8900060

Saputro, B. A., Adha, M., & Ulmi, S. A. (2024). Preserving harmony in Bali: An environmental philosophy approach to tackling over-tourism. *Digital Press Social Sciences and Humanities, 11*, 00004. https://doi.org/10.29037/digitalpress.411456

Silviana, A., Utama, Y. J., Ismail, N., & Arda, M. N. (2021). Land management policy in the coastal area based on the local wisdom. *AACL Bioflux, 14*(6), 3403–3415.

Sivakumar, M. (2021). Climate services' role in safeguarding pastoral disaster communities. *Revue Scientifique et Technique de l'OIE, 40*(2), 431–438. https://doi.org/10.20506/rst.40.2.3235

Suhardono, S., Lee, C.-H., & Suryawan, I. W. K. (2024). Trends in citizen influencing willingness to participate in marine debris management and social well-being in Bali metropolitan, Indonesia. *Urban Governance, 4*(4), 362–373. https://doi.org/10.1016/j.ugj.2024.12.005

Sujarwo, W., Caneva, G., & Zuccarello, V. (2020). Patterns of plant use in religious offerings in Bali (Indonesia). *Acta Botanica Brasilica, 34*(1), 40–53. https://doi.org/10.1590/0102-33062019abb0110

Sujarwo, W., Keim, A. P., Savo, V., Guarrera, P. M., & Caneva, G. (2015). Ethnobotanical study of Loloh: Traditional herbal drinks from Bali (Indonesia). *Journal of Ethnopharmacology, 169*, 34–48. https://doi.org/10.1016/j.jep.2015.03.079

Sumunar, D. R. S., Suparmini, S., & Setyawati, S. (2018). Masyarakat Desa Adat Tenganan Pegringsingan. *Jurnal Penelitian Humaniora, 22*(2), 111–124. https://doi.org/10.21831/hum.v22i2.19929

Theresia, M., & Budiastuti, S. (2020). *Agroforestry as climate change mitigation* (Vol. 2020, pp. 23–29). https://doi.org/10.11594/nstp.2020.0603

Wiessner, P. (2005). Norm enforcement among the Ju/'hoansi Bushmen: A case of strong reciprocity? *Human Nature, 16*(2), 115–145. https://doi.org/10.1007/s12110-005-1000-9

Wijana, N., Citrawathi, D. M., Mulyadiharja, S., Julyasih, K. S. M., & Dewi, N. P. S. R. (2022). Results of ethnoecological studies in the prospect of forest tourism development in Tenganan Pegringsingan traditional village, Karangasem, Bali. *Baltic Forestry, 28*(1). https://doi.org/10.46490/BF599

Wiryawan, I. W. (2024). Implementation of Balinese Tri Hita Karana concept for environmental conservation of cultural heritage of land consolidation arrangements. *International Journal of Conservation Science, 15*(1), 597–614. https://doi.org/10.36868/IJCS.2024.01.15

Zeigermann, U., & Böcher, M. (2020). Challenges for bridging the gap between knowledge and governance in sustainability policy—The case of OECD 'Focal Points' for policy coherence for development. *Forest Policy and Economics, 114*, 102005. https://doi.org/10.1016/j.forpol.2019.102005

Chapter 16
Agricultural Development Paradigm Shift in Indonesia: Promoting Farmer Corporations for Sustainability

Andi Ishak, Kurnia Suci Indraningsih, Carolina, Arditya Wicaksono, Febtri Wijayanti, Ria Jayanthi, Eko Wahyono, and Purnama Alamsyah

Abstract The agricultural sector is experiencing a significant transformation, prompting a shift from traditional development paradigms towards more sustainable and inclusive models. This chapter examines the establishment of farmer corporations through an agribusiness approach as a strategy to enhance agricultural sustainability. By organizing farmers into corporate structures, collective access to resources, knowledge and markets can be strengthened, improving productivity and profitability. Using the Context-Input-Process-Product (CIPP) framework, the chapter examines the key challenges faced by smallholder farmers. One of them is limited access to markets, financing and technology, factors that often lead to unsustainable practices. The agribusiness approach emphasizes the importance of value-added activities, such as processing and marketing, enabling farmers to increase their competitiveness

A. Ishak (✉) · K. S. Indraningsih · Carolina · A. Wicaksono · F. Wijayanti · R. Jayanthi · E. Wahyono · P. Alamsyah
National Research and Innovation Agency, South Jakarta, Indonesia
e-mail: andi060@brin.go.id

K. S. Indraningsih
e-mail: kurn028@brin.go.id

Carolina
e-mail: carolina@brin.go.id

A. Wicaksono
e-mail: ardi013@brin.go.id

F. Wijayanti
e-mail: febt002@brin.go.id

R. Jayanthi
e-mail: riaj002@brin.go.id

E. Wahyono
e-mail: ekow005@brin.go.id

P. Alamsyah
e-mail: purn008@brin.go.id

G. K. Panda et al. (eds.), *Socio-Ecological Crisis of Development*, Sustainability Solutions, https://doi.org/10.1007/978-3-032-22171-1_16

while preserving local identity. The chapter also examines the crucial role of policy support, institutional frameworks and capacity building in fostering the growth of farmer corporations. Training in business management, financial literacy and sustainable practices is crucial for empowering farmers during this transition. Drawing on case studies from various regions, the chapter highlights successful models of farmer corporations that have effectively adopted agribusiness strategies to achieve greater sustainability and resilience. Ultimately, this chapter offers insights and practical recommendations for policymakers, practitioners and researchers on promoting equitable and sustainable agricultural development through the integration of farmer corporations and agribusiness.

Keywords Agricultural development · CIPP · Farmer corporations · Agribusiness approach · Sustainability

16.1 Introduction

The agricultural sector plays a crucial role, as it provides food for the world and has become the backbone of economic development, particularly in Indonesia. Ironically, farmers, as the leading actors in the agricultural sector in rural areas, are still the dominant contributors to the number of poor people in Indonesia. It is recorded that 47.34% of poor households with heads of households are working in the agricultural sector (BPS-Statistics Indonesia, 2024). Various agricultural development programs in Indonesia continue to focus on enhancing food security, promoting agribusiness development, and improving farmer welfare. However, to date, agricultural development programs have not been successful in reducing poverty levels in rural areas, indicating that the government's ongoing efforts have not been optimal. As an illustration, poverty statistics data from BPS (2025) show that in 2000, the number of poor people in Indonesia was 38.7 million people, 68.22% of whom were in rural areas. In 2010, the number of people living in poverty was 31.02 million, with 64.25% residing in rural areas. By 2020, the number of poor people had decreased to 27.55 million, with 56.30% still residing in rural areas. In 2024, the number of people living in poverty is expected to decrease to 25.22 million, with 53.85% of them residing in rural areas. Based on these data, it can be stated that although the number tends to decrease, most people with low incomes in Indonesia are still in rural areas, which, of course, are very dependent on the agricultural sector.

Strengthening farmer institutions is needed to strengthen bargaining positions in market competition, provide capital and facilitate the integration of agricultural businesses from upstream to downstream (Darwanto et al., 2019). Strengthening the institutional capacity of farmers requires synergy between farmers and entrepreneurs, supported by government policies (Raharja et al., 2020). These government policies include intensive assistance to farmers through agricultural extension (Mariyono et al., 2021), facilitation of agricultural tools and machinery (Aldillah, 2016) and capital assistance for farmers (Harianto et al., 2019). Increasingly conducive government policies will increase farmers' entrepreneurial orientation and improve farming performance (Anisa et al., 2021). Meanwhile, partnerships with open and mutually beneficial business partners are expected to increase farmers' income (Sjaf et al., 2022).

The government's efforts to develop farmer institutions involve placing agricultural development programs within the recipient farmer groups that have been formed. It is hoped that the agricultural technology innovations developed can be disseminated more quickly to enhance food security and sustainable agricultural development (Sutardi et al., 2022). Therefore, it is necessary to integrate farmer institutions by establishing them with legal status (Harjanto et al., 2022). The formation of farmer groups began during the New Order era in the 1970s, aimed at implementing the national agricultural development program in Indonesia as a key agent of rural development. In the 1990s and 2000s, the formation of farmer groups began to be initiated by farmers independently. Following the regional autonomy era, various forms of farmer groups emerged as important partners in local development projects (Purnomo et al., 2022). Factors influencing the formation of farmer groups include government encouragement for development interests through top-down strategies (Pranadji et al., 2021) and the existence of shared economic interests such as increasing production and gaining access to capital or marketing assistance (Rahmadanih et al., 2018). However, farmer groups are still exclusive institutions (Rokhani et al., 2021). The reason for this exclusivity is that in the Regulation of the Minister of Agriculture Number 82/Permentan/OT.140/8/2013 concerning Guidelines for the Development of Farmer Groups and Farmer Group Associations, it is stipulated that the formation of farmer groups is based on the extent of agricultural land or the location of the farmer's domicile. The number is limited to between 20 and 25 people to facilitate assistance from agricultural extension workers in the field.

Agricultural development policies related to the development of farmer institutions in Indonesia have changed institutional form, territorial coverage, institutional legality and business capacity. Farmer institutions, in the form of farmer groups, began to be formed during the New Order era in the 1970s. They were then pushed into farmer group associations at the beginning of the reform era in the 2000s and subsequently evolved into farmer corporations since 2018. Figure 16.1 shows the shift in the form and orientation of these institutions.

The development of farmer corporation institutions within the framework of the agribusiness system, carried out through government programs, must be compatible with the socio-economic and cultural conditions of the local community. The economic system built on considering the socio-cultural and economic conditions of the community has been successfully practised in East and Southeast Asian countries (Block & Evans, 2005). Based on the description above, a theoretical framework can be described in the development of farmer corporations that integrate the roles of farmers, markets and government within the framework of the agribusiness system as follows (Fig. 16.2).

In developing farmer corporations, it is essential to focus on the objectives to be achieved, which are, in essence, to increase the added value and competitiveness

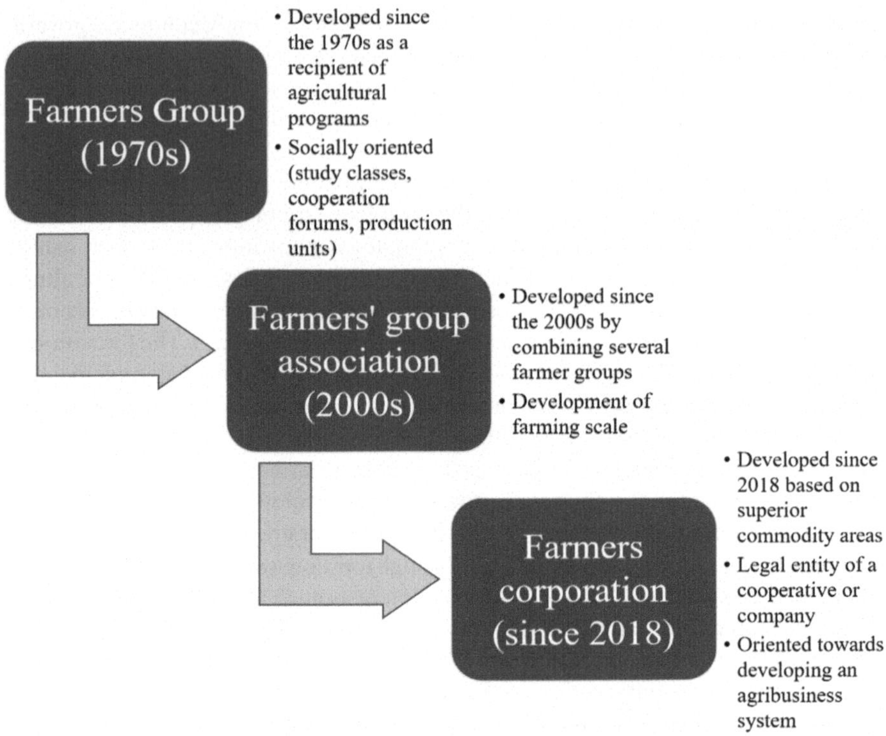

Fig. 16.1 Shifting forms of farmer institutions in Indonesia (1970–2018). *Source* Adapted from Munifah (2019)

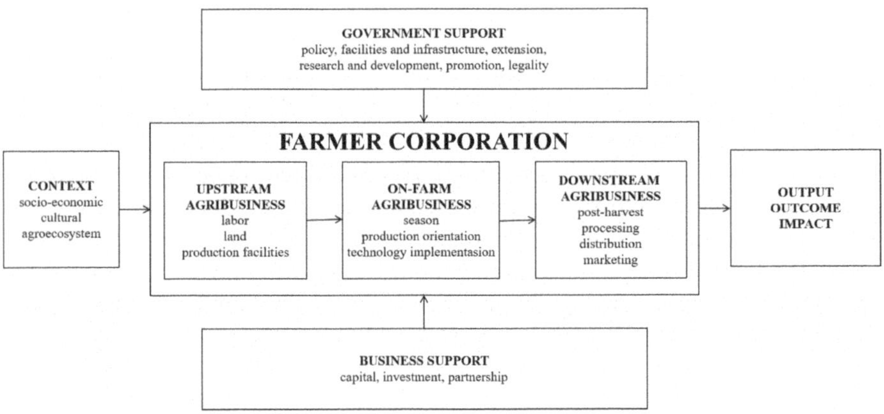

Fig. 16.2 Conceptual framework for developing farmer corporations in the agribusiness system. Adapted from Block and Evans (2005), Kementan (2018), Krinamurthi (2020)

of agricultural regions and commodities by strengthening the agribusiness system, thereby enhancing the efficiency and effectiveness of farming businesses that ultimately improve the welfare of rural communities. Strengthening the agribusiness system requires collaboration among farmer corporations, the government, and the business world or market players, who serve as supporters of the agribusiness system. The farmer corporations that are built cannot be separated from the suitability of the social, economic and cultural conditions of the community, as well as the local agroecosystem, so that farmers can easily participate in corporate development. Furthermore, a description is provided of the integration of the agribusiness system within each farmers' corporation and of how this integration relates to the future sustainability of the agribusiness system.

Global agricultural development has transitioned from state-led approaches to market-oriented and institution-driven models, emphasizing the role of farmer-led corporations in advancing sustainability. In Southeast Asia, nations such as Thailand, Vietnam and China have effectively integrated smallholder farmers into corporate structures, enhancing their productivity and competitiveness. Although Indonesia has embarked on similar reforms, the effectiveness of farmer corporations within institutional and agribusiness frameworks remains insufficiently examined. This study, therefore, aims to evaluate the development and performance of farmer corporations in Indonesia through the lens of the agribusiness system, employing the Context-Input-Process-Product (CIPP) evaluation model. The novelty of this research lies in its comprehensive assessment, which synergizes institutional and agribusiness perspectives across three pilot sites on the island of Java, Indonesia. By bridging global paradigms with localized practices, this study offers valuable insights into the discourse on sustainable agricultural transformation in developing countries, providing implications relevant to an international audience and policy stakeholders.

Agricultural development worldwide has shifted from top-down government programs to institution-led approaches that emphasize building local capacities, particularly through farmer-led organizations. While the concept of farmer-led corporations has been implemented in several Asian countries, the policy transition in Indonesia since 2018 has yet to be thoroughly evaluated in terms of its effectiveness. This study aims to fill this gap by examining farmer corporations in three pilot locations, analyzing them through a dual framework of institutional development and agribusiness system integration using the Context-Input-Process-Product (CIPP) evaluation model.

16.2 Study Area

This research was conducted in three farmer corporations located at the site of farmer corporation development for plantation and horticultural commodities on dry land. The three research locations are pilot farmer corporations in three provinces in Java (Fig. 16.3). The location was chosen on Java Island, considering that market access for farmer corporations is more open, as economic activities in Indonesia are still

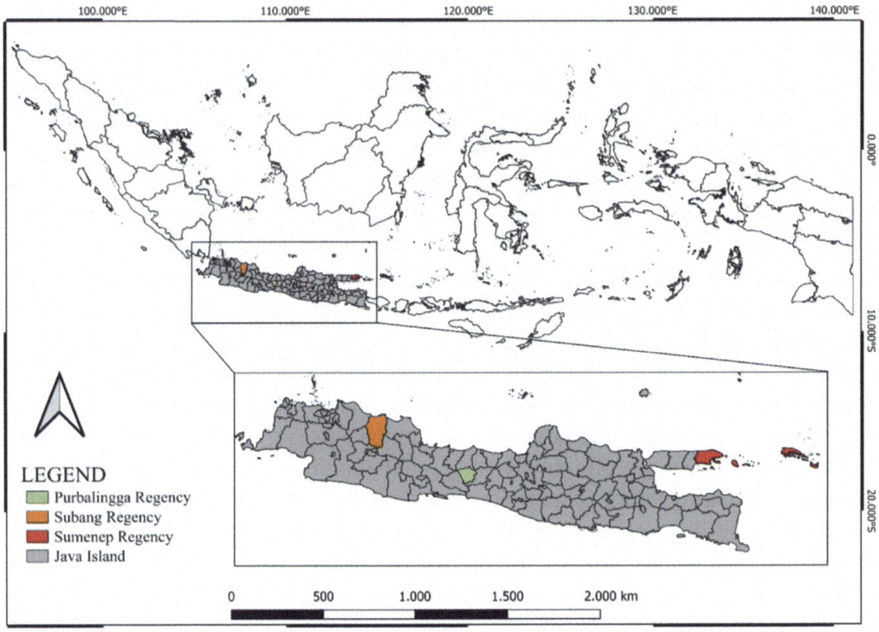

Fig. 16.3 Distribution of three research locations in Java

primarily centred on Java Island. With an area of approximately 7% of Indonesia's land area, Java Island is inhabited by 151.6 million people or 56.10% of the total population based on the results of the 2020 population census (Setkab, 2021).

The three farmer corporations have received UPLAND Project facilitation from the Ministry of Agriculture since 2021 and have been incorporated as cooperatives and limited companies in 2022 (Table 16.1). The three corporations are new institutions established through government program interventions.

16.3 Materials and Methods

This study employed a mixed-methods approach, combining quantitative evaluation using the Context-Input-Process-Product (CIPP) model with qualitative analysis of agribusiness system integration. The CIPP model was used to assess the performance of farmer corporations in three pilot locations—PCM, MJF and PIR—by evaluating 19 sub-indicators distributed across four dimensions. Each sub-indicator was measured using a structured questionnaire with a 4-point Likert scale (1 = strongly disagree to 4 = strongly agree), administered to 30 respondents per site, including farmers, administrators and agricultural officials. Mean scores were used to categorize each dimension as low, moderate or high. In parallel, qualitative data

Table 16.1 Research locations

	Farmers corporation	Corporate description	Information
1	Perwira Cipta Mandiri (PCM) cooperative	Location	Kejobong District, Purbalingga Regency, Central Java Province
		Year formed	2022
		Legal entity	Producer cooperative
		Number of members	64 farmers
		Featured commodities	Pepper, sugarcane
2	Midang Jaya Farm (MJF) cooperative	Location	Serangpanjang District, Subang Regency, West Java Province
		Year formed	2022
		Legal entity	Producer cooperative
		Number of members	70 farmers
		Featured commodities	Mangosteen
3	Permata Indah Rubaru (PIR) Limited Liability Company	Location	Rubaru District, Sumenep Regency, East Java Province
		Year formed	2022
		Legal entity	Limited liability company
		Featured commodities	Shallots

were collected through semi-structured interviews with key informants to examine the integration of agribusiness subsystems (upstream, on-farm, downstream). This integration was assessed based on the operational roles and linkages of each farmer corporation with input suppliers, production activities, processing and marketing channels. The combination of these methods enabled a comprehensive evaluation of both organizational effectiveness and the functional sustainability of agribusiness. The research methodology described above is illustrated in the following flow chart in Fig. 16.4.

The research methodology follows a structured, multi-phase approach to assess the sustainability of agribusiness systems. The study began with the selection of three representative sites: PCM, MJF and PIR. Data were collected through CIPP-based surveys, interviews and secondary sources. The CIPP (Context - Input - Process - Product) model was applied using a 4-point Likert scale to evaluate the effectiveness of each agribusiness scheme. Subsequently, the agribusiness system was analysed across three subsystems: upstream, on-farm and downstream. Data analysis combined quantitative methods—using mean score categorization—and qualitative synthesis to ensure comprehensive interpretation. Finally, sustainability was assessed

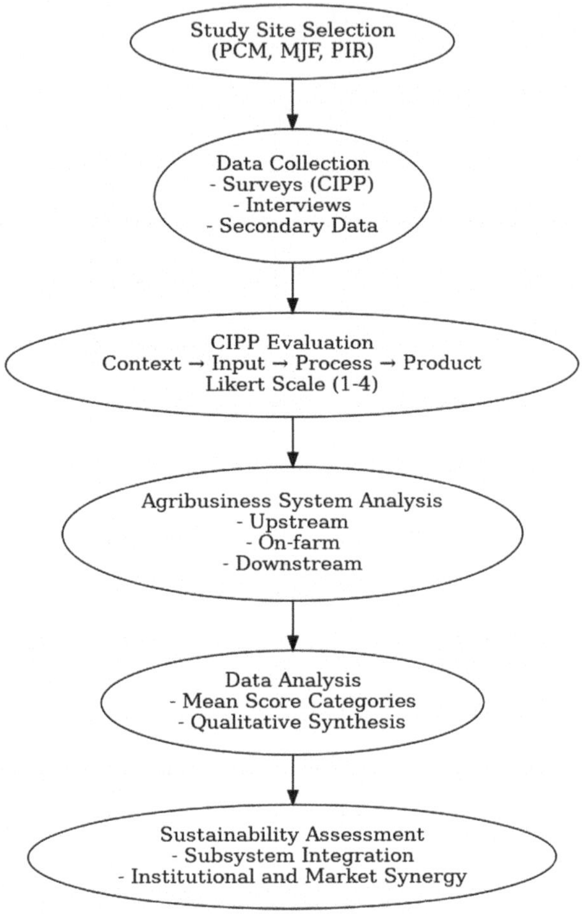

Fig. 16.4 Methodological flowchart

through subsystem integration and the synergy between institutional and market actors. The data collected in the study consisted of both primary and secondary data aimed at evaluating farmer corporations. Primary data were obtained from survey results (CIPP evaluation) and interviews (agribusiness system evaluation). The survey involved 30 people at each research location. The scope of activities is shown in Fig. 16.5.

The respondents interviewed were 30 people from each farmer corporation, consisting of farmers, farmer corporation administrators, district agriculture service personnel, agricultural extension workers and business actors involved in agribusiness activities. CIPP evaluation using scoring (Likert scale) on agreement with a value of 1–4, respectively, from strongly disagree, less agree, agree and strongly agree. Qualitative data collection through interviews with various sources focused on factors related to the performance of farmer corporations in agribusiness development. The data were analysed descriptively by classifying the average CIPP score

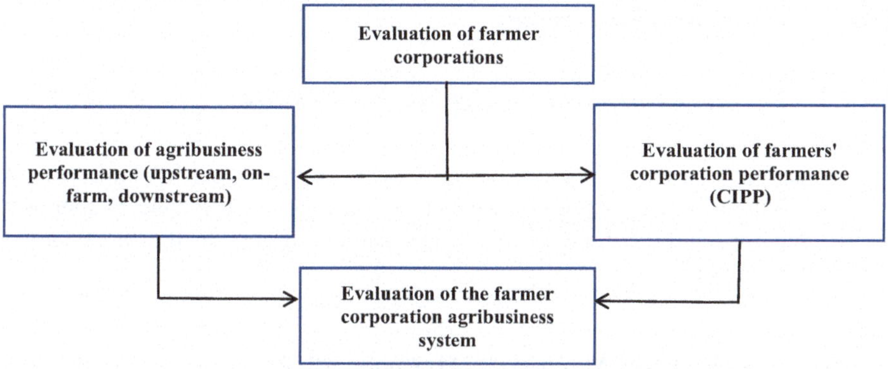

Fig. 16.5 Scope of farmer corporation evaluation

into three categories, namely, low (1 > Mi < 2), moderate (2 > Mi < 3) and high (3 > Mi < 4), where Mi is the mean of the respondents' answer scores. Sustainability in agribusiness development is a core value in evaluating farmer corporations. The sustainability evaluation is helpful for the prospects of developing farmer corporations in the future. Stufflebeam (2015) presents a framework for an evaluation using the CIPP concept (Fig. 16.6).

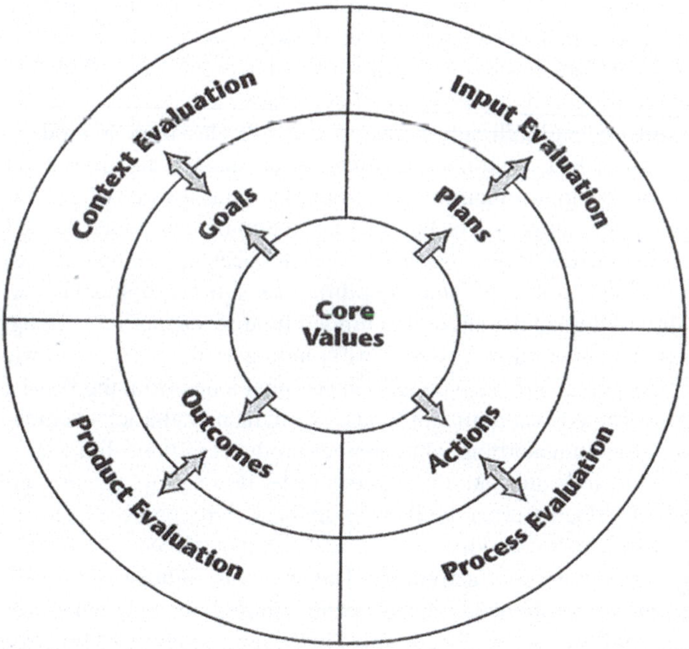

Fig. 16.6 Model evaluation using the CIPP concept (Stufflebeam, 2015)

16.4 Results and Discussion

16.4.1 Farmers Corporation Evaluation Results

The evaluation of farmer corporation performance below provides a brief overview of the evaluation of farmer corporation development using the CIPP concept approach and its relation to the formation of an agribusiness system in the three farmer corporations studied.

16.4.1.1 Perwira Cipta Mandiri (PCM) Cooperative

The results of the CIPP evaluation (Table 16.2) provide a comprehensive picture of the PCM performance profile. The highest value is in the Product indicator with a score of 3.18, followed by the Context indicator (3.06), Process (3.06) and Input (2.78). The results of this evaluation show high values for all CIPP indicators, except for the input indicator, which is classified as moderate. In aggregate, the results of the CIPP evaluation on PCM are classified as high with a value of 3.02.

Table 16.2 shows that, in the Context indicator, the sub-indicators "level of corporate perfection in planning preparation" and "level of completeness and clarity of guidelines" in organizational management are still in the moderate category, with average scores of 2.65 and 2.83, respectively. Within the sub-aspect related to establishing a farmer corporation, it is already classified as high, with a score exceeding 3. Therefore, improvements are needed in organizational management for the preparation of plans and the clarity of guidelines.

In the Input indicator, all sub-indicators are still classified as moderate (scores between 2.55 and 2.91), including human resource support, facilities and budget. In terms of corporate support facilities, for example, it has a value of 2.91, which falls within the moderate category. The facilities and equipment provided by the UPLAND Project for cooperatives are extensive, but have not been optimally utilized due to the relatively low production scale. The suboptimal use of this equipment is also related to the availability of human resources and limited funding capabilities. Budget support is critical in pepper marketing. The source of funding for the purchase of white pepper from farmers, especially those who rely on personal loans from the PCM Chairman, is related to the limited availability of funds. Consequently, the achievement of farmer corporation output in absorbing white pepper products remains limited.

In the Process indicator, attention needs to be directed to the sub-indicators of "success rate of reducing procedural errors in the activity process", "level of human resource utilization in the activity process" and "level of success of implementation of monitoring and evaluation of activities". The three sub-indicators are still classified as moderate and therefore need improvement, although the aggregate average value of the Process indicator is already classified in the high category (3.06). Product indicators have shown high performance (3.18). The impact achievement sub-indicator

Table 16.2 CIPP evaluation results at PCM

Indicators	Sub-indicators	Average of sub-indicators	Average of indicators
Context	The priority level of the objectives that have been set in establishing the corporation	3.32	3.06
	The level of opportunity to achieve corporate goals is based on the problems faced and the ability to implement activities	3.36	
	Level of corporate strength in achieving goals	3.16	
	Level of corporate perfection in planning and preparation	2.65	
	Level of completeness and clarity of guidelines	2.83	
Input	The level of availability of human resources for farmer corporations	2.86	2.78
	Level of availability and support of facilities for farmer corporations	2.91	
	Level of budget availability and support	2.55	
Process	The level of success of implementing socialization activities	3.38	3.06
	Level of conformity of the ongoing Process with the achievement of activity objectives	3.31	
	The level of accuracy of the Process with the guidelines and technical instructions that have been issued	3.03	
	Success rate of reducing procedural errors in the activity process	2.86	
	Level of human resource utilization in the activity process	2.93	
	Level of conformity of the activity implementation process with the use of available facilities	3.03	

(continued)

Table 16.2 (continued)

Indicators	Sub-indicators	Average of sub-indicators	Average of indicators
	Level of conformity of the implementation process with the use of the available budget	3	
	Degree of conformity of the ongoing Process with the use of available time	3.1	
	The level of success of the implementation of the activity assistance	3.03	
	The level of success of the implementation of monitoring and evaluation of activities	2.93	
Product	Achievement of output	3.05	3.18
	Achievement of outcome	3.07	
	Achievement of impact	3.43	
CIPP average value			3.02

Source Survey results (processed data)

has the highest average score (3.43) among all evaluated sub-indicators. This indicates that corporations have a positive impact on the community, offering benefits that can be felt by farmers in Purbalingga.

The purpose of establishing PCM is: (1) to make PCM an economic institution for farmers, (2) to create profitable businesses, (3) to help provide production facility needs for farmers and (4) to help market products by seeking product intensification and diversification at the best price in order to obtain optimal profits for farmers individually and in groups. PCM establishes business partnerships based on the principle of mutual symbiosis, or mutual benefit.

The cooperative received funding from various sources. Investment was secured through the Ministry of Agriculture's facilitation, which provided agricultural tools and machinery, as well as supporting facilities for white pepper processing. The cooperative received an investment of approximately 1.8 billion rupiah, which included assistance in the form of agricultural tools and machinery for pepper processing, such as soaking tubs, hullers, grading machines, color grading machines, automatic filling machines and blowers. The investment is not supported by assistance from the Ministry of Agriculture and the Regional Government, including training and product promotion.

In addition to government investment, cooperative members also receive people's business credit (KUR) funds from Bank Rakyat Indonesia by using the cooperative as a guarantor or guarantee provider. KUR from BRI is used for sugarcane cultivation, including capital for seeds and fertilizers. One loan period is 18 months. Credit payments are made at the time of sale of the sugarcane harvest. Sales are made directly by the cooperative to the sugar factory, and farmers receive the proceeds after deducting their bank debt. PCM member farmers borrow funds ranging from 10 to 15 million rupiah per person. Other sources of cooperative funding come from member savings. These savings consist of principal savings, mandatory savings and special savings. The total amount of cooperative savings at the time of the study was around 541 million rupiah, most of which was used to purchase white pepper.

Pepper is a commodity that the Kejobong community has cultivated since ancient times, so farmers have experience in post-harvest handling with varying qualities. This is caused by collectors who buy farmers' pepper at the same price for all qualities, resulting in the quality and price of white pepper being low. The same thing happens with sugarcane purchases by collectors using the "tebas" system, where a single price applies to all qualities.

The cooperative has established market networks with various parties, both at the local and national levels. For local, pepper and sugarcane marketing is not a problem. PCM has a strong enough network with various parties to build a pepper and sugarcane agribusiness system. PCM collaborates with traders to secure raw material supplies of the expected quality. Partnerships have been established between farmer cooperatives and private parties who care about farmer corporate programs. Java Agritech in Semarang (export to Japan), PT. SUPA Surya Niaga in Surabaya, and PT. Haniori in Jakarta. Before being sold, white pepper is processed first to a maximum water content of 16% with facilities owned by PCM.

Meanwhile, for sugarcane marketing, PCM collaborates with sugarcane collectors who harvest farmers' sugarcane and transport it to the PT Rajawali Sugar factory in Cirebon, West Java, with which PCM has a partnership. The purchase price is determined transparently among farmers, collectors and PCM based on sales receipts at the sugar factory, ensuring that farmers are not disadvantaged. Ten sugarcane collectors from two districts in Central Java, namely, Purbalingga and Banjarnegara (neighbouring Purbalingga Regency), partner with PCM. Sugarcane collectors who have harmed farmers by applying a slashing system at low prices can no longer manipulate prices due to the partnership between farmer corporations and buyers.

PCM has successfully developed an integrated agribusiness system for pepper and sugarcane commodities.

16.4.1.2 Midang Jaya Farm (MJF) Cooperative

No different from PCM, the results of the performance evaluation on MJF with CIPP show that the Input indicator is categorized as moderate with an average value of 2.91, while the other three indicators, namely, Context (3.18), Process (3.37) and Product (3.32) are categorized as high, with an average value for all indicators being categorized as high with a score of 3.20 (Table 16.3). Several sub-aspects that require improvement based on the results of the CIPP evaluation are located in the Context aspect, specifically "level of corporate strength in achieving goals" (2.94) and "level of completeness and clarity of guidelines" (2.77). Likewise, in the Input aspect, in the sub-aspect “level of availability and supporting facilities for farmer cooperatives” (2.90) and “level of budget availability and support” (2.41). In the Process indicator, improvements are needed in the sub-indicators “level of conformity of the implementation process with the use of the available budget” (2.83) and “degree of conformity of the ongoing process with the use of available time” (2.97). The evaluation results in Table 16.3 indicate that several sub-indicators require consideration for improving MJF performance, including the need to implement clear guidelines for cooperative operational activities, enhance facilities and secure funding to support the integration of agribusiness systems from upstream, on-farm and downstream operations.

The purpose of establishing MJF is to create an independent, competitive and sustainable agricultural business through effective corporate land management, utilizing existing community resources and institutional opportunities in an optimal manner. The goal is to enhance the added value of mangosteen commodities that already have branding, enabling them to compete at regional, national and international levels, both in their fresh form and in other processed forms.

Through the UPLAND Project, MJF has been facilitated with the assistance of warehouses, four-wheeled vehicles and three-wheeled vehicles. The cooperative focuses its efforts on marketing mangosteen. However, the cooperative’s limited capital means that not all of its members’ mangosteen harvests can be absorbed. In the last harvest season, MJF was only able to buy 2–3 tons of farmers’ harvests per day. In 2023, the cooperative marketed 91 tons of mangosteen from production in Subang Regency, specifically in Serangpanjang District, which accounted for 4.3% of the total production. As a result, the cooperative was only able to buy and market 2.1% of the mangosteen production. It is estimated that MJF requires approximately 5 billion rupiah to purchase all the mangosteen in Serangpanjang District, as the farmers’ purchasing system operates on a cash (cash and carry) basis. Adequate capital is the key to increasing the income of mangosteen farmers, particularly in mitigating the effects of price uncertainty. Business competition in purchasing mangosteen occurs between the cooperative and the collectors. However, the cooperative’s purchasing system is different from the collector’s purchasing system. Cooperatives buy using a weighing system (prices are based on fruit quality or grade), while traders buy using a slashing system.

Mangosteen fruit that is harvested by farmers and purchased by MJF is sorted into five grades based on size and quality (Table 16.4). Grade 1 is fruit categorized as export-worthy quality, also known as Super 1 quality. Perfectly round shape, no defects on the skin surface and the ears or calyx are still complete. The weight and diameter of the fruit are also important determinants of quality. The proportion of super-quality mangosteen is as expected, given its high profitability in terms of price compared to mangosteens included in other grades. Buyers can purchase export-quality mangosteen for IDR 50,000 to IDR 60,000 per kilogram. During the last harvest season, MJF obtained only around 30% of Super 1 quality from all mangosteens purchased at the farmer level. The rest are grades 2, 3, even 4 and 5. Therefore, it is necessary to improve cultivation technology to improve the quality of mangosteen produced by farmers.

Table 16.3 CIPP evaluation results at MJF

Indicators	Sub-indicators	Average of sub-indicators	Average of indicators
Context	The priority level of the objectives that have been set in establishing the corporation	3.49	3.18
	The level of opportunity to achieve corporate goals is based on the problems faced and the ability to implement activities	3.50	
	Level of corporate strength in achieving goals	2.94	
	Level of corporate perfection in planning and preparation	3.19	
	Level of completeness and clarity of guidelines	2.77	
Input	The level of availability of human resources for farmer corporations	3.42	2.91
	Level of availability and support of facilities for farmer corporations	2.90	
	Level of budget availability and support	2.41	
Process	The level of success of implementing socialization activities	3.83	3.37
	Level of conformity of the ongoing Process with the achievement of activity objectives	3.80	
	The level of accuracy of the Process with the guidelines and technical instructions that have been issued	3.60	
	Success rate of reducing procedural errors in the activity process	3.20	
	Level of human resource utilization in the activity process	3.27	
	Level of conformity of the activity implementation process with the use of available facilities	3.03	
	Level of conformity of the implementation process with the use of the available budget	2.83	
	Degree of conformity of the ongoing Process with the use of available time	2.97	
	The level of success of the implementation of the activity assistance	3.57	
	The level of success of the implementation of monitoring and evaluation of activities	3.57	
Product	Achievement of output	3.09	3.32
	Achievement of outcome	3.23	
	Achievement of impact	3.64	
CIPP average value			3.20

Source Survey results (processed data)

Table 16.4 Quality and price of mangosteen purchased by MJF

Mangosteen grade	Price at the farmer level (Rp/kg)
SP 1 (Super 1)	18.000–20.000
SP 2 (Super 2)	10.000–15.000
SP 3 (Super 3)	7.000–8.000
Cherry	5.000–6.000
Former sorting	2.000–3.000

Source Interview results

MJF has anticipated the lack of funds through cooperation with BUMDesa. In 2023, MJF borrowed 100 million rupiah from BUMDesa, utilizing a profit-sharing system based on the profits generated. Limited funds are an obstacle to partnership with buyers. MJF has not dared to enter into a partnership because it is unable to prepare mangosteen fruit according to the contract due to limited funds. Another problem is the packing house (PH) that is not yet owned. PH functions as a sorting place based on quality and size. Fruit shipments from MJF will be re-sorted by buyers at their PH, which causes many fruits to become sorted, thus decreasing their prices.

Based on the results of the agribusiness system evaluation, MJF has not adopted the mangosteen agribusiness system as a whole, encompassing the upstream agribusiness subsystem, on-farm operations and downstream agribusiness. MJF only plays a role in the downstream agribusiness subsystem, specifically in post-harvest and marketing, which contributes to the weak ability of farmer corporations to streamline the supply chain. The upstream agribusiness subsystem and on-farm activities related to plant cultivation are not under the cooperative's control. Capital constraints and PH facilities are challenges in developing agribusiness marketing of mangosteen products, especially in establishing partnerships with buyers.

16.4.1.3 Permata Indah Rubaru (PIR) Limited Liability Company

PIR is a farmer corporation facilitated by the UPLAND Project for the development of shallot agribusiness in Sumenep Regency, with the support of modern processing tools and machines, warehouses, training for farmers and product promotion. The agribusiness system, utilizing shallot production centres in Sumenep Regency, especially in Rubaru District, has great potential for development with this facilitation. The establishment of PIR aims to improve the welfare of shallot farmers sustainably. Table 16.5 shows the average CIPP score for the PIR performance assessment.

The highest score is in the Context indicator, with a score of 3.36, followed by Product (3.27), Process (3.16) and Input (3.12). The average score for all indicators is high, at 3.23. The evaluation results of all CIPP indicators indicate that farmers' responses to the presence of PIR are generally positive. This is because PIR has been able to absorb some of the shallot production for processing into various products, such as fried shallots, shallot flour and shallot paste. Despite limited capital, the presence of PIR has been able to attract buyers. Through promotions carried out

Table 16.5 CIPP evaluation results at PIR

Indicators	Sub-indicators	Average of sub-indicators	Average of indicators
Context	The priority level of the objectives that have been set in establishing the corporation	3.57	3.36
	The level of opportunity to achieve corporate goals is based on the problems faced and the ability to implement activities	3.53	
	Level of corporate strength in achieving goals	3.39	
	Level of corporate perfection in planning and preparation	3.13	
	Level of completeness and clarity of guidelines	3.20	
Input	The level of availability of human resources for farmer corporations	3.07	3.12
	Level of availability and support of facilities for farmer corporations	3.15	
	Level of budget availability and support	3.12	
Process	The level of success of implementing socialization activities	3.22	3.16
	Level of conformity of the ongoing Process with the achievement of activity objectives	3.19	
	The level of accuracy of the Process with the guidelines and technical instructions that have been issued	3.15	
	Success rate of reducing procedural errors in the activity process	3.07	
	Level of human resource utilization in the activity process	3.04	
	Level of conformity of the activity implementation process with the use of available facilities	3.19	
	Level of conformity of the implementation process with the use of the available budget	3.15	
	Degree of conformity of the ongoing Process with the use of available time	3.15	
	The level of success of the implementation of the activity assistance	3.22	
	The level of success of the implementation of monitoring and evaluation of activities	3.22	
Product	Achievement of output	3.08	3.27
	Achievement of outcome	3.23	
	Achievement of impact	3.50	
CIPP average value			3.23

Source Survey results (processed data)

by the Ministry of Agriculture and the Sumenep Regency Government, a marketing contract has been successfully signed with Ben Helen Trading, the Netherlands, for a period of 5 years (2023–2028) with a contract value of $400,000 for packaged fried shallot products.

In addition to the export market, PIR has also pioneered the marketing of packaged fried shallots to modern markets, including supermarkets, minimarkets and hypermarkets, as well as traditional markets in Surabaya and Jakarta. This distribution channel enables the provision of larger and more consistent product quantities, reaching a broader customer base. Consumers in the modern market tend to look for products with guaranteed quality and more aesthetically pleasing and hygienic packaging. Marketing is also conducted through various e-commerce platforms, including Tokopedia, Bukalapak and Shopee, to reach a broader spectrum of customers.

Despite having developed a downstream agribusiness subsystem, in the form of processing and marketing shallot products in Sumenep Regency, PIR is still constrained in developing an integrated agribusiness system that spans from upstream to downstream on-farm operations. On the upstream side, shallot cultivation requires relatively large funds, especially for the provision of seeds, while farmers' financial capabilities vary. PIR is still unable to provide shallot seeds due to limited corporate funds. At the same time, farmers remain reluctant to access credit schemes from banks, such as the People's Business Credit (KUR), which the Government subsidizes. Farmers circumvent the provision of shallot seeds by allocating the remaining harvest from the previous planting season. Lack of collateral is the main obstacle for farmers in accessing credit, in addition to the mismatch between the shallot harvest cycle and the credit payment schedule.

On the on-farm side, shallot cultivation at the farmer level is constrained by the increasing prices of fertilizer and pesticides. Climate change is also a problem for farmers because the weather is unpredictable. The prolonged rainy season causes shallots to rot quickly during harvest. PIR has coordinated with farmer groups to prepare warehouse facilities, implement storage technology and enhance the quality of shallot production and post-harvest management through various training and mentoring programs for farmers.

16.4.2 Integration of Agribusiness Systems Through Farmer Corporation Development

Evaluation of the development of farmer corporations in three UPLAND Project locations has provided lessons that the dynamics of performance and institutions in each corporation show encouraging things. According to the CIPP evaluation results, the average value of all indicators falls within the high category for the three farmer corporations. This shows that farmers have responded well to the establishment of farmer corporations in their environment. Farmer corporations have succeeded in increasing added value, bargaining power and farmer income by marketing superior agricultural products in each location through partnerships with local, national and export partners. The agribusiness system within the three farmer corporations is briefly outlined in Table 16.6.

In the case of PCM, the upstream and downstream agribusiness systems in commodity development have been integrated. However, in MJF and PIR, the integration of the agribusiness system has not been fully formed. In the development of pepper and sugarcane in Purbalingga Regency, PCM can enter the upstream agribusiness subsystem by providing capital and agricultural production facilities, implementing cultivation and harvest planning in the on-farm subsystem, and processing and marketing in the downstream agribusiness subsystem. Meanwhile, MJF in Subang Regency plays a role only in the upstream agribusiness subsystem, specifically in mangosteen marketing. Likewise, PIR plays a role in the processing and marketing of shallots in Sumenep Regency. Although the agribusiness system that has been formed has not been integrated from upstream to downstream within a single farmer corporation institution, MJF and PIR can build synergy with farmers and farmer groups within the farmer corporation environment and have developed upstream agribusiness and cultivation activities on their respective farms.

Several studies have shown that strengthening farmer institutions can significantly enhance their bargaining power in market competition (Darwanto et al., 2019; Yeny et al., 2022). The role of the government is not only related to the establishment of institutions as agents for distributing programs, but also requires institutional assistance to ensure sustainability within the agribusiness system, thereby improving performance and entrepreneurial orientation (Anisa et al., 2021). Mutually beneficial market partnerships will enhance farmers' bargaining power in the market, leading to increased farmer income (Sjaf et al., 2022).

The farmer corporation, initiated through the UPLAND Project of the Ministry of Agriculture since 2021, represents a form of institutional innovation in the development of an agribusiness system within the farmer environment, with a sustainable

Table 16.6 Integration of agribusiness systems in farmer corporations

Agribusiness subsystem	PCM	MJF	PIR
Upstream	Funds for purchasing white pepper; provision of production facilities (pepper seeds, sugarcane seeds, fertilizers and pesticides for sugarcane cultivation)	–	–
On-farm	Technical improvements in sugarcane cultivation	–	–
Downstream	Post-harvest and processing of white pepper, marketing of white pepper and sugarcane	Post-harvest and marketing of mangosteen	Post-harvest, processing and marketing of shallots

Source Analysed data results

orientation. This approach differs from previous methods, such as those through farmer groups or farmer group associations, which did not emphasize the upstream-downstream perspective of agribusiness. Silvestre and Ţîrcă (2019) stated that institutional innovation must lead to transformation in individuals, organizations, supply chains and communities, ultimately contributing to a sustainable future. Sustainable innovation seeks to integrate social, economic and environmental aspects without being solely oriented towards increasing profits, but rather towards solving problems faced by the community.

This study makes a unique contribution by combining the CIPP evaluation model with an analysis of agribusiness subsystems to assess the performance of farmer corporations. This methodological integration has not been extensively applied in the agricultural development literature of Indonesia. The uniqueness lies in evaluating institutional performance alongside agribusiness integration to determine sustainability prospects. The discussion of each farmer corporation (PCM, MJF, PIR) aligns directly with the study's objectives: PCM exemplifies full integration of agribusiness subsystems, demonstrating optimal model application; MJF and PIR illustrate partial integration, revealing specific challenges such as capital limitations and infrastructural needs. These comparative insights clarify how institutional form and agribusiness linkage affect corporate effectiveness and sustainability. Thus, the chapter not only meets its evaluative aims but also provides a scalable model for improving agricultural institutions in similar contexts globally.

The integration of agribusiness subsystems in PCM and the emerging synergy in MJF and PIR reflect a paradigm shift from fragmented smallholder models to consolidated, market-responsive institutional frameworks. This chapter demonstrates that such a shift is not only feasible but necessary for achieving economic resilience and social equity in the agricultural sector, making it a relevant contribution to discussions on transforming agriculture in developing economies.

16.5 Conclusions

Strong farmer institutions are needed in the development of such agribusiness. Since the 1970s, the Ministry of Agriculture has developed farmer groups as agents receiving agricultural development programs. The development of the scale of farming businesses began to play a significant role, leading to the formation of farmer group associations since the 2000s. However, increasing the scale of business has not enabled the optimal development of agribusiness activities in rural areas, as they are still focused on a single village area. The Ministry of Agriculture then initiated farmer corporations as institutional structures for farmers, based on a single superior commodity development area. Farmer corporations are also legal entities, enabling them to form partnerships with business actors involved in agricultural commodities to develop an optimal agricultural commodity agribusiness system.

Research conducted in three farmer corporations initiated through the UPLAND Project, funded by the Ministry of Agriculture, reveals that the downstream agribusiness subsystem is crucial for developing agribusiness that can enhance farmers' bargaining power and income within an agricultural commodity agribusiness system. Integration between upstream agribusiness subsystems, on-farm agribusiness and downstream agribusiness has been successfully established in the PCM farmer corporation in Purbalingga Regency, Central Java, for pepper and sugarcane farming. Meanwhile, in the MJF farmer corporation in Subang Regency, West Java and PIR in Sumenep Regency, East Java, they have not been able to integrate all agribusiness subsystems in the development of mangosteen commodities in MJF and shallots in PIR.

This research provides empirical evidence on how institutional innovation through farmer corporations can operationalize integrated agribusiness systems, particularly in the Context of rural economic transformation. The findings underscore the importance of structured institutional models supported by state facilitation and market partnerships in promoting sustainability and improving smallholder livelihoods. Methodologically, the study advances the use of the CIPP framework in evaluating institutional performance in the agricultural sector, offering a replicable model for similar assessments in developing regions. Globally, this research contributes to ongoing discourse on inclusive agribusiness development and smallholder integration into value chains—a theme relevant to achieving SDG 1 (No Poverty), SDG 2 (Zero Hunger) and SDG 12 (Responsible Consumption and Production). By linking national agricultural policy with grassroots institutional outcomes, the study offers insights that can inform global efforts to transform rural agrifood systems, particularly in the Global South.

Acknowledgements We want to thank all the Managers of the farmer corporations (PCM, MJF and PIR), farmer members of the farmer corporations, collectors, field officers and the Agriculture Offices of Purbalingga, Subang and Sumenep Regencies for their willingness to provide information through surveys and interviews related to the research topic. We would also like to thank the Director of Agricultural Irrigation, Ministry of Agriculture and the UPLAND Project Manager of the Ministry of Agriculture, who have facilitated funding for the farmer corporation evaluation research activities in collaboration with the National Research and Innovation Agency.

Funding This research was funded by the Development of Integrated Farming Systems in Upland Areas (UPLAND) Project, Ministry of Agriculture, Fiscal Year 2024.

Potential conflicts of interest On behalf of all authors, the corresponding author declares that there are no conflicts of interest.

Informed Consent Informed consent was obtained from all individual participants included in the study.

Authors' Contribution All authors have contributed equally to this work.

Data Availability Statement The datasets used and/or analysed during the current study are available from the corresponding author on reasonable request.

AI Disclosure Statement The authors declare that Artificial Intelligence (AI) tools were used in a limited capacity to support this manuscript, specifically for language refinement,paraphrasing, and sentence structuring in accordance with international academic standards. All analyses, interpretations, and conclusions are the sole responsibility of the authors. AI was not used to replace any part of the scientific process, including data collection, analysis, or validation.

References

Aldillah, R. (2016). Agricultural mechanization and its implications for food production acceleration in Indonesia. *Forum Penelitian Agro Ekonomi, 34*(2), 163–171.

Anisa, A., Kusnadi, N., & Rachmina, D. (2021). The Government's role impacts the entrepreneurial orientation of rice farmers. *IOP Conference Series: Earth and Environmental Science, 807*, 032072 (IOP Publishing). https://doi.org/10.1088/1755-1315/807/3/032072

Block, F., Evans, P.: (2005). 22 The State and the economy. 505. *The handbook of economic sociology, 505*. https://d1wqtxts1xzle7.cloudfront.net/51070776/_Neil_J._Smelser__Richard_Swedberg__The_Handbook_Economic_sociology-libre.pdf?1482830849=&responsecontent-disposition=inline%3B+filename%3DNeil_J_Smelser_Richard_Swedberg_The_Han.pdf&Expires=1774796034&Signature=Ip5ywqiYYwGR~Q~aK1lNF6-DdZHD5fWB4BrgBgPiAM6NUaN-4xAaRR3MYUldW3W0VtUmrWHZ6zYBCOlTfsDYRlfxzusGUMD4Z~kDDD1p~byNt9I6r6gKsefdFEpEEvTLGAjGALRPUenPtNd10GyBb2Wim~cuT-4KrBhGUhiXUHIg8OYXBdN96UH7bqGEptserdGEGNpCnVudELuqQa5C99I1KQp7y~mdpZ1w92oqEANQSAele--4pOY6Hv~jiosfec32kg1UakGNfOJ96T-Fo05zFZnZfswqHX1bvGQgKIc-zc2E~mDpY7Rr3Ge6NKiDaoPrnfCP41XWxJGAYoTHA__&Key-Pair-Id=APKAJLOHF5GGSLRBV4ZA#page=518

BPS-Statistics of Purbalingga Regency. (2024). Purbalingga Regency in Figures 2024. Katalog: 1102001.3303. ISSN 0215-6075. BPS-Statistics of Purbalingga Regency.

BPS-Statistics Indonesia. (2024). Penghitungan dan Analisis Kemiskinan Makro Indonesia 2024. Volume 16, 2024. Katalog: 3205022. ISSN 2502-7484. Statistics Indonesia.

BPS-Statistics Indonesia. (2025). Jumlah Penduduk Miskin menurut Wilayah. Statistics Indonesia. https://www.bps.go.id/id/statistics-table/2/MTgzIzI=/number-of-poor-people-by-region.html.

Darwanto, Roessali, W., Woyanti, N., Salam, A. N., & Santosa, P. B. (2019). Sharia microfinance institutions financing model for strengthening agricultural sector. *IOP Conference Series: Earth and Environmental Science, 292*, 012016 (IOP Publishing). https://doi.org/10.1088/1755-1315/292/1/012016

Harianto, Hutagaol, M. P., & Widhiyanto, I. (2019). Sources and effects of credit accessibility on smallholder paddy farms performance: An empirical analysis of Government subsidized credit program in Indonesia. *International Journal of Economics and Financial Issues, 9*(5), 1–10. https://doi.org/10.32479/ijefi.8388

Harjanto, P., Fahmid, I. M., Ali, S. & Demmallino, E. B. (2022). Institutional development of farmers through agricultural area-based corporations in Indonesia. *IOP Conference Series: Earth and Environmental Science, 1114*, 012039 (IOP Publishing). https://doi.org/10.1088/1755-1315/1114/1/012039

Krinamurti, B. (2020). Pengertian Agribisnis. Puspita Swara.

Mariyono, J., Waskito, J., Suwandi, T., Kuntariningsih, A., Latifah, E., & Suswati, E. (2021). Farmer field school: Non-formal education to enhance livelihoods of Indonesian farmer communities. *Community Development, 52*(2), 153–168. https://doi.org/10.1080/15575330.2020.1852436

Munifah, S. (2019). Transformasi Kelembagaan Petani Menuju Korporasi Petani. Badan Penyuluhan dan Pengembangan SDM Pertanian—Kementerian Pertanian. https://fungsional.pertanian.go.id/ujikompjf/assets/file/elearning/elearning_66_5d95a4dd4da6c.pdf

Pangan, B.K.: (2018). Laporan Kinerja Badan Ketahanan Pangan tahun 2017. *Jakarta: Kementerian Pertanian*. https://badanpangan.go.id/storage/app/media/Evalap/Lakin%20BKP%202017.pdf

Pranadji, T., & Anugrah, I. S. (2021, November). Turning point the concept of rural development in Indonesia from top-down to bottom-up strategy. *IOP Conference Series: Earth and Environmental Science, 892*, 012079 (IOP Publishing). https://doi.org/10.1080/1755.1315/892/1/012079

Purnomo, M., Hadiwiyono, E., Andriatmoko, N. D., Pariasa, I. I., Kustanti, A., & Faust, H. (2022). Transformation role of Forest Farmer Group's (FFGs) in Indonesia from development agent to brokering project. *SAGE Open, 12*(1), 21582440221082144. https://doi.org/10.1177/21582440221082144

Raharja, S., Marimin, Machfud, Papilo, P., Safriyana, Massijaya, M. Y., Asrol, M., & Darmawan, M. A. (2020). Institutional strengthening model of oil palm independent smallholder in Riau and Jambi Provinces, Indonesia. *Heliyon, 6*(2020), e03875. https://doi.org/10.1016/j.heliyon.2020.e03875

Rahmadanih, Bulkis, S., Arsyad, M., Amrullah, A., & Viantika, N.M.: (2018, May). Role of farmer group institutions in increasing farm production and household food security. *In IOP Conference Series: Earth and Environmental Science, 157*(1), 012062 (IOP Publishing) Rahmadanih et al 2018 IOP Conf. Ser.: Earth Environ. Sci. 157 012062. https://doi.org/10.1088/1755-1315/157/1/012062

Rokhani, Fauziyah, D., Supriono, A., Hariyati, Y., Raharto, S., Hapsari, T. D., et al. (2021). Factors affecting the participation of sugarcane and tobacco farmers in farmer groups, associations and cooperatives in Indonesia. *Caraka Tani: Journal of Sustainable Agriculture, 36*(2), 340–354. https://doi.org/10.20961/carakatani.v36i2.46817

Setkab. (2021). Hasil Sensus Penduduk 2020; BPS: Meski Lambat, Ada Pergeseran Penduduk Antarpulau. https://setkab.go.id/hasil-sensus-penduduk-2020-bps-meski-lambat-ada-pergeseran-penduduk-antarpulau/.

Silvestre, B. S., & Ţîrcă, D. M. (2019). Innovations for sustainable development: Moving toward a sustainable future. *Journal of Cleaner Production, 208*, 325–332. https://doi.org/10.1016/j.jclepro.2018.09.244

Sjaf, S., Arsyad, A. A., Mahardika, A. R., Gandi, R., Elson, L., Hakim, L., et al. (2022). Partnership 4.0: Smallholder farmer partnership solutions. *Heliyon, 8*(2022), e12012. https://doi.org/10.1016/j.heliyon.2020.e12012

Stufflebeam, D. L. (2015). *CIPP evaluation model checklist: A tool for applying the CIPP model to assess projects and programs*. Western Michigan University Evaluation Center. http://rszarf.ips.uw.edu.pl/ewalps/dzienne/cipp-model-stufflebeam2015.pdf

Sutardi, Apriyana, Y., Rejekiningrum, P., Alifia, A. D., Ramadhani, F., Darwis, V., et al. (2022). The transformation of rice crop technology in Indonesia: Innovation and sustainable food security. *Agronomy, 13*(1). https://doi.org/10.3390/agronomy13010001

Yeny, I., Garsetiasih, R., Suharti, S., Gunawan, H., Sawitri, R., Karlina, E., et al. (2022). Examining the socio-economic and natural resource risks of food estate development on peatlands: A strategy for economic recovery and natural resource sustainability. *Sustainability, 14*(7), 3961. https://doi.org/10.3390/su14073961

Chapter 17
Mapping Urban Green Spaces for Neighborhood Sustainability Using Urban Neighborhood Green Index: A Geospatial Case Study in Midnapore Town, West Bengal, India

Kishor Dandapat, Prasanna Mahata, Biplab Mahali, Sheuli Biswas, and Gopal Krishna Panda

Abstract Urban green spaces are a part of every city. Urban green spaces keep the environment clean and help sustainability. Modern urbanization changes the environment. Modern urbanization reduces spaces and threatens human well-being. The municipality of Medinipur is an urban center, in West Bengal. The municipality of Medinipur loses spaces. To study the spaces of the municipality of Medinipur, we used the Urban Greenness Index (UNGI). Our analysis of the UNGI, by sector showed differences in the quality. Spread of the green spaces. Sector 25 recorded the UNGI. The UNGI in sector 25 points to spaces. Sector 15 is an area. Recorded the lowest average UNGI. The UNGI in sector 15 points to a lack of spaces. Sector 15 also showed the difference in the UNGI values between semi public lands. The UNGI differences, in sector 15 show that the green spaces are not even. When I looked at the data, I saw that sector 13 which's mostly lands showed the least variation, in the UNGI values; that shows a spread of the green spaces. The study highlighted the areas that lack the spaces and that could use more green spaces. Districts 9 and 24 had UNGI values of 0.710 and 0.711. The standard deviations of 0.137 and 0.141 showed

K. Dandapat (✉)
Assistant Professor, Department of Geography, Jogamaya Devi College (Affiliated to University of Calcutta), Kolkata, West Bengal, India
e-mail: kishordandapat@gmail.com

P. Mahata · S. Biswas
Department of Geography, Sadhu Ram Chand Murmu University, Jhargram, West Bengal, India

P. Mahata
The University of Burdwan, Bardhaman, West Bengal, India

B. Mahali
Department of Geography, University of Calcutta, Kolkata, West Bengal, India

G. K. Panda
Department of Geography, School of Tribal Resource Management, Kalinga Institute of Social Science, Deemed to be University, Odisha, Bhubaneswar, India

G. K. Panda et al. (eds.), *Socio-Ecological Crisis of Development*, Sustainability Solutions, https://doi.org/10.1007/978-3-032-22171-1_17

differences, in the quality of the spaces. The mean scores were similar. This highlights the importance of considering both average UNGI values and their variability when assessing urban green spaces. By identifying areas with insufficient green spaces and those with potential for improvement, our analysis can inform targeted green-space initiatives and urban planning strategies to improve the city's overall sustainability and quality of life. These results can guide policymakers in prioritizing areas for green infrastructure development and optimizing resource allocation for maximum impact.

Keywords Urban green spaces · Medinipur municipality · UNGI · Sustainability

17.1 Introduction

Urbanization is radically transforming the world and cities. The global urbanization rate is projected to increase from 55% in 2018 to 68% over the next 30 years, and global population growth will result in an increase of approximately 2.5 billion people living in urban areas by 2050. Rapid urbanization in today's world has significantly altered the environment, leading to a decline in green spaces and posing a substantial risk to human well-being. As cities expand and urban populations grow, natural habitats are being lost, and the benefits that green spaces provide, such as clean air, recreation, and mental health support, are being compromised. In recent years, there is increased demand for urban green space (UGS) in urban areas. Urban green spaces offer numerous environmental benefits, including mitigating pollution impacts and alleviating high temperatures during summer months (Baklanov et al., 2008) and also green infrastructure has been given increasing attention in urban planning and its policies, especially due to its ecological, environmental, and social contributions to planning and designing high-quality UGS. Urban green spaces (UGS) include parks, gardens, and recreational areas, as well as informal spaces such as riverbanks, promenades, green spaces around historical monuments, railway rights-of-way, and areas of native vegetation. They can also include urban areas such as brownfield sites and abandoned or wild gardens (Venn & Niemelä, 2004). Several studies show that people living near green spaces tend to be happier and healthier than those without access to them (Bounoua et al., 2020; Lyytimäki & Sipilä, 2009; Mudau et al., 2020; Wolch et al., 2014). These green spaces offer an ideal place to escape the stresses of daily life and recharge.

The city of Medinipur, though small and rapidly developing, is important both administratively and for the greater metropolitan area. This small city, rich in culture and boasting numerous tourist attractions thanks to its history, is gradually urbanizing. This gradual urbanization is due to improvements in infrastructure, educational institutions, the healthcare sector, and the growth of commercial activity. The municipality faces several challenges, including uncontrolled urban growth, the saturation of existing infrastructure, and environmental problems such as the reduction of green spaces and inadequate waste management. Though, Medinipur is a unique example to study the change of the town in the different secondary cities of India, which now can be called urban, hence, the town becomes a valuable source for understanding the complexities of urban growth, socio-economic transitions, and environmental impacts.

Aside from the fact that urban greening is a crucial means of combating climate change, it appears that sustainable living is gradually becoming an essential feature of people's modern lifestyles (Cai et al., 2023; Hou & Wu, 2021; Hwang et al., 2020; Van et al., 2020). There is a plethora of potentials to create and promote green areas that could not only enhance the sporting activities of the people but also be beneficial both to the climate and the city's beautification. Parks, where people very often stand, sit, or relax, are the jewels in the urban crown. Besides, the development of green corridors along roads and rivers can extend these benefits, thus giving rise to a healthy environment and more attractive cities.

Green spaces are studied using a variety of methods, including subjective techniques such as surveys (Gorzelany et al., 2023) and audits, as well as objective methods such as geographic information systems (GIS) (Sokhi & Ruangrit, 2004) and the Normalized Difference Vegetation Index (NDVI). NDVI measures greenness and vegetation density, which helps detect variations in plant health. In addition, there are indices such as the Urban Neighborhood Greenness Index (UNGI) (Gupta et al., 2012), the Neighborhoods Under Construction Greenness Index (BNGI), the Vegetation Cover Index (VCI), and the Green View Index (GVI) (Yang et al., 2009) that first provide information about the composition of green areas and then about their usability. Additionally, tools such as the Multidimensional Balance Index (MBI) and the Green Accessibility Index (GAI) (Zhang et al., 2020) are used to determine the quantity and productivity of green spaces. Their use allows researchers to assess and monitor changes in urban green spaces, helping them identify areas for improvement and promote sustainable urban development. The Normalized Urban Greenness Index (NUGI) was used to measure green spaces in Medinipur Municipality, identifying areas with high-quality vegetation and those in need of improvement. The UNGI based on the types of urban vegetation and the characteristics of built-up areas provides the necessary information for the urban planning. The UNGI is a comprehensive framework for the evaluation and improvement of urban green spaces that similarly employs the four main parameters used in the study by namely, green space per capita, proximity to green space, area percentages, and types of green spaces.

17.2 Geographical Locations of the Study Area

The study area includes Medinipur Municipality, which is a major urban agglomeration within Medinipur district, West Bengal. It falls between 22°26′N to 22°29′N latitudes and 86°59′E to 87°03′E longitudes with an area of approximately 17.78 km^2 (see Fig. 17.1). The region enjoys beneficial transportation links as it is well connected to larger neighboring towns of Medinipur and Kharagpur. This includes the district headquarters and forms an important administrative, commercial and residential

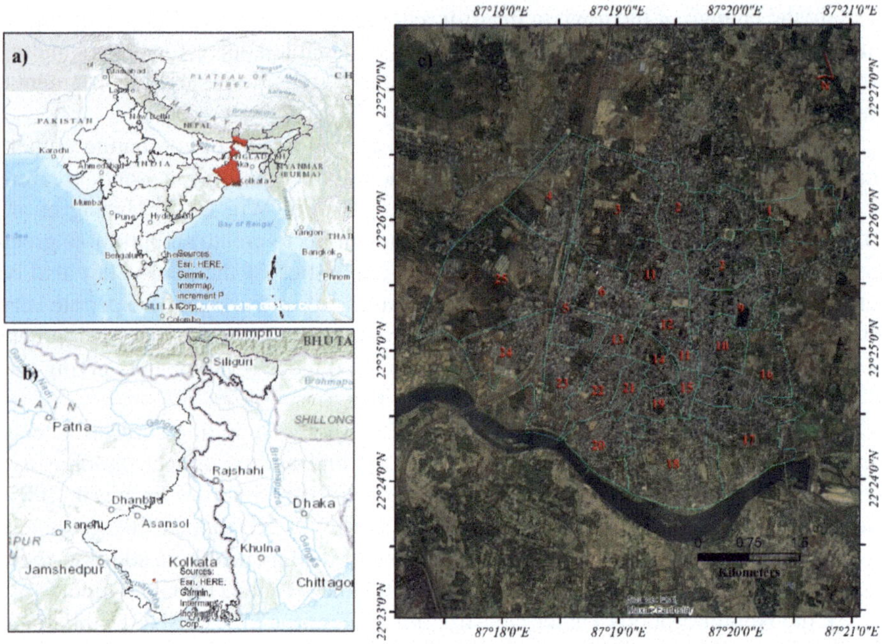

Fig. 17.1 Location map of the study area

center in the region. The general climate of the region is hot tropical monsoon. The summers are very hot while winters are mild. The soils are diverse in the area-alluvial soils along the Kangsabati River and lateritic soils elsewhere in the region. Vegetation consists of eucalyptus and sal forests. The region hosts a number of wildlife, particularly elephants, which invade human settlements in nearby villages. According to the 2011 census, Midnapore municipality had a population of 169,264, of which males were 84,977 and females were 84,287. Population of children in the age group of 0–6 year was 15,172. Effective literacy rate was more than 88.99%. Compared to the 2001 census, where the population was 153,349, the average literacy rate was 75%, higher than the national average of 59.5%. Male literacy rate was 80% and that of females was 71%, with about 10% of the population under six years of age.

17.3 Objectives of the Study

The objectives of the study in Medinipur Municipality are to assess the quality of green spaces near built-up areas; analyze the spatial distribution of green spaces; and measure greenness using remote sensing images by computing the percentage area of green, thereby informing urban planning and policy decisions with a view toward ensuring sustainable development and improving the lives of residents.

Table 17.1 Description of remote sensing images used for assessing UNGI

Input	Satellite/sensor	Resolution	Type	Date of acquisition
Image-1	IRS P6 LISS—IV	5.8 m	Multispectral	09.04.2025

17.4 Database and Methodology

17.4.1 Data Used

For the assessment of neighborhood green spaces, this study has made use of satellite data from Indian Remote Sensing IRS P6 LISS IV, dated 9 April 2025 (see Table 17.1). This data will allow for quantitative assessment of vegetation and consequently categorize it into four classes: dense vegetation, intermediate/low vegetation or grass, open spaces, and built-up areas. This sensor offers a very high spatial resolution of 5.8 m and three spectral bands that help in identifying the vegetation with its classes more efficiently. Open-building data from Google Earth Engine were also used for determining the typology of neighborhoods and building heights, along with reference information. Image subsets for 25 wards within Medinipur Municipality were generated using a boundary map derived from the sector land-use map.

17.4.2 Function of UNGI

A neighborhood is defined as a spatial unit that possesses homogeneous features, such as ethnicity, housing, or development type, and it can also define the area or quality of green space around built environments. The Urban Neighborhood Green Index defines the spatial distribution of urban green space within a neighborhood. Neighborhood greenness is defined by two key features: (1) green cover, typically measured by the percentage of vegetation cover in an area, and (2) type of vegetation, as identified by dense cover, low-lying vegetation, or grass. In addition, built-up environment features of a city, in terms of proximity to green spaces, density of buildings, and structural height of buildings, are taken into account. Proximity to green space can enhance property value and environmental quality, while increased built-up density and structure height can be used as surrogate measures of population density. These elements are combined to calculate the UNGI, providing an overall measure of neighborhood green quality (see Figs. 17.2 and 17.3).

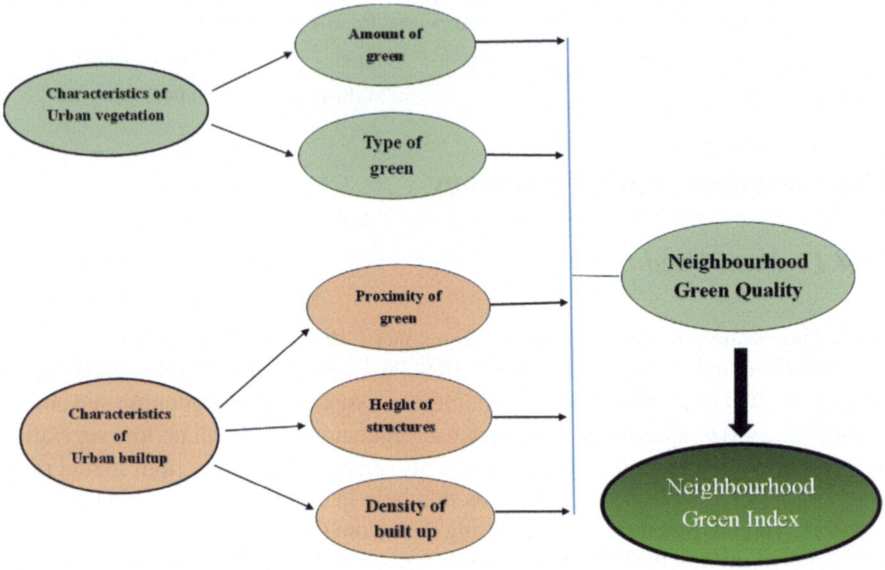

Fig. 17.2 A conceptual framework for urban neighborhood green index (UNGI)

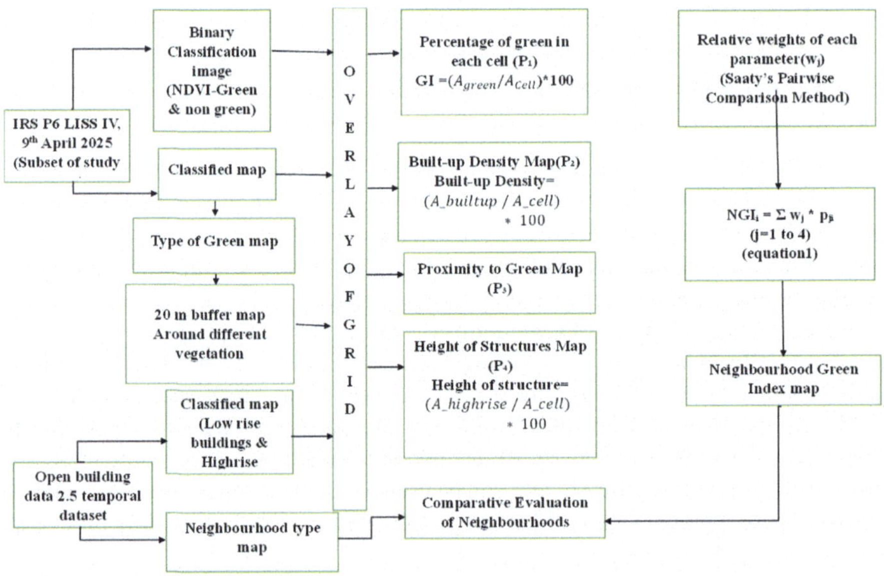

Fig. 17.3 Methodology flow chart

17.4.3 Derivation of Green Index Map

The percentage of vegetation in each cell was calculated using a binary classification (areas with vegetation and without vegetation) on IRS LISS IV multispectral data from Medinipur. NDVI was generated using red and near-infrared bands, with negative values classified as built-up areas and positive values as vegetation. A 20 m × 20 m grid was overlaid on the binary image, and the percentage of green in each cell was calculated (see Fig. 17.4). Cells were then classified into four green quality classes: low (less than 25%), moderate (25–50%), high (50–75%), and very high (more than 75%), with corresponding values of 0.25, 0.5, 0.75, and 1, respectively (see Table 17.2). This categorization allows for relative comparison between neighborhoods, providing a simple and effective way to evaluate green quality.

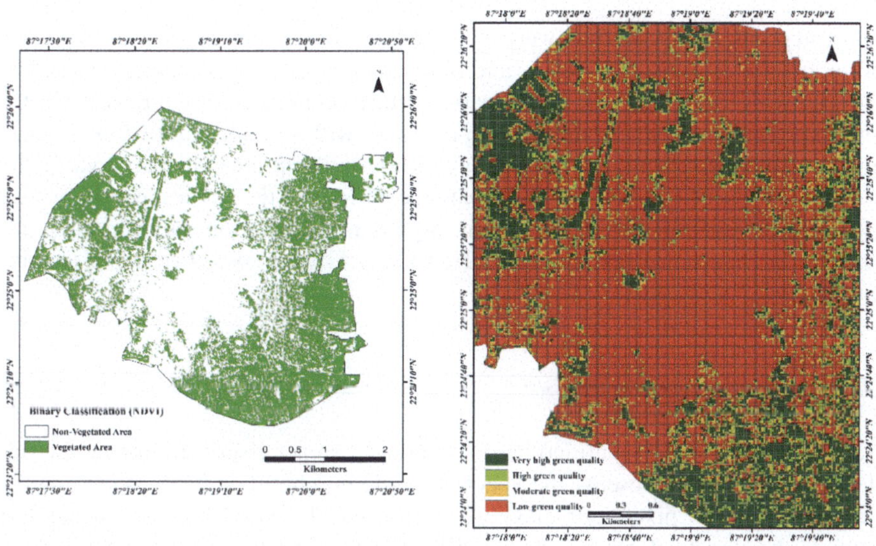

Fig. 17.4 Binary classification of image (NDVI) and preparation of green index map (percentage of green in each cell)

Table 17.2 Criteria for classifying percentage green

% of green in each pixel	Parameter p1 value	Quality
Below 25%	0.25	Low green quality
25–50%	0.5	Moderate green quality
50–75%	0.75	High green quality
Above 75%	1	Very high green quality

17.4.4 Preparation of Proximity to Green and Built-Up Density Map

The proximity to green map was prepared by first generating a green cover map. This was achieved by classifying IRS P6 LISS IV imagery using the Maximum Likelihood Classifier, a parametric method of classification. The classification process involved several stages: data preparation, gathering training samples, applying the classification algorithm, and evaluating its accuracy. The generated green map then enabled the creation of a proximity map, which categorizes features by distance to green areas.

17.4.4.1 Classification System

Four categories of urban green were classified: dense vegetation, grass or low vegetation, open non-vegetated spaces, and built-up areas. Dense vegetation appears as dark red with irregular shapes and coarse texture on false color composite (FCC). Grass/low vegetation shows dull red to pink tones with smooth texture. Open spaces appear bright white with smooth texture, while built-up areas display cyan color with coarse texture and block-like appearance. However, the classes identified were slightly different: low dense vegetation, high dense vegetation, built-up area, and water body (instead of open spaces). These classes were identified using training datasets.

17.4.4.2 Training Data Sets

Training datasets were identified based on image characteristics, following the rule of 30 m, where n represents the spectral bands number. For LISS IV's 3 bands, a minimum of 90 pixels per class were selected. To avoid bias and accurately estimate the statistical parameters, small and homogeneous training samples were selected. The histogram plots of these samples were unimodal, showing the purity and normality of the training dataset. As mentioned by, this supported reliable classification outcomes. Care was taken to balance sample size and homogeneity, allowing for accurate representation of each class.

17.4.4.3 Image Classification

The IRS P6 LISS IV image was classified using identified training datasets into the four basic urban green classes described earlier. As a result, the final green cover map represents all four classified categories (see Fig. 17.5a).

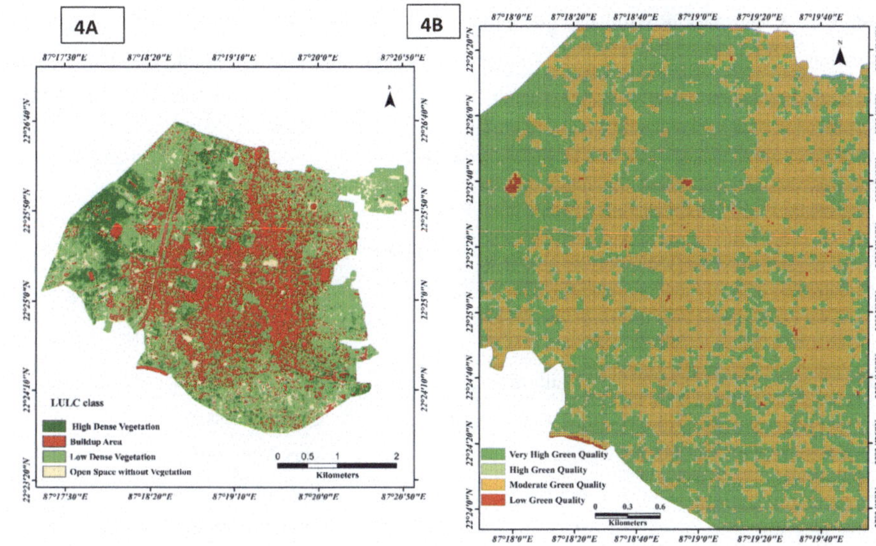

Fig. 17.5 **a** Type of green map (LULC Class), **b** proximity to green map

17.4.4.4 Assessment the Accuracy

Stratified random sampling was carried out to evaluate the classification accuracy by comparing it with reference data. The classified image achieved an overall accuracy of 86.6%, which is higher than the minimum threshold of 85% recommended by Anderson et al.

17.4.4.5 Generation of Buffer and Proximity to Green Map

A 20 m buffer was created around vegetated zones, excluding built-up areas, and classified into high-, moderate-, and low-quality green spaces corresponding to dense vegetation, grass/low vegetation, and open spaces, respectively. To generate the proximity to green map, a 20 m × 20 m grid was overlaid on the buffered image, incorporating these green quality classes (see Fig. 17.5b) defined according to specific criteria outlined in Table 17.3.

17.4.4.6 Preparation of Built-Up Density Map

After identifying un-vegetated areas, built-up areas were classified based on density and building height. The classified image was then used to quantify open areas suitable for potential vegetation. A supervised classification was performed on the LISS-4 image using predefined classes. A 20 m × 20 m grid cell was applied to

Table 17.3 Criteria for proximity to green map

Percentage (%) of buffer pixel	Value of parameter	Quality classes
>50% of area of cell in buffer around dense vegetation	1	Very high green quality
<50% and >20% of area of cell in buffer around dense vegetation	0.75	High green quality
Buffer around low/grass vegetation	0.5	Moderate green quality
Buffer around open spaces	0.25	Low green quality
Other areas	0.25	Low green quality

Table 17.4 Criteria for built-up density map

% of Build-up pixels	Parameter value	Quality classes
<25%	1	Very high green quality
>25% to <50%	0.75	High green quality
>50% to <75%	0.5	Moderate green quality
>75%	0.25	Low green quality

analyze the percentage of built-up area per cell, with output parameters classified according to the criteria in Table 17.4 (Fig. 17.6).

17.4.5 Derivation of Heights of Structure Map

The analysis used norms from the Medinipur Buildings Byelaws & Regulations to evaluate land-use variations and built-form concerns using spatial analysis. In the absence of Cartosat-2 data, open building data from Google Earth Engine were used as the data source for assessing building heights. Buildings were classified into two classes: low-rise (height <10 m) and high-rise (height >10 m). A 20 m × 20 m grid cell framework was applied to calculate the percentage of high-rise structures within each cell, with the output parameters delineated as per Table 17.5 (Fig. 17.7).

17.4.6 Derivation of Weights

AHP uses Saaty's pairwise comparison to estimate the weights of parameters. The procedure involves comparing each parameter with all other parameters on a 1–9 scale and thus creates a pairwise comparison matrix.

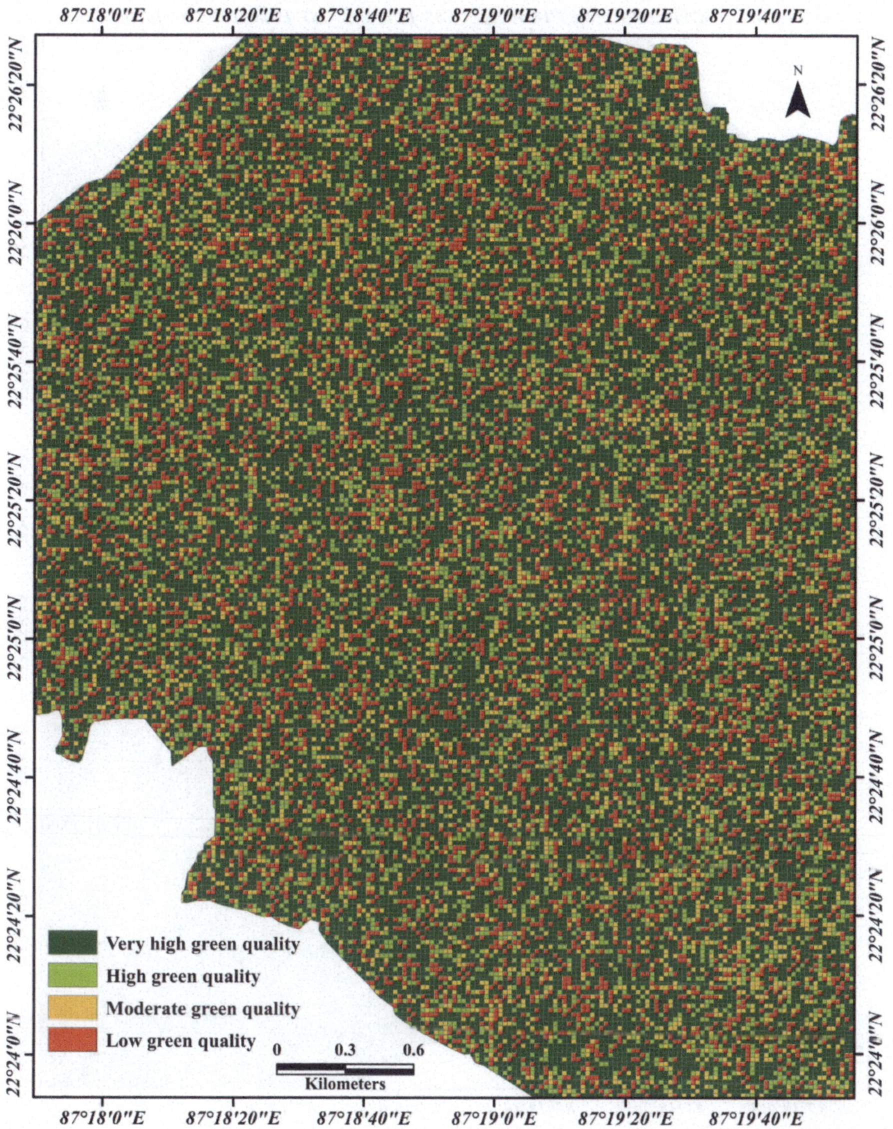

Fig. 17.6 Built-up density map of Medinipur municipality

Table 17.5 Criteria for height of structures map

Percentage of high-rise pixels	Value of the parameter	Quality classes
<25%	1	Very high green quality
>25% to <50%	0.75	High green quality
>50% to <75%	0.5	Moderate green quality
>75%	0.25	Low green quality

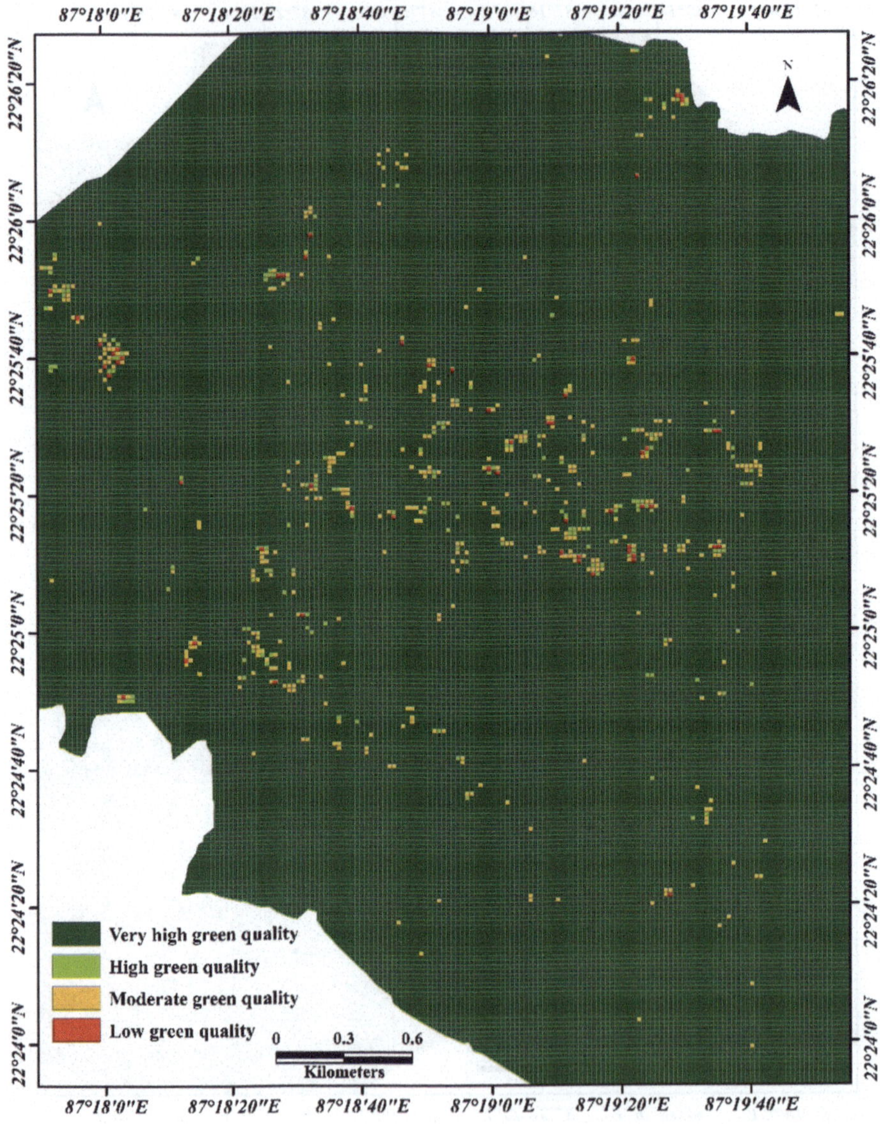

Fig. 17.7 Height of structure map of Medinipur municipality

17.4.7 *Generation of UNGI Map and Comparative Evaluation*

The consequent weights are used to compute the UNGI applying the derived layers. The UNGI value for each cell is computed as:

$$UNGI_i = \sum_{j=1}^{4} \left(W_j \times P_{ij}\right), \quad \text{for } i = 1 \text{ ton}$$

where UNGI_i is the Urban Neighborhood Green Index in the i-th cell ($i = 1, \ldots, n$), W_j is the relative weight of the j-th parameter ($j = 1, \ldots, 4$), and P_{ij} is the value of the j-th parameter in the i-th cell.

A neighborhood type map for comparative neighborhood assessment was created from the Open Building 2.5D Temporal Dataset with code implemented in Google Earth Engine. From characteristics such as built-up density, development type, and building height, the study area was differentiated into four main neighborhood types: namely, open space (non-residential), low-rise low-density neighborhood (LRLDN), low-rise high-density neighborhood (LRHDN), and high-rise low-density neighborhood (HRLDN) (see Fig. 17.8).

17.5 Result and Discussion

The quality of green index (GI) and Urban Neighborhood Green Index (UNGI) across four types of neighborhoods: open spaces, low-rise low-density (LRLD), high-rise low-density (HRLD), and low-rise high-density (LRHD). Open spaces stand out with the highest proportion of very high-quality green areas (66.33% GI and 67.32% UNGI), showing that these zones provide the best green quality compared to all other neighborhoods. In low-rise low-density (LRLD) neighborhoods, the largest share is in low-quality areas for GI (58.26%), but UNGI distribution is spread more toward moderate quality (46.65%), indicating uneven green quality (see Table 17.6). High-rise low-density (HRLD) neighborhoods are dominated by low-quality areas (59.55% GI, 50.62% UNGI), with only limited representation in high and very high-quality categories, reflecting poor green-space management in such settings (see Table 17.6). Low-rise high-density areas represent the neighborhood type characterized by the most significant imbalance in green infrastructure quality, as shown in Table 17.6: 92.83% of GI is classified as low quality, while only 1.91–1.86% is considered high or very high quality. From Table 17.6, open spaces are indicated as containing the highest percentage of very high-quality green area, followed by a ranking that makes such a neighborhood the most sustainable environment. The other two cases-low-rise low-density and high-rise low-density-are dominated by low-quality green space, with minimal area covered at higher quality classes. This is particularly critical for LRHD neighborhoods, where more than 90% of GI falls into the low-quality category. Taken together, these results clearly show the worsening quality of green areas as the density of the neighborhoods increases and point out the need for a better approach in green planning for densely populated urban areas.

These results reflect a sharp decline in the quality of green space from open spaces to densely built-up neighborhoods. Open spaces are the most sustainable green environments, while low-rise high-density areas are the most vulnerable; the

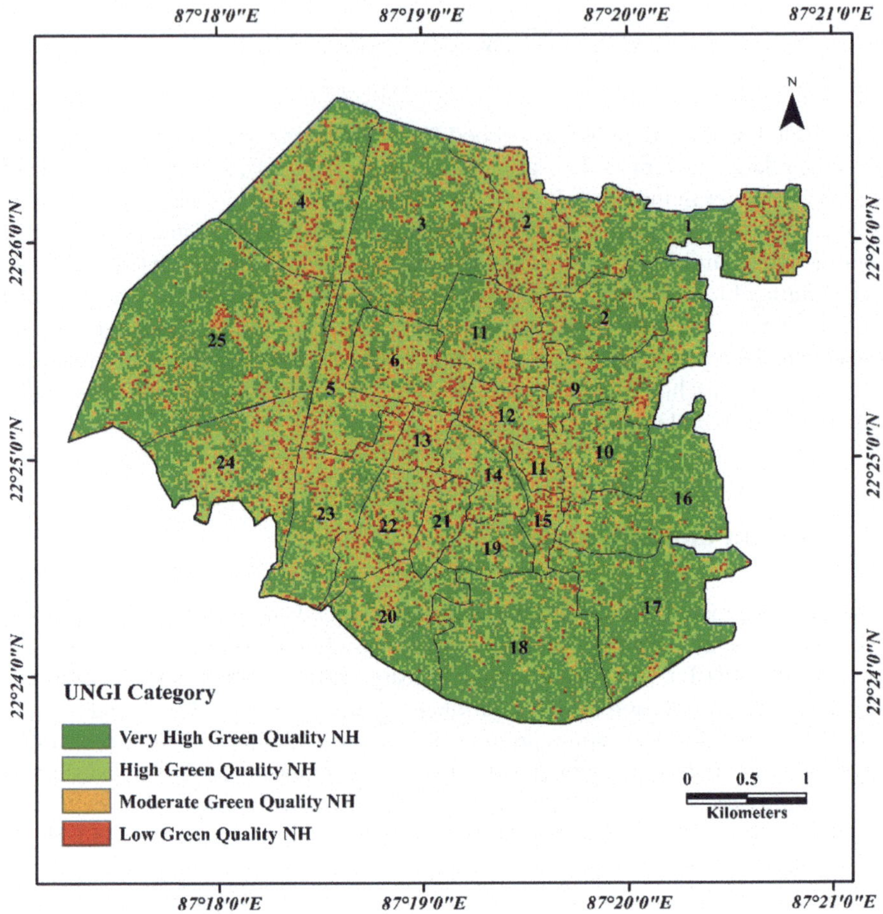

Fig. 17.8 Derived urban neighborhood green index map

dominant type of GI present in those areas is of low quality. This trend indicates that densely developed urban neighborhoods need increased planning and management for green spaces, especially in the LRHD regions.

Indeed, the results point to significant heterogeneity across the district in terms of the quality of green space. More than 50% of the area is categorized as a high-quality green space in districts 16, 17, and 18, while districts 12, 13, 14, 22, and 23 show serious deficiencies, as more than 90% of their area consists of low-quality green spaces. District 13 totally lacks high-quality green spaces-100% are of low quality. See Table 17.7. Moreover, a majority of wards, including 5, 6, 11, and 21, are dominated by predominantly low-quality green spaces, ranging from over 70–85%, suggesting very limited ecological benefits. In contrast, wards 1, 9, 10, 15, 19, 20, and 25 show a more balanced distribution, even though low-quality green spaces are dominant. Even though wards 16–18 could play the role of ecological hubs

Table 17.6 Area wise distribution of various neighborhoods

Area of green quality	Open spaces (n = 6540)		Low-rise low-density NH (n = 13,278)		High-rise low-density NH (n = 20,034)		Low-rise high-density NH (n = 9833)	
	GI	UNGI	GI	UNGI	GI	UNGI	GI	UNGI
	(%)		(%)		(%)		(%)	
Area of low quality	3.67	6.56	58.26	21.46	59.55	50.62	92.83	60.11
Area of moderate quality	10.00	18.25	11.25	46.65	15.61	23.87	3.41	27.59
Area of high quality	20.33	7.87	14.25	18.25	12.42	20.63	1.91	9.80
Area of very high quality	66.33	67.32	16.24	13.64	12.42	4.88	1.86	2.50
Total	100	100	100	100	100	100	100	100

given the availability of green areas, most of the districts suffer from insufficiency or extreme insufficiency of green infrastructure. See Table 17.7. In this way, this irregular distribution calls for targeted interventions to improve green-space quality in the most vulnerable districts and points to a more equitable distribution of environmental benefits throughout the city.

Table 17.8 shows that the majority of the wards are dominated by high-quality green spaces at about 50–65% and by medium quality at 20–30%. Low-quality green spaces are less than 1% in all the districts, indicating a very good overall environmental sustainability. The most favorable results are obtained for Districts 16, 17, and 18, where more than 47–55% of the areas were of a very high-quality green space, and their medium-quality shares were less than 10%. Districts 10, 20, and 25 also present very good environmental conditions, with 27–39% of very high-quality green spaces, due to the high share of high-quality green spaces. On the other hand, in Districts 12, 13, 14, and 22, the percentage of very high-quality green space was less than 5%, indicating that they were entirely covered with high- and medium-quality areas, reflecting poor ecological richness. Wards 5, 6, 11, 21, and 23 have more balanced conditions, but they were relatively rich in medium-quality green spaces, around 29–31%, indicating possibilities for improvement measures. Generally speaking, the information shows that green spaces are sufficiently distributed in the sectors with highly prevailing high-quality areas and minimal low-quality green spaces. While Districts 16–18 and 25 have a high abundancy of green spaces, the presence of very high-quality areas is restricted in Districts 12, 13, 14, and 22, while slight improvements can be brought forward to ensure an equitable urban green environment throughout the city.

Table 17.7 Ward wise percentage of green index in Medinipur municipality

Ward No.	Very high green quality	High green quality	Moderate green quality	Low green quality
1	34.22	9.74	9.74	46.30
2	21.23	10.19	10.77	57.81
3	12.68	6.57	7.68	73.07
4	17.22	8.46	10.09	64.23
5	4.77	3.23	7.56	84.45
6	5.71	3.64	4.58	86.07
9	23.75	15.91	18.76	41.58
10	29.90	14.91	16.34	38.85
11	9.84	6.51	9.25	74.39
12	1.50	1.62	2.66	94.22
13	0.00	0.00	0.00	100.00
14	1.73	1.73	3.12	93.42
15	19.58	9.58	14.38	56.46
16	59.59	14.06	10.94	15.41
17	54.39	16.59	12.62	16.40
18	52.89	19.33	14.08	13.70
19	21.28	11.75	11.49	55.48
20	27.56	11.91	11.91	48.62
21	7.91	5.72	6.06	80.30
22	2.88	3.18	3.56	90.38
23	7.69	4.80	6.03	81.48
24	12.73	8.60	12.11	66.56
25	32.84	12.47	12.36	42.33

Table 17.8 Ward wise percentage of UNGI in Medinipur municipality

Ward No.	Very high-quality green NH	High-quality green NH	Moderate quality green NH	Low quality green NH
1	26.31	52.52	21.12	0.06
2	18.65	56.27	24.94	0.15
3	19.59	62.89	17.47	0.05
4	22.06	55.44	22.30	0.20
5	8.43	61.58	29.18	0.81
6	7.97	61.48	29.80	0.75
9	19.61	55.91	23.66	0.82
10	27.27	52.49	20.17	0.07
11	11.67	58.96	29.05	0.32
12	2.08	60.51	35.80	1.62
13	0.19	60.64	38.79	0.38
14	2.89	67.67	29.33	0.12
15	20.54	48.55	30.71	0.21
16	47.60	43.51	8.85	0.04
17	54.71	37.33	7.93	0.03
18	52.27	39.06	8.58	0.08
19	24.74	54.84	20.29	0.13
20	30.94	52.12	16.77	0.17
21	9.95	59.70	30.02	0.34
22	4.55	63.68	31.39	0.38
23	11.32	62.21	25.83	0.64
24	17.98	57.91	23.88	0.23
25	38.82	47.49	13.28	0.41

17.6 Conclusion

Urbanization is expanding at an unprecedented rate, especially in developing countries, where rapid population growth and economic development are putting significant pressure on available land resources. One of the most pressing challenges of urban growth is the loss of green and natural spaces, which directly affects environmental quality, human health, and the quality of life in cities. In this context, the Urban Neighborhood Greenness Index (UNGHI) emerges as a highly effective and practical tool for decision-making regarding the planning, selection and management of urban green spaces.

The UNGHI methodology is very useful because the UNGHI methodology goes beyond mapping vegetation cover. The UNGHI methodology gives an measurable way to check the quality and spread of spaces, at the neighborhood level. I have found that the UNGI is built to be simple, easy to use and clear, for planners, decision makers, and local authorities. The UNGI works even if the user does not know much about GIS or RS. The UNGI does not need the GIS and RS tools that require a lot of knowledge. The feature of the index that makes it such a strong connection between science and policy is that it acts as a bridge from the technical research to the implementation of policies.

The Urban Neighborhood Green Index (UNGI) helps government officials and city planners find the places where green projects work best; that is why the Urban Neighborhood Green Index (UNGI) matters. We noticed that the city has green spaces spread unevenly. Some neighborhoods have parks, trees and open land. Other neighborhoods are packed with buildings. Have no green. With the help of UNGI, the city planners can evaluate these disparities and decide on how to allocate the funds for the various green initiatives such as parks, community gardens, roadside plantings, or vertical greening. This index thus becomes a tool for advocating for environmental justice, ensuring all residents, regardless of where they live, have access to excellent green infrastructure.

Also the UNGI adapts well can be used for land uses and setups. We think of a complex, a school, a shopping center and an industrial area. Each of these places has possibilities and needs, for the vegetation cover. The index determines best vegetation levels, for each type. The index mixes nature with what the city needs. The many functions give the UNGI its status as a measurement tool. The many functions also give the UNGI the role of a planning system that supports design.

One key part of the UNGI is the assessment of spaces both public green spaces and private green spaces. We see that the UNGI looks at spaces such, as parks, playgrounds and community gardens. We also see that the UNGI looks at spaces such as community gardens in residential areas university campuses and commercial complexes. The UNGI maps neighborhood vegetation. Gathers input, from stakeholders. The UNGI brings spaces and private green spaces together in one network. A collaborative approach is essential for the development of the city's long-term environmental plans.

The UNGI lets people watch the growth and progress of greening initiatives. Urban greening initiatives need monitoring to make sure urban greening initiatives meet the expected results. In our work, we have seen the UNGI help the local authorities verify if the actions really benefit the spaces at the level. The local authorities can then adapt the strategies as needed. The feedback system of the UNGI strengthens the UNGI as a planning tool that can respond to change. In the end, the UNGI method is an effective decision support system, for urban planning. The UNGI method lets elected officials and urban planners evaluate, compare and improve the quality of spaces at the district level. Beyond the benefits the UNGI method shows environmental commitment. The UNGI method promotes distribution of spaces. The UNGI method strengthens resilience, to climate change. The UNGI method helps create more pleasant environments. We see that cities keep growing. We see that using easy to use but indices will be critical to reach the two goals of development and sustainability. We also see that UNGI lets policymakers and planners—those who do not have expertise in remote sensing—design greener healthier more sustainable cities (Erdoğan & Yücel, 2023; Frontiers, 2023; Gupta et al., 2012).

AI Discloser This chapter is the result of the authors' own research, analysis, and writing. At no stage whether in developing the ideas, drafting the text, or editing the final manuscript were generative AI tools used. All sections, including literature discussion, data interpretation, and figure preparation, were carried out using established scholarly methods and conventional research software. The work reflects the authors' direct engagement with the subject matter and is entirely the product of human intellectual effort.

References

Baklanov, A., Korsholm, U., Mahura, A., Petersen, C., & Gross, A. (2008). Enviro-HIRLAM: Online coupled modelling of urban meteorology and air pollution. *Advances in Science and Research, 2*, 41–46. https://doi.org/10.5194/asr-2-41-2008

Bounoua, L., Fathi, N., El Berkaoui, M., El Ghazouani, L., & Messouli, M. (2020). Assessment of sustainability development in urban areas of Morocco. *Urban Science, 4*(2), 18. https://doi.org/10.3390/urbansci4020018

Cai, G., Zou, B., Chi, X., He, X., Guo, Y., & Jiang, W. et al. (2023). Neighbourhood spatio-temporal impacts of SDG 8.9: The case of urban and rural exhibition-driven tourism by multiple methods. *Land, 12*(2), 368. https://doi.org/10.3390/land12020368

Erdoğan, R., & Yücel, G. (2023). Assessment of urban green space quality at the neighborhood scale: A case study of Northern Nicosia. *Sustainability, 15*(18), 13880.

Frontiers in Environmental Science. (2023). The role of urban green spaces in climate adaptation and human well-being. *Frontiers in Environmental Science.*

Gorzelany, J., Noszczyk, T., Kukulska-Kozieł, A., & Hernik, J. (2023). Urban green spaces management during the COVID-19 pandemic: Experiences from Kraków, Poland. *Land Degradation and Development, 34*, 423–440. https://doi.org/10.1002/ldr.4469

Gupta, K., Kumar, P., Pathan, S. K., & Sharma, K. P. (2012). Urban neighborhood green index—A measure of green spaces in urban areas. *Landscape and Urban Planning, 105*(3), 325–335. https://doi.org/10.1016/j.landurbplan.2012.01.003

Hou, H., & Wu, H. (2021). Tourists' perceptions of green building design and their intention of staying in green hotel. *Tourism and Hospitality Research, 21*(1), 115–128. https://doi.org/10.1177/1467358420963379

Hwang, J., Kim, I., & Gulzar, M. A. (2020). Understanding the eco-friendly role of drone food delivery services: Deepening the theory of planned behavior. *Sustainability, 12*(4), 1440. https://doi.org/10.3390/su12041440

Lyytimäki, J., & Sipilä, M. (2009). Hopping on one leg—the challenge of ecosystem disservices for urban green management. *Urban Forestry & Urban Greening, 8*(4), 309–315. https://doi.org/10.1016/j.ufug.2009.09.003

Mudau, N., Mwaniki, D., Tsoeleng, L., Mashalane, M., Beguy, D., & Ndugwa, R. (2020). Assessment of SDG indicator 11.3.1 and urban growth trends of major and small cities in South Africa. *Sustainability, 12*(17), 7063. https://doi.org/10.3390/su12177063

Sokhi, B., & Ruangrit, V. (2004). Remote sensing and GIS for urban green space analysis—A case study of Jaipur city, Rajasthan. *Journal of Institute of Town Planners India, 1*(2), 55–67.

Van, K., Korman, T. M., Nicholson, S., Troutbeck, R., Lister, D. M., & Woolley, I. (2020). Case report: Japanese encephalitis associated with chorioretinitis after short-term travel to Bali, Indonesia. *The American Journal of Tropical Medicine and Hygiene, 103*(4), 1691–1693. https://doi.org/10.4269/ajtmh.19-0330

Venn, J. S., & Niemelä, K. J. (2004). Ecology in a multidisciplinary study of urban green space: The URGE project. *Boreal Environment Research, 9*(6), 479–489.

Wolch, J. R., Byrne, J., & Newell, J. P. (2014). Urban green space, public health, and environmental justice: The challenge of making cities 'just green enough.' *Landscape and Urban Planning, 125*, 234–244. https://doi.org/10.1016/j.landurbplan.2014.01.017

Yang, J., Zhao, L., McBride, J., & Gong, P. (2009). Can you see green? Assessing the visibility of urban forests in cities. *Landscape and Urban Planning, 91*(2), 97–104. https://doi.org/10.1016/j.landurbplan.2008.12.004

Zhang, J., Yue, W., Fan, P., & Gao, J. (2020). Measuring the accessibility of public green spaces in urban areas using web map services. *Applied Geography*. https://doi.org/10.1016/j.apgeog.2020.102381

Chapter 18
Climate Resilience Development Policy: An Overview of Indonesia's Strategies and Practices

Intan Hapsari Surya Putri, Badrun Susantyo, Herma Juniati, Tri Bastuti Purwantini, and Tri Margono

Abstract Climate resilience development policies in Indonesia are crucial in addressing the nation's vulnerability to climate change impacts, such as sea-level rise, extreme weather, and increased disaster risk. This study examines the structure, thematic priorities, and implementation of these policies at national and sub-national levels through document analysis. Key frameworks such as the nationally determined contributions (NDCs) and the national medium-term development plan (RPJMN) are analyzed to identify recurring themes, policy objectives, and gaps in consistency and implementation. The study also examines adaptation efforts in urban areas as an example of how national policies are translated into concrete actions tailored to the regional context. These efforts highlight the challenges and opportunities in implementing climate strategies amidst diverse environmental and socio-economic conditions. Governance structures, disaster risk management, and integrating resilience into development planning are key elements in bridging national objectives with local realities. The research findings provide valuable insights into the alignment between national policy frameworks and sub-national practices, while also suggesting pathways for improving policy coherence and effectiveness. By addressing these dynamics, this study offers recommendations to strengthen climate resilience and ensure sustainable development across all levels of government in Indonesia.

I. H. S. Putri (✉) · B. Susantyo · H. Juniati · T. B. Purwantini · T. Margono
National Research and Innovation Agency (BRIN), Jakarta, Indonesia
e-mail: inta017@brin.go.id; intanhsp@gmail.com

B. Susantyo
e-mail: badr004@brin.go.id

H. Juniati
e-mail: herm018@brin.go.id

T. B. Purwantini
e-mail: trib005@brin.go.id

T. Margono
e-mail: trim004@brin.go.id

G. K. Panda et al. (eds.), *Socio-Ecological Crisis of Development*, Sustainability Solutions, https://doi.org/10.1007/978-3-032-22171-1_18

Keywords Climate resilience · Policy analysis · Adaptation strategies · Sustainable development · Governance

18.1 Introduction

The effect of climate change is and can continue to be regarded as the most fundamental and challenging threat to the world and its people. And relative to the rest of the world, developing countries, such as Indonesia, face disproportionately higher risks. This can be illustrated by global warming, severe storms, coastal flooding, and loss of biodiversity, and the negative growth impact of these on the already developing and economically vulnerable Indonesia is as stated in the report (IPCC, 2014). Indonesia is an economically vulnerable developing country. The country's unique geography, characterized by its sprawling archipelago, extensive and diverse ecosystems, and high coastal area to land ratio, translates to a high sensitivity to cascading climate risks (Case et al., 2007; Djalante & Thomalla, 2012; Hadi, 2011).

Indonesian climate resilience policies are crucial in addressing the potential impacts of climate change. As a tropical archipelagic nation, Indonesia is susceptible to the consequences of climate change, such as rising sea levels, extreme weather patterns, and the depletion of natural and economically crucial resources like agriculture and maritime zones (ADB, 2021; Dharmawan, 2021; Zikra et al., 2015). The IPCC (2014) defines climate resilience as the extent to which a social, economic, and environmental system can withstand and/or adapt to climate change while reducing the associated losses. As such, climate resilience or climate adaptive capacity has to be planned at the national as well as sub-national levels, which Indonesia has tried to achieve through various policies. One such policy is the climate resilience development (CRD) policy which is a vertically integrated policy at the national level. The document seeks to provide frameworks to various ministries and institutions in developing programs aimed at addressing the impacts of climate change in the most vulnerable sectors, such as maritime and agriculture, health, and water (Bappenas, 2020).

Resilience policies designed to address climate changes continue to face challenges, especially in integrating local policies to fulfill the goals set by the national policies. As noted by Chenani et al. (2021), the lack of inter-institutional coordination, inadequate budgets, and regional-level disparity contribute to the stalled execution of climate resilience policies. This creates a disconnect between central-level policies and local realities, as pointed out by Vázquez (2017), who noted that the policies, although comprehensive, face implementation challenges in socially and environmentally diverse regions. The key issue is the lack of coherence between national frameworks and sub-national implementation. The CRD and RPJMN documents, as the rest of the country, face considerable gaps between design and implementation due to institutional fragmentation, limited financial resources, and uneven regional capabilities.

Moreover, and especially in the case of Indonesia, climate resilience policies are hindered by the disparate sectoral interests and the "low hanging fruits" dependence of agriculture and maritime affairs that are climate vulnerable. In Haasnoot et al. (2024) explanation, "adaptive planning" of urban development in the context of climate change needs to be multisectoral, considering the integration and interactions of the sectors involved, rather than the silo approach.

Also, the bottom-up climate change adaptation policies are crucial in determining the overall climate resilience of a jurisdiction. The regional action plan for climate change adaptation (RAD-API) executed in several cities like Semarang illustrates how overarching national policies on climate resilience can be locally translated into focused and context-specific RAD-API. In the case of Semarang, the RAD-API emphasizes adaptation planning which integrates disaster resilient infrastructure, safe and sustainable urban water cycle management, and climate change awareness-raising (Semarang City Government, 2020).

Moreover, attention is given to the need for integration of urban development with policies on climate change. In the study by Ribot and Peluso (2003), the issue of access to natural resources, which is part of climate change adaptation, is often in conflict with the formal ownership of land and other resources. This is even more complex in the inner city, where the policies of urban planning and the development of infrastructure need to deal with the adverse effects of climate change such as flooding, variable weather, and food insecurity. In summary, despite having a climate policy which is as detailed as any in the world, the greatest problem in the case of Indonesia is the inability to execute policy in practice. The major challenges consist of limited inter-institutional coordination, regional differences in capacity, and the disjuncture between policies and the ground reality. This is why this study aims to explore more rational pathways for climate change policies with special emphasis on the strengthening of integration between national and sub-national policies, as well as community-based actions to improve climate change adaptation in Indonesia. This research is important because it tries to address the missed policy–practice gap. Closing the gap between strategic planning and action is essential for Indonesia and other developing countries dealing with similar governance challenges. The analysis conducted in this research advances arguments in adaptive governance as well as fulfilling the concern of policymakers regarding the need for climate resilience.

Because of this vulnerability, Indonesia reframed the development paradigm for the country in line with climate resilience. One of the important development policies in this respect is the climate resilience development policy, which integrates climate adaptation into the national planning framework for the country (Bappenas, 2020). The policy outlines a vision that moves away from siloed climate change adaptation responses to climate change resilience governance and development at all tiers and in all areas of governance. This includes economic, social, environmental, and infrastructure domains. There is a marked change toward the governance of long-term risks, especially anticipatory adaptation, systemic resilience, and deep transformation of institutions.

This is also seen in the enhanced NDC of Indonesia, which comprises targets and plans for mitigation and adaptation of unprecedented scale. These NDCs, paired with sectoral strategies in energy, forestry, and agriculture, aim to cut greenhouse gas emissions and enhance adaptive capacity by 2030 at the national and sub-national levels. This positions Indonesia as a crucial case of the global South. While making attempts to bridge the gap between development and climate goals, the country is faced with an array of complexity, disparity, and resource challenges.

Despite the policy advancements made, the gaps between the two extremes remain bridged. Implementation of secondary strategies at the sub-national level, which are more concrete and effective in nature, are lacking. Such barriers are a result of lacking political will, finances, and insufficient institutional cross-sectoral planning. Even in more favorable conditions, these barriers are most pronounced in the remote and "forgotten" parts of the territory, which face the greatest challenge to adaptation.

There are several notable reasons this whole research is particularly important. Firstly, it is part of the ever-growing scholarship on the adaptive governance of climate change and the institutionalization of climate change resilience into development policy. This study offers empirical evidence on the workings of multi-level climate governance by analyzing Indonesia's efforts to link national frameworks with local implementation. Secondly, it fills an important policy gap: how national climate commitments can be effectively translated into sub-national actionable plans. Given the global trend to "mainstream" adaptation within planning frameworks, Indonesia's approach to this challenge provides helpful insights and advocacy lessons on adaptive governance of climate change as well as on enduring systemic barriers. Thirdly, this research is about the broader theorized debates on resilience building in developing contexts where the institutional capacity, political economy, and socio-ecological diversity of the region are meshed in intricate ways.

This chapter also contributes to the expanding literature on comparative climate policy analysis by enhancing the understanding of Indonesia within the context of climate-resilient development. As other countries in the Global South adopt a similar approach of attempting to align climate objectives with development aspirations under resource constraints, the Indonesian experience can serve as practical policy guidance for international policy diffusion (Ampaire et al., 2017; Biesbroek et al., 2013; Chowdhury et al., 2022; Panditharatne, 2016). This research holds both applied and scholarly value. Consequently, this study provides a preliminary assessment of Indonesia's climate resilience development framework, focusing on the recently established policy framework and its governance implementation. It aims to clarify the institutional, political, and socio-technical barriers to climate-resilient development in Indonesia in order to highlight the broader implications for climate adaptation and sustainable development scholarship and practice.

This analysis combines in-depth assessments of the country's national and sub-national climate resilience policies and their internal coherence. It translates strategic policies such as the climate resilience development (CRD) policy and the national medium-term development plan (RPJMN) 2020–2024 into instruments at the sub-national level, including the Semarang City regional action plan for climate change adaptation (RAD-API) and the regional medium-term development plan (RPJMD), as well as within the Semarang City framework. This research is globally significant as it sheds light on the disaggregation of climate adaptation policies within the context of a developing country, a notable gap in the climate governance literature (Chenani et al., 2021; IPCC, 2014; Moser & Ekstrom, 2010). This research addresses the gap in the global climate adaptation discourse by focusing on regional scale challenges. It enhances the comparative debates on climate policies and curricula by providing 12 SDG and net-zero country new policies as well as educational insights. It offers valuable perspectives for other vulnerable countries to reconcile their national obligations with sub-national realities. This accomplishment particularly highlights the divide between climate resilience governance and the implementation of vertical and horizontal policies, which is crucial as development becomes conditioned upon addressing cross-border complexities of climate change.

This pivotal study is the first to evaluate the impact of climate resilience policies related to national policies and policies on climate change adaptation on the regional action plan for climate change adaptation (RAD-API) and the regional medium-term development plan (RPJMD) in Semarang City, Indonesia. The guiding question of the study is as follows: do national and sub-national climate governance frameworks and legislations function in harmony, and what are the actual challenges of such implementation? This chapter contributes to the dominant narrative of the book by elucidating the integration of climate change resilience within the sustainable development planning process, thereby enriching vertical policy integration, cross-institutional coordination, and flexible governance planning for cities in the Global South.

Despite these improvements, the core issue that remains is the persistent difference between the creation of a national policy and its execution on the ground at the sub-national level. This gap constitutes not only a governance problem but more importantly, a research problem, as it diminishes the effectiveness of the climate adaptation agenda in Indonesia and the country's overall sustainable development objectives. Articulating this gap is important because it brings to the fore the structural issues, such as inadequate inter-institutional collaboration, uneven capability and resource distribution at the regional level, and the urgent need for improved policy integration both vertically and laterally. This gap is of much concern to the present research because it sheds light on the need to align national frameworks more closely with local realities. The document-based comparative method used in this research is appropriate because it identifies patterns, gaps, and connections between policy levels and hence, is able to generate plausible, well-informed recommendations. This emphasizes the uniqueness of the research, explains the objectives in a coherent manner, and sets the stage to make a substantial contribution to the discourse on malaria governance as well as to policy formulation in vulnerable regions to climate change.

Therefore, this study has three objectives: (1) to examine the integration of climate change policies and climate change adaptation policies at the national and sub-national levels; (2) to assess the implementation of policy integration, including the governance mechanisms, financial resources, and technical assistance available; (3) to devise methods for improving cross-tier policy integration and mobilizing sub-national actors in climate change adaptation efforts. The study is clear in its objectives, allowing for a coherent explanation of the methodology adopted.

18.2 Methods

This study relies on an analysis of documents policy methodology to trace and assess policies on climate resilience development at the national and sub-national levels. This method was chosen as the policy documents are the most pivotal means through which climate resilience policies are articulated, coordinated, and enforced in Indonesia. The documents' analysis in this study is intended to assess the alignment of framework of the country policies with local and in particular, Semarang City, efforts on adaptation.

The analysis uses a policy review method in comparative approach focusing on two levels (Fig. 18.1). First, a within-city approach is directed to the development planning documents of Semarang City and these are the regional action plan for climate change adaptation (RAD-API) and the regional medium-term development plan (RPJMD) 2021–2026. This analysis seeks to assess the extent to which climate resilience strategies are embraced and prioritized in local planning as well as the extent to which these documents respond to the vulnerabilities and to what extent adaptation measures are put in place.

Second, a cross-level review compares local documents with the national, and in particular the climate resilience development (CRD) policy as well as the national medium-term development plan (RPJMN) 2020–2024. This step in the analysis seeks to assess the extent to which national objectives are translated to sub-national implementation focusing on the alignment of the national-level guidance and local strategies in place to facilitate the guidance.

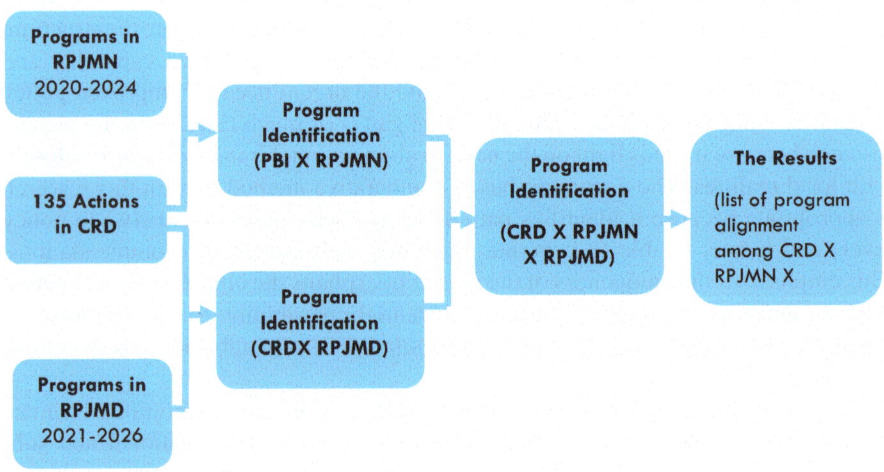

Fig. 18.1 The method of comparison among the CRD, RPJMN, and RPJMD documents. *Source* Authors (2025)

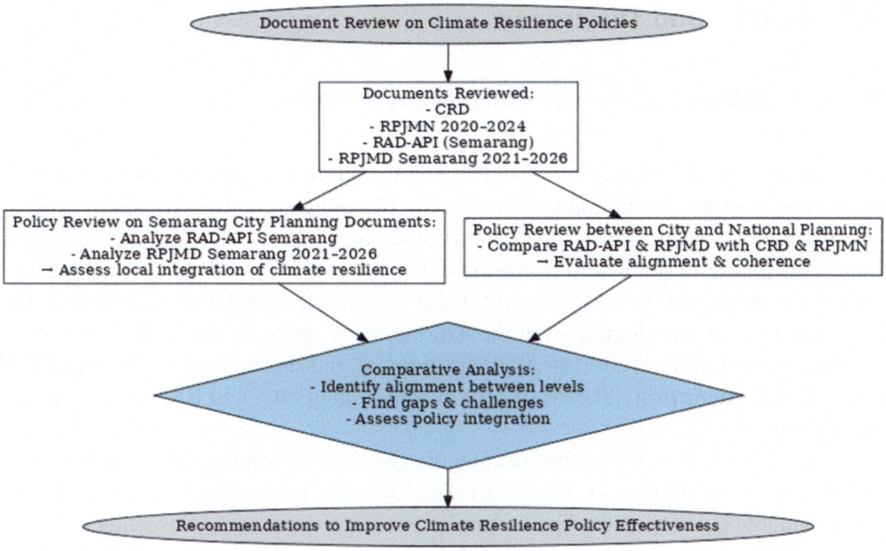

Fig. 18.2 Methodological flowchart of climate resilience policy review. *Source* Authors (2025)

The analytical process occurs in three stages (Fig. 18.2):

- Document Identification and Selection: Identifying relevant national and sub-national policy documents that directly discuss and implement strategies for climate resilience.
- Content Review and Thematic Coding: The documents are systematically reviewed and coded thematically in accordance with resilience policy objectives and measures documents. Thematic coding is used for policy objectives and thematic measures for comparability across documents.
- Comparative Assessment: Identification of the coherence and integration of policies across levels of governance to assess consistencies, gaps, and contradictions.

The intention of this approach is to not merely map the policies that focus on climate resilience, but also assess the practical integration of those policies at different levels of governance. The research aim is to analyze national strategies and their incorporation into local planning, the existing gaps of which, along with institutional and governance obstacles, are not effectively implemented. The primary intention is to offer insights and suggestions in order to increase the coherence of policies at different levels for improving climate resilience in Indonesia.

18.3 Results and Discussion

18.3.1 Linkages Between Planning Documents

There are two broad categories of development planning documents in Indonesia: general and sectoral or thematic. The general planning scheme includes long-term (twenty years), medium-term (five years), and short-term (one year) development plans. This planning document accommodates comprehensive development aspects (strategic planning) at the national, provincial, and district/city levels. In this case, the development planning document at the national level includes the RPJPN, RPJMN, and RKP, hereinafter referred to as the provincial and district/city RPJP, the RPJP of provincial/regency/city, RPJM of provincial/regency/city, and RKP of provincial/regency/city.

Meanwhile, the sectoral planning category regulates urban development based on specific sectors or themes or aspects of development, for example, climate change and sustainable development. The sectoral planning documents in this study focus on climate change documents, namely the climate resilience development (CRD) policy at the national level and the regional action plan for climate change adaptation (RAD-API) at the city level. The preparation of a comprehensive development planning document certainly has links with various other planning documents. In this preparation, aspects of sectoral planning documents have been mainstreamed in comprehensive planning documents. Figure 18.3 illustrates the position and relationship between these planning documents.

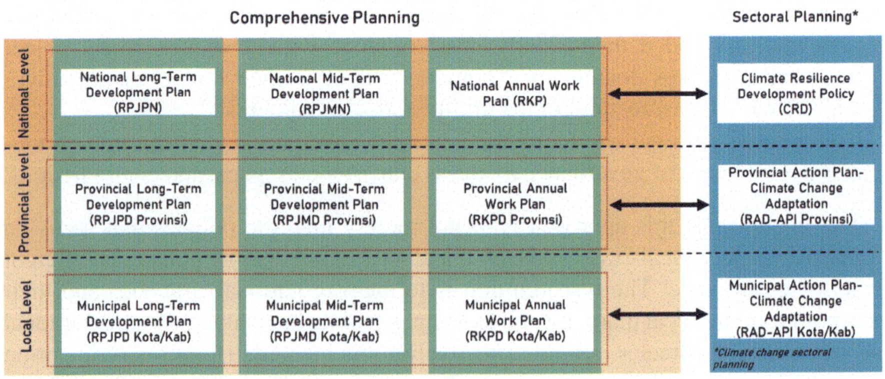

Fig. 18.3 Linkages between planning documents. *Source* Authors (2025)

18.3.2 Climate Resilience Development (CRD)

As a guideline for handling climate change, the Ministry of National Development Planning/Bappenas has launched a climate resilience development (CRD) policy document to create national climate resilience. Through Presidential Regulation No. 18 of 2020, CRD has become one of National Priorities (PN) 6 in the 2020–2024 National Medium-Term Development Plan (RPJMN). The CRD is intended as a reference for the parties in implementing the development priorities listed in the 2020–2024 RPJMN and the subsequent national development planning framework. In this case, especially for preparing climate-resilient programs and activity plans, guidelines for the distribution of authority for agencies to avoid duplication related to climate resilience efforts in priority sectors.

As stated in the CRD document, in Indonesia itself, climate resilience focuses on four affected sectors: the marine and coastal sector, the agricultural sector, the water sector, and the health sector. This sector division aims to guide and provide reference locations and interventions to increase climate resilience to the ministries/agencies involved. First, the climate resilience activities of the marine subsector are related to increasing shipping safety, tiny fishing vessels in the face of the threat of increasing wave height, which reduces the safe zone of the ship's shipping. Improving shipping safety can be through increasing ship capacity, increasing access to safety information for fishermen, and so on.

Meanwhile, climate resilience activities in the coastal subsector are related to protection against potential coastal inundation and tidal flooding caused by rising sea levels in coastal areas with a high level of vulnerability. Inundation in coastal areas impacts loss and damage to infrastructure for settlements and other strategic areas, such as aquaculture or ponds (damage to pond structures, loss of fish due to flooding, and death of fish due to too saline water). Coastal climate resilience actions focus on increasing the readiness of coastal settlements and cultivation areas on the coast, which support the stability of the coastal areas.

Second, climate resilience activities in the water sector focus on protecting water availability and preventing and overcoming drought to avoid water scarcity. Actions for climate resilience that can be taken are conservation in the upstream watersheds, construction of water storage infrastructure, and others. In addition to decreasing water availability and drought, flooding is also one of the climate hazards, especially in terms of frequency of occurrence and depth of flooding. Climate resilience action also considers the vulnerability of the region to flooding and how to overcome it.

Third, climate resilience activities in the agricultural sector focus on handling the effects of climate change on food crop production, in this case, rice. The impact of climate change in question is the potential for changes in plant growth by increasing temperatures and changes in rainfall (climate that is no longer suitable), which results in a decrease in production yields. In addition to changes in plant physiology, decreased production yields can also cause by drought, flooding, and disturbance of plant pest organisms (OPT). Adaptation actions in the agricultural sector include fulfilling crop water needs, controlling flooding in rice fields, and others through the

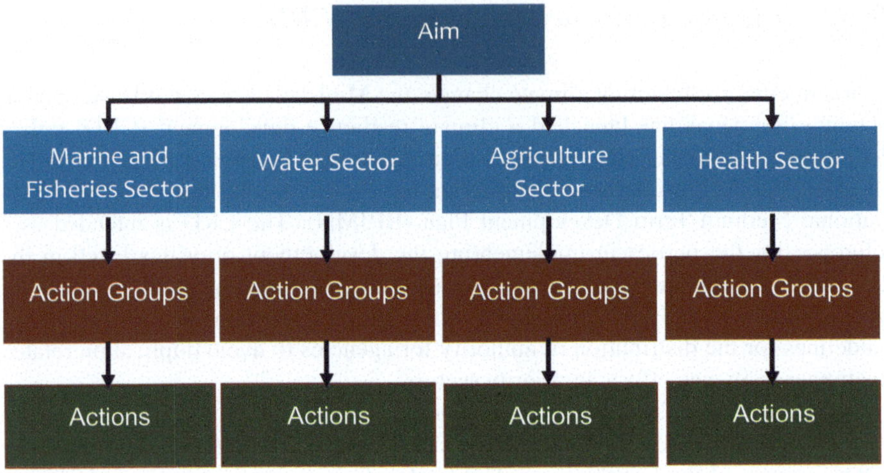

Fig. 18.4 Cascading of climate resistant development documents (CRD). *Source* Authors (2025)

application of climate smart agriculture that supports production stability and food security.

Then, the fourth sector is the health sector (Fig. 18.4). Climate resilience activities in the health sector focus on preventing and overcoming extraordinary events (KLB) of diseases which are influenced by climate, namely dengue hemorrhagic fever (DHF), malaria, and pneumonia. Some actions for climate resilience in the health sector include improving health facilities and services, preventing and controlling disease outbreaks, and improving public health and the environment.

The CRD document's cascading consists of objectives, four affected sectors translated into action groups, and actions as the smallest unit (Fig. 18.5). The four sectors breakdown into core activities and supporting activities, which indicate priority activities that need to be implemented in an area. In general, the total number of actions listed in the CRD is 135 actions (Table 18.1). Water sector has the most actions compare to other sectors, about 34 actions in total as shown in Fig. 18.4. For more details, Table 18.1 describes number of actions in each sector and its core and supporting activities.

18.3.3 *RAD-API/Regional Action Plan for Climate Change Adaptation*

RAD-API Semarang City was prepared concerning several relevant laws and regulations regarding climate change and several climate change studies conducted previously. The existence of RAD-API as a direction in preparing development planning documents that are resistant to climate change. To minimize risks and anticipate

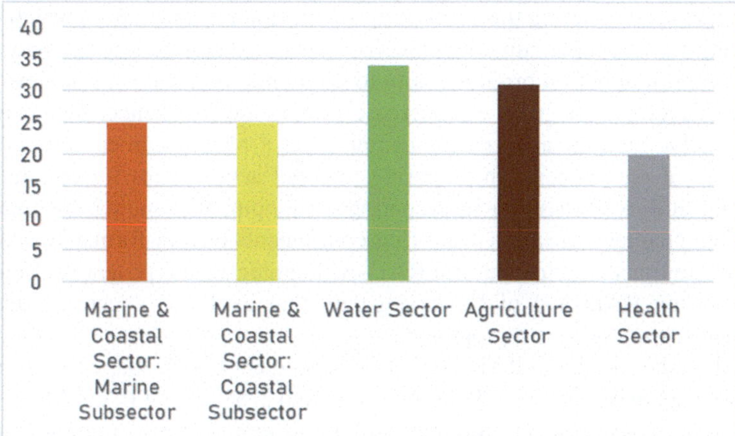

Fig. 18.5 Number of CRD actions by sector. *Source* Authors (2025) (compiled from CRD document)

Table 18.1 Number of CRD actions by sector

Sector	Number of actions
Marine and coastal sector core activities: marine subsector	8
Marine and coastal sector support activities: marine subsector	17
marine and coastal sector core activities: coastal subsector	13
Marine and coastal sector support activities: coastal subsector	12
Water sector core activities	18
Water sector support activities	16
Agricultural sector core activities	14
Agricultural sector support activities	17
Health sector core activities	6
Health sector support activities	14
Total action	**135**

Source Authors (2025) (Compiled from CRD document)

the impacts of climate change, the Semarang City government requires a series of adaptation strategies, which are further translated into climate change adaptation programs that involve all stakeholders in Semarang City. The adaptation strategies and programs arranged in the study frame of the regional action plan for climate change adaptation (RAD-API) for the Semarang City to increase community resilience and have a city development system that is increasingly adaptive to climate change. Determination of climate change adaptation actions at the city level looks at climate phenomena that occur such as changes in rainfall, surface temperatures, and sea-level rise. Adaptation actions are structured to overcome or minimize these impacts, considering the concept of transformative adaptation, which consists of institutions, urban planning, urban ecosystems, and community behavior.

The adaptation action programs aim to guarantee or secure the achievement of the main development goals and increase community resilience, both physically, economically, socially, and environmentally. As a plan document that is a reference for local governments in preparing regional action plans for climate change adaptation and as a direction in preparing development planning documents that are climate change-resistant. Of the four components of transformative adaptation, then they are translated into several RAD-API strategies and programs. Figure 18.6 is a cascading RAD-API of Semarang City that explains objectives, adaptation components, strategies, and programs.

Based on the cascading RAD-API of Semarang City, each adaptation component has its strategy and program. All the strategies and programs of RAD-API Semarang City, which amounted to 11 strategies and 37 programs, the adaptation component of urban planning has the most number of strategies and programs compared to other adaptation components, as shown in Fig. 18.7. This indicates that the urban planning component that focuses on adaptive regional planning and the development of climate-resilient facilities and infrastructure is one of the concerns of RAD-API Semarang City. In more detail, Table 18.2 describes the components, strategies, and numbers of the Semarang City RAD-API program.

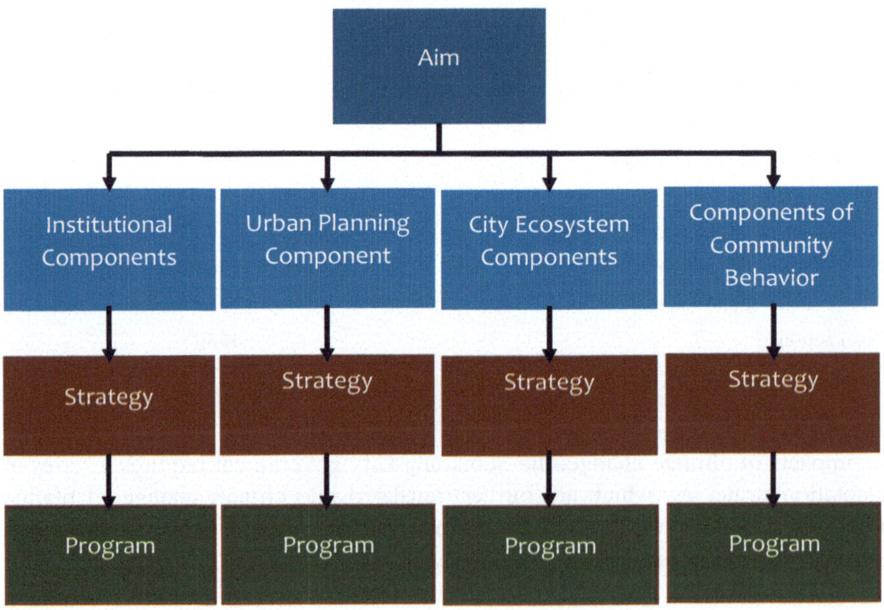

Fig. 18.6 RAD-API document cascading Semarang city. *Source* Authors (2025) (compiled from RAD-API Semarang city)

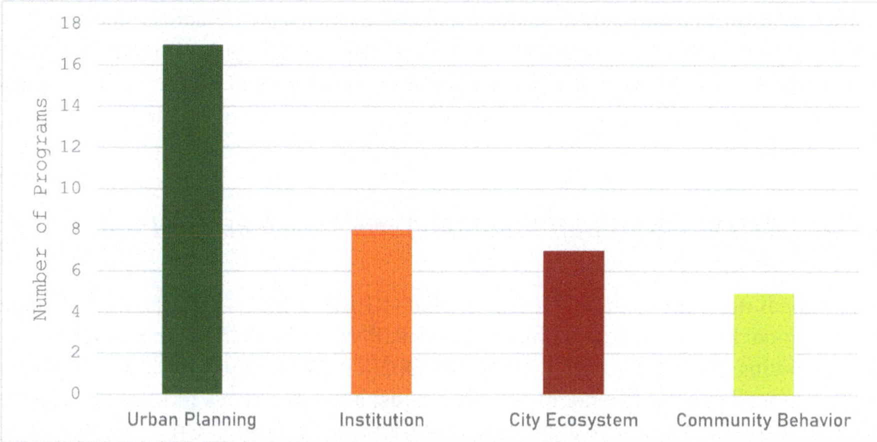

Fig. 18.7 Diagram number of RAD-API programs in each component. *Source* Authors (2025) (compiled from RAD-API Semarang city)

Table 18.2 Number of RAD-API programs by strategies and component

Component	Strategies	Number of programs
Urban planning	Optimizing regional arrangements that are adaptive to climate change	7
	Developing climate-resilient facilities and infrastructure	6
	Developing information systems related to climate change	2
	Optimizing the means of meeting the community's clean water needs	2
Institution	Strengthening regulations related to climate change adaptation	2
	Develop an innovative financing system to encourage the implementation of climate change adaptation actions	6
City ecosystem	Develop an innovative financing system to encourage the implementation of climate change adaptation actions	2
	Developing agricultural systems that are adaptive to climate change	5
Community behavior	Develop a health care system that is pro-active toward the phenomenon of climate change	1
	Raising public awareness regarding climate change	3
	Gender integration in climate change adaptation activities	1

Source Authors (2025) (Compiled from RAD-API Semarang City)

Each program has an achievement indicator to measure the achievement of the Semarang City RAD-API program. From the 37 RAD-API programs, there are 51 indicators that are aligned with climate resistant development. Table 18.3 provides examples of RAD-API indicators that are aligned with climate resistant development.

18.3.4 RPJMN/National Medium-Term Development Plan

The national medium-term development plan (RPJMN) 2020–2024 is the last stage of the national long-term development plan (RPJPN) 2005–2025 so it becomes essential. Following the 2005–2025 RPJPN, the 2020–2024 medium-term development target is to realize an independent, advanced, just and prosperous Indonesian society through accelerated development in various fields by emphasizing the establishment of a solid economic structure based on competitive advantage in various regions supported by qualified and competitive human resources.

There are 4 (four) pillars of the RPJMN 2020–2024, which are the mandate of the 2005–2025 RPJPN to achieve the main objectives of the last period's national development plan. The four pillars translated into seven development agendas, including priority programs, priority activities, and priority projects under the cascading of the RPJMN. The objectives of the RPJMN IV for 2020–2024 are in line with the sustainable development goals (SDGs). The 17 goals (goals) in the sustainable development goals (SDGs) and their indicators have been accommodated in 7 development agendas. The following are the 7 RPJMN development agendas:

Agenda 1: Strengthening economic resilience for quality and equitable growth

Agenda 2: Regional Development for reducing inequality

Agenda 3: Upgrading Human Resource qualification and competitiveness

Agenda 4: Mental Revolution and Cultural Development

Agenda 5: Strengthening The Infrastructure for supporting Economic Development and Basic Needs

Agenda 6: Building Living Environment, Increasing Disaster Resilience and Climate Change

Agenda 7: Strenthening Politics, Law, Defence and Security Stabilization and Public Service Transformation

Table 18.3 Examples of Semarang city RAD-API indicators that are in line with CRD

No.	Program	Indicator
1	Development of information systems related to climate impacts and phenomena	Percentage of population who can access information systems regarding climate phenomena, such as rainfall, temperature, and sea-level rise
2	Development of an early warning system for disasters and driving safety	Percentage of population who can access predictive and real-time information services on urban road conditions every time/when extreme climate events occur
3	Establishment of a zoning system and control of development in coastal areas	Availability of zoning maps for built-up areas on the coast
4	Infrastructure development to increase the resilience of coastal areas	City resilience index
5	Development of settlements that are adaptive to climate change	Number of houses that are adaptive to climate change such as floating houses and stilt houses
		Percentage of settlements that are adaptive to climate change
6	Application of the concept of environmentally friendly buildings (green building) in urban areas	Percentage of buildings with the predicate of green buildings
7	Development of climate-resilient irrigation facilities and systems	Percentage of climate-resilient irrigation facilities and systems
8	Development of an early warning system for disasters and driving safety	Percentage of population who can access predictive and real-time information services on urban road conditions every time/when extreme climate events occur
9	Preparation of regulations on building and environmental arrangements in zones prone to tidal inundation	Number of regulations governing building and environmental arrangements in rob inundation-prone zones
10	Development and development of tourism areas that are comfortable and resilient to disasters	Percentage of tourist areas that apply the principle of disaster resilience
11	Optimizing tourism potential that accommodates the phenomenon of climate change	Number of tourist destinations/tourism activities related to the phenomenon of climate change
12	Mapping of disaster-prone areas	Availability of disaster-prone maps based on ecoregion
13	Development and development of tourism areas that are comfortable and resilient to disasters	Percentage of tourist areas that apply the principle of disaster resilience
14	Provision of contingency funds for disaster-affected asset owners	Number of disaster-affected asset owners who can access contingency funds
15	Provision of a climate-related disaster contingency fund system for the community	Number of people who have disaster contingency funds due to climate change

(continued)

Table 18.3 (continued)

No.	Program	Indicator
16	Development of microfinance alternatives for climate change adaptation	Number of people who can access microfinance for climate change adaptation
17	Development of the government business entity cooperation (PPP) system for infrastructure financing	Number of infrastructure built with the PPP system
		Number of business entities participating in infrastructure development
18	Development of innovation in establishing tax disincentive incentive mechanisms based on regional ecosystem services to increase local revenue (PAD)	The percentage increase in PAD originating from the development of innovation in the establishment of tax disincentive incentive mechanisms based on regional ecosystem services
19	Development of microfinance alternatives for climate change adaptation	Number of people who can access microfinance for climate change adaptation
20	Conservation and increase of catchment area and rainwater catchment area	Percentage of area of urban conservation area
21	Development of climate-resilient irrigation facilities and systems	Percentage of climate-resilient irrigation facilities and systems
22	Development of agricultural activities with the smart agriculture system	Percentage of smart farmland
23	Utilization of social media networking systems	Number of people who access information on climate change impacts
		City emergency response system coverage area
24	Dissemination of information related to gender-specific health threats due to climate change	Number of educational activities related to gender in climate change adaptation
25	Involvement of women in climate change adaptation activities	Number of climate change adaptation activities involving women

Source Authors (2025) (Compiled from RAD-API Semarang City)

The first agenda implies that increasing innovation and investment quality is the main capital to encourage higher, sustainable, and prosperous economic growth fairly and equitably. The second agenda focuses on regional development to increase economic growth and fulfill essential services by harmonizing development plans and space utilization, skillful and character.

Meanwhile, the fourth development agenda focuses on developing culture and national character, which has a central position within the national development framework to create an advanced, modern, superior, competitive, and able to compete with other nations. Strengthening infrastructure aims to support economic activity and encourage equitable distribution of national development as the fifth development agenda, where the government ensures that infrastructure development will be based

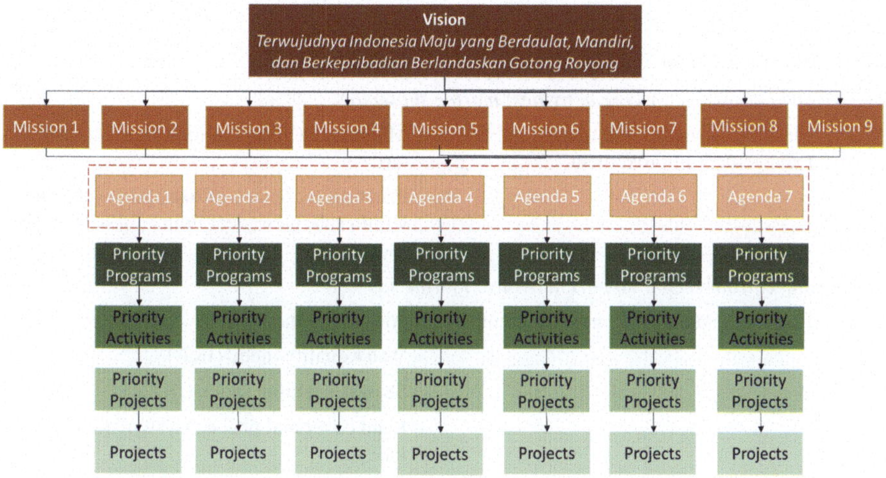

Fig. 18.8 Cascading RPJMN 2020–2024 document. *Source* Authors (compiled from RPJMN 2020–2024 document)

on regional needs and advantages. The SDGs objectives regarding climate are also manifested in the sixth agenda of the RPJMN by focusing on national development that needs to pay attention to the carrying capacity of natural resources and the capacity of the environment, disaster vulnerability, and climate change (Fig. 18.8).

As a measure of the achievement of the RPJMN priority programs, each program has an achievement indicator. From the entire RPJMN program, there are indicators that are in line with the actions for climate resilience development. Table 18.4 is an example of indicators in the RPJMN priority programs that are aligned with climate resistant development based on each of the core and supporting activities.

18.3.5 RPJMD/Regional Medium-Term Development Plan

The RPJMD of Semarang City is a regional development planning document for the Semarang City for 5 (five) years, starting from 2021 to 2026. The RPJMD document is a document that is prepared based on the vision and mission of the regional head in one period and consists of strategies to develop planning programs to realize the region's vision and mission. Development planning also involves various elements of stakeholders for the utilization and allocation of existing resources. The RPJMD document is prepared by referring to the long-term development plan (RPJPD) of the Semarang City and taking into account other planning documents at the national and provincial levels and other strategic documents in the Semarang City.

The RPJMD for 2021–2026 elaborates the development stages of the fourth or final period of the RPJPD for Semarang City 2005–2025. The theme of RPJPD's development for this period is one of the references for regional heads in compiling the

Table 18.4 Examples of RPJMN indicators aligned with CRD

No.	Priority program	Indicator
Marine and coastal sector core activities: marine subsector		
1	Maritime, fisheries and marine management improvement	Marine area conservation (14.5.1*) (million ha)
		Proportion of catch of fish species that are within safe biological limits (14.4.1*) (%) (percent)
Marine and coastal sector support activities: marine subsector		
2	Environmental quality improvement	Air quality index (KPI)
		Water quality index (IKA)
		Seawater quality index (IKAL)
		Land cover and peat ecosystem quality index
3	Improving disaster and climate resilience	Percentage of potential GDP loss impact of disaster (%)
		Percentage of reduction in potential GDP loss for sectors affected by climate hazards (%)
		Speed of public information delivery (minutes)
Marine and coastal sector core activities: coastal subsector		
4	Basic service infrastructure	Percentage of households occupying housing with building resilience (roof, floor, wall) (%)
		Percentage of households occupying housing with access to proper and safe sanitation (wastewater) (%)
		Percentage of households that still practice open defecation (%)
		Percentage of households occupying dwellings with access to well-managed waste (%)
		Percentage of households occupying housing with access to proper drinking water (%)
		Percentage of households occupying housing with access to safe drinking water (%)
		Water storage volume per capita (m^3/capita)
		Number of modernized irrigation areas

(continued)

Table 18.4 (continued)

No.	Priority program	Indicator
		The area of irrigated rice and non-rice agricultural land increased (ha)
		Provinces with reduced disaster risk in disaster risk areas
		Provision of raw water for drinking water needs, industry, and superior areas (m^3/s)
		Road accident fatality ratio per 10,000 vehicles (against 2010 baseline) (%)
Marine and coastal sector support activities: coastal subsector		
6	Strengthening entrepreneurship, micro, small and medium enterprises (MSMEs), and cooperatives	National entrepreneurship ratio (%)
		Contribution of SMEs to GDP (%)
		Cooperative contribution to GDP (%)
Water sector core activities		
7	Increasing water quantity/security to support economic growth	Water productivity (US $/m^3)
Water sector support activities		
8	Environmental quality improvement	Air quality index
		Water quality index
		Seawater quality index
		Land cover and peat ecosystem quality index
Agricultural sector core activities		
9	Increasing the availability, access, and quality of food consumption	Expected food pattern score (2.2.2(c))
		Energy adequacy rate (2.1.2(a)) (kcal/day)
		Protein adequacy rate (g/capita/day)
		Prevalence of undernourishment (PoU)
		Prevalence of population with moderate or severe food insecurity (food insecurity experience scale/FIES)
Agricultural sector support activities		
10	Strengthening entrepreneurship, micro, small and medium enterprises (MSMEs), and cooperatives	National entrepreneurship ratio (%)
		Contribution of SMEs to GDP (%)
		Cooperative contribution to GDP (%)
Health sector core activities		
11	Access and quality of health services improvement	Percentage of accredited first-rate health facilities
		Percentage of accredited hospitals

(continued)

Table 18.4 (continued)

No.	Priority program	Indicator
Health sector support activities		
12	Strengthening the implementation of social protection	Proportion of population covered by social security program
		Proportion of poor and vulnerable households receiving government social assistance

Source Authors (Compiled from RPJMN 2020–2024 document)

vision and mission of the Semarang City for 2021–2026. The vision and mission in the RPJMD document are the vision and mission of the elected regional head in that one leadership period. The vision of Semarang City for 2021–2026 is "The Realization of a Greater Semarang City Based on Pancasila in the Frame of the Unitary State of the Republic of Indonesia with Unity in Diversity." This vision implies that the Semarang City will be ready to face global challenges, more heterogeneous urban problems, higher demands for public services, and still pay attention to environmental insights.

Mission 1: Improving the Quality and Capacity of Superior and Productive Human Resources to Achieve Prosperity and Social Justice

Mission 2: Improving the Potential of a Competitive Local Economy and Stimulating Industrial Development, Based on Research and Innovation Based on Democratic Principles

Mission 3: Ensuring the Freedom of the Community to Perform Worship, Fulfillment of Basic Rights and Protection of Social Welfare and Human Rights for the Community in a Just

Mission 4: Improving Quality Infrastructure that is Environmentally Friendly to Support City Progress

Mission 5: Carrying out Dynamic Government Bureaucratic Reforms and Developing Legal Products in accordance with Pancasila Values within the Framework of the Unitary State of the Republic of Indonesia

The first mission implies that development is prioritized on improving the quality of human resources with high levels of education and health, empowered communities, skilled workers, and the spirit of youth development. The second mission focuses on development that is prioritized on increasing regional economic competitiveness based on local economic potential, enthusiasm for innovation, and support for industry players in marketing regional goods and services and support in business partnerships. Meanwhile, the third mission is prioritized on increasing social empowerment, providing protection and social security for the entire community of Semarang City, gender mainstreaming, and women's empowerment and child protection to realize social welfare for the entire community. Optimizing the development of quality urban facilities and infrastructure for the entire community and the development of regional connectivity while still paying attention to the concept of environmentally sound and sustainable development by controlling pollution and environmental damage and maintaining biodiversity is the focus of development in the fourth mission. Finally, the fifth mission focuses on governance, prioritizing strengthening bureaucratic reform by optimizing information technology, improving internal management, simplifying regulations, and increasing regional financial capacity to improve the quality and reach of public services for the entire community. Optimizing the development of quality urban facilities and infrastructure for the entire community and the development of regional connectivity, while still paying attention to the concept of environmentally sound and sustainable development by controlling pollution and environmental damage and maintaining biodiversity is the focus of development in the fourth mission. Finally, the fifth mission focuses on governance, prioritizing strengthening bureaucratic reform by optimizing information technology, improving internal management, simplifying regulations, and increasing regional financial capacity to improve the quality and reach of public services for the entire community.

Based on the cascading RPJMD in Fig. 18.9, the five missions that have been described have their objectives. The five RPJMD missions were then reduced to six goals and 19 development targets. The objectives and targets are the strategy formulation stage, which shows the highest priority level in the regional medium-term development planning, which will become the basis for preparing the overall regional development performance architecture. At a smaller cascading level, there are programs, activities, and sub-activities that focus on the implementation of Semarang City development.

As a measure of the achievement of the Semarang City RPJMD program, each program has an achievement indicator. From the entire RAD-API program, there are indicators that are in line with the actions for climate resilience development. Table 18.5 is an example of RPJMD indicators that are aligned with climate resistant development based on each of the core and supporting activities.

18.3.6 City Government Policies Related to Climate Change, Development Planning, and Spatial Planning

The RPJMD Semarang City 2021–2026 is prioritized to realize the vision of the Semarang City through accelerating development in various fields by emphasizing the construction of a strong social, cultural, and economic life structure based on

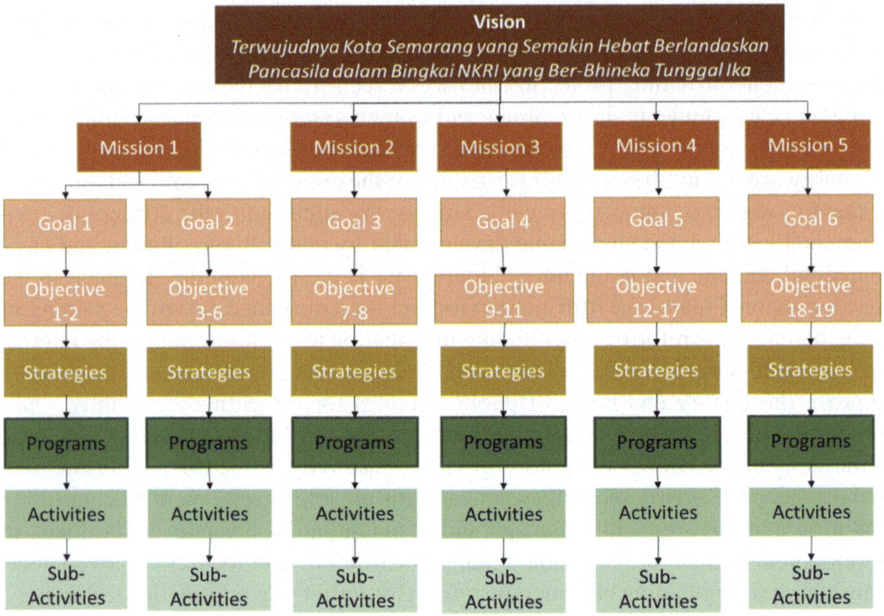

Fig. 18.9 Cascading RPJMD documents for Semarang city. *Source* Authors (2025) (compiled from RPJMD Semarang city document)

competitive advantages in various regions supported by quality and competitive human resources. The policy focus is to realize the quality of Semarang City's human resources, realize good governance and a democratic and responsible political life, realize regional independence and competitiveness, realize sustainable regional spatial planning and infrastructure, and realize the community's social welfare.

Concerning regional documents, the preparation of the RPJMD has also been linked to the Semarang City RTRW 2011–2031. The preparation of the RPJMD takes into account the RTRW document, which contains the pattern of regional zoning development. This is to be a guideline in formulating development policies to develop regional growth centers in the Semarang City based on the region's potential and to determine the location of development programs related to the use of space in the Semarang City.

Meanwhile, several government policy documents related to the RPJMD of the Semarang City and the RAD-API of the Semarang City consist of the mayor's regulation and the Semarang City regional regulation. In addition, there are also study documents or supporting studies related to certain themes such as water and energy, urban resilience, and transportation. The three types of regulations are then divided into documents, namely urban planning documents, sectoral planning documents, and technical/guidance/study documents. As for related study documents, such as water as leverage documents, study on potential development of new renewable energy, study on low carbon city, resilient city strategies, and study on thematic

Table 18.5 Example of Semarang city RPJMD indicators that are in line with CRD

No.	Program	Indicator
Water sector support activities		
1	Construction services development program	Construction service business license issued (%)
2	Water resources management program (SDA)	Level of implementation of flood and tidal control facilities and infrastructure management (%)
3	Spatial planning implementation program	Spatial planning documents prepared (%)
Agricultural sector core activities		
4	Agricultural infrastructure provision and development program	Agricultural productivity per ha per year (tons)
5	Regional research and development program	Percentage of innovation and R&D results implemented in regional planning and development (%)
6	Agricultural disaster management and control program	Percentage of agricultural disasters in the form of pest attacks and climate change impacts that can be controlled (%)
Water sector core activities		
7	Biodiversity management program (Kehati)	Percentage of green open space (%)
8	Drinking water supply system management and development program	Percentage of households served by SPAM (%)
Marine and coastal sector support activities: coastal subsector		
9	Disaster management program	Improvement of infrastructure for disaster preparedness and management
10	Housing development program	Implementation of housing needs facilitation (%)
11	Aquaculture management program	Aquaculture production (tons)
12	Empowerment program for medium, small and micro enterprises (MSMEs)	Percentage of the number of micro business actors who received facilitation of business assistance, business partnerships, institutional development, access to financing (%)
13	Capture fisheries management program	Capture fisheries production (tons)
Marine and coastal sector core activities: coastal subsector		
14	Settlement development program	Percentage of slum area under 10 ha in the city being handled (%)
15	Building and environmental arrangement program	Organized occupancy utilization (%)
16	Residential area program	Percentage of livable houses (%)
Marine and coastal sector support activities: marine subsector		
17	Family welfare empowerment and improvement program (KS)	Coverage of active family members (%)

(continued)

Table 18.5 (continued)

No.	Program	Indicator
Kelantan and coastal sector core activities: marine subsector		
18	Capture fisheries management program	Capture fisheries production (tons)

Source Authors (2025) (Compiled from RPJMD Semarang City document)

village development. In the disaster sector, the Semarang City government also has several supporting documents, including as follows:

- Semarang City Flood Contingency Plan 2017
- Semarang City Disaster Management Plan 2018
- Semarang City Disaster Risk Assessment 2018
- Semarang City Earthquake Contingency Plan in 2021.

There are 12 Semarang City major regulatory documents related to the RPJMD and RAD-API of Semarang City in this study. The Semarang mayor's regulatory documents include the following:

- Semarang Mayor Regulation Number 3a of 2018 concerning Regional Policies and Strategies for the Development of the Drinking Water Supply System (KSDP SPAM) of Semarang City 2016–2021
- Semarang City Mayor Regulation Number 24 of 2019 concerning Green Buildings
- Regulation of the Mayor of Semarang City Number 18 of 2015 concerning the Semarang City Disaster Information System
- Semarang Mayor Regulation Number 14 of 2019 concerning Local Food Development in Semarang City
- Regulation of the Mayor of Semarang Number 24 of 2021 concerning the Urban Agriculture Cultivation Movement in the Semarang City.

On the other hand, 10 city regulatory documents are in accordance with the RPJMD and RAD-API of Semarang City. The Semarang City regulation documents are as follows:

- Semarang City Regional Regulation Number 5 of 2015 concerning the Master Plan of Semarang City Tourism Development for 2015–2025
- Semarang City Regional Regulation No. 7 of 2010 concerning the Arrangement of Green Open Space (RTH)
- Semarang City Regional Regulation Number 13 of 2010 concerning Implementation of Disaster Management in Semarang City
- Semarang City Regional Regulation No. 8 of 2016 concerning Tree Management in Public Green Open Spaces, Roads and Parks.

The relationship between the Semarang City RPJMD development plan document 2021–2026 and the Semarang City RAD-API with regulatory documents and other study documents can be seen in Fig. 18.10.

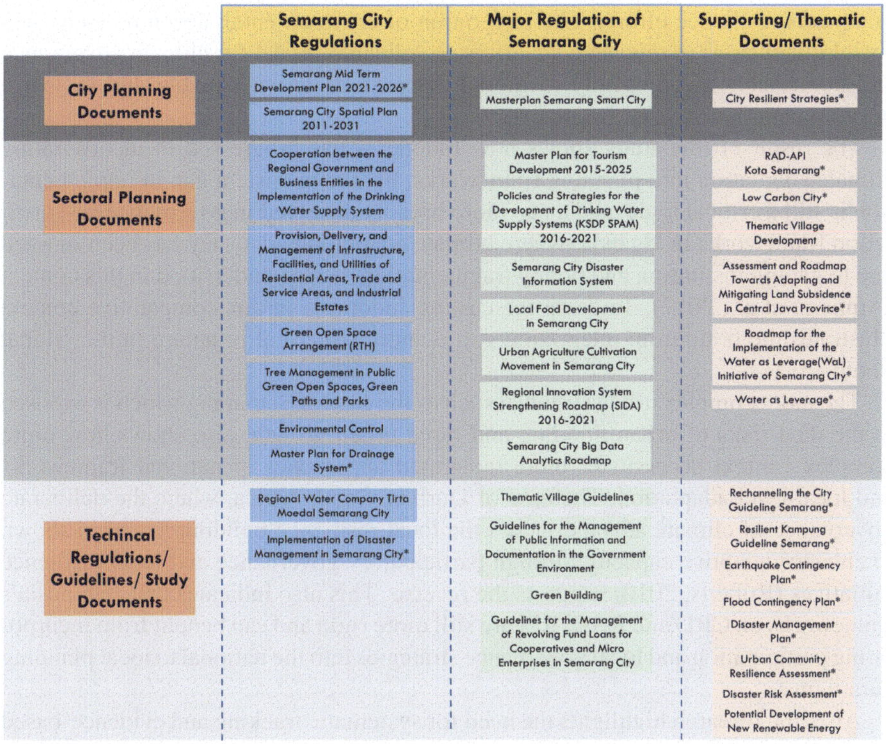

Fig. 18.10 City government policies related to climate change, development planning, and spatial planning. *Source* Authors (2025)

This study is strongly aligned with the core theme of socio-ecological crisis of development: community resilience and sustainable solutions, as it addresses the intersection of environmental governance, development planning, and social vulnerability within the context of climate change. Through the lens of Indonesia's national and sub-national planning frameworks—particularly the CRD, RPJMN, RAD-API, and RPJMD—the research illustrates how institutional responses to climate threats are being integrated into urban systems that directly affect communities. Semarang City, as a case study, reflects how environmental crises like sea-level rise and flooding are managed not only through infrastructure but through participatory planning, health responses, and targeted support for vulnerable populations, including gender and livelihood-sensitive interventions.

Moreover, the research provides a critical framework for understanding community resilience not just as a policy aspiration but as a systemic process that links governance structures with localized adaptive practices. By mapping how adaptation components, such as early warning systems, green infrastructure, and community behavior change, are embedded into regional planning, the study fills a gap in the existing literature that often overlooks how top-down planning connects with grassroots resilience efforts. The integration of environmental, developmental, and social dimensions in this analysis offers a replicable model for climate governance and contributes meaningfully to global discussions on sustainable and inclusive development in crisis-prone contexts.

The result of this study shows how Indonesia has progressed in incorporating climate adaptation into planning frameworks, but the persistent gap in implementation is still a critical issue, and a gap more broadly observed in governance and institution frameworks in Uganda, where climate adaptation advocacy has been mooted due to gaping institution and policy fragmentation, is also understood in this context (Ampaire et al., 2017). To grasp the case of Indonesia in this comparative context illustrates the bottom-up integration gap is more structural in nature in the Global South.

The more complex urban systems, such as the case of Semarang which is exposed to the dual risks of urban flooding and rapid urban growth, also shows how more complex context necessitates more integrated approaches of national frameworks and local level adaptation. The case of Durban, South Africa, where the deliberate governance of climate adaptation and the local resilience building has been shown to enhance adaptive capacity through participatory governance and local resilience initiatives (Roberts, 2010) suggests the reverse. This also indicates that Indonesia's frameworks on CRD and RAD-API are still more rigid and can benefit from incorporating participation and local governance strategies into the national to local planning paradigm.

Also, the research highlights the need for systematic tracking and evidence-based evaluation mechanisms. Adaptation strategies, so far, have proved to be largely policy-based, unsupported by data. More sophisticated global research illustrates that marked vulnerability and adaptation metrics tracking is fundamental to ensuring that policies have controllable and impactful results over time and provide pathways for prudent foresight (Birkmann et al., 2022). Utilizing such frameworks could aid Indonesia in refining the RAD-API and RPJMD to ensure that the adaptation measures proposed go beyond mere advocacy to effective and demonstrable results in vulnerability.

This research advances the discussion in Indonesia, particularly the need to track and evaluate policies, by employing international comparative analysis. Even while underscoring that the outcomes of Indonesia's climate resilience strategies are of global significance, the study also highlights a need for contextualized, participatory, and evidence-based approaches to advance the policy-practice divide.

18.4 Conclusion

The impacts of climate change on Indonesia are immense, and thus, strong actions are needed at both the national and local levels. This research investigated the implementation of climate resilience policies in the Semarang City area, focusing on the upper-tier national documents, such as the national medium-term development plan

(RPJMN) 2020–2024 and climate resilience development (CRD), in comparison to local documents such as the Semarang City regional medium-term development plan (RPJMD) 2021–2026 and the regional action plan for climate change adaptation (RAD-API).

The foundation of the country's comprehensive climate resilience policies is still struggling at the local level. The commitment made by Semarang City in RAD-API is impressive, as it aims to weave retrofitted climate adaptation measures into key sectors such as urban and rural planning, infrastructural development, public health, and farming. Yet, nine years later, major gaps in implementational strategy coherence, institutional alignment, and financing still exist.

The Semarang RPJMD has included objectives for climate resilience, which, however, will be more effective if they focus on integration with RAD-API and national guidelines. Furthermore, the integration of local and sectoral regulations and plans needs to be aligned for effective implementation. Suggested recommendations are as follows: (1) bridging gaps in national and local climate resilience frameworks, (2) promoting cross-agency collaboration, (3) mobilizing climate adaptation financing, and (4) partnering on climate change education and advocacy programs.

This research demonstrates the need to formulate and implement context-specific climate change adaptation strategies, while at the same time, calling for greater integration and coordination of strategies at the national and local levels. Meeting these issues head on will put Semarang, like the rest of Indonesian cities, in a better position to manage climate threats while pursuing sustainable development. More efforts need to be devoted to research to uncover implementation practices from elsewhere. Advanced from this, climate resilience will entail routine assessment of policies, flexible and coordinated governance, and cross-sectoral collaboration to attain the desired climate socio-economic development of Indonesia.

The outcomes of this study bear on both practical policy and theoretical considerations. This study fills the existing gap on vertical policy integration—more practiced than researched—policy in the context of Indonesia's national and sub-national climate resilience integration frameworks, and adds to the literature on policy integration at different levels of governance. This research also has a wider scientific value and may be applied in other contexts facing the challenges of multi-tiered governance systems in climate change. Indonesia forms a part of the Global South, and the case study speaks to larger dilemmas climate change poses to its developing countries. Therefore, it has relevance to audiences beyond Indonesia interested in practical approaches to climate-resilient development. In addition, the experiences of Semarang's adaptation actions and their integration with national objectives offer lessons to international policymakers, scholars, and practitioners in climate change and urbanization who, in particular, are concerned with sustainable city development.

Authors Contribution All authors have contributed equally to this work.

Funding The research was not supported by any funding.

Potential Conflicts of Interest On behalf of all authors, the corresponding author states that there is no conflict of interest.

Data Availability Statement The datasets analyzed during the current study are available from the corresponding author on reasonable request.

AI Disclosure This chapter was developed with the assistance of artificial intelligence (AI) tools ChatGPT and Grammarly to enhance the writing, organization, and clarity of the content. AI was utilized in the drafting, editing, and formatting stages to improve coherence, grammar, and readability. The authors take full responsibility for the accuracy, originality, and integrity of the content. Any AI-generated text has been reviewed and validated to ensure compliance with academic and ethical standards.

References

ADB. (2021). *Climate risk country profile: Indonesia.* https://www.adb.org/sites/default/files/publication/700411/climate-risk-country-profile-indonesia.pdf

Ampaire, E. L., Jassogne, L., Providence, H., Acosta, M., Twyman, J., Winowiecki, L., & van Asten, P. (2017). Institutional challenges to climate change adaptation: A case study on policy action gaps in Uganda. *Environmental Science and Policy, 75*, 81–90. https://doi.org/10.1016/j.envsci.2017.05.013

Bappenas. (2020). *Kebijakan pembangunan berketahanan iklim.* Kementerian Perencanaan Pembangunan Nasional/Badan Perencanaan Pembangunan Nasional (Bappenas). https://perubahaniklim.bappenas.go.id

Biesbroek, G. R., Klostermann, J. E. M., Termeer, C. J. A. M., & Kabat, P. (2013). On the nature of barriers to climate change adaptation. *Regional Environmental Change, 13*(5), 1119–1129. https://doi.org/10.1007/s10113-013-0421-y

Birkmann, J., Jamshed, A., McMillan, J. M., Feldmeyer, D., Totin, E., Solecki, W., Ibrahim, Z. Z., Roberts, D., Kerr, R. B., Poertner, H. O., Pelling, M., Djalante, R., Garschagen, M., Leal Filho, W., Guha-Sapir, D., & Alegría, A. (2022). Understanding human vulnerability to climate change: A global perspective on index validation for adaptation planning. *Science of the Total Environment, 803*. https://doi.org/10.1016/j.scitotenv.2021.150065

Case, M., Ardiansyah, F., & Spector, E. (2007). Climate change in indonesia implications for humans and nature. *Energy, 1*(1), 1–13.

Chenani, E., Yazdanpanah, M., Baradaran, M., Azizi-Khalkheili, T., & Mardani Najafabadi, M. (2021). Barriers to climate change adaptation: Qualitative evidence from southwestern Iran. *Journal of Arid Environments, 189*. https://doi.org/10.1016/j.jaridenv.2021.104487

Chowdhury, Md. A., Hasan, Md. K., & Islam, S. L. U. (2022). Climate change adaptation in Bangladesh: Current practices, challenges and the way forward. *The Journal of Climate Change and Health, 6*, 100108. https://doi.org/10.1016/j.joclim.2021.100108

Dharmawan, I. W. S. (2021). Mitigation and adaptation of climate change disaster. *IOP Conference Series Earth and Environmental Science, 874*(1), 12005. https://doi.org/10.1088/1755-1315/874/1/012005

Djalante, R., & Thomalla, F. (2012). Disaster risk reduction and climate change adaptation in Indonesia: Institutional challenges and opportunities for integration. *International Journal of Disaster Resilience in the Built Environment, 3*(2), 166–180. https://doi.org/10.1108/17595901211245260

Hadi, P. (2011). Integrated climate action: Linking mitigation and adaptation to make Indonesian cities resilient. In K. Otto-Zimmermann (Ed.), *Resilient cities 2: Cities and adaptation to climate change* (pp. 349–357). Springer. https://doi.org/10.1017/CBO9781107415324.004

Haasnoot, M., Di Fant, V., Kwakkel, J., & Lawrence, J. (2024). Lessons from a decade of adaptive pathways studies for climate adaptation. *Global Environmental Change, 88*, 102907. https://doi.org/10.1016/j.gloenvcha.2024.102907

IPCC. (2014). Climate change 2014: Mitigation of climate change. In *Working group III contribution to the fifth assessment report of the intergovernmental panel on climate change*. https://doi.org/10.1017/CBO9781107415416

Moser, S. C., & Ekstrom, J. A. (2010). A framework to diagnose barriers to climate change adaptation. *Proceedings of the National Academy of Sciences of the United States of America, 107*(51), 22026–22031. https://doi.org/10.1073/pnas.1007887107

Panditharatne, C. (2016). Institutional barriers in adapting to climate change: A case study in Sri Lanka. *Ocean and Coastal Management, 130*, 73–78. https://doi.org/10.1016/j.ocecoaman.2016.06.003

Ribot, J. C., & Peluso, N. L. (2003). A theory of access. *Rural Sociology, 68*(2), 153–181. https://doi.org/10.1111/j.1549-0831.2003.tb00133.x

Roberts, D. (2010). Prioritizing climate change adaptation and local level resilience in Durban, South Africa. *Environment and Urbanization, 22*(2), 397–413. https://doi.org/10.1177/0956247810379948

Semarang City Government. (2020). Regional action plan for climate change adaptation (RAD API) of Semarang City. Semarang: Semarang City Government.

Vázquez, L. M. (2017). Implementation challenges of climate change adaptation initiatives in coastal lagoon communities in the Gulf of Mexico. *Maritime Studies, 16*(1). https://doi.org/10.1186/s40152-017-0068-2

Zikra, M., Suntoyo, O. P. K. R., & Lukijanto, L. (2015). Climate change impacts on Indonesian coastal areas. *Procedia Earth and Planetary Science, 14*, 57. https://doi.org/10.1016/j.proeps.2015.07.085

The manufacturer's authorised representative in the EU is Springer Nature Customer Service Centre GmbH, Europaplatz 3, 69115 Heidelberg, Germany. If you have any concerns regarding our products, please contact ProductSafety@springernature.com

Printed and bound by CPI Group (UK) Ltd, Croydon, CR0 4YY
21/09/2026
02220010-0018